PROTEIN CHAPERONES AND PROTECTION FROM NEURODEGENERATIVE DISEASES

WILEY SERIES ON PROTEIN AND PEPTIDE SCIENCE

VLADIMIR N. UVERSKY, Series Editor

Metalloproteomics • Eugene A. Permyakov

Instrumental Analysis of Intrinsically Disordered Proteins: Assessing Structure and Conformation • Vladimir Uversky, Sonia Longhi

Protein Misfolding Diseases: Current and Emerging Principles and Therapies • Marina Ramirez-Alvarado, Jeffery W. Kelly, Christopher M. Dobson

Calcium Binding Proteins • Eugene A. Permyakov and Robert H. Kretsinger

Protein Chaperones and Protection From Neurodegenerative Diseases • Stephan N. Witt

PROTEIN CHAPERONES AND PROTECTION FROM NEURODEGENERATIVE DISEASES

Edited By

Stephan N. Witt

WILEY

A JOHN WILEY & SONS, INC., PUBLICATION

Published by John Wiley & Sons, Inc., Hoboken, New Jersey.
Published simultaneously in Canada.

For general information on our other products and services or for technical support, please contact our Customer Care Department within the United States at (800) 762-2974, outside the United States at (317) 572-3993 or fax (317) 572-4002.

Wiley also publishes its books in a variety of electronic formats. Some content that appears in print may not be available in electronic formats. For more information about Wiley products, visit our web site at www.wiley.com.

Library of Congress Cataloging-in-Publication Data:
Protein chaperones and protection from neurodegenerative diseases / [edited by] Stephan Witt.
p.; cm.—(Wiley series on protein and peptide science)
Includes bibliographical references and index.
ISBN 978-0-470-56907-8 (cloth)
1. Molecular chaperones. 2. Nervous system—Degeneration. I. Witt, Stephan. II. Series: Wiley series in protein and peptide science.
[DNLM: 1. Molecular Chaperones. 2. Neurodegenerative Diseases—genetics. QU 55.6]
QP552.M64P76 2011
612.8—dc22

2010053393

Printed in Singapore

oBook ISBN: 978-1-118-06390-3
ePDF ISBN: 978-1-118-06388-0
ePub ISBN: 978-1-118-06389-7

10 9 8 7 6 5 4 3 2 1

CONTENTS

PREFACE

Proteins and peptides are the major functional components of the living cell. They are involved in all aspects of the maintenance of life. Their structural and functional repertoires are endless. They may act alone or in conjunction with other proteins, peptides, nucleic acids, membranes, small molecules and ions during various stages of life. Dysfunction of proteins and peptides may result in the development of various pathological conditions and diseases. Therefore, the protein/peptide structure-function relationship is a key scientific problem lying at the junction point of modern biochemistry, biophysics, genetics, physiology, molecular and cellular biology, proteomics, and medicine.

The *Wiley Series on Protein and Peptide Science* is designed to supply a complementary perspective from current publications by focusing each volume on a specific protein- or peptide-associated question and endowing it with the broadest possible context and outlook. The volumes in this series should be considered required reading for biochemists, biophysicists, molecular biologists, geneticists, cell biologists and physiologists as well as those specialists in drug design and development, proteomics and molecular medicine with an interest in proteins and peptides. I hope that each reader will find in the volumes within this book series interesting and useful information.

First and foremost I would like to acknowledge the assistance of Anita Lekhwani of John Wiley & Sons, Inc. throughout this project. She has guided me through countless difficulties in the preparation of this book series and her enthusiasm, input, suggestions and efforts were indispensable in bringing the *Wiley Series on Protein and Peptide Science* into existence. I would like to take this opportunity to thank everybody whose contribution in one way or another has helped and supported this project. Finally, special thank you goes to my wife, sons and mother for their constant support, invaluable assistance, and continuous encouragement.

VLADIMIR N. UVERSKY

INTRODUCTION

Owing to advances in DNA sequencing and the deciphering of the human genome in 2000, the genes that cause many of the neurodegenerative diseases that plague mankind, such as Alzheimer's, amyotrophic lateral sclerosis (ALS), Huntington's, Parkinson's disease (PD), prion diseases, and others, have been identified. Intriguingly, many of the identified genes code for proteins that misfold or aggregate, meaning that incorrectly folded proteins cause disease. Misfolded and aggregated proteins are thought to slowly accumulate in neurons with age and are probably toxic for the following reasons. First, an aggregated protein has no biochemical activity; thus, if an essential protein aggregates, essential biochemical activity is lost. Loss of activity of an essential enzyme devastates cells and organs, and this "loss-of-function" phenotype causes the pathobiology in some diseases. Second, aggregated proteins can disrupt essential cellular processes and damage membranes. Thus, certain aggregated proteins display a toxic "gain-of-function" and this causes the pathobiology in some diseases. Of course, disease may result from a combined loss-of-function and toxic gain-of-function. Toxic protein aggregates also cause disease in other organs, especially in the kidney, liver, and pancreas.

Many of the neurodegenerative diseases discussed in this book are caused by proteins that, in addition to forming soluble oligomeric species, also form amyloid fibers. Amyloid fibers are insoluble, protease-resistant chains of repeating molecules of one type of protein. The soluble protein monomer typically adopts a β-sheet-rich conformation; the β-sheet-rich monomers gradually bind to each other and create a highly ordered fiber that elongates as more monomers add to the ends. Depending on the protein that makes up the amyloid, amyloid fibers may be toxic in one disease but protective in another. Given that many of the disease-causing proteins discussed in this book form an array of different soluble and insoluble aggregates and amyloid, it is a great challenge to identify the toxic, disease-causing protein species. It is an equally great challenge to decipher the various mechanisms by which toxic aggregates kill cells.

How molecular chaperones protect neurons from toxic protein aggregates is the focus of this book. Molecular chaperones are proteins that protect cells by inhibiting the formation of toxic aggregates or by breaking up toxic aggregates into smaller units that are not toxic. In some inherited neurodegenerative diseases, a particular mutation triggers a particular protein to aggregate, and cells may become so burdened with the aggregated protein that even the natural chaperone system cannot dissolve away the aggregates. In other neurodegenerative diseases, the chaperones themselves may be mutated or damaged by oxidants, and this leads to the accumulation of toxic aggregated proteins. Although the field of molecular chaperones is relatively mature, it is still important to define this term. Webster's dictionary defines a *chaperone* as . . .*a person, esp. an older or married woman, who accompanies young unmarried people in public or is present at their parties, dances, etc. to supervise their behavior*. Although this definition does not exactly describe a *molecular* chaperone, elements of this definition are valid. I define a *molecular chaperone* as . . .*a protein that reversibly binds to an unfolded, misfolded, or aggregated substrate protein and through cycles of binding and release helps the substrate protein attain its native, active conformation, which it otherwise would not attain*. A molecular chaperone briefly accompanies its substrate protein and makes sure that the substrate makes no inappropriate inter- and intramolecular interactions. At one level, this is very much similar to the chaperone that makes sure that her charges do not misbehave!

This book gives a state-of-the-art account of the diverse biological roles that molecular chaperones play in neurodegenerative diseases. Vladimir Uversky (Chapter 1) provides an excellent background for the various types of chaperones, and he also discusses how many chaperones possess intrinsically disordered protein domains that serve to impart unique functionality to chaperones. Tomohiro Nakamura and Stuart Lipton (Chapter 2) discuss nitrosative stress and how it leads to protein misfolding and neurotoxicity. Nitric oxide (NO) oxidizes PDI (a chaperone in the endoplasmic reticulum), parkin (a ubiquitin ligase), and Drp1 (a protein involved in mitochondrial fission) and the oxidation of these proteins causes protein aggregation, mitochondrial dysfunction, and neuronal damage. This is an example of how damaging a chaperone harms cells. Marta Martinez-Vicente and Esther Wong (Chapter 3) discuss the importance of Chaperone-Mediated Autophagy (CMA) to Alzheimer's, Huntington's, and Parkinson's, focusing on PD. CMA is a selective type of autophagy that relies on molecular chaperones to proteolytically degrade the PD-related protein α-synuclein. Makoto Hashimoto and colleagues (Chapter 4) follow up on PD and discuss the yin and yang of α-synuclein, the related proteins β- and γ-synuclein, and small heat shock proteins. By yin and yang, I mean that the three synucleins can be beneficial to cells when they act like chaperones, but they can also aggregate into toxic oligomeric forms that lack chaperone activity, and these aggregates are referred to as *anti-chaperones*. The biological consequence of a protein's chaperone-to-anti-chaperone conversion is discussed. Anne Bertolotti (Chapter 5) discusses the biochemistry of the proteasome and how proteasome activity can be inhibited by polyglutamine expanded proteins, such as the mutant huntingtin protein associated with Huntington's disease. Ann offers a provocative idea that certain proteasomal chaperones facilitate the conversion of the soluble mutant huntingtin protein into toxic misfolded and aggregated forms. These proteasomal chaperones may thus have anti-chaperone activity in that they induce another protein to aggregate. Following up with another polyQ disease, Andrew Lieberman and William Pratt (Chapter 6) discuss how the chaperones Hsp70 and Hsp90 regulate the degradation of the polyglutamine variant of the androgen receptor; this variant causes Kennedy's disease, or spinal and bulbar muscular atrophy, a slowly progressive degenerative disorder that affects only men. James Shorter (Chapter 7) discusses how the yeast molecular chaperone Hsp104, which has no ortholog in mammalian cells, might be used in gene therapy experiments to rid human neurons of toxic protein aggregates that form in PD. The Hsp104 chaperone has extremely powerful disaggregase activity and can dissolve yeast prions. Chapters 8 and 9 cover prion proteins, which cause diseases like Creutzfeldt–Jakob, fatal familial insomnia, and Gerstmann–Sträussler–Scheinker in humans. Douglas Cyr and colleagues (Chapter 8) discuss examples in which amyloid formation is benign or cytoprotective in disease model systems, describe how an Hsp40 molecular chaperone promotes the formation of amyloid-like aggregates as a protective mechanism in prion toxicity, and highlight cellular pathways that promote amyloid assembly for a functional role in cell biology. Daniel Masison (Chapter 9) discusses how Hsp70 and its cofactors influence the propagation of yeast prions. The yeast model system could shed light on how Hsp70 proteins affect prion formation in the human brain. Marjatta Son and Jeffrey Elliott (Chapter 10) discuss the role of the Copper Chaperone for Sod1 (CCS) in the devastating disease ALS. CCS is a copper chaperone for the protein superoxide dismutase (SOD1); mutations in SOD1 are linked to ALS. CCS donates copper to SOD1, promotes the maturation of SOD1, and affects the subcellular localization of SOD1. CCS overexpression in G93A SOD1 or G37R SOD1 mice causes the most significant acceleration of mutant SOD1-linked familial ALS reported to date. Michal Zolkiewski and Hui-Chuan Wu (Chapter 11) discuss a disease called dystonia and how mutations in the gene torsinA bring about this disease. The torsinA protein has properties of a chaperone, and ubiquitinated aggregates of this protein occur in cells of individuals with this disease. David Gross, Ronald Klein, and I (Chapter 12) discuss modulation of the heat shock response to boost the concentration of protective chaperones, gene therapy, and osmolytes and pharmaceutical chaperones. Osmolytes are sugars, amino acids, and polyols. Pharmaceutical chaperones are low molecular mass organic compounds that rescue the mistrafficking of mutant enzymes in cells. Such compounds are used for certain liver diseases and lipid storage diseases such as Fabry's and Gaucher's diseases, which have neurodegenerative components, but also have great potential as treatments for Huntington's disease and PD.

Neurodegenerative diseases are "sporadic," which means there is no known cause, or "familial," which means the disease is inherited and a defective gene causes the disease. Evidence is mounting that sporadic neurodegenerative diseases are due to the accumulation of toxic protein aggregates, but why aggregates occur in some people but not in others of the same age, and even from the same family, is a mystery. Scientists speculate that individuals who suffer from sporadic neurodegenerative diseases have a genotype (genetic susceptibility) that confers sensitivity to environmental toxins, bacterial and viral infections, or unknown stressors that act to trigger a slow, progressive neurodegeneration. Symptoms occur late in life (>65 years), although the triggering event probably occurred early in life. Familial neurodegenerative diseases occur early in life (~30 – 35 year) and are devastating. As an example, only about 5% of all PD cases are inherited, whereas the rest are sporadic. Sporadic PD typically occurs after 65 years of age, and aggregates of the protein α-synuclein are thought to kill off dopaminergic neurons. Familial PD generally occurs much earlier in life and is due to a mutation in one of the several PD-associated genes, only one of which is α-synuclein. The various familial forms of PD present with similar symptoms, and toxic protein aggregates are thought to cause the gradual demise of dopaminergic neurons in these various PDs.

The chapters in this book show that scientists are revealing new insights into the mechanisms by which aberrantly folded proteins damage cells, and that in the near future new therapeutics, perhaps many of them based on molecular and pharmacological chaperones, will become available to individuals who suffer from neurodegenerative diseases. I am grateful to the authors for investing their time and talent into this project. I also thank my wife, son and daughter for their encouragement and support.

STEPHAN N. WITT

CONTRIBUTORS

Anne Bertolotti, MRC Laboratory of Molecular Biology, Hills Road, Cambridge CB2 0QH, UK

Douglas M. Cyr, Department of Cell and Developmental Biology, School of Medicine, University of North Carolina, 516 Taylor Hall, Chapel Hill, NC 27599-7090, USA

Jeffrey L. Elliott, Department of Neurology, University of Texas, Southwestern Medical Center, 5323 Harry Hines BLVD, Dallas, TX 75390, USA

Masayo Fujita, Laboratory for Chemistry and Metabolism, Tokyo Metropolitan Institute for Neuroscience, Fuchu, Tokyo 183-8526, Japan

David S. Gross, Department of Biochemistry and Molecular Biology, Louisiana State University Health Sciences Center, 1501 Kings Highway, Shreveport, LA 71130-3932, USA

Makoto Hashimoto, Laboratory of Chemistry and Metabolism, Tokyo Metropolitan Institute for Neuroscience, Musashidai, Fuchu, Tokyo 183-8526, Japan

Ronald L. Klein, Department of Pharmacology, Toxicology and Neuroscience, Louisiana State University Health Sciences Center, 1501 Kings Highway, Shreveport, LA 71130-3932, USA

Andrew P. Lieberman, Department of Pathology, University of Michigan Medical School, Ann Arbor, MI 48109, USA

Stuart A. Lipton, Del E. Webb Center for Neuroscience, Aging, and Stem Cell Research, Sanford-Burnham Medical Research Institute, 10901 North Torrey Pines Road, La Jolla, CA 92037, USA; Departments of Neurosciences and Psychiatry, University of California at San Diego, 9500 Gilman Drive, La Jolla, CA 92039, USA; Molecular Neurobiology Laboratory, The Salk Institute for Biological Studies, La Jolla, CA, USA; Departments of Molecular and Integrative Neurosciences, and Molecular and Experimental Medicine, The Scripps Research Institute, La Jolla, CA, USA

Daniel C. Masison, Laboratory of Biochemistry and Genetics, National Institute of Diabetes Digestive and Kidney Diseases, National Institutes of Health, Building 8, Room 225, Bethesda, MD 20892-0830, USA

Katie J. Wolfe, Department of Cell and Developmental Biology, School of Medicine, University of North Carolina, 516 Taylor Hall, Chapel Hill, NC 27599-7090, USA

Tomohiro Nakamura, Del E. Webb Center for Neuroscience, Aging, and Stem Cell Research, Sanford-Burnham Medical Research Institute, 10901 North Torrey Pines Road, La Jolla, CA 92037, USA

William B. Pratt, Department of Pharmacology, University of Michigan Medical School, Ann Arbor, MI 48109, USA

Akio Sekigawa, Laboratory for Chemistry and Metabolism, Tokyo Metropolitan Institute for Neuroscience, Fuchu, Tokyo 183-8526, Japan

Kazuanri Sekiyama, Laboratory for Chemistry and Metabolism, Tokyo Metropolitan Institute for Neuroscience, Fuchu, Tokyo 183-8526, Japan

James Shorter, Department of Biochemistry and Biophysics, University of Pennsylvania School of Medicine, 805b Stellar-Chance Laboratories, 422 Curie Boulevard, Philadelphia, PA 19104, USA

Marjatta Son, Department of Neurology, University of Texas, Southwestern Medical Center, 5323 Harry Hines, Dallas, TX 75390, USA

Daniel W. Summers, Department of Cell and Developmental Biology, School of Medicine, University of North Carolina, 516 Taylor Hall, Chapel Hill, NC 27599-7090, USA

Vladimir N. Uversky, Department of Molecular Medicine, College of Medicine, University of South Florida, 12901 Bruce B. Downs Blvd, MDC3540, Tampa, FL 33612, USA

Marta Martinez-Vicente, Vall d'Hebron Research Insititue, Barcelona, Spain

Stephan N. Witt, Department of Biochemistry and Molecular Biology, Louisiana State University Health Sciences Center, 1501 Kings Highway Shreveport, LA 71130-3932, USA

Esther Wong, Department of Developmental and Molecular Biology, Marion Bessin Liver Research Center, Albert Einstein College of Medicine, 1300 Morris Park Avenue, Chanin B.504, Bronx, NY 10461, USA

Hui-Chuan Wu, Department of Biochemistry, Kansas State University, Manhattan, KS 66506, USA

Michal Zolkiewski, Department of Biochemistry, Kansas State University, Manhattan, KS 66506, USA

1

INTRINSICALLY DISORDERED CHAPERONES AND NEURODEGENERATION

Vladimir N. Uversky

Department of Molecular Medicine, College of Medicine, University of South Florida, Tampa, FL, USA

1.1 INTRODUCTION

This chapter is dedicated to the description of the intrinsically disordered chaperones and their roles in neurodegenerative diseases. Three major concepts, namely, intrinsically disordered proteins (IDPs), chaperones, and neurodegeneration are briefly introduced below.

1.1.1 Intrinsically Disordered Proteins

1.1.1.1 Concept. Evidence is rapidly accumulating that many protein regions and even entire proteins lack stable tertiary and/or secondary structure in solution, existing instead as dynamic ensembles of interconverting structures. These naturally flexible proteins are known by different names, including intrinsically disordered (Dunker *et al.*, 2001), natively denatured (Schweers

Protein Chaperones and Protection from Neurodegenerative Diseases, First Edition.
Edited by Stephan N. Witt.

et al., 1994), natively unfolded (Uversky *et al.*, 2000; Weinreb *et al.*, 1996), intrinsically unstructured (Tompa, 2002; Wright and Dyson, 1999), and natively disordered proteins (Daughdrill *et al.*, 2005). By "intrinsic disorder," it is meant that the protein exists as a structural ensemble, either at the secondary or at the tertiary level. In other words, in contrast to ordered proteins whose 3D structure is relatively stable and Ramachandran angles vary slightly around their equilibrium positions with occasional cooperative conformational switches, IDPs or intrinsically disordered regions (IDRs) exist as dynamic ensembles in which the atomic positions and backbone Ramachandran angles vary significantly over time with no specific equilibrium values and typically undergo noncooperative conformational changes. To some extent, conformational behavior and structural features of IDPs and IDRs resemble those of nonnative states of "normal" globular proteins, which may exist in at least four different conformations: ordered, molten globule, premolten globule, and coil-like (Fink, 2005; Uversky, 2003b; Uversky and Ptitsyn, 1994, 1996b). Using this analogy, IDPs and IDRs might contain collapsed disorder (i.e., where intrinsic disorder is present in a molten globular form) and extended disorder (i.e., regions where intrinsic disorder is present in the form of a random coil or premolten globule) under physiological conditions *in vitro* (Daughdrill *et al.*, 2005; Dunker *et al.*, 2001; Uversky, 2003b).

1.1.1.2 Experimental Techniques for IDP Detection. The disorder in IDPs has been detected by several physicochemical methods elaborated to characterize protein self-organization. The list includes, but is not limited to, X-ray crystallography (Ringe and Petsko, 1986), NMR (nuclear magnetic resonance) spectroscopy (Bracken *et al.*, 2004; Daughdrill *et al.*, 2005; Dyson and Wright, 2002, 2004, 2005a, b), near-UV circular dichroism (CD) (Fasman, 1996), far-UV CD (Adler *et al.*, 1973; Provencher and Glockner, 1981; Uversky *et al.*, 2000; Woody, 1995), optical rotatory dispersion (ORD) (Adler *et al.*, 1973; Uversky *et al.*, 2000), FTIR (Fourier transform infrared spectroscopy)(Uversky *et al.*, 2000), Raman spectroscopy and Raman optical activity (Smyth *et al.*, 2001), different fluorescence techniques (Receveur-Brechot *et al.*, 2006; Uversky, 1999), numerous hydrodynamic techniques (including gel filtration, viscometry, small angle X-ray scattering (SAXS), small angle neutron scattering (SANS), sedimentation, and dynamic and static light scattering) (Receveur-Brechot *et al.*, 2006; Uversky, 1999), rate of proteolytic degradation (Fontana *et al.*, 1997, 2004; Hubbard *et al.*, 1994; Markus, 1965; Mikhalyi, 1978), aberrant mobility in sodium dodecyl sulfate (SDS)-gel electrophoresis (Iakoucheva *et al.*, 2001; Tompa, 2002), low conformational stability (Privalov, 1979; Ptitsyn, 1995; Ptitsyn and Uversky, 1994; Uversky, 1999; Uversky and Ptitsyn, 1996a), H/D exchange (Receveur-Brechot *et al.*, 2006), immunochemical methods (Berzofsky, 1985; Westhof *et al.*, 1984), interaction with molecular chaperones (Uversky, 1999), electron microscopy or atomic force microscopy (Miyagi *et al.*, 2008; Receveur-Brechot *et al.*, 2006), and the charge state analysis of electrospray ionization mass spectrometry (Frimpong *et al.*, 2010; Kaltashov and Mohimen, 2005). For more detailed reviews on methods used to detect intrinsic disorder, see Bracken *et al.* (2004), Daughdrill *et al.* (2005), Longhi and Uversky (2010), Receveur-Brechot *et al.* (2006), and Uversky (2002a).

1.1.1.3 Sequence Peculiarities of IDPs and Predictors of Intrinsic Disorder. IDPs and IDRs differ from structured globular proteins and domains with regard to many attributes, including amino acid composition, sequence complexity, hydrophobicity, charge, flexibility, and type and rate of amino acid substitutions over evolutionary time. For example, IDPs are significantly depleted in a number of so-called order-promoting residues, including bulky hydrophobic (I, L, and V) and aromatic amino acid residues (W, F, and Y), which would normally form the hydrophobic core of a folded globular protein, and also possess low content of C and N residues. On the other hand, IDPs were shown to be substantially enriched in the so-called disorder-promoting amino acids: A, R, G, Q, S, P, E, and K (Dunker *et al.*, 2001; Radivojac *et al.*, 2007; Romero *et al.*, 2001; Williams *et al.*, 2001). Many of the differences mentioned were utilized to develop numerous disorder predictors, including PONDR® (Li *et al.*, 1999; Romero *et al.*, 2001), CH-plot (Uversky *et al.*, 2000), NORSp (Liu and Rost, 2003), GlobPlot (Linding *et al.*, 2003a, b), FoldIndex© (Prilusky *et al.*, 2005), IUPred (Dosztanyi *et al.*, 2005), and DisoPred (Jones and Ward, 2003; Ward *et al.*, 2004a, b) to name a few. It is important to remember that comparing several predictors on an individual protein of interest or on a protein data set can provide additional insight regarding the predicted disorder if any exists.

1.1.1.4 Natural Abundance of IDPs and Their Biological Functions. Application of various disorder predictors to different proteomes revealed that intrinsic disorder is highly abundant in nature and the overall amount of disorder in proteins increases from bacteria to archaea to eukaryota, with over a half of the eukaryotic proteins containing long predicted IDRs (Dunker *et al.*, 2000; Oldfield *et al.*, 2005; Ward *et al.*, 2004b). One explanation for this trend is a change in the cellular requirements for certain protein functions, particularly cellular signaling. In support of this hypothesis, an analysis of a eukaryotic signal protein database indicated that the majority of known signal transduction proteins were predicted to contain significant regions of disorder (Dunker *et al.*, 2002a).

Although IDPs fail to form unique 3D structures under physiological conditions, they are known to carry out a great number of important biological functions, a fact that was recently confirmed by several comprehensive studies (Daughdrill *et al.*, 2005; Dunker *et al.*, 1998, 2001, 2002a, b, 2005; Dunker and Obradovic, 2001; Dyson and Wright, 2005b; Tompa, 2002, 2005; Tompa and Csermely, 2004; Tompa *et al.*, 2005; Uversky, 2002a, b 2003b; Uversky *et al.*, 2000, 2005; Vucetic *et al.*, 2007; Wright and Dyson, 1999; Xie *et al.*, 2007a, b). Furthermore, sites of posttranslational modifications (acetylation, hydroxylation, ubiquitination, methylation, phosphorylation, etc.) and proteolytic attack are frequently associated with regions of intrinsic disorder (Xie *et al.*, 2007a). The functional diversity provided by IDRs was suggested to complement functions of ordered protein regions (Vucetic *et al.*, 2007; Xie *et al.*, 2007a, b).

Another very important feature of the IDPs is their unique capability to fold under a variety of conditions (Dunker *et al.*, 2002a, 2005; Dunker and Obradovic, 2001; Dyson and Wright, 2002, 2005b; Fink, 2005; Iakoucheva *et al.*, 2002; Tompa, 2002; Uversky, 2002a, b; Uversky *et al.*, 2000, 2005; Wright and Dyson, 1999). In fact, the folding of these proteins can be brought about by interaction with other proteins, nucleic acids, membranes, or small molecules. It can also be driven by changes in the protein environment. The resulting conformations could be either relatively noncompact (i.e., remain substantially disordered) or tightly folded.

In a living organism, proteins participate in complex interactions, which represent the mechanistic foundation of the organism's physiology and function. Regulation, recognition, and cell signaling involve the coordinated actions of many players. To achieve this coordination, each participant must have a valid identification that is easily recognized by the other players. For proteins, these identification features are often located within IDRs (Dunker *et al.*, 2005; Uversky *et al.*, 2005). Despite (or may be due to) their high flexibility, IDPs are involved in regulation, signaling, and control pathways in which interactions with multiple partners and high-specificity/low-affinity interactions are often required (Dunker *et al.*, 2005; Uversky *et al.*, 2005).

IDPs have specific functions that can be grouped into four broad classes: (i) molecular recognition; (ii) molecular assembly; (iii) protein modification; and (iv) entropic chain activities (Dunker *et al.*, 2002a). Recently, the crucial role of intrinsic disorder in the action of RNA and protein chaperones was emphasized by showing that IDRs in these complex machines can function as molecular recognition elements that act as solubilizers by locally loosening the structure of the kinetically trapped folding intermediates (Tompa and Csermely, 2004).

1.1.2 Chaperones

1.1.2.1 Concept. Generally, a polypeptide chain of a protein contains all the information required to achieve the functional conformation (Anfinsen, 1973; Crick, 1958). Although this principle is generally correct for many foldable proteins, the information contained in some proteins is not sufficient to guarantee them the gain of functionally active structure. Such proteins cannot fold spontaneously and require the help of molecular chaperones. According to Ellis, molecular chaperones represent "a class of cellular proteins whose function is to ensure that the folding of certain other polypeptide chains and their assembly into oligomeric structures occur correctly" (Ellis, 1987). Chaperones are an important part of the cellular quality control system maintaining an intricate balance between protein synthesis and degradation and protecting cells from devastating consequences of uncontrolled protein aggregation. In addition to chaperones, this system includes the ubiquitin–proteasome system and the autophagy–lysosome system. Molecular chaperones protect cells from apoptosis induced by toxic oligomers. There are several mechanisms by which chaperones fight devastating consequences of misfolding and aggregation. These mechanisms can be grouped into three major classes of action: prevention, reversal, and elimination. At the prevention stage, chaperones bind to unfolded stretches in proteins and keep them in a folding-competent state while preventing aggregation. In the reversal mechanism, chaperones act as disaggregating and unfolding machines that help dissolve aggregates and give a misfolded protein a second chance for correct folding. At the elimination step, chaperones target misfolded proteins for degradation by the ubiquitin–proteasome system and/or the autophagy–lysosome system.

1.1.2.2 Functional Classification of Chaperones. The principal heat-shock proteins (HSPs) that have chaperone activity belong to five conserved classes: HSP33, HSP60, HSP70, HSP90, HSP100, and the small heat shock proteins (sHsps). On the basis of their mechanism of action, molecular chaperones have been divided into three functional subclasses. "Folding" chaperones (e.g., DnaK and GroEL in prokaryotes, and Hsp60 and Hsp70 as well as the HspB group of Hsps including Hsp27 and HspB1 in eukaryotes) rely on adenosine triphosphate (ATP)-dependent conformational changes to mediate the net refolding/unfolding of their substrates. "Holding" chaperones (e.g., Hsp33 and Hsp31) bind partially folded proteins and maintain these substrates on their surface to await availability of "folding" chaperones. "Disaggregating" chaperones constitute the third class of chaperones (e.g., ClpB in prokaryotes and Hsp104 in eukaryotes), which promote the solubilization of proteins that have become aggregated as a result of stress.

According to their expression mechanisms, molecular chaperones are classified as inducible and constitutively expressed. Both types of chaperones act by selective binding of solvent-exposed hydrophobic segments of nonfolded polypeptides, and through multiple binding–release cycles bring about the folding, transport, and assembly of the target polypeptides (Bukau *et al.*, 2006; Hartl and Hayer-Hartl, 2002; Slepenkov and Witt, 2002b). Some chaperones are ATPases; that is, they use free energy from ATP binding and/or hydrolysis to perform work on their substrates.

The concentration of inducible chaperones, also known as *HSPs*, increases as a response to the stress conditions. Some of the illustrative examples of inducible chaperones are sHsps (e.g., αA-crystallin (HspB4), αB-crystallin (HspB5), Hsp27 (HspB1), and Hsp22 (HspB8); family of Hsp40; Hsp70 chaperones and their regulators-co-chaperones HDJ1, HDJ2, BAG1 (Bcl-2–associated athanogene), HSPBP1, Hip, Hop, and CHIP (carboxyl terminus of Hsc70-interacting protein); HspC group of Hsp including Hsp90, Grp94, Hsp104, and Hsp110. These molecular chaperones prevent and reverse the misfolding and aggregation of proteins, which occurs as a consequence of the stress (Lindquist, 1986; Lindquist and Craig, 1988).

On the other hand, constitutively expressed chaperones, also known as the heat shock cognate proteins (HSCs), facilitate protein translation, help newly synthesized proteins to fold, promote assembly of proteins into functional complexes, and assist translocation of proteins into cellular compartments such as mitochondria and chloroplasts (Hartl and Hayer-Hartl, 2002; Young *et al.*, 2004). In the HSP70 family of proteins, in addition to the inducible Hsp70 form, there is a constitutively expressed form, the HSC (Hsc70), which has 85% identity with human Hsp70 and binds to nascent polypeptides to facilitate its correct folding. Hsc70 also acts as an ATPase participating in the disassembly of clathrin-coated vesicles during transport of membrane components through the cell (Goldfarb *et al.*, 2006).

Irrespective of being inducible or constitutively expressed, molecular chaperones evolve to protect proteins from misfolding and aggregation. An important feature of chaperones is that although they assist the noncovalent folding/unfolding and the assembly/disassembly of other macromolecular structures, they do not occur in these structures when the latter are performing their normal biological functions. Generally, molecular chaperones have no effect on protein folding rate. Of course, apparent folding and assembly rates can be increased by elimination of nonproductive oligomer/aggregate formation. Furthermore, by binding to partially folded species and preventing their aggregation, chaperones increase the yield of functional folded/assembled proteins. However, these actions do not affect the intramolecular folding rates. On the other hand, there is a last class of protein helpers that assist protein folding and are not present in the final folded/assembled functional form of a protein substrate. Therefore, these helpers known as *foldases* belong to the family of chaperones. Contrary to the typical chaperones considered so far, foldases evolve to catalyze the folding process by directly accelerating the protein folding rate-limiting steps. Among well-known foldases are eukaryotic protein disulfide isomerase (Goldberger *et al.*, 1963; Hatahet *et al.*, 2009; Nagradova, 2007), peptidyl-prolyl *cis/trans*-isomerase (Fischer *et al.*, 1984; Nagradova, 2007), and lipase-specific foldases, Lifs, found in the periplasm of gram-negative bacteria (Jorgensen *et al.*, 1991; Nagradova, 2007). Finally, there is a large class of the so-called intramolecular chaperones, which are specific protein regions, which are essential for protein folding but not required for protein function. Often, these N-terminal or C-terminal extensions are removed after the protein is folded by autoprocessing or by specific exogenous proteases (Chen and Inouye, 2008). On the basis of their roles in protein folding, intramolecular chaperones were classified into two categories. Type I category includes those intramolecular chaperones that assist tertiary structure formation and mostly are produced as the N-terminal sequence extension of the protein carrier. Type II category contains intramolecular chaperones that are not directly involved in tertiary structure formation but guide the assembly of quaternary structure to form the functional protein complex and are mostly located at the C-terminus of the protein carrier (Chen and Inouye, 2008).

1.1.3 Neurodegeneration

1.1.3.1 Concept. The term *neurodegeneration* is derived from the Greek word νευρο-, *néuro-*, "nerval" and a Latin verb *dēgenerāre*, "to decline" or "to worsen." Therefore, neurodegenerative diseases are a large class of human maladies, which includes various acquired neurological diseases with distinct phenotypic and pathologic symptoms, all characterized by the pathological conditions in which cells of the brain and/or spinal cord are lost. As the death of neurons increases, affected brain regions begin to shrink: by the final stage of Alzheimer's disease (AD), damage is widespread and the brain tissue has shrunk significantly; in prion disease, the brain undergoes damage known as *spongiform change* or *spongiosis* because when the tissue is examined under a microscope, it looks like a sponge, with many tiny holes.

Neurodegeneration is a slow process that begins long before the patient experiences any symptoms. It can take months or even years before visible outcomes of this degeneration are felt and diagnosed: in the case of AD, damage to the brain begins 10–20 years before any problems are evident. The progression through various AD stages may last from 8 to 10 years, whereas in Huntington disease (HD), death occurs approximately 18 years from the time of onset. Symptoms are usually noticed when many cells die or fail to function and a part of the brain begins to cease functioning properly. For example, the symptoms of Parkinson's disease (PD) become apparent after more than ~70% dopaminergic neurons die in a specific area of the midbrain known as *substantia nigra*.

As neurons are not readily regenerated, their deterioration over time leads to dysfunction and disabilities. Neurodegeneration, in principle, can affect various peripheral and central areas of the nervous system resulting in the great variability of the disease manifestations. Generally, neurodegenerative diseases can be divided into three groups according to their phenotypic effects: (i) conditions causing problems with movements; (ii) conditions affecting memory and leading to dementia; (iii) conditions affecting both movement and cognitive abilities; and (iv) conditions causing problems with peripheral nervous system.

Illustrative examples of movement neurodegenerative disorders include PD (characterized by symptoms originating from the neuronal loss in substantia nigra such as resting tremor on one (or both) side(s) of the body; generalized slowness of movement (bradykinesia); stiffness of limbs (rigidity); and gait or balance problems (postural dysfunction)); multiple system atrophy (MSA, characterized by several clinical features of PD); Kennedy disease (also known as spinal and bulbar muscular atrophy (SBMA) or X-linked spinal muscular atrophy since it affects the motor neurons of males only and characterized by muscle weakness); and various forms of ataxia (characterized by a failure of muscle coordination due to pathology arising in the spinocerebellar tract of the spinal cord).

Cognitive neurodegeneration is illustrated by AD and prion diseases (Creutzfeldt–Jakob disease (CJD), Gerstmann–Sträussler–Scheinker (GSS) disease, fatal familial insomnia, and kuru). Some of the movement/cognition-affecting neurodegenerative diseases are neurodegeneration with brain iron accumulation type 1 (NBIA1, characterized by rigidity, dystonia, dyskinesia, and choreoathetosis (Malandrini *et al.*, 1996; Sugiyama *et al.*, 1993; Swaiman, 1991; Taylor *et al.*, 1996)), together with dysarthria, dysphagia, ataxia, and dementia (Dooling *et al.*, 1974; Jankovic *et al.*, 1985; Swaiman, 1991); dementia with Lewy bodies (DLB, characterized by neuropsychiatric changes, often with marked fluctuations in cognition and attention, hallucinations, and parkinsonism (Galpern and Lang, 2006)); and HD (characterized by clinical effects on motor, cognitive, and psychological functions (Melone *et al.*, 2005)).

Illustrative examples of conditions with predominant involvement of the peripheral nervous system with minimal central nervous system involvement include pure autonomic failure (also known as *Bradbury–Eggleston syndrome* characterized by orthostatic hypotension leading to dizziness and fainting, visual disturbances, neck pain, chest pain, fatigue and sexual dysfunction (Hague *et al.*, 1997)), and Lewy body dysphagia (characterized by swallowing abnormalities caused by the localized Lewy body accumulation in both dorsal vagal motor nucleus and the nucleus ambiguus (Jackson *et al.*, 1995)).

1.1.3.2 Molecular Mechanisms of Neurodegeneration. Although neurodegenerative diseases are characterized by an extremely wide range of clinical symptoms resulting from dysfunction of different areas of the central and the peripheral nervous systems, the unifying mechanism of all these pathologies is the deterioration of specific regions of the nervous system caused by the highly specific and localized death of neurons. At the molecular level, many factors can induce neuronal death. Some of these factors are protein misfolding and aggregation, oxidative damage, mitochondrial dysfunction and impaired bioenergetics, disruption of neuronal Golgi apparatus and transport, and failure of cell protective mechanisms including chaperone system and impaired protein degradation machinery (e.g., proteasomal proteolysis and autophagy–lysosome system).

1.1.3.2.1 Protein Misfolding and Aggregation: Neurodegenerative Diseases as Proteinopathies and Amyloidoses. For a long time, a link between AD, PD, prion diseases, HD, and several other neurodegenerative disorders was elusive. However, recent advances in molecular biology, immunopathology, and genetics indicated that these diseases might share a common pathophysiologic mechanism, where derangement of a specific protein processing, functioning, and/or folding takes place. Therefore, neurodegenerative disorders represent a set of proteinopathies, which can be classified and grouped on the basis of the causative proteins. In fact, from this viewpoint, neurodegenerative disorders represent a subset of a broader class of human diseases known as *protein conformational* or *protein misfolding diseases*. These disorders arise from the failure of a specific peptide or protein to adopt its native functional conformational state. The obvious consequences of misfolding are protein aggregation (and/or fibril formation), loss of function, and gain of toxic function. Some proteins have an intrinsic propensity to assume a pathologic conformation, which becomes evident with aging or at persistently high concentrations. Interactions (or impaired interactions) with some endogenous factors (e.g., chaperones, intracellular or extracellular matrixes, other proteins, and small molecules) can change conformation of a pathogenic protein and increase its propensity to misfold. Misfolding can originate from point mutation(s) or result from an exposure to internal or external toxins, impaired posttranslational modifications (phosphorylation, advanced glycation, deamidation, racemization, etc.), an increased probability of degradation, impaired trafficking, lost binding partners, or oxidative damage. All these factors can act independently or in association with one another.

Many of the neurodegenerative diseases are in fact protein deposition diseases. In other words, they are associated with the formation of extracellular amyloid fibrils or intracellular inclusions with amyloid-like characteristics. Protein deposition diseases can be sporadic (idiopathic, 85%), hereditary (familial or genetically inherited, 10%), or even transmissible, as in the case of prion diseases (5%) (Chiti and Dobson, 2006). In the first case, neurodegeneration develops spontaneously, without obvious alterations in the patient's DNA (although genetic differences may act as risk factors). In the second case, neurodegeneration is caused by mutation(s) in specific gene(s). Although these diseases are very different clinically, they share similar molecular mechanisms where a specific protein or protein fragment changes from its natural soluble form into insoluble fibrils. It has been pointed out that prior to fibrillation, amyloidogenic polypeptides may be rich in β-sheet, α-helix, β-helix, or contain both α-helices and β-sheets. They may be globular proteins with rigid 3D structure or belong to the class of natively unfolded (or intrinsically unstructured) proteins (Uversky and Fink, 2004). Despite these differences, the fibrils from different pathologies display many common properties, including a core cross-β-sheet structure in which continuous β-sheets are formed, with β-strands running perpendicular to the long axis of the fibrils (Sunde *et al.*, 1997). This β-pleated sheet structure of fibrils constitutes the basis of the unusual resistance of all kinds of amyloid to degradation and, therefore, the progressive deposition of the material (Westermark, 2005). Furthermore, all fibrils have similar twisted, rope-like structures that are typically 7–13 nm wide (Serpell *et al.*, 2000; Sunde and Blake, 1997) and consist of a number of protofilaments (typically 2–6), each about 2–5 nm in diameter (Serpell *et al.*, 2000). Alternatively, protofilaments may associate laterally to form long ribbons that are 2–5 nm thick and up to 30 nm wide (Bauer *et al.*, 1995; Pedersen *et al.*, 2006; Saiki *et al.*, 2005).

Although amyloid-like fibrils are frequently observed in several neurodegenerative diseases and although the importance of specific amyloidogenic proteins in etiology of corresponding diseases was established by multiple genetic and pathological studies, there is no unifying model explaining toxicity of these deposits. In fact, several different mechanisms of toxicity have been proposed on the basis of the monomeric/polymeric nature of the proposed toxic species. Let us consider the role of α-synuclein in the pathology of PD as an illustrative example, for which at least three different mechanisms of neurotoxicity were discussed (Waxman and Giasson, 2009). An increase in intracellular abundance of monomeric α-synuclein has been considered as a potential cause of neuronal toxicity. This hypothesis is supported by the fact that 50% or 100% increase in α-synuclein expression caused by the duplication or triplication of the α-synuclein gene is known to result in familial forms of PD or DLB (Ross *et al.*, 2008). Furthermore, increased α-synuclein expression was reported in specific brain areas or types of neurons in individuals with sporadic PD (Dachsel *et al.*, 2007) as well as in brains of model animals as a result of toxic insult (Goers *et al.*, 2003; Manning-Bog *et al.*, 2002). In another model, specific oligomeric and protofibrillar forms of α-synuclein have been proposed as potent toxic species. Here, α-synuclein oligomers were proposed to form pores on intracellular membranes such as the plasma membrane and may increase cation permeability (Ding *et al.*, 2002; Lashuel *et al.*, 2002; Volles *et al.*, 2001). Finally, it was emphasized that the fibrillation of α-synuclein and formation of large intracytoplasmic inclusions that can cause the dysfunction and the demise of neurons or oligodendrocytes (Waxman and Giasson, 2009). These inclusions may act as "sinks," recruiting other necessary, cellular proteins from their normal cellular functions (Waxman and Giasson, 2009). They may affect proteasome function (Lindersson *et al.*, 2004) and can impair cellular functions by obstructing normal cellular trafficking (including disruption of endoplasmic reticulum (ER) and Golgi apparatus), by disrupting cell morphology, by impairing axonal transport, and by trapping cellular components (e.g., mitochondria) (Waxman and Giasson, 2009). Of course, the discussed mechanisms of α-synuclein toxicity based on the different polymeric forms from small oligomers to amyloid fibrils are not necessarily mutually exclusive because the presence of any polymeric form of α-synuclein is abnormal and may be problematic for the normal activities of cells, thereby resulting in neurodegeneration (Waxman and Giasson, 2009).

1.1.3.2.2 Mitochondrial Dysfunction and Impaired Bioenergetics. Mitochondria, in addition to being a source of ATP, perform pivotal biochemical functions necessary for homeostasis and represent a convergence point for both extracellular and intracellular death signals. Mitochondrial dysfunction has been described in several neurodegenerative diseases including AD, PD, HD, and amyotrophic lateral sclerosis (ALS) (Moreira *et al.*, 2010). For example, in AD brains, the impaired activity of three tricarboxylic acid cycle complexes, pyruvate dehydrogenase, isocitrate dehydrogenase, and α-ketoglutarate dehydrogenase, was observed (Bubber *et al.*, 2005) together with the reduced respiratory chain activities in complexes I, III, and IV (Valla *et al.*, 2006) and the presence of alterations in mitochondria morphology and distribution (Wang *et al.*, 2008). In PD, mitochondria were demonstrated to be one of the direct targets of α-synuclein-triggered toxicity, which that caused reduced mitochondrial complex I activity and increased production of reactive oxygen species (ROS; Devi *et al.*, 2008). Furthermore, in both sporadic and familial forms of PD, reported mitochondrial abnormalities include impaired functioning of the mitochondrial electron transport chain, aging-associated damage to mitochondrial DNA, impaired calcium buffering, and anomalies in mitochondrial morphology and dynamics (Banerjee *et al.*, 2009; Gibson *et al.*, 2010). Reductions in the activities of complexes II, III, and IV have been observed in the caudate and putamen of HD patients (Browne *et al.*, 1997). Finally, in ALS, the presence of mutant Cu/Zn superoxide dismutase (SOD1) within motor neurons was shown to cause alterations of the mitochondrial respiratory chain (Dupuis *et al.*, 2004), specifically affecting performance of the mitochondrial complexes II and IV (Zimmerman *et al.*, 2007).

1.1.3.2.3 Oxidative Damage. There are several factors that put the brain at risk from oxidative damage (Fatokun *et al.*, 2008). Some of these factors include high oxygen consumption (20% of the total basal O_2 consumption of the body), critically high levels of both iron and ascorbate, relatively low levels of antioxidants (e.g., catalase), a tendency to accumulate metals with age, and low regenerative capacity (Barnham *et al.*, 2004; Gaeta and Hider, 2005; Halliwell, 2006). Furthermore, microglia, the resident immune cells of the brain, produce superoxide and H_2O_2 upon activation; they also produce cytokines that can enhance production of ROS and NO (Halliwell, 2006). Astrocytes equally produce cytokines through which they can be activated to generate NO from iNOS (Halliwell, 2006). The microglia and astrocytes are therefore major mediators of inflammatory processes in the brain (Duncan and Heales, 2005). Some cytochromes *P*450 are also a source of ROS in certain brain regions (Gonzalez, 2005).

Therefore, it is not surprising that although the etiology, symptoms, and disease localization are not the same for neurodegenerative diseases, oxidative stress is recognized as an important pathway leading to neuronal death and is implicated in many neurodegenerative diseases including AD, PD, HD, ALS, and Friedreich's ataxia (FA) (Barnham *et al.*, 2004; Fatokun *et al.*, 2008; Qureshi and Parvez, 2007). In AD, the major sources of oxidative stress and free radical production are copper and iron when bound to Aβ, and the various forms of Aβ in the AD brain are commonly found to be oxidatively modified (Barnham *et al.*, 2004). In PD, resulting from selective degeneration of neuromelanin-containing neurons, most notably the nigral dopaminergic neurons, the catechol dopamine can generate H_2O_2 and the oxidative stress could come from a failure to regulate dopamine–iron biochemistry (Barnham *et al.*, 2004). In ALS, the mutations in SOD are known to lead to a toxic gain of function promoting a pro-oxidant activity of SOD generating ROS (Barnham *et al.*, 2004). FA originates because of an abnormal GAA trinucleotide expansion within the gene encoding the mitochondrial protein frataxin, causing frataxin deficiency. Iron therefore accumulates in the mitochondria, promoting oxidative stress that leads to cardiomyopathy and neurodegeneration (Barnham *et al.*, 2004).

1.1.3.2.4 Disruption of Neuronal Golgi Apparatus and Impaired Transport. The Golgi apparatus plays a central role in the transport, processing, and sorting of proteins. The complex consists of stacks of parallel cisternae and vesicles that carry molecular "cargo" from one cisterna to the next by the coordinated fission of vesicles from the lateral edge of one cisterna and fusion to the next cisterna (Rambourg and Clermont, 1997). Interactions between amyloidogenic proteins and any one or more proteins involved in the maintenance of the structure of the Golgi apparatus might disrupt its structure and function. Golgi apparatus fragmentation was reported in ALS, corticobasal degeneration, AD, PD, CJD, and in spinocerebellar ataxia type 2 (SCA2). In mice model of familial ALS, fragmentation of the Golgi apparatus of spinal cord motor neurons and aggregation of mutant protein were detected months before the onset of paralysis (Gonatas *et al.*, 2006). In a cellular PD model, cells with prefibrillar α-synuclein aggregates had fragmented Golgi apparatus and showed trafficking impairment. These results strongly suggested that the fragmentation of the Golgi apparatus is an early event that occurs before the appearance of the fibrillar α-synuclein forms (Gosavi *et al.*, 2002).

1.1.3.2.5 Impaired Protein Degradation Machinery. The proteasome, in collaboration with a sophisticated ubiquitin system used for marking target proteins, selectively degrades short-lived regulatory proteins as well as abnormal proteins that must be eliminated from cells. The lysosome-linked autophagy system is a bulk protein degradation system designed to eliminate cytoplasmic constituents and to play a prominent role in starvation response and quality control of organelles in cells. The majority of characteristic proteinaceous inclusions in AD, PD, ALS, and frontotemporal lobar degeneration (FTLD) are ubiquitin-positive (Alves-Rodrigues *et al.*, 1998; Lim, 2007). This clearly suggests that impaired proteasomal proteolysis is the main mechanism for the accumulation of ubiquitinated proteins and inclusion body formation in many neurodegenerative diseases (Matsuda and Tanaka, 2010).

Furthermore, since ubiquitination is recently recognized as a mechanism relevant to the autophagy–lysosome system, the fact that specific inclusions in neurodegenerative diseases are ubiquitinated may reflect the impairment of this degradation system too. In fact, the autophagosome sequesters cytosolic material nonspecifically and therefore for a long time the autophagic degradation was considered as a nonselective process. However, recent studies clearly showed that several subcellular structures such as mitochondria and protein aggregates are degraded by selective autophagy and that ubiquitin is involved in this process (Ishihara and Mizushima, 2009; Kirkin *et al.*, 2009). Later, the impairment of the autophagy system in neurons was shown to cause neurodegeneration and ubiquitin-positive inclusion formation in mice (Hara *et al.*, 2006; Komatsu *et al.*, 2006).

1.1.3.2.6 Chaperone System Dysfunctions. Maintaining the appropriate intracellular complement of functional proteins depends on the robust, well-organized, and self-regulated protein quality control system that maintains a balance between protein synthesis and degradation and is capable of a targeted response if an imbalance occurs where misfolded, aggregated, or otherwise damaged proteins accumulate (Bukau *et al.*, 2006; Leidhold and Voos, 2007; McClellan *et al.*, 2005; Witt, 2010). This system tags misfolded and aggregated proteins for refolding by molecular chaperones or degradation by protein degradation machinery such as the ubiquitin-dependent proteasome system or the lysosome-linked autophagy system (Goldberg, 2003). The first line of defense against protein misfolding and aggregation are molecular chaperones. Although, under normal conditions, any protein can spontaneously misfold and aggregate, the "nonstress" concentration of such misfolded, aggregated, or amyloid proteins is negligible and these potentially toxic species are efficiently eliminated by the quality control system. However, several conditions are known to promote protein misfolding and aggregation. This includes the classic environmental stresses such as heat and cold, heavy metals, toxic chemical compounds, UV radiation, the synthesis of proteins with mutations, and age-related decrements in the protein quality control system itself. The enhanced misfolding and aggregation result in the abuse and potential failure of the quality control system. In its turn, the failure of this protein quality control system to fulfill its functions or malfunction of either one or both of its components generates the potential for tissue-specific buildup of protein aggregates termed *amyloid* and is related to the development of neurodegenerative or "conformational" diseases (Gao and Hu, 2008). More details of chaperone action in neurodegeneration together with the description of the role of intrinsic disorder in their activities are given in the next section of this chapter.

1.2 INTRINSICALLY DISORDERED CHAPERONES IN NEURODEGENERATION

As mentioned above, molecular chaperones play a number of important roles in fighting protein misfolding and aggregation and therefore in protecting neurons from the cytotoxic effects of misfolded/aggregated species. This neuroprotection involves a highly coordinated and orchestrated action of multiple players. Therefore, there is an entire net of macromolecular chaperones and their helpers, co-chaperones. The detailed description of individual chaperones and their role in neuroprotection are covered in subsequent chapters of this book. Earlier, it has been emphasized that the importance of intrinsic disorder for the function of chaperones can be underlined by the analysis of the abundance of predicted intrinsically disordered residues in chaperones (Tompa and Csermely, 2004). This analysis revealed a high proportion of such regions in protein chaperones, 36.7% residues of which fall into disordered regions and 15% fall within disordered regions longer than 30 consecutive residues (Tompa and Csermely, 2004). The major goal of this section is to show that many neuroprotective chaperones/co-chaperones are either completely disordered or possess long disordered regions and to emphasize that intrinsic disorder plays a crucial role in their action. Corresponding information is provided for the Hsp70 system, the Hsp90 system, several sHsps, and members of the synuclein family.

1.2.1 The Hsp70 System

1.2.1.1 Major Players. Hsp70 is a 70-kDa molecular machine that is able to interact with exposed hydrophobic amino acids in various polypeptides, hydrolyzes ATP, directs its substrates into a variety of distinct fates, and therefore acts at multiple steps in a protein's life cycle, including its folding, trafficking, remodeling, and degradation (Bukau *et al.*, 2006; Frydman, 2001; Genevaux *et al.*, 2007; Mayer and Bukau, 2005; Patury *et al.*, 2009). Since Hsp70 is able to bind promiscuously, it is considered now as a core chaperone for the proteome (Erbse *et al.*, 2004; Rudiger *et al.*, 1997a, b) and a central mediator of protein homeostasis. The activity of Hsp70 is known to be modulated by a number of co-chaperones, which bind to the core chaperone and influence its functions. Among the most important Hsp70 co-chaperones are various J-domain proteins (e.g., HDJ1 and HDJ2), the number of nucleotide exchange factors (NEFs such as GrpE, Bag1, Hsp110, and HspBP1), and several tetratricopeptide repeat (TPR) co-chaperones (e.g., Hip). Mammalian cells contain a large net of various Hsp70s and their decorating proteins: there are approximately 13 Hsp70s, >40 J-domain proteins, at least 4 distinct types of NEFs, and dozens of proteins with TPR domains. Since at any given time, an individual Hsp70 molecule can only interact with a single representative of each major co-chaperone class, this means that tens of thousands of possible chaperone–co-chaperone complexes might be formed in the cell (Patury *et al.*, 2009).

Finally, there are also several co-chaperones connecting Hsp70 to the Hsp90 and proteasomal degradation pathways. For example, the Hsp-organizing protein (Hop) mediates interactions between Hsp70 and Hsp90 (Scheufler *et al.*, 2000). Hsp70 and Hsp90 also bind to a protein co-chaperone CHIP (Ballinger *et al.*, 1999; Connell *et al.*, 2001), which is a member of the family of E3 ubiquitin ligases. CHIP ubiquitinates unfolded proteins bound to Hsp70 and Hsp90, and these tagged proteins are degraded by the proteasome. Therefore, CHIP links Hsp70 and Hsp90 chaperones to the proteasomal degradation pathway (Witt, 2010).

1.2.1.2 Hsp70. Hsp70 is a highly abundant (~1–2% of total cellular protein) and highly conserved protein, with ~50% sequence identity between prokaryotic and mammalian family members. Many organisms express multiple Hsp70s (e.g., 13 in humans), and members of this class of chaperones are found in all the major subcellular compartments (Patury *et al.*, 2009). Hsp70 is composed of three major domains: an ~44–kDa N-terminal nucleotide-binding domain (NBD, residues 1–388), an ~15-kDa substrate-binding domain (SBD, residues 393–537), and an ~10-kDa C-terminal α-helical, "lid" domain (residues 538–638). All three domains are important for the function of Hsp70. NBD competitively binds ATP and adenosine diphosphate (ADP) and can slowly hydrolyze ATP (McCarty *et al.*, 1995). SBD binds target peptide via the hydrophobic substrate-binding cleft. NBD and SBD are connected by a hydrophobic linker that is crucial for the functional association of two domains: when ATP is bound to NBD, the SBD and NBD exhibit coupled motion, suggesting their tight association (Bertelsen *et al.*, 2009; Schuermann *et al.*, 2008). The position of the lid domain regulates the accessibility of the peptide-binding site. In the ATP-bound form, the lid domain remains open, which facilitates transient interactions with substrates. Following ATP hydrolysis, a conformational change releases the SBD, resulting in closure of the lid and an ~10-fold increase in the affinity for substrate (Slepenkov and Witt, 2002a; Wittung-Stafshede *et al.*, 2003). An important feature of the ATP binding to Hsp70 is that this chaperone binds ATP tightly ($K_d = 1$ nM) but hydrolyses it very slowly ($k_{hy}^{app} = 3 \times 10^{-4}$ s^{-1} at 25°C) (Russell *et al.*, 1998). Another important feature of these chaperones is that the nucleotide modulates their peptide binding and release; in the absence of co-chaperones, ADP-bound DnaK binds and releases peptides over a timescale of minutes or even hours, whereas ATP-bound DnaK binds and releases peptides over a timescale of seconds or even milliseconds (Slepenkov and Witt, 2002a; Wittung-Stafshede *et al.*, 2003). Overall, HSP70s function in a dynamic cycle of binding and releasing polypeptide substrate coupled to a cycle of ATP binding and hydrolysis by the intramolecular ATPase. In the ATP-bound state, HSP70s exhibit fast kinetics and low affinity for polypeptide substrates, whereas in the ADP-bound state, this chaperone exhibits slow kinetics and high affinity for polypeptide substrates. These cycling states are highly regulated by at least five different co-chaperones: Hsp40, Hip, Hop, Bag-1 (Bcl2-associated athanogene 1 or RAP46), and CHIP (Shi *et al.*, 2007).

The substrates of Hsp70 proteins are unrelated in sequence and structure and represent a large spectrum of folding conformers of foldable proteins. In addition, Hsp70 interacts indiscriminately with nonnative polypeptides such as a broad spectrum of heat-denatured proteins (Wegele *et al.*, 2004). The broad substrate specificity of Hsp70s implies a rather degenerative binding motif. It was shown by NMR and X-ray crystallography that DnaK binds its substrate in an extended conformation (Wegele *et al.*, 2004).

The data above clearly show that action of HSP70s involves a lot of dynamics. To test how this functional dynamics correlates with the intrinsic disorder status, DnaK sequence was analyzed by several disorder predictors. DnaK was chosen as the illustrative member of Hsp70 family. Figure 1.1 represents the distribution of the intrinsic disorder propensity within the DnaK sequence evaluated by PONDR VSL2 and illustrates that all three domains of the protein contain extensive amounts of intrinsic disorder. The mean disorder propensity within these domains is arranged in the following order: NBD (0.36 ± 0.13) $<$ SBD (0.49 ± 0.19) $<$ lid (0.84 ± 0.08), with the C-terminal lid domain predicted to be entirely disordered. This high intrinsic propensity of the protein for intrinsic disorder is crucial for its function. It is also reflected in the fact that for a very long time, the 3D structure of a full-length Hsp70 chaperone has not been resolved and all structural information about this important protein was derived from the analysis of its separate domains obtained from partial proteolysis (Slepenkov and Witt, 2002a, 2002b). Furthermore, even when structures comprising both the NBD and the SBD have become available, none of these structures was compatible with any of the others. The location where the SBD docks to the NBD differs by tens of Angstroms (Bertelsen *et al.*, 2009).

Recently, the solution conformations for the full-length, *Escherichia coli* DnaK (1–638) and for a truncation (1–605) for the chaperone bound to substrate peptide (NRLLLTG) and ADP were determined by NMR techniques (Bertelsen *et al.*, 2009). The analysis revealed that although the NBD, SBD, and linker move relatively independently of each other, the motion of SBD with respect

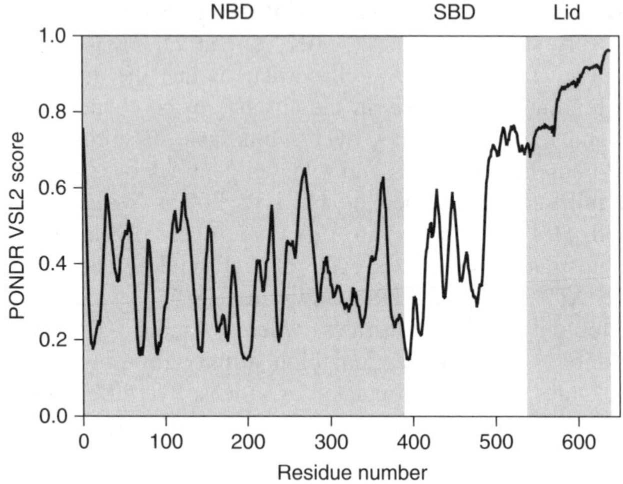

Figure 1.1 The distribution of the intrinsic disorder propensity within the DnaK sequence evaluated by PONDR VSL2. Positions of the three domains, NBD (residues 1–388), SBD (residues 389–537), and lid (538–638) are also shown. All domains contain a significant amount of disorder, with the C-terminal lid domain predicted to be entirely disordered.

to NBD is restricted to a cone of $\sim 70^{\circ}$ opening angle. However, within this cone, there was a preferred orientation of SBD with respect to NBD that can be defined with a relatively high precision (Bertelsen *et al.*, 2009). Importantly, NBD–SBD linker residues (379–397) were shown to possess a large amount of flexibility for these residues, and the lack of dispersion in NMR chemical shifts suggested that this flexible linker had a random coil conformation (Bertelsen *et al.*, 2009). On the other hand, SBD and lid domains were shown to move together as a single rigid unit in the ADP–peptide state, and residues 606–638 were disordered in solution (Bertelsen *et al.*, 2009).

1.2.1.3 Hsp70 Co-chaperones

1.2.1.3.1 J-Domain Proteins. The members of the J protein family, also known as the Hsp40 family (or DnaJ-related co-chaperones), are highly diverse and range in size from 116 amino acids (DnaJC19) to 2243 amino acids (DnaJC13 or Rme-8) (Gibbs and Braun, 2008). In humans, there are over 44 J proteins with the only common feature of this family being a conserved ~70 amino acid signature region known as a *J domain* (Zhao *et al.*, 2008). Although J proteins are known to modulate the Hsp70 ATP catalytic activity via their conserved J domain, all these proteins have multiple additional domains with various functionalities, clearly reflecting the functional diversity of this family probably related to the client protein recognition (Gibbs and Braun, 2008). Therefore, J proteins serve as regulators of the ATPase activity and substrate-binding specificity of Hsp70s. Hsp40 proteins are classified into three main subfamilies (*A–C*, also referred to as types I–III) (Cheetham and Caplan, 1998; Mayer *et al.*, 2001; Ohtsuka and Hata, 2000). Subfamily *A* consists of proteins with the four domains: the highly conserved α-helical N-terminal domain, referred to as the *J domain* (Greene *et al.*, 1998; Karzai and McMacken, 1996; Laufen *et al.*, 1999; Suh *et al.*, 1999; Szabo *et al.*, 1994), a glycine/phenylalanine-rich region that is disordered and likely to be responsible for flexibility (Karzai and McMacken, 1996; Szyperski *et al.*, 1994), the central cysteine-rich domain that includes four repeats of the motif C*XX*C*X*G*X*G (where *X* is any amino acid) and folds in a zinc-dependent manner with two repeats bound to one zinc ion (Banecki *et al.*, 1996; Martinez-Yamout *et al.*, 2000; Szabo *et al.*, 1996), and the C-terminal domain that forms a β-sheet structure and is involved in the dimerization of Hsp40 (Sha *et al.*, 2000). The Cys-rich and C-terminal domains are involved in substrate binding and presentation (Banecki *et al.*, 1996; Li *et al.*, 2003; Lu and Cyr, 1998). Subfamily *B* contains proteins that lack the Cys-rich domain, and subfamily *C* has only the J domain, which is not necessarily located at the N terminus (Cheetham and Caplan, 1998; Mayer *et al.*, 2001; Ohtsuka and Hata, 2000).

Results of the disorder prediction for illustrative members of J proteins from classes *A* (human DnaJ homolog subfamily A member 1), *B* (human DnaJ homolog subfamily B member 1), and *C* (human DnaJ homolog subfamily C member 21) are shown in Figure 1.2. All three co-chaperones are predicted to be highly disordered, possessing averaged disorder scores of 0.60 ± 0.20, 0.54 ± 0.23, and 0.74 ± 0.28, respectively. The fact that all three proteins contain long disordered regions is further confirmed by the lack of the resolved 3D structures of the full-length proteins.

1.2.1.3.2 TPR Co-Chaperone Hip. Hip (Hsp70-interacting protein) is a 369-amino acid cytosolic protein that is composed of an N-terminal region (residues 1–100, which is responsible for protein homo-oligomerization (Velten *et al.*, 2000)), a central, TPR domain (residues 114–215), followed by a highly charged region (residues 230–272), and a C-terminal region (residues 283–369) containing GGMP repeats and a Sti1 domain (heat shock chaperonin-binding motif) (Shi *et al.*, 2007). Hip plays a crucial role in the Hsp70 cycle. In fact, the initial Hsp70 interaction with a polypeptide substrate is achieved through chaperone cooperation with a member of the Hsp40 family, the binding of which stimulates the ATPase activity of Hsp70 and generates the high affinity, ADP-bound state. However, in the presence of Hsp40 alone, the ADP state of Hsp70 is unstable, and the newly formed Hsp70-substrate complex may dissociate prematurely (Hohfeld *et al.*, 1995). The TPR domain of Hip binds to the HSP70 ATPase domain and stabilizes the ADP-bound state of Hsp70, thus stabilizing the chaperone–substrate complex. TPR domain and flanking highly charged region are required for Hip to bind the HSP70 ATPase domain (Hohfeld *et al.*, 1995). Figure 1.3a presents disorder prediction for human Hip co-chaperone and shows that this protein is predicted to be intensively disordered.

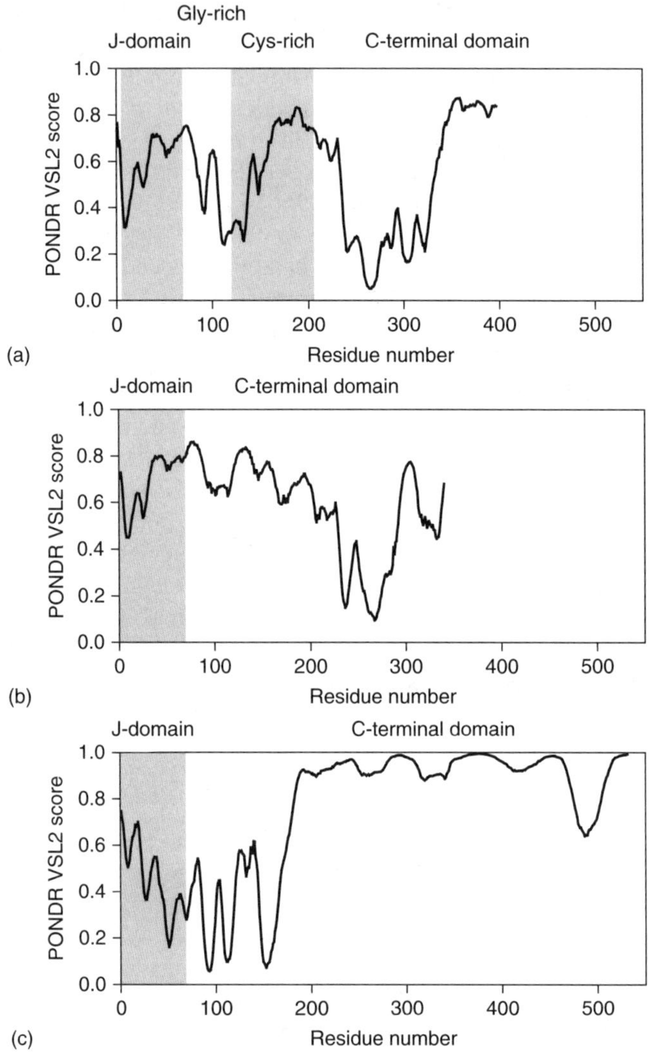

Figure 1.2 Intrinsic disorder in illustrative members of the J proteins (Hsp40s) co-chaperones from the classes *A* ((a) human DnaJ homolog subfamily A member 1), *B* ((b) human DnaJ homolog subfamily B member 1), and *C* ((c) human DnaJ homolog subfamily C member 21). The localizations of major domains in these Hsp40 proteins are also indicated.

1.2.1.3.3 NEF Co-Chaperone Hsp110. Hsp110 proteins constitute a heterogeneous family of abundant molecular chaperones, which are found exclusively in the cytosol of eukaryotic organisms and are evolutionarily related to the Hsp70 family (Easton *et al.*, 2000). Hsp100 was shown to reside in a large molecular complex that includes Hsp70 and Hsp25 (Wang *et al.*, 2000). The N-terminal

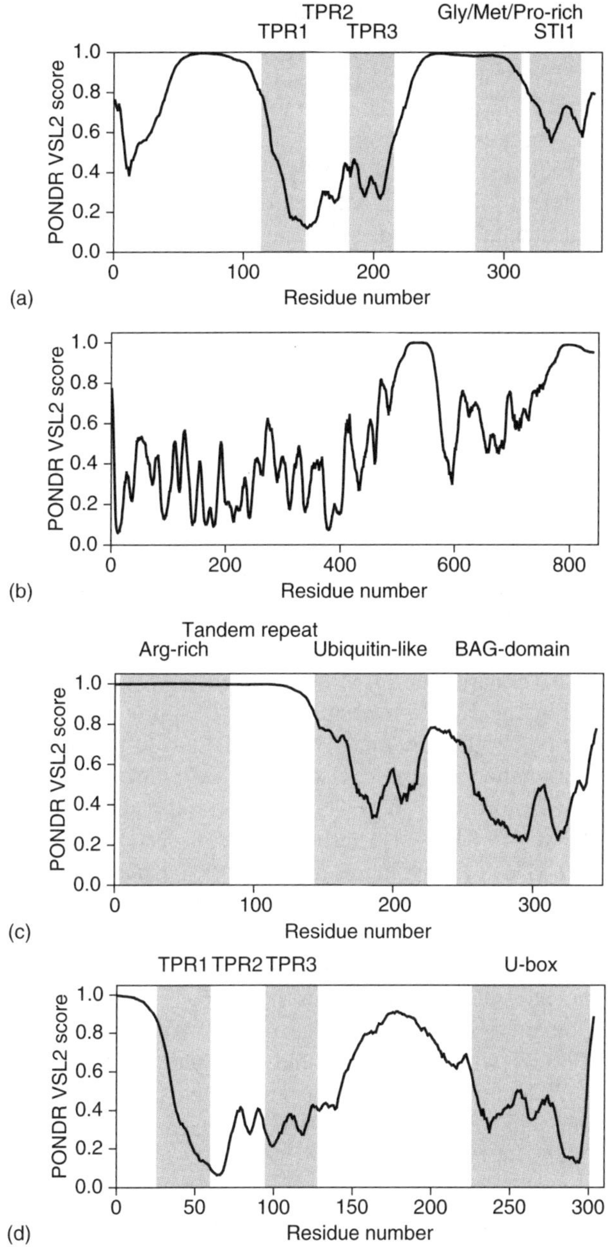

Figure 1.3 Predicted intrinsic disorder in the co-chaperones of the Hsp70 machinery: (a) Disorder prediction for human Hip co-chaperone; (b) disorder in the NEF co-chaperone, human Hsp70-related protein APG-2 (human Hsp110); (c) intrinsic disorder prediction in the human BAG family molecular chaperone regulator 1L (BAG-1L); and (d) predicted disorder in the human CHIP also known as the STIP1 homology and U box-containing protein 1. The localizations of major functional domains in these proteins (when known) are also indicated.

ATPase domain of Hsp110 proteins possesses significant amino acid sequence homology with the Hsp70 proteins, whereas sequence homology between these proteins in their C-terminal domains is very low and hardly recognizable. The C-terminal domain of Hsp100s is considerably longer than that of classical Hsp70 proteins, partially because of a highly negatively charged insertion characteristic for Hsp110 proteins (Oh *et al.*, 1999). Members of the Hsp110 family are known to be efficient "holdases"—they prevent the aggregation and assist the refolding of heat-denatured model substrates in the presence of Hsp70 chaperones and their co-chaperones (Raviol *et al.*, 2006a). It has been shown recently that the Hsp110 from the yeast, Sse1p, acts as an efficient NEF for the yeast cytosolic Hsp70s, Ssa1p, and Ssb1p (Raviol *et al.*, 2006b). Figure 1.3b shows that human Hsp70-related protein APG-2 (human Hsp110) possesses a significant amount of intrinsic disorder, especially in its C-terminal domain. In fact, the mean disorder score for this protein is 0.52 ± 0.28, whereas its last 380 residues are characterized by the disorder score of 0.76 ± 0.19.

1.2.1.3.4 NEF Co-chaperone BAG1. BAG-1 is a multifunctional protein implicated in the modulation of a large variety of cellular processes ranging from transcriptional regulation, to the regulation of apoptosis, to the control of cell migration (Alberti *et al.*, 2003; Doong *et al.*, 2002; Takayama and Reed, 2001). In relation to the chaperone system, BAG-1 is intimately involved in the regulation of Hsp70 chaperone proteins in the eukaryotic cytosol and nucleus, thereby modulating the Hsp70-mediated protein folding and degradation pathways (Alberti *et al.*, 2003). In fact, BAG-1 is known to stimulate nucleotide exchange on mammalian cytosolic Hsc70 (Hohfeld and Jentsch, 1997). There are at least four isoforms of BAG-1: BAG-1L (apparent molecular mass of 52 kDa), BAG-1M (46 kDa; also termed HAP46, RAP46), BAG-1 (34 kDa), and BAG-1S (29 kDa) (Takayama *et al.*, 1998). The major difference between these isoforms is in their N-terminal domains, which differ in their content of several structural elements, the presence or absence of which brings functional diversity to distinct isoforms (Alberti *et al.*, 2003). In addition to the conserved BAG domain located at the C-terminus of BAG-1s, all isoforms contain the ubiquitin-like domain, which serves as an integral sorting signal to stimulate an interaction of BAG-1 with the proteasome and therefore provide a unique link between the Hsp70 and ubiquitin–proteasome systems (Alberti *et al.*, 2003). Furthermore, human cells contain several BAG-1-related proteins: BAG-2, BAG-3 (CAIR-1; Bis), BAG-4 (SODD), BAG-5, and BAG-6 (Scythe, BAT3), which in addition to the conserved BAG domain required for binding and regulation of Hsc70, possess various functional domains that mediate their targeting to diverse partner proteins and subcellular compartments (Takayama and Reed, 2001). The plethora of biological functions ascribed to BAG-1 can be understood by taking into account the fact that this protein is highly disordered. Intrinsic disorder distribution within the sequence of the human BAG family molecular chaperone regulator 1L (BAG-1L) is shown in Figure 1.3c.

1.2.1.3.5 Carboxyl Terminus of Hsc70 Interacting Protein (CHIP). Similar to BAG-1, CHIP (carboxyl terminus of Hsc70 interacting protein) contains both a chaperone-binding site and a domain implicated in the regulation of the ubiquitin–proteasome system. CHIP interacts with binding sites for TPR-containing co-chaperones Hsc70 and Hsp90 via a tandem of three TPR motifs at its amino terminus, whereas at its carboxyl terminus, this protein contains a U-box, which is structurally related to RING finger domains found in many ubiquitin ligases (Jackson *et al.*, 2000). CHIP by itself possesses ubiquitin ligase activity and, in coordination with ubiquitin-conjugating enzymes of the Ubc4/5 family, mediates ubiquitin attachment to protein substrates bound by Hsc70 and Hsp90 (Ballinger *et al.*, 1999; Murata *et al.*, 2001). Since Hsc70 contains the nonoverlapping binding sites for BAG-1 and CHIP, these two factors can simultaneously associate with the chaperone (Ballinger *et al.*, 1999). Furthermore, in the ternary BAG-1/Hsc70/CHIP complex, CHIP mediates the attachment of a polyubiquitin chain to BAG-1 promoting the association of the chaperone complex with the proteasome, thereby providing the mechanism of the chaperone-assisted degradation pathway regulation (Alberti *et al.*, 2002). In application to neurodegeneration, it has been recently established that CHIP targets the toxic α-synuclein oligomers for degradation (Tetzlaff *et al.*, 2008). Figure 1.3d represents the results of disorder prediction for human CHIP, also known as the *STIP1 homology*, and U-box-containing protein 1. Figure 1.3d shows that although TPR-containing domain and U-box domain are predicted to be mostly ordered, they are connected by a long, highly disordered linker. The flexibility of this linker very likely helps in decoupling CHIP interactions with BAG-1 and Hsp70.

1.2.1.3.6 Co-Chaperone Hop. Hop (Hsp70/Hsp90-organization protein, also known as stress-induced phosphoprotein 1, STIP1) does not act as chaperone by itself (Bose *et al.*, 1996). However, it is involved in the organization of Hsp70/Hsp90 complex via its three TPR domains serving as nonoverlapping binding sites for both Hsp70 and Hsp90. In fact, the EEVD-containing C-termini of Hsp70 and Hsp90 bind specifically to the Hop TPR domains, TPR1, and TPR2a, respectively (Scheufler *et al.*, 2000). The connection of and the interplay between the Hsp70 and Hsp90 chaperone machineries is of crucial importance for cell viability. Although originally Hop was considered as a linker protein that brings and holds together Hsp70 and Hsp90 (Smith *et al.*, 1993), the functional repertoire of this co-chaperone is essentially broader, since Hop is involved in regulation of activities of these two chaperones (Odunuga *et al.*, 2004). The chaperone activities of both Hsp70 and Hsp90 are dependent on their ability to bind and hydrolyze ATP. These two chaperones are constantly recycled between the ADP- and ATP-bound forms. The ATPase activity of either Hsp70 or Hsp90 can be divided mechanistically into two stages: ATP hydrolysis and ADP/ATP (nucleotide) exchange. Hop serves as one such modulator. Hsp40 enhances the binding of Hsp70 to preexisting Hop–Hsp90 complex by stimulating the conversion of Hsp70-ATP to Hsp70-ADP (Hernandez *et al.*, 2002). Human Hop binds to Hsp70 with low affinity, but the strength of interaction increases in the presence of Hsp90 (Hernandez *et al.*, 2002). Despite noticeable sequence homology between human Hop and its yeast homolog Sti1 (37% identity), there are fundamental differences between these two proteins in the regulation of the mammalian Hsp90 system compared to the yeast complex (Wegele *et al.*, 2003). In the mammalian system Hop has no influence on the ATPase activity of the Hsp70 or Hsp90 component (Wegele *et al.*, 2003), whereas Sti1 is a noncompetitive inhibitor of yHsp90 (Richter *et al.*, 2003) and a potent activator of yHsp70 (Wegele *et al.*, 2003).

Hop exists as a dimeric molecule in solution and binds as a dimer to dimeric Hsp90 (Carrigan *et al.*, 2004). Structurally, Hop is defined by the presence of nine TPR motifs (which are loosely conserved repeats of roughly 34 amino acids known to mediate protein–protein interactions) clustered into three TPR domains, each consisting of three TPR motifs. A TPR motif shows a helix-turn-helix structure and subsequent TPR motifs are ordered in antiparallel α-helices (Das *et al.*, 1998). Each TPR domain is able to form a structural module that directs protein–protein interactions and has been recruited by different proteins and adapted for various protein–protein interaction functions (Blatch and Lassle, 1999). Owing to its TPR domains, Hop participates in the formation of several Hsp70/Hsp90-unrelated complexes, for example, serving as a receptor for prion proteins (Odunuga *et al.*, 2004). The results of disorder prediction in human Hop (STIP1) are shown in Figure 1.4.

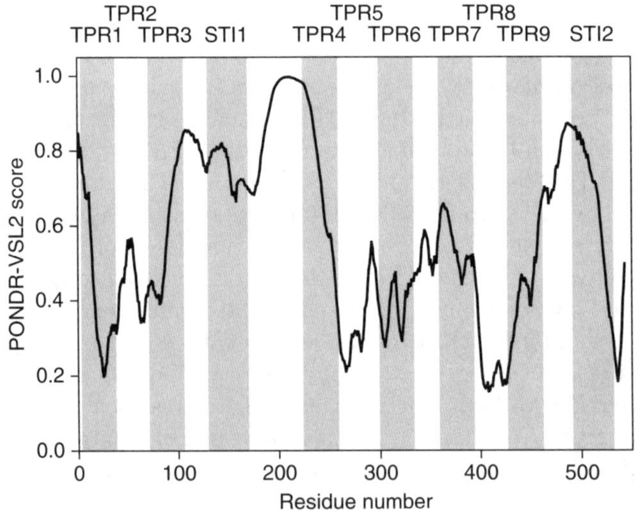

Figure 1.4 The distribution of the intrinsic disorder propensity within the sequence of the human co-chaperone Hop (Hsp70/Hsp90-organization protein, also known as stress-induced-phosphoprotein 1, STIP1) evaluated by PONDR VSL2. The localizations of major domains in this co-chaperone are also indicated.

1.2.2 The Hsp90 Chaperone System

1.2.2.1 Major Players. Although the Hsp70 machinery is one of the most frequently used folding systems in the cell, which is responsible for the correct folding of a wide variety of protein substrates, some proteins are processed by Hsp70 and then transferred to the Hsp90 machinery. In this case, the scaffold protein Hop connects elements of the Hsp70 and Hsp90 machineries to form the "intermediate complex." The Hsp70 component dissociates and, at the same time, p23 and prolyl isomerases enter the complex. After that, the substrate is released from this "final complex." After binding to Hop, Hsp90 is able to reenter the chaperone cycle (Wegele *et al.*, 2004). There is some evidence that Hsp90 is also able to act independently of Hsp70. Each chaperone-folding pathway can either lead to folded, functional proteins or to degradation. Neither Hsp70 nor Hsp90 acts alone. The activity of both chaperones is precisely regulated by a number of co-chaperones. Interestingly, some of the co-chaperones (e.g., Hop and CHIP) are able to interact with both Hsp70 and Hsp90. Other co-chaperones are specific for the individual chaperone machinery: Hsp70 exclusively interacts with J proteins, Hip, Hsp110, and BAG-1, whereas co-chaperone p23, a signal-transduction-related protein Cdc37/p50 (which is required for the Hsp90 substrate-specific folding activity), prolyl isomerases FKPB51, FKPB52, and Cyp40 (also known as *immunophilins*), prolyl isomerase-related protein XAP5, phosphatase PP5, and the Hsp90 ATPase regulator Aha1 are specific cofactors of Hsp90 (Wegele *et al.*, 2004). Overall, the Hsp90 system is a complex machinery, the uniqueness of which is defined by its close collaboration with Hsp70 and the large number of cofactors.

1.2.2.2 Hsp90. Hsp90 is one of the most abundant proteins in unstressed cells accounting for 1–2% of total soluble cell protein (Lai *et al.*, 1984; Welch and Feramisco, 1982). In eukaryotes, cytoplasmic Hsp90 is essential for viability under all conditions. There are two genes encoding cytosolic Hsp90 homologues in mammalian cells. For example, the human Hsp90α shows 85% sequence identity to Hsp90β (Hickey *et al.*, 1989). ATP hydrolysis is crucial for Hsp90 function *in vivo*. However, ATP binding to Hsp90 is generally weak with a dissociation constant in the high micromolar range (Prodromou *et al.*, 1997). The ATPase activity of human Hsp90 is barely detectable, with a k_{cat} of $0.089 \pm 0.004\ \text{min}^{-1}$ and a K_m of $840 \pm 60\ \mu\text{M}$ (McLaughlin *et al.*, 2002). Therefore, as in the case of Hsp70, the ATPase cycle of Hsp90 is modulated by partner proteins that act in complex with Hsp90 *in vivo*.

Hsp90 is known to regulate a number of specific targets. Among the established substrates or "client proteins" of Hsp90 are transcription factors, such as steroid hormone receptors and p53, as well as some proto-oncogenic serine/threonine and tyrosine kinases, such as Raf and Src in higher eukaryotes (for reviews see Buchner (1999), Picard (2002), and Pratt and Toft (2003)).

In neurodegenerative disorders associated with protein aggregation, Hsp90 is known to regulate the heat shock response (Barral *et al.*, 2004; Klettner, 2004;

Soti *et al.*, 2005). Inhibition of Hsp90 activates heat shock factor 1 (HSF1) to induce the production of the chaperones Hsp70 and Hsp40, which promote disaggregation and protein degradation. Under nonstressed conditions, Hsp90 binds to HSF1 and maintains the transcription factor in a monomeric state (Zou *et al.*, 1998). Stress-induced inhibition of Hsp90 releases HSF1 from the Hsp90 complex, leading to its trimerization, activation, and translocation to the nucleus where it initiates a heat shock response (Zou *et al.*, 1998).

Structurally, Hsp90 is an elongated dimer (Maruya *et al.*, 1999; Richter *et al.*, 2001), which, in higher eukaryotes, exists either as $\alpha-\alpha$ or $\beta-\beta$ homodimers or as $\alpha-\beta$ heterodimers (Nemoto *et al.*, 1996; Perdew *et al.*, 1993). The quaternary structure is important for the ATPase activity and associated conformational changes. There are three major domains in Hsp90: a highly conserved N-terminal ATPase domain; a middle domain, which is potentially involved in binding of the substrate proteins; and a C-terminal dimerization domain, which is essential for Hsp90 function and provides the binding site for a subset of Hsp90 co-chaperones, containing TPR domains (Chadli *et al.*, 2000; Chen *et al.*, 1998; Maruya *et al.*, 1999; Prodromou *et al.*, 1997). In eukaryotic Hsp90s, the amino-terminal nucleotide-binding domain is connected to the remainder of the protein by a highly charged and protease-sensitive segment that is variable both in length and composition between different species and between different isoforms in the same species (Wegele *et al.*, 2004).

Because of the intrinsic conformational flexibility of the intact protein, for a long time, atomic resolution crystal structures have only been solved for individual structural domains of Hsp90. Recently, solution structure of the first nucleotide-free eukaryotic Hsp90 (apo-Hsp90) from pig brain was analyzed using the combination of small-angle X-ray scattering and single-particle cryo-electron microscopy (cryo-EM). This analysis revealed the intrinsic flexibility of the full-length eukaryotic apo-Hsp90 and showed that apo-Hsp90 exists in a conformational equilibrium between two open states, transitions between these, the fully open and the semi-open states, require large movements of the N-terminal domain and middle domain around two flexible hinge regions (Bron *et al.*, 2008). Figure 1.5a illustrates the disorder distribution within the sequence of human Hsp90α, whereas Figure 1.5b shows the disorder status in human Hsp90β. Data of this analysis show that in agreement with structural studies functional domains are separated by long disordered regions.

1.2.2.3 Co-chaperone p23. p23 (also known as prostaglandin E synthase 3, telomerase-binding protein p23, and progesterone receptor complex p23 in human or Sba1 in yeast) is a small protein with chaperone activity (Bose *et al.*, 1996). Similar to Hop, p23 interacts with the N-terminal ATPase domain of Hsp90 (Wegele *et al.*, 2003). This interaction is dependent on ATP binding (Prodromou *et al.*, 2000; Sullivan *et al.*, 2002; Young and Hartl, 2000) and inhibits the intrinsic ATPase activity of Hsp90 (Panaretou *et al.*, 2002). In yeast, amino-terminal dimerization of Hsp90 noticeably increases the affinity of p23 for yeast Hsp90.

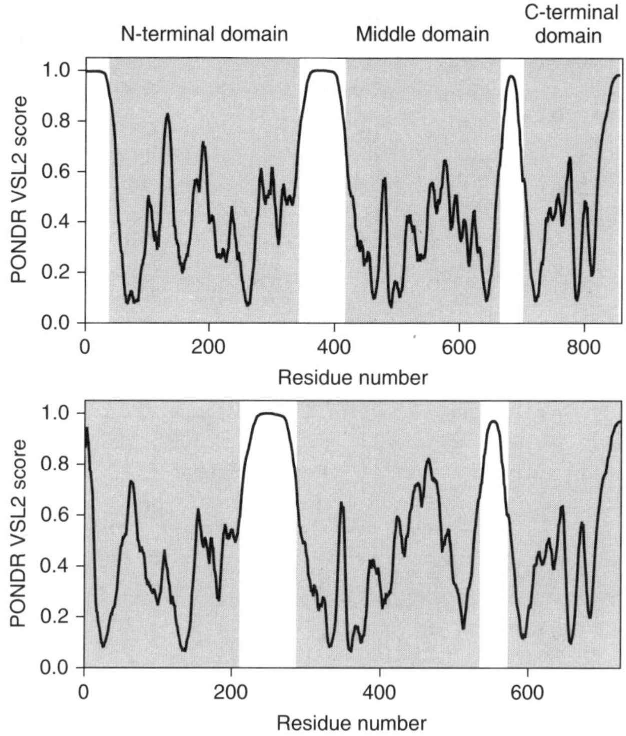

Figure 1.5 The disorder distribution within the sequence of human Hsp90α (a), and the disorder status in human Hsp90β (b). In agreement with structural studies, functional domains are separated by long disordered regions.

The p23 interaction is counteracted by Hop, which prevents amino-terminal dimerization of yeast Hsp90 and therefore binding of p23 to yeast Hsp90 (Wegele *et al.*, 2003).

The N-terminal region of p23 contains a CS domain, which is an ~100-residue protein–protein interaction module named after CHORD-containing proteins and SGT1 (Shirasu *et al.*, 1999). The CS domain has a compact antiparallel β-sandwich fold consisting of seven β-strands (Garcia-Ranea *et al.*, 2002). In the crystal structure of human p23 (Weaver *et al.*, 2000), the C-terminal tail (residues 91–160) is unresolved, which implies that it may be unstructured. This disordered tail occupies almost one half of the protein and is highly enriched in Asp and Glu residues. Although this flexible tail is not needed for the binding of p23 to Hsp90, it is necessary for optimum active chaperoning activity of p23 in assays measuring inhibition of heat-induced protein aggregation (Weaver *et al.*, 2000; Weikl *et al.*, 1999). In agreement with this structural analysis, Figure 1.6a shows a very high level of predicted intrinsic disorder in the C-terminal half of human p23.

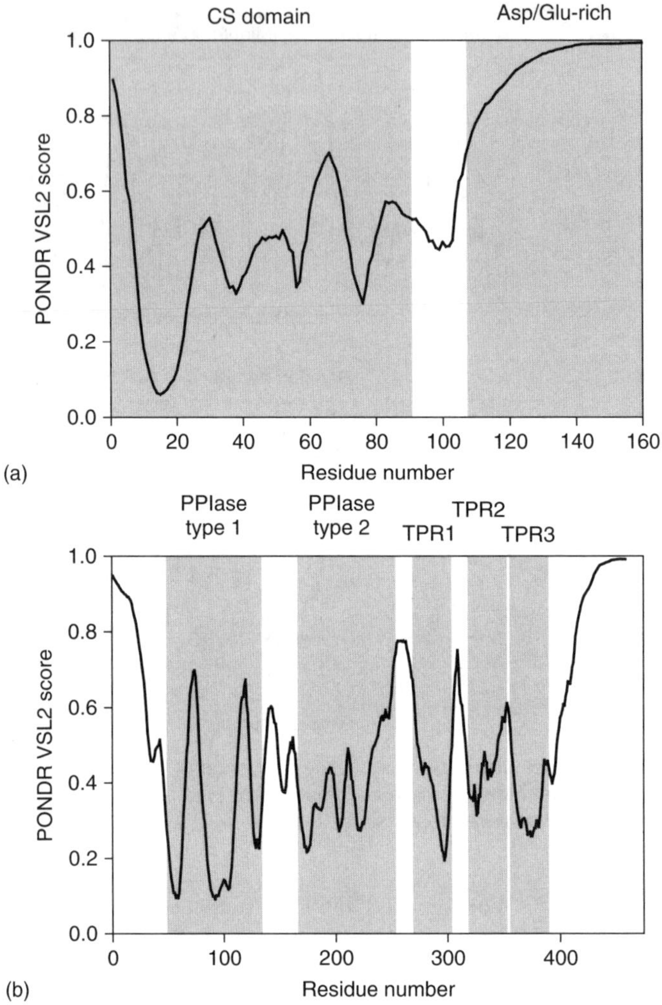

Figure 1.6 Predicted intrinsic disorder in Hsp90 co-chaperones and accessory proteins. Disorder distribution in human p23 co-chaperone (a) and in human FKBP52, also known as FK506-binding protein 4, HSP-binding immunophilin (HBI), 52 kDa FK506-binding protein, FKBP59, and p59 protein (b).

1.2.2.4 Prolyl Isomerases/Immunophilins. The immunophilins are ubiquitous and conserved proteins that have peptidylprolyl isomerase (PPIase) activity, suggesting that they may play a role in protein folding in the cell (Schmid, 1993). Functionally, immunophilins are divided into two classes: the cyclophilins (CsA-binding proteins) and the FKBP (FK506/rapamycin-binding proteins) (Galat, 1993). High molecular mass immunophilins possess several TPR domains and a calmodulin-binding domain in their C-terminal half (Pratt and Toft, 1997). The binding of immunophilins to Hsp90 via TPR domains is conserved in plants and in the animal kingdom (Owens-Grillo *et al.*, 1996), suggesting that this is a basic function of the high molecular mass immunophilins.

High molecular mass prolyl isomerases/immunophilins are crucial for the effective action of the Hsp90 machinery, where together with p23 they are involved in the release of the substrate protein from the Hsp70–Hop complex, and contribute to the formation of the "final complex." In more detail, this mechanism looks as follows. The substrate protein bound to Hsp70 is brought into contact with Hsp90 via Hop. Hop and Hsp70 as parts of the "intermediate complex" are exchanged for a prolyl isomerase/immunophilin and p23 to yield the "final complex." Upon maturation from the "intermediate complex" to the "final complex," the substrate is transferred from Hsp70 to Hsp90 (Wegele *et al.*, 2003). In addition to this crucial role in the Hsp70–Hsp90 chaperone cycle, immunophilins are involved in a number of very important biological processes, for example, in hormonal activation. For example, in the absence of hormone, the glucocorticoid receptor (GR), which is a hormone-activated transcription factor that requires hormonally driven movement to its site of action within the nucleus, is found in the cytosolic fraction of cells as a mixture of complexes. The common feature of all of these complexes is that they all contain GR and Hsp90. However, each of these heterogeneous complexes contains only one molecule of either FKBP52, FKBP51, Cyp40, or PP5, which are unified by being the TPR domain-containing members of the prolyl isomerase/immunophilin family (Davies *et al.*, 2002).

In FKBP52, there are three globular domains followed by a C-terminal portion containing a predicted calmodulin-binding domain (Callebaut *et al.*, 1992). The N-terminal domain possesses the highest homology (49%) with the well-characterized low molecular mass prolyl isomerase FKBP12 (Callebaut *et al.*, 1992). This domain has PPIase activity *in vitro* (Chambraud *et al.*, 1993) and possesses the dimerization site (Wiederrecht *et al.*, 1992). Domain II, which virtually has no PPIase activity (Chambraud *et al.*, 1993), is less homologous to FKBP12 (28%) and contains a consensus nucleotide-binding sequence (Callebaut *et al.*, 1992). Domain III, deletion of which abrogates FKBP52 binding to Hsp90, comprises three TPR domains (Radanyi *et al.*, 1994). The functionality of predicted calmodulin-binding domain at the C-terminus of FKBP52 was supported by the specific retention of this protein by calmodulin-Sepharose in the presence of calcium (Massol *et al.*, 1992).

Figure 1.6b shows that human FKBP52 (also known as FK506-binding protein 4, HSP-binding immunophilin (HBI), 52-kDa FK506-binding protein, FKBP59, and p59 protein) contains a noticeable amount of disorder. In fact, both ends of the protein are highly disordered. Furthermore, although FKBP52 contains several structured domains (PPIase type 1, PPIase type 2, and three TPR domains), all of them are separated by highly flexible linkers, providing the unique functional plasticity to the protein, where each domain can act independently from its neighbors.

1.2.3 Small Heat Shock Proteins

sHsps constitute a structurally divergent family of stress proteins characterized by the presence of the α-crystallin domain, a conserved sequence of 80–100 amino acid residues (Boelens *et al.*, 1998; Carver, 1999; Derham and Harding, 1999; Kim *et al.*, 1998; van Montfort *et al.*, 2001b). The sHSPs are molecular chaperones, storing aggregation-prone proteins as folding competent intermediates and conferring enhanced stress resistance on cells by suppressing aggregation of denatured or nonfolded proteins.

The core of sHsps is the conserved α-crystallin domain, which is typically located in the middle of sHSPs, being flanked by two extensions. The α-crystallin domains of sHsps share a common structure in all members of this family consisting of a seven-stranded, IgG-like β-sandwich with topology identical to p23 (Kim *et al.*, 1998; van Montfort *et al.*, 2001a,b). The poorly conserved N-terminal region varies in sequence and length and influences oligomer construction and chaperone activity. The highly flexible and variable C-terminal extension stabilizes quaternary structure and enhances protein/substrate complex solubility (Carver, 1999). In human sHsps, α-crystallin domain modulates both the structural integrity and the function. This domain is similar to a major lens protein α-crystallin that is composed of two similar subunits, αA- and αB-crystallins. α-Crystallin is highly abundant in lens cells, where it comprises as much as 40% of the cytoplasmic protein and is typically assembled into a heterogeneous mixture of large complexes (Derham and Harding, 1999). Electron microscopic analysis of a 32-subunit complex of αB-crystallin reveals a micelle-like hollow globular structure with outside dimensions of ∼190 Å and inside dimensions of ∼100 Å (Haley *et al.*, 1998).

The molecular mass of various sHSPs in different species ranges from 12 to 43 kDa, and even within a single species, most organisms express multiple sHsps in a cell-specific and developmentally regulated pattern (Kappe *et al.*, 2003). For example, the number of sHsp genes in the known eukaryotic genomes ranges from 2 in yeast to 12 in *Drosophila melanogaster* (Michaud *et al.*, 2002), 16 in *Caenorhabditis elegans* (Candido, 2002), and 19 in *Arabidopsis thaliana* (Scharf *et al.*, 2001). Furthermore, *Arabidopsis thaliana* genome contains additional distantly related 25 genes coding for proteins that contain one or more α-crystallin domains (Scharf *et al.*, 2001). In humans, there are 10 sHSPs, many of which are constitutively present at high levels and implicated in various diseases (Franck *et al.*, 2004; Kappe *et al.*, 2003). These 10 human sHsps are Hsp27/HspB1, HspB2, HspB3, αA-crystallin/HspB4, αB-crystallin/HspB5, Hsp20/HspB6, cvHsp/HspB7, H11/HspB8, HspB9, and a sperm tail protein known as *outer dense fiber protein 1* (ODF1) (Kappe *et al.*, 2003). Their genes are dispersed over nine chromosomes, suggesting their ancient origin (Kappe *et al.*, 2003).

The sHsps occur as homo- or heteromeric complexes, comprising about 2–40 subunits. These globular complexes are often polydisperse and dynamic, readily exchanging subunits, and bind a wide range of cellular substrates. sHsps serve mostly as "holdases," prevent the *in vitro* aggregation of unfolding proteins, which can be transferred to ATP-dependent chaperones, such as Hsp70, and refolded (Haslbeck and Buchner, 2002). sHSP–substrate binding capacity is known to be enhanced by structural changes that expose hydrophobic surfaces that are normally occluded in the native sHsp oligomeric structure (van Montfort *et al.*, 2001a). Putative substrate binding sites may become available through dissociation of sHsp oligomers to dimers as a result of the dynamic equilibrium of sHsp subunits between oligomeric and suboligomeric species or through more subtle environmentally induced (e.g., high temperature) changes in sHSP tertiary structure (Haslbeck *et al.*, 2005; van Montfort *et al.*, 2001a). Therefore, one proposed mechanism of action of sHsps involves breaking down the large oligomer into smaller subunits, exposing hydrophobic surfaces in the α-crystallin domain, which enables binding of the unfolded substrate, followed by reassembly into large soluble complexes aided by sequence extensions (Stamler *et al.*, 2005). sHsps protect against several cellular stressors (Arrigo *et al.*, 2002; Latchman, 2002) and therefore their expression can be upregulated by various forms of stress (Davidson *et al.*, 2002; Michaud *et al.*, 2002). For example, increased levels of several human sHsps were found in neurodegenerative disorders (Krueger-Naug *et al.*, 2002) and in certain tumors (Ciocca and Vargas-Roig, 2002).

Many of the sHsps are multifunctional proteins. For example, in addition to serving as an ATP-independent chaperone involved in protein folding, human Hsp27 (also known as HspB1) is involved in interactions with various cell structures and is implicated in architecture of the cytoskeleton, cell migration, metabolism, cell survival, growth/differentiation, mRNA stabilization, and tumor progression (Kostenko and Moens, 2009). Furthermore, a variety of stimuli induce phosphorylation of this protein at serine residues 15, 78, and 82, which is crucial for its subsequent activity (Kostenko and Moens, 2009). This functional diversity of sHsps is translated into their exceptional structural plasticity and flexibility. For example, human HspB8 (also known as Hsp22), which was shown to decrease or prevent aggregation of Huntingtin fragments and $A\beta_{1-40}$ of the Dutch type, is a highly flexible protein belonging to the group of IDPs (Shemetov *et al.*, 2008). Recently, a comprehensive search for the specific positions of a sHsp that interact directly with partially denatured substrates revealed that although all three domains of the chaperone, the N-terminal arm, the α-crystallin domain, and the C-terminal arm, are able to interact with the substrate, the N-terminal extension plays the most important role in the substrate binding. Several substrates were shown to form strong contacts with multiple residues of this region, the intrinsically disordered nature of which helps in adopting diverse geometries of interaction sites necessary for the interaction with different substrate proteins. This property of the N-terminal arm is critical for the ability of sHsps to protect efficiently many different substrates (Jaya *et al.*, 2009). Figure 1.7 reports on the disorder status of human sHsps and shows that many of these chaperones are highly disordered.

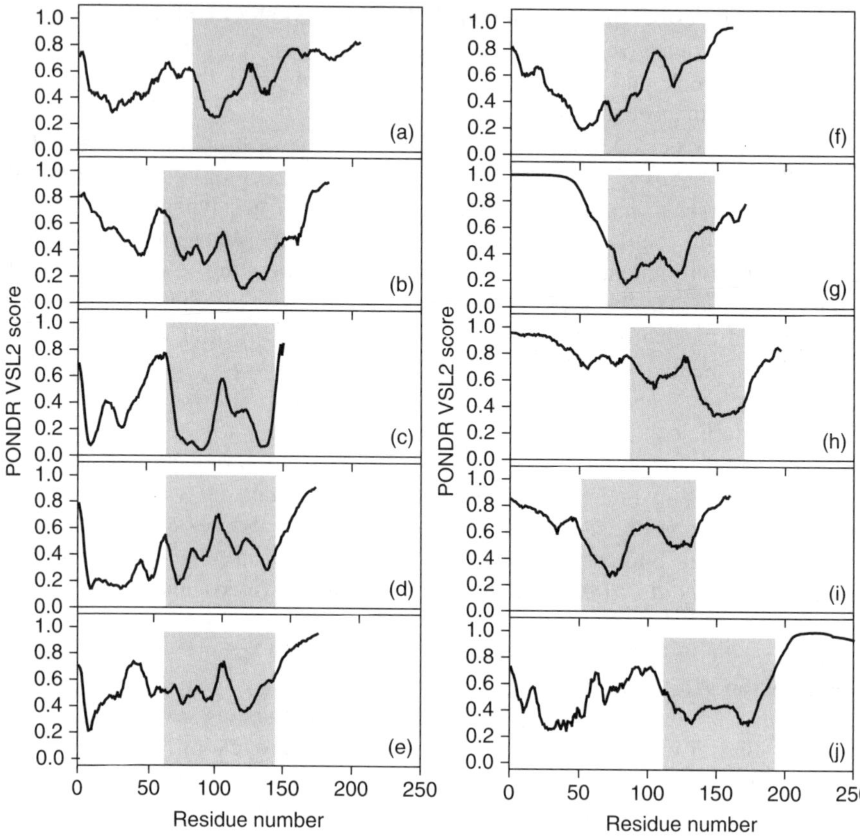

Figure 1.7 The distribution of PONDR VSL2 predicted intrinsic disorder in sequences of human small heat shock proteins, sHsps: Hsp27/HspB1 (a), HspB2 (b), HspB3 (c), αA-crystallin/HspB4 (d), αB-crystallin/HspB5 (e), Hsp20/HspB6 (f), cvHsp/HspB7 (g), H11/HspB8 (h), HspB9 (i), and outer dense fiber protein 1 (ODF1) (j). In each plot, the localization of the α-crystallin domain is indicated as gray area.

1.2.4 Synucleins

Synucleins belong to a family of closely related presynaptic proteins that arise from three distinct genes, described currently only in vertebrates (Clayton and George, 1999). This family includes α-synuclein, which is also known as the non-amyloid component precursor protein (NACP) or synelfin (Jakes *et al.*, 1994; Maroteaux *et al.*, 1988; Ueda *et al.*, 1993); β-synuclein, also referred to as phosphoneuro-protein 14 or PNP14 (Jakes *et al.*, 1994; Nakajo *et al.*, 1993; Tobe *et al.*, 1992); and γ-synuclein, also known as *breast cancer-specific gene 1* or BCSG1 and persyn (Buchman *et al.*, 1998a,b; Ji *et al.*, 1997; Lavedan *et al.*, 1998; Ninkina *et al.*, 1998). All three proteins belong to the family of IDPs (Uversky *et al.*, 2002a; Yamin *et al.*, 2005), with α-synuclein being one of the most well-characterized IDPs (Uversky, 2003a, 2007, 2008, 2009; Uversky and Eliezer, 2009; Uversky *et al.*, 2001).

α-Synuclein is an abundant presynaptic brain protein, whose misfolding, aggregation, and fibrillation are implicated as critical factors in several neurodegenerative diseases. α-Synucleins from different organisms possess a high degree of sequence conservation. For example, mouse and rat α-synucleins are identical throughout the first 93 residues, whereas human and canary proteins differ from them by only two residues (Clayton and George, 1998). At least three α-synuclein isoforms are produced in humans by alternative splicing (Beyer, 2006). The best known isoform is α-synuclein-140 which is the whole and the major transcript of the protein. Two other isoforms, α-synuclein-126 and α-synuclein-112, are produced by AS resulting from the in-frame deletion of exons 3 and 5, respectively. Exon 3 localizes at the N-terminal of the protein and codes for amino acid residues 41–54, whereas exon 5 is located at the C-terminal domain of the protein, coding for residues 103–130. The whole transcript of human α-synuclein, a protein composed of 140 amino acid residues, can be divided into three regions:

1. Residues 1–60 form the N-terminal region. It includes the sites of three familial PD mutations and contains four 11-amino acid imperfect repeats with a highly conservative hexameric motif (KTKEGV). The N-terminal region is predicted to form amphipathic α-helices, typical of the lipid-binding domain of apolipoproteins (Clayton and George, 1998; George *et al.*, 1995).
2. Residues 61–95 constitute the central region and comprise the highly amyloidogenic NAC sequence (NAC stays for *N*on-*A*β*C*omponent of AD amyloid) (Han *et al.*, 1995; Ueda *et al.*, 1993). NAC contains three additional KTKEGV repeats and represents a second major intrinsic constituent of Alzheimer's plaques, amounting to about 10% of these inclusions (Ueda *et al.*, 1993). An 11-amino-acid segment within the central part of the NAC domain (corresponding to residues 73–83 of α-synuclein) is missing in β-synuclein.
3. The highly charged C-terminal region is constituted by residues 96–140. This part of α-synuclein is highly enriched in acidic residues and prolines, suggesting that it adopts a disordered conformation. Three highly conserved tyrosine residues, which are considered as a family signature of α- and β-synucleins, are located in this region. This region is mostly missing in γ-synuclein.

α-Synuclein exhibits a 40% homology with members of the 14-3-3 chaperone protein family (Ostrerova *et al.*, 1999). The 14-3-3 proteins constitute a family of protein chaperones that are particularly abundant in the brain, similar to α-synuclein. The 14-3-3 family of proteins consists of five different isoforms that share extensive sequence homology, both among the different isoforms and between similar isoforms in different species (Broadie *et al.*, 1997; Layfield *et al.*, 1996). 14-3-3 Proteins appear to be involved in diverse cellular functions, mostly via the regulation of protein kinases (Aitken, 1995; Aitken *et al.*, 1995). They bind to ligands at sites containing phosphoserine residues. Binding of 14-3-3 to phosphorylated Raf-1 stabilizes it in an active conformation (Tzivion *et al.*, 1998). 14-3-3 binds to a phosphorylated epitope of protein kinase Cε (PKCε) and stabilizes PKCε in an inactive conformation that is unable to translocate to the membrane (Meller *et al.*, 1996). 14-3-3 also binds to phosphorylated death agonist BAD, a very distant BCL2 family member and a pro-apoptotic oncogene that remains inactive when sequestered in the cytosol. The interaction of 14-3-3 with BAD was shown to stabilize maintenance of BAD in a cytoplasmic localization (Zha *et al.*, 1996).

Deposition of α-synuclein has been implicated in the pathogenesis of several neurodegenerative disorders, known as *synucleinopathies*. Synucleinopathies share common pathologic proteinaceous lesions that are composed of aggregated α-synuclein and are deposited in the selectively vulnerable populations of neurons and glia (Galvin *et al.*, 2001; Goedert, 1999; Spillantini and Goedert, 2000; Trojanowski and Lee, 2003). The term *synucleinopathies* was introduced in 1998 (i.e., just one year after the discovery of α-synuclein deposition in PD) when it was recognized that filamentous α-synuclein deposits might represent a common hallmark linking MSA with PD and DLB (Spillantini *et al.*, 1998b). In addition to these three diseases, the current list of the synucleinopathies includes (but is not limited to) neurodegeneration with brain iron accumulation, type I (also known as *adult neuroaxonal dystrophy* or Hallervorden–Spatz diseases (HSD)); pure autonomic failure and several Lewy body disorders; and diffuse Lewy body disease (DLBD), the Lewy body variant of Alzheimer's disease (LBVAD) (Arawaka *et al.*, 1998; Gai *et al.*, 1998; Lucking and Brice, 2000; Okazaki *et al.*, 1961; Spillantini *et al.*, 1998a, b 1997; Takeda *et al.*, 1998; Trojanowski *et al.*, 1998; Wakabayashi *et al.*, 1997, 1998). Furthermore, even before the detection of α-synuclein as the major Lewy body (LB) component in PD, the peptide derived from the central hydrophobic region of this protein (residues 61–95), known as NAC, was found to represent a second major intrinsic constituent of the AD senile plaques (Han *et al.*, 1995; Ueda *et al.*, 1993). Intriguingly, subsequent work failed to confirm the presence of NAC in amyloid plaques (Bayer *et al.*, 1999). Growing evidence associates the onset and progression of clinical symptoms as well as the degeneration of affected brain regions in these neurodegenerative disorders with the formation of abnormal filamentous aggregates containing α-synuclein. Therefore, it has been concluded that all aforementioned disorders are brain amyloidoses unified by pathological intracellular inclusions of aggregates having the α-synuclein protein as a key component (Galvin *et al.*, 2001; Goedert, 1999; Lundvig *et al.*, 2005; Spillantini *et al.*, 1998a, b; Spillantini and Goedert, 2000; Trojanowski and Lee, 2003; Wakabayashi *et al.*, 1997). Some key facts linking α-synuclein aggregation with the pathogenesis of different synucleinopathies are outlined below. It is believed that understanding why α-synuclein pathology develops in these apparently unrelated conditions may shed light on the mechanisms operating in different synucleinopathies (Goedert, 2001).

It has recently been established that in addition to the traditional α-synuclein-containing LBs and LNs, the development of PD and DLB is accompanied by the appearance of novel α-, β-, and γ-synuclein-positive lesions at the axon terminals of hippocampus (Galvin *et al.*, 1999). These pathological vesicular-like lesions located at the presynaptic axon terminals in the hippocampal dentate, hilar, and CA2/3 regions have been co-stained by antibodies to α- and β-synucleins, whereas antibodies to γ-synuclein detect previously unrecognized axonal spheroid-like inclusions in the hippocampal dentate molecular layer (Galvin *et al.*, 1999). This broadens the concept of neurodegenerative "synucleinopathies" by implicating β- and γ-synucleins, in addition to α-synuclein, in the onset/progression of these two diseases. Additionally, abnormal expression of γ-synuclein has recently been reported in some breast tumors (Ninkina *et al.*, 1998). Using Northern blots and *in situ* hybridization, it has been shown that a high percentage of malignant breast tumors, but not benign breast tumors or normal breast tissue, express γ-synuclein mRNA (Ninkina *et al.*, 1998). In addition, a direct link between γ-synuclein overexpression and increased invasiveness of breast tumor cells has been demonstrated (Ninkina *et al.*, 1999).

Human β-synuclein is a 134-amino acid neuronal protein showing 78% identity to α-synuclein. The α- and β-synucleins share a conserved C-terminus with three identically placed tyrosine residues. However, β-synuclein is missing 11 residues within the specific NAC region (Clayton and George, 1998; Lucking and Brice, 2000). The activity of β-synuclein may be regulated by phosphorylation (Nakajo *et al.*, 1993). This protein, like α-synuclein, is expressed predominantly in the brain; however, in contrast to α-synuclein, β-synuclein is distributed more uniformly throughout the brain (Nakajo *et al.*, 1994; Shibayama-Imazu *et al.*, 1993). Besides the central nervous system, β-synuclein was also found in Sertoli cells of the testis (Nakajo *et al.*, 1996; Shibayama-Imazu *et al.*, 1998), whereas α-synuclein was found in platelets (Hashimoto *et al.*, 1997).

The third member of the human synuclein family is the 127-aa γ-synuclein, which shares 60% similarity with α-synuclein at the amino acid sequence level (Clayton and George, 1998; Lucking and Brice, 2000). This protein specifically lacks the tyrosine-rich C-terminal signature of α- and β-synucleins (Clayton and George, 1998). γ-Synuclein is abundant in spinal cord and sensory ganglia (Buchman *et al.*, 1998a, b). Interestingly, this protein is more widely distributed within the neuronal cytoplasm than α- and β-synucleins, being present throughout the cell body and axons (Buchman *et al.*, 1998b). It was also found in metastatic breast cancer tissue (Ninkina *et al.*, 1998) and epidermis (Ninkina *et al.*, 1999).

Despite the facts that α-synuclein was estimated to account for as much as 1% of the total protein in soluble cytosolic brain fractions (Iwai *et al.*, 1995) and that it is assumed to play a crucial role in the pathogenesis of several neurodegenerative disorders, the precise function of this protein remains mainly elusive. α-Synuclein is expressed in a number of neuronal and non-neuronal cell types including cortical neurons, dopaminergic neurons, noradrenergic neurons, endothelial cells, and platelets (Abeliovich *et al.*, 2000; Hashimoto *et al.*, 1997;

Li *et al.*, 2002; Tamo *et al.*, 2002). Interestingly, torpedo synuclein was reported to localize within the nucleus and presynaptic nerve terminals (Maroteaux *et al.*, 1988); however, most subsequent studies have shown α-synuclein localization only within nerve terminals in the central nervous system (Clayton and George, 1998, 1999; Lavedan, 1998). Although the precise function of α-synuclein remains unknown, this localization, in addition to the close association of this protein with vesicular structures, has led to the hypothesis that it may regulate vesicular release and/or turnover and synaptic function in the central nervous system (Clayton and George, 1998, 1999; Davidson *et al.*, 1998; Lavedan, 1998; Ueda *et al.*, 1993). In agreement with this hypothesis, mice lacking α-synuclein, being superficially normal, exhibited alterations in transmitter release from dopaminergic terminals in striatum, following paired electrical stimulation and in locomotor responses after amphetamine administration (Abeliovich *et al.*, 2000). Additional observations suggest that α-synuclein may play a role in neuronal plasticity responses because its avian homolog synelphin is upregulated in zebra finch brain at a critical period of song learning (George *et al.*, 1995), and rat synuclein-1 is upregulated during brain development (Hsu *et al.*, 1998; Petersen *et al.*, 1999) and in cultured neonatal sympathetic neurons after nerve growth factor treatment (Stefanis *et al.*, 2001). α-Synuclein was shown to act as a high-affinity inhibitor of phospholipase D2, which hydrolyzes phosphatidylcholine to phosphatidic acid and may be involved in vesicle trafficking in the secretory pathway (Chen *et al.*, 1997; Jenco *et al.*, 1998). Overall, functions ascribed to α-synuclein include binding fatty acids and physiological regulation of certain enzymes, transporters, and neurotransmitter vesicles, as well as roles in neuronal survival (Dev *et al.*, 2003).

It has been shown that α-synuclein can act as a molecular chaperone (Chandra *et al.*, 2005), cellular levels of which in both substantia nigra and frontal cortex were shown to be significantly increased as a response to the toxic insult (Manning-Bog *et al.*, 2002). These toxicant-induced changes in the expression of α-synuclein were characterized by a very peculiar time course where levels of the protein in the mice brain were consistently enhanced at two days after paraquat administrations and returned to basal control values within seven days posttreatment (Manning-Bog *et al.*, 2002). A similar time course of α-synuclein upregulation has also been reported for mice treated with the parkinsonism-inducing neurotoxicant 1-methyl-4-phenyl-1,2,3,6-tetrahydropyridine (MPTP) (Vila *et al.*, 2000). These findings suggested that α-synuclein plays a role in overcoming the consequences of the toxic insult. Furthermore, overexpression of α-synuclein was shown to actually delay cell death caused by toxic agents and protect cells against the apoptotic stimuli (da Costa *et al.*, 2000; Lee *et al.*, 2001).

Methionine oxidation was proposed to play a role in the potential α-synuclein function as a chaperone. In fact, since the addition of methionine-oxidized α-synuclein inhibited fibrillation of the nonoxidized form, it was suggested that the methionine residues in α-synuclein may be used by the cells as a natural scavenger of ROS, and therefore, α-synuclein can serve as a redox chaperone (Uversky *et al.*, 2002b). This hypothesis was based on the facts that

(i) methionine can react with essentially all of the known oxidants found in normal and pathological tissues; (ii) α-synuclein is a very abundant brain protein; (iii) the concentration of α-synuclein can increase significantly as a result of the neuronal response to toxic insult (Manning-Bog *et al.*, 2002); and (iv) methionine sulfoxide residues in proteins can be cycled back to their native methionines by methionine sulfoxide reductase (Levine *et al.*, 1996), a process that might protect other functionally essential residues from oxidative damage (Reddy and Bhagyalakshmi, 1994). However, the antifibrillation role of the methionine-oxidized α-synuclein was strongly compromised in the presence of certain heavy metals, such as lead, aluminum, zinc, titanium, and others (Yamin *et al.*, 2003). Therefore, in the presence of the enhanced concentrations of such industrial pollutants, toxic insult-induced upregulation of α-synuclein may no longer play a protective role; rather, it may represent a risk factor, leading to metal-triggered fibrillation of the methionine-oxidized protein (Yamin *et al.*, 2003).

Comprehensive structural analysis of human α-, β-, and γ-synucleins revealed that these three proteins are almost completely unfolded under the physiological conditions *in vitro* (Uversky *et al.*, 2002a). Figure 1.8 supports this experimental data-based conclusion and shows that all three human synucleins are predicted to be completely disordered.

1.3 THE ACTION MECHANISMS OF INTRINSICALLY DISORDERED CHAPERONES

On the basis of their molecular mechanism of action, chaperones have been divided into several functional subclasses: the "(un)folding" chaperones that utilize the ATP-dependent conformational changes to promote unfolding and subsequent refolding of their substrates; the "holding" chaperones that hold partially folded substrates in the folding-competent state, preventing their aggregation while waiting for the available "folding" chaperones; the "disaggregating" chaperones that are responsible for the solubilization of aggregated proteins; the "foldases" that catalyze the folding process by directly accelerating the protein folding rate-limiting steps; and the "redox" chaperones that help prevent the consequences of the oxidative damage. Illustrative members of all these chaperone classes were discussed in this chapter to show that all of them clearly belong to the family of IDPs, being either completely disordered or containing functionally important long disordered regions. Some potential modes of the intrinsic disorder involvement in the function of some of these proteins are briefly considered below.

1.3.1 Intrinsically Disordered Holding Chaperones

Many proteins in their misfolded or partially (un)folded conformation(s) are sticky and therefore possess increased propensity to aggregate. The efficiency of an aggregation process depends dramatically on protein concentration. Therefore,

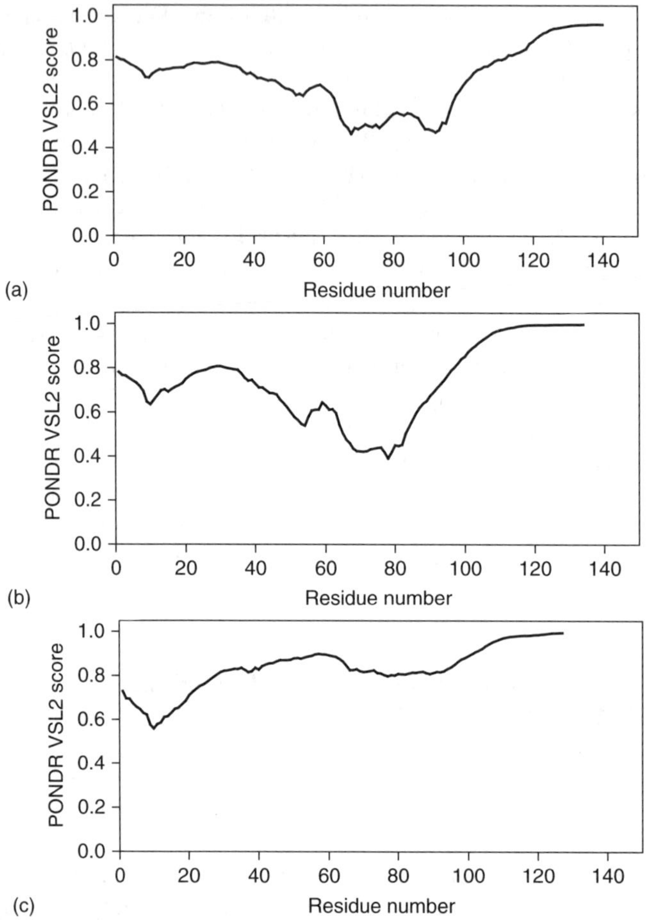

Figure 1.8 Intrinsic disorder in human α- (a), β- (b), and γ-synucleins (c) as evaluated by PONDR VSL2.

the major task of holding chaperones or holdases is to bind misfolded or partially (un)folded substrates, to decrease the pool of free molecules available for non-productive and potentially toxic aggregation, and to hold misfolded/aggregated proteins until they can be disaggregated and refolded by the ATP-dependent chaperones. As mentioned above, sHsps, which normally exist as oligomers that are polydisperse and change size and organization on exposure to stress and when interacting with substrate (Basha *et al.*, 2004b; Ehrnsperger *et al.*, 1997; Horwitz, 1992; Lee *et al.*, 1997; Studer and Narberhaus, 2000), serve as holding chaperones. Three possible modes of chaperone action have been suggested for sHsps:

1. The large oligomeric form remains intact as it binds the substrate on its surface (Kim *et al.*, 2003);

2. The large oligomer breaks down into smaller subunits (that may be monomers, dimers, or larger assemblies), which exposes hydrophobic surfaces enabling binding of unfolded substrate, and subsequently reassembles into large soluble complexes that are then handled by ATP-dependent refolding machinery (Haslbeck *et al.*, 1999; Stamler *et al.*, 2005; van Montfort *et al.*, 2001a); and
3. sHsp molecules are intercalated into large insoluble protein aggregates, which enables subsequent disaggregation and refolding by the ATP-dependent refolding machinery (Basha *et al.*, 2004a; Cashikar *et al.*, 2005; Haslbeck *et al.*, 2005; Mogk *et al.*, 2003).

It is also possible that all three mechanisms are realized in nature and that the holding chaperones use different approaches to work with the different targets. Irrespective of the mechanism of action, intrinsic disorder is crucial for holding chaperones. Model substrates protected *in vitro* as well as proteins found associated with sHsps *in vivo* include proteins of a wide range of molecular masses, pI values, and structures, with no obvious common characteristics (Basha *et al.*, 2004a; Haslbeck *et al.*, 2004a). This astonishing promiscuity of sHsps and their ability to interact with very different targets "frozen" at different folding stages rely on the flexibility of the chaperones' binding sites. In fact, the intrinsically disordered nature of these binding sites helps them to accommodate different substrate proteins by adopting structurally diverse binding platforms (Jaya *et al.*, 2009). The fact that the surface of a holding chaperone possesses multiple substrate-binding sites determines the peculiar interaction mechanism where sHsps bind coils and secondary structural elements by wrapping them around the α-crystallin domain (Stamler *et al.*, 2005).

Furthermore, intrinsically disordered N- and C-terminal arms of sHsps play a number of roles important for the sHsp structure and function. They are involved in the regulation of the oligomeric state of these chaperones, stabilize their quaternary structure, regulate chaperone activity, and enhance solubility of the protein/substrate complex (Carver, 1999). For example, the evolutionarily variable N-terminal arms of sHSP subunits, which are often unresolved in the crystal structures of sHsps (Sharma *et al.*, 1997; van Montfort *et al.*, 2001b) and are disordered in cryo-EM images of α-crystallins (Haley *et al.*, 2000), are essential for interaction with substrates, and sHSPs with an N-terminal arm that is naturally short or truncated by mutagenesis most often lack chaperone function (Fu *et al.*, 2005; Haslbeck *et al.*, 2004b; Leroux *et al.*, 1997; Stromer *et al.*, 2004). Similarly, the C-terminal arms of sHsps, which protrude from the domain core of the molecule, are flexible and solvent exposed (Carver *et al.*, 1992, 1995). These highly flexible C-terminal extensions represent a conserved feature of mammalian sHsps (Caspers *et al.*, 1995). Although they share no sequence similarity, the C-terminal arms of the mammalian sHsps have several common characteristics, being polar, having no ordered structure, and being conformationally flexible. Importantly, these arms are not involved in direct binding to the substrate. However, they act as polar solubilizing agents for the relatively hydrophobic sHsp proteins and the sHsp–substrate complexes (Lindner *et al.*, 2000). If the polarity and flexibility of the extensions are disrupted, as in the hydrophobic mutant of αA-crystallin, the stability and chaperone activity of the protein are reduced significantly (Smulders *et al.*, 1996). Because of the flexibility of the C-terminal arms of the mammalian sHsps, their solubilizing function was suggested to arise from an entropic contribution to the free energy of the solution state (Lindner *et al.*, 2000). These extensions were also proposed to serve as entropic "spacers," which prevent sHsp–substrate complexes from aggregation (Lindner *et al.*, 2000). In other words, disordered segments may act via the entropic exclusion effect, a long-range repulsive force, which prevents molecules from approaching each other (Tompa and Csermely, 2004). Therefore, similar to the disordered regions of microtubule-associated proteins (Mukhopadhyay and Hoh, 2001), neurofilaments (Brown and Hoh, 1997), and nucleoporins (Rout *et al.*, 2000), highly flexible extensions of holding chaperones work as entropic brushes that, because of the thermally driven motion, maintain spacing between the chaperone oligomers.

1.3.2 Intrinsically Disordered (Un)Folding Chaperones

Kinetically trapped misfolded substrates are stuck in a local conformational energy minimum. Chaperones were proposed to assist folding by randomly disrupting the misformed bonds via repeated cycles of binding and release, allowing the substrate to resume search in the conformational space toward the global energy minimum (Tompa and Csermely, 2004). Numerous data indicate that ATP-dependent molecular chaperones use free energy from ATP binding and/or hydrolysis to minimize the concentrations of misfolded and nonproductively aggregated proteins in the cell. One of the models of such chaperone action is that they serve as unfoldases that use free energy from ATP binding and/or hydrolysis to unfold misfolded proteins to yield productive folding intermediates (Hubbard and Sander, 1991; Rothman, 1989; Rothman and Kornberg, 1986; Slepenkov and Witt, 2002b). These reestablished productive folding intermediates can fold spontaneously to the native state. In this mechanism, ATP-dependent chaperones (e.g., Hsp70 proteins) are able to lower the activation energy barrier for the transitions from the misfolded species to the productive intermediates without changing the microscopic rate constant for the folding reaction. Acting as an unfoldase, ATP-dependent chaperones reverse the nonproductive reactions leading to the misfolded species, and therefore keep the polypeptide chain in a folding-competent state (Slepenkov and Witt, 2002b). The unfoldase activity was studied in detail for several members of the Hsp70 family, such as DnaK (Slepenkov and Witt, 2002b); the mitochondrial heat shock protein, mtHsp70, which was shown to unfold preproteins in an ATP-dependent mechanism before their import into mitochondria (Voisine *et al.*, 1999); and GroEL (Shtilerman *et al.*, 1999). The critical prerequisites for the ATP-driven chaperone to serve as an unfoldase mediating proper unfolding/refolding of many different proteins in the cell are (Sharma *et al.*, 2009):

1. the propensity to specifically bind misfolded substrates with a higher affinity than their natively refolded products;
2. the capability to recruit the energy of ATP hydrolysis to favor unfolding of the bound misfolded substrate;
3. the ability to dissociate timely from the newly unfolded product, allowing the latter to refold spontaneously into a low-affinity chaperone product.

For GroEL and several other ATP-dependent chaperones, the proposed mechanism of action involves active unfolding of misfolded proteins–substrates. Importantly, to serve as an unfoldase in protein homeostasis, these ATP-dependent chaperones must collaborate with co-chaperones: GroEL works as an unfoldase in the complex with GroES, whereas Hsp70s are known to work with a J-domain co-chaperone and an NEF.

There are three potential mechanisms for the unfoldase activity of chaperones: entropic pulling model, mechanical unfolding via forcible stretching, and the entropy transfer model. The peculiarities of these mechanisms are briefly outlined below.

In Hsp70s, which consist of an N-terminal ATPase domain and a C-terminal substrate-binding domain also containing a specific lid, hydrolysis of ATP and ADP/ATP exchange in the nucleotide-binding domain control the functional properties of the substrate-binding domain by allosteric cross talk. Hsp70 interacts transiently with short extended peptide segments of the substrate made of about seven non-bulky hydrophobic residues, ideally flanked by positive charges (Rudiger *et al.*, 1997b). ATP-liganded Hsp70 exhibits low affinity for misfolded substrates and fast rates of substrate binding and release (Palleros *et al.*, 1993; Schmid *et al.*, 1994). In contrast, ADP·Hsp70 is characterized by a 100-fold higher affinity for substrate and by a very slow rate of substrate release (Mayer *et al.*, 1999; Siegenthaler and Christen, 2006). During the chaperone cycle, Hsp70s alternate between the low and the high affinity state, under the control of their co-chaperones.

Binding of Hsp70 to misfolded protein accelerates ATP hydrolysis by chaperone up to two orders of magnitude (Han and Christen, 2003; Laufen *et al.*, 1999). This substrate effect is further amplified by the J-domain-containing Hsp40 co-chaperones (Laufen *et al.*, 1999). For example, DnaJ stimulates ATP hydrolysis in the substrate-bound ATP·DnaK molecules and thus promotes the formation of considerably more stable [ADP·DnaK·substrate] complexes. These long-lived chaperone–substrate complexes then act as entropic pulling species: the dangling bound chaperone molecule actively unfolds the misfolded regions that flank the chaperone-binding sites in the substrate (Sharma *et al.*, 2009) (De Los Rios *et al.*, 2006; Goloubinoff and De Los Rios, 2007). GrpE accelerates the release of ADP and rebinding of ATP, triggering the "unlocking" of DnaK from its substrate. The unlocked chaperone may dissociate from an unfolded peptide loop, which may spontaneously refold into a more native structure. The final result of this cycle is the productive transient unfolding of a stably misfolded polypeptide region.

GroEL binds nonnative proteins by means of a ring of hydrophobic residues that line the entrance to the central cavity of its heptameric ring (Braig *et al.*, 1994; Fenton *et al.*, 1994). When GroEL, ATP, and the GroES co-chaperonin come together, massive structure changes double the GroEL cavity volume and occlude its hydrophobic binding surface (Roseman *et al.*, 1996; Xu *et al.*, 1997). In fact, before the ATP and GroES that bind the binding sites of GroEL are located 25 Å from each other, whereas on addition of ATP and GroES, the apical domain of each GroEL subunit twists upward and outward so that the binding sites move apart to a position 33 Å from one another. As a result, neighboring binding sites move apart by 8 Å and nonneighboring sites by larger increments, up to 20 Å. These large-scale movements provide the means for the mechanical unfolding of the misfolded substrate protein which, being tethered to these sites, will be forcibly stretched and partially unfolded (Lorimer, 1997; Xu and Sigler, 1998). These mechanical unfolding of the misfolded substrate protein relieves it from the misfolded state, thus restoring its capability for normal folding. Incompletely folded proteins undergo further iterations until they achieve the native state (Shtilerman *et al.*, 1999; Xu and Sigler, 1998).

Similar to the holdases discussed above, intrinsic disorder and flexibility of binding sites of ATP-dependent unfoldases add unique versatility to the recognition process since such regions can bind several different partners, enabling an enhanced speed of interaction and uncoupling specificity from binding strength (Tompa and Csermely, 2004). For example, the intrinsic disorder in the C-terminal tail of GroEL was proposed to be necessary for the ability of this chaperone to bind a wide range of unrelated, misfolded substrates (Braig *et al.*, 1994).

Another potential role of intrinsic disorder in the unfol dase activity of chaperones is described by a so-called entropy transfer model (Tompa and Csermely, 2004): the ability of disordered segments of chaperone for rapid and transient binding of the substrate and the subsequent local ordering may "pay" the thermodynamic cost of local substrate unfolding. In other words, the entropy transfer model implies the ordering of the chaperone with a concomitant unfolding of the substrate. Here, the binding-induced folding of a disordered, nonspecific binding segment of a chaperone may promote the local unfolding of the misfolded segment, and the energy required for the local unfolding of the substrate may be covered by the binding and folding of the chaperone (Tompa and Csermely, 2004).

1.3.3 Intrinsically Disordered Disaggregating Chaperones

Misfolded and partially folded proteins are often trapped in oligomeric/aggregated forms since these nonproductive aggregated states are energetically favorable and, therefore, the half-times for the return to folding-competent intermediates are typically very long. The stability of such aggregates of misfolded polypeptides is attributed to the formation of numerous small hydrophobic surfaces with increased propensity to form β-sheet-enriched structures, which tend to cooperatively associate with one another to achieve less exposure to water, thus maintaining the misfolded polypeptides tightly entangled (Sharma *et al.*,

2009). This highly cooperative nature of intermolecular interactions between the misfolded polypeptide chains restrains the local random molecular motions which, being allowed, would eventually release individual misfolded polypeptides from the aggregate, thus giving them a chance for their native spontaneous refolding. Since aggregated proteins are potentially toxic and since the chance of their spontaneous clearance is very low, nature has elaborated a complex protective system, a network of chaperones, which can recognize misfolded proteins, prevent their aggregation, unfold misfolded regions, and disaggregate preformed aggregates. Both holding and unfolding chaperones can disaggregate nonproductive aggregates. In the cytoplasm of animal cells, unfolding and disaggregation may be achieved by Hsp70/Hsp40/NEF alone. In lower organisms, a special and very powerful disaggregating machinery has evolved. For example, yeast heat-shock protein 104 (Hsp104), plant Hsp101, and their bacterial homolog caseinolytic peptidase B (ClpB) are molecular chaperones that have the ability to solubilize almost any protein that becomes aggregated after severe stress. Although these chaperones are not found in humans, Hsp104 is of considerable interest because of its ability to dissolve numerous types of aggregates (see Chapter 7).

Unfoldases can disaggregate proteins using entropic pulling, mechanical unfolding via forced stretching, and the entropy transfer mechanisms. For example, as discussed in a recent review (Sharma *et al.*, 2009), the energy of ATP hydrolysis serves to "lock" Hsp70 onto a loop at the surface of stable protein aggregates. Multiple ADP-liganded Hsp70 molecules tightly attached to loops of the same aggregated substrate polypeptide cooperate in applying stretching forces by entropic pulling (De Los Rios *et al.*, 2006). A single Hsp70, or more effectively, several Hsp70 molecules, locked to loops in substrate polypeptide chains, recruit random Brownian motions to pull apart aggregated proteins and thereby distend the loop segments caught up in aggregates. The gain in entropy resulting from the increased motility of the Hsp70·loop complexes diffusing away from the aggregate may therefore overcome the aggregate-stabilizing energy (Hubbard and Sander, 1991; Rothman, 1989). In the entropy transfer mechanism, binding and folding of a disordered chaperone fragment can induce local unfolding and local disaggregation of the misfolded segment from the tightly packed aggregate. Even "passive" holdases can cause disaggregation via the entropic exclusion effects of their disordered extensions. Here, interaction of sHsp molecules with aggregated proteins may prevent and minimize intermolecular hydrophobic interactions among the aggregating polypeptides. Then, the chaperone-bound polypeptides can be "brushed away" from the aggregated species because of the entropic brush activity of the highly flexible extensions of holding chaperones that are in the constant thermally driven motion maintaining spacing between the chaperone oligomers.

1.4 CONCLUDING REMARKS

It is difficult to overestimate the role of protein chaperones in protein homeostasis. These guardians of the cell are intimately involved at all the stages of the protein's life, starting from its birth (chaperones are awaiting the newly synthesized polypeptides exiting the ribosome and help them avoid misfolding and aggregation before the spontaneous folding into a native structure), through the maturation and productive adulthood (chaperones help in assembly of multichain complexes and in protein translocation through membranes), in the norm and in the stress (many types of stress and toxic insults can induce transient unfolding of proteins which can then misfold and aggregate, unless prevented by chaperones), and till the death (chaperones are involved in various protein's clearance mechanisms). Chaperones help proteins to fold, prevent them from misfolding and aggregation, work with misfolded and aggregated species to promote their disaggregation, productive folding, or degradation. Obviously, failure of this protective system results in the appearance of proteotoxic species that will eventually induce a strong inflammatory response, apoptosis, and tissue loss (Hinault *et al.*, 2006).

As shown in this chapter, intrinsic disorder plays a number of important roles in the action of protein chaperones. IDRs determine the promiscuity of chaperones, acting as pliable molecular recognition elements. They help to fold misfolded chains, participate in disaggregation and local unfolding of aggregated and misfolded species, and increase the solubility and foldability of proteins. The abundance of IDRs and the versatility of their functions are crucial for the success of protein chaperones.

ACKNOWLEDGMENT

This work was supported in part by the grants R01 LM007 688-01A1 and GM071 714-01A2 from the National Institute of Health, the grant EF 0 849 803 from the National Science Foundation, and the Program of the Russian Academy of Sciences for the "Molecular and Cellular Biology." We gratefully acknowledge the support of the IUPUI Signature Centers Initiative.

REFERENCES

Abeliovich A, Schmitz Y, Farinas I, Choi-Lundberg D, Ho WH, Castillo PE, Shinsky N, Verdugo JM, Armanini M, Ryan A, *et al.* Mice lacking alpha-synuclein display functional deficits in the nigrostriatal dopamine system. *Neuron* 2000;**25**:239–252.

Adler AJ, Greenfield NJ, Fasman GD. Circular dichroism and optical rotatory dispersion of proteins and polypeptides. *Methods Enzymol* 1973;**27**:675–735.

Aitken A. 14-3-3 proteins on the MAP. *Trends Biochem Sci* 1995;**20**:95–97.

Aitken A, Jones D, Soneji Y, Howell S. 14-3-3 proteins: biological function and domain structure. *Biochem Soc Trans* 1995;**23**:605–611.

Alberti S, Demand J, Esser C, Emmerich N, Schild H, Hohfeld J. Ubiquitylation of BAG-1 suggests a novel regulatory mechanism during the sorting of chaperone substrates to the proteasome. *J Biol Chem* 2002;**277**:45920–45927.

Alberti S, Esser C, Hohfeld J. BAG-1—a nucleotide exchange factor of Hsc70 with multiple cellular functions. *Cell Stress Chaperones* 2003;**8**:225–231.

Alves-Rodrigues A, Gregori L, Figueiredo-Pereira ME. Ubiquitin, cellular inclusions and their role in neurodegeneration. *Trends Neurosci* 1998;**21**:516–520.

Anfinsen CB. Principles that govern the folding of protein chains. *Science* 1973;**181**:223–230.

Arawaka S, Saito Y, Murayama S, Mori H. Lewy body in neurodegeneration with brain iron accumulation type 1 is immunoreactive for alpha-synuclein. *Neurology* 1998;**51**:887–889.

Arrigo AP, Paul C, Ducasse C, Sauvageot O, Kretz-Remy C. Small stress proteins: modulation of intracellular redox state and protection against oxidative stress. *Prog Mol Subcell Biol* 2002;**28**:171–184.

Ballinger CA, Connell P, Wu Y, Hu Z, Thompson LJ, Yin LY, Patterson C. Identification of CHIP, a novel tetratricopeptide repeat-containing protein that interacts with heat shock proteins and negatively regulates chaperone functions. *Mol Cell Biol* 1999;**19**:4535–4545.

Banecki B, Liberek K, Wall D, Wawrzynow A, Georgopoulos C, Bertoli E, Tanfani F, Zylicz M. Structure-function analysis of the zinc finger region of the DnaJ molecular chaperone. *J Biol Chem* 1996;**271**:14840–14848.

Banerjee R, Starkov AA, Beal MF, Thomas B. Mitochondrial dysfunction in the limelight of Parkinson's disease pathogenesis. *Biochim Biophys Acta* 2009;**1792**:651–663.

Barnham KJ, Masters CL, Bush AI. Neurodegenerative diseases and oxidative stress. *Nat Rev Drug Discov* 2004;**3**:205–214.

Barral JM, Broadley SA, Schaffar G, Hartl FU. Roles of molecular chaperones in protein misfolding diseases. *Semin Cell Dev Biol* 2004;**15**:17–29.

Basha E, Lee GJ, Breci LA, Hausrath AC, Buan NR, Giese KC, Vierling E. The identity of proteins associated with a small heat shock protein during heat stress *in vivo* indicates that these chaperones protect a wide range of cellular functions. *J Biol Chem* 2004a;**279**:7566–7575.

Basha E, Lee GJ, Demeler B, Vierling E. Chaperone activity of cytosolic small heat shock proteins from wheat. *Eur J Biochem* 2004b;**271**:1426–1436.

Bauer HH, Aebi U, Haner M, Hermann R, Muller M, Merkle HP. Architecture and polymorphism of fibrillar supramolecular assemblies produced by *in vitro* aggregation of human calcitonin. *J Struct Biol* 1995;**115**:1–15.

Bayer TA, Jakala P, Hartmann T, Havas L, McLean C, Culvenor JG, Li QX, Masters CL, Falkai P, Beyreuther K. Alpha-synuclein accumulates in Lewy bodies in Parkinson's disease and dementia with Lewy bodies but not in Alzheimer's disease beta-amyloid plaque cores. *Neurosci Lett* 1999;**266**:213–216.

Bertelsen EB, Chang L, Gestwicki JE, Zuiderweg ER. Solution conformation of wild-type E. coli Hsp70 (DnaK) chaperone complexed with ADP and substrate. *Proc Natl Acad Sci U S A* 2009;**106**:8471–8476.

Berzofsky JA. Intrinsic and extrinsic factors in protein antigenic structure. *Science* 1985;**229**:932–940.

Beyer K. alpha-Synuclein structure, posttranslational modification and alternative splicing as aggregation enhancers. *Acta Neuropathol (Berl)* 2006;**112**:237–251.

Blatch GL, Lassle M. The tetratricopeptide repeat: a structural motif mediating protein-protein interactions. *Bioessays* 1999;**21**:932–939.

Boelens WC, Croes Y, de Ruwe M, de Reu L, de Jong WW. Negative charges in the C-terminal domain stabilize the alphaB-crystallin complex. *J Biol Chem* 1998;**273**:28085–28090.

Bose S, Weikl T, Bugl H, Buchner J. Chaperone function of Hsp90-associated proteins. *Science* 1996;**274**:1715–1717.

Bracken C, Iakoucheva LM, Romero PR, Dunker AK. Combining prediction, computation and experiment for the characterization of protein disorder. *Curr Opin Struct Biol* 2004;**14**:570–576.

Braig K, Otwinowski Z, Hegde R, Boisvert DC, Joachimiak A, Horwich AL, Sigler PB. The crystal structure of the bacterial chaperonin GroEL at 2.8 A. *Nature* 1994;**371**:578–586.

Broadie K, Rushton E, Skoulakis EM, Davis RL. Leonardo, a Drosophila 14-3-3 protein involved in learning, regulates presynaptic function. *Neuron* 1997;**19**:391–402.

Bron P, Giudice E, Rolland JP, Buey RM, Barbier P, Diaz JF, Peyrot V, Thomas D, Garnier C. Apo-Hsp90 coexists in two open conformational states in solution. *Biol Cell* 2008;**100**:413–425.

Brown HG, Hoh JH. Entropic exclusion by neurofilament sidearms: a mechanism for maintaining interfilament spacing. *Biochemistry* 1997;**36**:15035–15040.

Browne SE, Bowling AC, MacGarvey U, Baik MJ, Berger SC, Muqit MM, Bird ED, Beal MF. Oxidative damage and metabolic dysfunction in Huntington's disease: selective vulnerability of the basal ganglia. *Ann Neurol* 1997;**41**:646–653.

Bubber P, Haroutunian V, Fisch G, Blass JP, Gibson GE. Mitochondrial abnormalities in Alzheimer brain: mechanistic implications. *Ann Neurol* 2005;**57**:695–703.

Buchman VL, Adu J, Pinon LG, Ninkina NN, Davies AM. Persyn, a member of the synuclein family, influences neurofilament network integrity. *Nat Neurosci* 1998a;**1**:101–103.

Buchman VL, Hunter HJ, Pinon LG, Thompson J, Privalova EM, Ninkina NN, Davies AM. Persyn, a member of the synuclein family, has a distinct pattern of expression in the developing nervous system. *J Neurosci* 1998b;**18**:9335–9341.

Buchner J. Hsp90 & Co. - a holding for folding. *Trends Biochem Sci* 1999;**24**:136–141.

Bukau B, Weissman J, Horwich A. Molecular chaperones and protein quality control. *Cell* 2006;**125**:443–451.

Callebaut I, Renoir JM, Lebeau MC, Massol N, Burny A, Baulieu EE, Mornon JP. An immunophilin that binds M(r) 90,000 heat shock protein: main structural features of a mammalian p59 protein. *Proc Natl Acad Sci U S A* 1992;**89**:6270–6274.

Candido EP. The small heat shock proteins of the nematode Caenorhabditis elegans: structure, regulation and biology. *Prog Mol Subcell Biol* 2002;**28**:61–78.

Carrigan PE, Nelson GM, Roberts PJ, Stoffer J, Riggs DL, Smith DF. Multiple domains of the co-chaperone Hop are important for Hsp70 binding. *J Biol Chem* 2004;**279**:16185–16193.

Carver JA. Probing the structure and interactions of crystallin proteins by NMR spectroscopy. *Prog Retin Eye Res* 1999;**18**:431–462.

Carver JA, Aquilina JA, Truscott RJ, Ralston GB. Identification by 1H NMR spectroscopy of flexible C-terminal extensions in bovine lens alpha-crystallin. *FEBS Lett* 1992;**311**:143–149.

Carver JA, Esposito G, Schwedersky G, Gaestel M. 1H NMR spectroscopy reveals that mouse Hsp25 has a flexible C-terminal extension of 18 amino acids. *FEBS Lett* 1995;**369**:305–310.

Cashikar AG, Duennwald M, Lindquist SL. A chaperone pathway in protein disaggregation. Hsp26 alters the nature of protein aggregates to facilitate reactivation by Hsp104. *J Biol Chem* 2005;**280**:23869–23875.

Caspers GJ, Leunissen JA, de Jong WW. The expanding small heat-shock protein family, and structure predictions of the conserved "alpha-crystallin domain". *J Mol Evol* 1995;**40**:238–248.

Chadli A, Bouhouche I, Sullivan W, Stensgard B, McMahon N, Catelli MG, Toft DO. Dimerization and N-terminal domain proximity underlie the function of the molecular chaperone heat shock protein 90. *Proc Natl Acad Sci U S A* 2000;**97**:12524–12529.

Chambraud B, Rouviere-Fourmy N, Radanyi C, Hsiao K, Peattie DA, Livingston DJ, Baulieu EE. Overexpression of p59-HBI (FKBP59), full length and domains, and characterization of PPlase activity. *Biochem Biophys Res Commun* 1993;**196**:160–166.

Chandra S, Gallardo G, Fernandez-Chacon R, Schluter OM, Sudhof TC. Alpha-synuclein cooperates with CSPalpha in preventing neurodegeneration. *Cell* 2005;**123**:383–396.

Cheetham ME, Caplan AJ. Structure, function and evolution of DnaJ: conservation and adaptation of chaperone function. *Cell Stress Chaperones* 1998;**3**:28–36.

Chen YJ, Inouye M. The intramolecular chaperone-mediated protein folding. *Curr Opin Struct Biol* 2008;**18**:765–770.

Chen YG, Siddhanta A, Austin CD, Hammond SM, Sung TC, Frohman MA, Morris AJ, Shields D. Phospholipase D stimulates release of nascent secretory vesicles from the trans-Golgi network. *J Cell Biol* 1997;**138**:495–504.

Chen S, Sullivan WP, Toft DO, Smith DF. Differential interactions of p23 and the TPR-containing proteins Hop, Cyp40, FKBP52 and FKBP51 with Hsp90 mutants. *Cell Stress Chaperones* 1998;**3**:118–129.

Chiti F, Dobson CM. Protein misfolding, functional amyloid, and human disease. *Annu Rev Biochem* 2006;**75**:333–366.

Ciocca DR, Vargas-Roig LM. Hsp27 as a prognostic and predictive factor in cancer. *Prog Mol Subcell Biol* 2002;**28**:205–218.

Clayton DF, George JM. The synucleins: a family of proteins involved in synaptic function, plasticity, neurodegeneration and disease. *Trends Neurosci* 1998;**21**:249–254.

Clayton DF, George JM. Synucleins in synaptic plasticity and neurodegenerative disorders. *J Neurosci Res* 1999;**58**:120–129.

Connell P, Ballinger CA, Jiang J, Wu Y, Thompson LJ, Hohfeld J, Patterson C. The co-chaperone CHIP regulates protein triage decisions mediated by heat-shock proteins. *Nat Cell Biol* 2001;**3**:93–96.

Crick FH. On protein synthesis. *Symp Soc Exp Biol* 1958;**12**:138–163.

Dachsel JC, Lincoln SJ, Gonzalez J, Ross OA, Dickson DW, Farrer MJ. The ups and downs of alpha-synuclein mRNA expression. *Mov Disord* 2007;**22**:293–295.

da Costa CA, Ancolio K, Checler F. Wild-type but not Parkinson's disease-related ala-53-> Thr mutant alpha-synuclein protects neuronal cells from apoptotic stimuli. *J Biol Chem* 2000;**275**:24065–24069.

Das AK, Cohen PW, Barford D. The structure of the tetratricopeptide repeats of protein phosphatase 5: implications for TPR-mediated protein-protein interactions. *EMBO J* 1998;**17**:1192–1199.

Daughdrill GW, Pielak GJ, Uversky VN, Cortese MS, Dunker AK. Natively disordered proteins. In: Buchner J, Kiefhaber T, editors. *Handbook of protein folding*. Weinheim, Germany: Wiley-VCH, Verlag GmbH & Co. KGaA; 2005. pp. 271–353.

Davidson WS, Jonas A, Clayton DF, George JM. Stabilization of alpha-synuclein secondary structure upon binding to synthetic membranes. *J Biol Chem* 1998;**273**:9443–9449.

Davidson SM, Loones MT, Duverger O, Morange M. The developmental expression of small HSP. *Prog Mol Subcell Biol* 2002;**28**:103–128.

Davies TH, Ning YM, Sanchez ER. A new first step in activation of steroid receptors: hormone-induced switching of FKBP51 and FKBP52 immunophilins. *J Biol Chem* 2002;**277**:4597–4600.

De Los Rios P, Ben-Zvi A, Slutsky O, Azem A, Goloubinoff P. Hsp70 chaperones accelerate protein translocation and the unfolding of stable protein aggregates by entropic pulling. *Proc Natl Acad Sci U S A* 2006;**103**:6166–6171.

Derham BK, Harding JJ. Alpha-crystallin as a molecular chaperone. *Prog Retin Eye Res* 1999;**18**:463–509.

Dev KK, Hofele K, Barbieri S, Buchman VL, van der Putten H. Part II: alpha-synuclein and its molecular pathophysiological role in neurodegenerative disease. *Neuropharmacology* 2003;**45**:14–44.

Devi L, Raghavendran V, Prabhu BM, Avadhani NG, Anandatheerthavarada HK. Mitochondrial import and accumulation of alpha-synuclein impair complex I in human dopaminergic neuronal cultures and Parkinson disease brain. *J Biol Chem* 2008;**283**:9089–9100.

Ding TT, Lee SJ, Rochet JC, Lansbury PT Jr. Annular alpha-synuclein protofibrils are produced when spherical protofibrils are incubated in solution or bound to brain-derived membranes. *Biochemistry* 2002;**41**:10209–10217.

Dooling EC, Schoene WC, Richardson EP Jr. Hallervorden-Spatz syndrome. *Arch Neurol* 1974;**30**:70–83.

Doong H, Vrailas A, Kohn EC. What's in the 'BAG'?—a functional domain analysis of the BAG-family proteins. *Cancer Lett* 2002;**188**:25–32.

Dosztanyi Z, Csizmok V, Tompa P, Simon I. IUPred: web server for the prediction of intrinsically unstructured regions of proteins based on estimated energy content. *Bioinformatics* 2005;**21**:3433–3434.

Duncan AJ, Heales SJ. Nitric oxide and neurological disorders. *Mol Aspects Med* 2005;**26**:67–96.

Dunker AK, Brown CJ, Lawson JD, Iakoucheva LM, Obradovic Z. Intrinsic disorder and protein function. *Biochemistry* 2002a;**41**:6573–6582.

Dunker AK, Brown CJ, Obradovic Z. Identification and functions of usefully disordered proteins. *Adv Protein Chem* 2002b;**62**:25–49.

Dunker AK, Cortese MS, Romero P, Iakoucheva LM, Uversky VN. Flexible Nets: the roles of intrinsic disorder in protein interaction networks. *FEBS J* 2005;**272**:5129–5148.

Dunker AK, Garner E, Guilliot S, Romero P, Albrecht K, Hart J, Obradovic Z, Kissinger C, Villafranca JE. Protein disorder and the evolution of molecular recognition: theory, predictions and observations. *Pac Symp Biocomput* 1998;**3**:473–484.

Dunker AK, Lawson JD, Brown CJ, Williams RM, Romero P, Oh JS, Oldfield CJ, Campen AM, Ratliff CM, Hipps KW, *et al.* Intrinsically disordered protein. *J Mol Graph Model* 2001;**19**:26–59.

Dunker AK, Obradovic Z. The protein trinity—linking function and disorder. *Nat Biotechnol* 2001;**19**:805–806.

Dunker AK, Obradovic Z, Romero P, Garner EC, Brown CJ. Intrinsic protein disorder in complete genomes. *Genome Inform Ser Workshop Genome Inform* 2000;**11**:161–171.

Dupuis L, Gonzalez de Aguilar JL, Oudart H, de Tapia M, Barbeito L, Loeffler JP. Mitochondria in amyotrophic lateral sclerosis: a trigger and a target. *Neurodegener Dis* 2004;**1**:245–254.

Dyson HJ, Wright PE. Insights into the structure and dynamics of unfolded proteins from nuclear magnetic resonance. *Adv Protein Chem* 2002;**62**:311–340.

Dyson HJ, Wright PE. Unfolded proteins and protein folding studied by NMR. *Chem Rev* 2004;**104**:3607–3622.

Dyson HJ, Wright PE. Elucidation of the protein folding landscape by NMR. *Methods Enzymol* 2005a;**394**:299–321.

Dyson HJ, Wright PE. Intrinsically unstructured proteins and their functions. *Nat Rev Mol Cell Biol* 2005b;**6**:197–208.

Easton DP, Kaneko Y, Subjeck JR. The hsp110 and Grp1 70 stress proteins: newly recognized relatives of the Hsp70s. *Cell Stress Chaperones* 2000;**5**:276–290.

Ehrnsperger M, Graber S, Gaestel M, Buchner J. Binding of non-native protein to Hsp25 during heat shock creates a reservoir of folding intermediates for reactivation. *EMBO J* 1997;**16**:221–229.

Ellis J. Proteins as molecular chaperones. *Nature* 1987;**328**:378–379.

Erbse A, Mayer MP, Bukau B. Mechanism of substrate recognition by Hsp70 chaperones. *Biochem Soc Trans* 2004;**32**:617–621.

Fasman GD. *Circular dichroism and the conformational analysis of biomolecules*. New York: Plenem Press; 1996.

Fatokun AA, Stone TW, Smith RA. Oxidative stress in neurodegeneration and available means of protection. *Front Biosci* 2008;**13**:3288–3311.

Fenton WA, Kashi Y, Furtak K, Horwich AL. Residues in chaperonin GroEL required for polypeptide binding and release. *Nature* 1994;**371**:614–619.

Fink AL. Natively unfolded proteins. *Curr Opin Struct Biol* 2005;**15**:35–41.

Fischer G, Bang H, Mech C. Determination of enzymatic catalysis for the cis-trans-isomerization of peptide binding in proline-containing peptides. *Biomed Biochim Acta* 1984;**43**:1101–1111.

Fontana A, de Laureto PP, de Filippis V, Scaramella E, Zambonin M. Probing the partly folded states of proteins by limited proteolysis. *Fold Des* 1997;**2**: R17–R26.

Fontana A, de Laureto PP, Spolaore B, Frare E, Picotti P, Zambonin M. Probing protein structure by limited proteolysis. *Acta Biochim Pol* 2004;**51**:299–321.

Franck E, Madsen O, van Rheede T, Ricard G, Huynen MA, de Jong WW. Evolutionary diversity of vertebrate small heat shock proteins. *J Mol Evol* 2004;**59**:792–805.

Frimpong AK, Abzalimov RR, Uversky VN, Kaltashov IA. Characterization of intrinsically disordered proteins with electrospray ionization mass spectrometry: conformational heterogeneity of alpha-synuclein. *Proteins* 2010;**78**:714–722.

Frydman J. Folding of newly translated proteins *in vivo*: the role of molecular chaperones. *Annu Rev Biochem* 2001;**70**:603–647.

Fu X, Zhang H, Zhang X, Cao Y, Jiao W, Liu C, Song Y, Abulimiti A, Chang Z. A dual role for the N-terminal region of Mycobacterium tuberculosis Hsp16.3 in self-oligomerization and binding denaturing substrate proteins. *J Biol Chem* 2005;**280**:6337–6348.

Gaeta A, Hider RC. The crucial role of metal ions in neurodegeneration: the basis for a promising therapeutic strategy. *Br J Pharmacol* 2005;**146**:1041–1059.

Gai WP, Power JH, Blumbergs PC, Blessing WW. Multiple-system atrophy: a new alpha-synuclein disease? *Lancet* 1998;**352**:547–548.

Galat A. Peptidylproline cis-trans-isomerases: immunophilins. *Eur J Biochem* 1993;**216**:689–707.

Galpern WR, Lang AE. Interface between tauopathies and synucleinopathies: a tale of two proteins. *Ann Neurol* 2006;**59**:449–458.

Galvin JE, Lee VM, Trojanowski JQ. Synucleinopathies: clinical and pathological implications. *Arch Neurol* 2001;**58**:186–190.

Galvin JE, Uryu K, Lee VM, Trojanowski JQ. Axon pathology in Parkinson's disease and Lewy body dementia hippocampus contains alpha-, beta-, and gamma-synuclein. *Proc Natl Acad Sci U S A* 1999;**96**:13450–13455.

Gao X, Hu H. Quality control of the proteins associated with neurodegenerative diseases. *Acta Biochim Biophys Sin (Shanghai)* 2008;**40**:612–618.

Garcia-Ranea JA, Mirey G, Camonis J, Valencia A. p23 and HSP20/alpha-crystallin proteins define a conserved sequence domain present in other eukaryotic protein families. *FEBS Lett* 2002;**529**:162–167.

Genevaux P, Georgopoulos C, Kelley WL. The Hsp70 chaperone machines of Escherichia coli: a paradigm for the repartition of chaperone functions. *Mol Microbiol* 2007;**66**:840–857.

George JM, Jin H, Woods WS, Clayton DF. Characterization of a novel protein regulated during the critical period for song learning in the zebra finch. *Neuron* 1995;**15**:361–372.

Gibbs SJ, Braun JE. Emerging roles of J proteins in neurodegenerative disorders. *Neurobiol Dis* 2008;**32**:196–199.

Gibson GE, Starkov A, Blass JP, Ratan RR, Beal MF. Cause and consequence: Mitochondrial dysfunction initiates and propagates neuronal dysfunction, neuronal death and behavioral abnormalities in age-associated neurodegenerative diseases. *Biochim Biophys Acta* 2010;**1802**:122–134.

Goedert M. Filamentous nerve cell inclusions in neurodegenerative diseases: tauopathies and alpha-synucleinopathies. *Philos Trans R Soc Lond B Biol Sci* 1999;**354**:1101–1118.

Goedert M. Alpha-synuclein and neurodegenerative diseases. *Nat Rev Neurosci* 2001;**2**:492–501.

Goers J, Manning-Bog AB, McCormack AL, Millett IS, Doniach S, Di Monte DA, Uversky VN, Fink AL. Nuclear localization of alpha-synuclein and its interaction with histones. *Biochemistry* 2003;**42**:8465–8471.

Goldberg AL. Protein degradation and protection against misfolded or damaged proteins. *Nature* 2003;**426**:895–899.

Goldberger RF, Epstein CJ, Anfinsen CB. Acceleration of reactivation of reduced bovine pancreatic ribonuclease by a microsomal system from rat liver. *J Biol Chem* 1963;**238**:628–635.

Goldfarb SB, Kashlan OB, Watkins JN, Suaud L, Yan W, Kleyman TR, Rubenstein RC. Differential effects of Hsc70 and Hsp70 on the intracellular trafficking and functional expression of epithelial sodium channels. *Proc Natl Acad Sci U S A* 2006;**103**:5817–5822.

Goloubinoff P, De Los Rios P. The mechanism of Hsp70 chaperones: (entropic) pulling the models together. *Trends Biochem Sci* 2007;**32**:372–380.

Gonatas NK, Stieber A, Gonatas JO. Fragmentation of the Golgi apparatus in neurodegenerative diseases and cell death. *J Neurol Sci* 2006;**246**:21–30.

Gonzalez FJ. Roles of cytochromes P450 in chemical toxicity and oxidative stress: studies with CYP2E1. *Mutat Res* 2005;**569**:101–110.

Gosavi N, Lee HJ, Lee JS, Patel S, Lee SJ. Golgi fragmentation occurs in the cells with prefibrillar alpha-synuclein aggregates and precedes the formation of fibrillar inclusion. *J Biol Chem* 2002;**277**:48984–48992.

Greene MK, Maskos K, Landry SJ. Role of the J-domain in the cooperation of Hsp40 with Hsp70. *Proc Natl Acad Sci U S A* 1998;**95**:6108–6113.

Hague K, Lento P, Morgello S, Caro S, Kaufmann H. The distribution of Lewy bodies in pure autonomic failure: autopsy findings and review of the literature. *Acta Neuropathol (Berl)* 1997;**94**:192–196.

Haley DA, Bova MP, Huang QL, McHaourab HS, Stewart PL. Small heat-shock protein structures reveal a continuum from symmetric to variable assemblies. *J Mol Biol* 2000;**298**:261–272.

Haley DA, Horwitz J, Stewart PL. The small heat-shock protein, alphaB-crystallin, has a variable quaternary structure. *J Mol Biol* 1998;**277**:27–35.

Halliwell B. Oxidative stress and neurodegeneration: where are we now? *J Neurochem* 2006;**97**:1634–1658.

Han W, Christen P. Mechanism of the targeting action of DnaJ in the DnaK molecular chaperone system. *J Biol Chem* 2003;**278**:19038–19043.

Han H, Weinreb PH, Lansbury PT Jr. The core Alzheimer's peptide NAC forms amyloid fibrils which seed and are seeded by beta-amyloid: is NAC a common trigger or target in neurodegenerative disease? *Chem Biol* 1995;**2**:163–169.

Hara T, Nakamura K, Matsui M, Yamamoto A, Nakahara Y, Suzuki-Migishima R, Yokoyama M, Mishima K, Saito I, Okano H, *et al.* Suppression of basal autophagy in neural cells causes neurodegenerative disease in mice. *Nature* 2006;**441**:885–889.

Hartl FU, Hayer-Hartl M. Molecular chaperones in the cytosol: from nascent chain to folded protein. *Science* 2002;**295**:1852–1858.

Hashimoto M, Yoshimoto M, Sisk A, Hsu LJ, Sundsmo M, Kittel A, Saitoh T, Miller A, Masliah E. NACP, a synaptic protein involved in Alzheimer's disease, is differentially regulated during megakaryocyte differentiation. *Biochem Biophys Res Commun* 1997;**237**:611–616.

Haslbeck M, Braun N, Stromer T, Richter B, Model N, Weinkauf S, Buchner J. Hsp42 is the general small heat shock protein in the cytosol of Saccharomyces cerevisiae. *EMBO J* 2004a;**23**:638–649.

Haslbeck M, Buchner J. Chaperone function of sHsps. *Prog Mol Subcell Biol* 2002;**28**:37–59.

Haslbeck M, Ignatiou A, Saibil H, Helmich S, Frenzl E, Stromer T, Buchner J. A domain in the N-terminal part of Hsp26 is essential for chaperone function and oligomerization. *J Mol Biol* 2004b;**343**:445–455.

Haslbeck M, Miess A, Stromer T, Walter S, Buchner J. Disassembling protein aggregates in the yeast cytosol. The cooperation of Hsp26 with Ssa1 and Hsp104. *J Biol Chem* 2005;**280**:23861–23868.

Haslbeck M, Walke S, Stromer T, Ehrnsperger M, White HE, Chen S, Saibil HR, Buchner J. Hsp26: a temperature-regulated chaperone. *EMBO J* 1999;**18**:6744–6751.

Hatahet F, Ruddock LW, Ahn K, Benham A, Craik D, Ellgaard L, Ferrari D, Ventura S. Protein disulfide isomerase: a critical evaluation of its function in disulfide bond formation. *Antioxid Redox Signal* 2009;**11**:2807–2850.

Hernandez MP, Sullivan WP, Toft DO. The assembly and intermolecular properties of the hsp70-Hop-hsp90 molecular chaperone complex. *J Biol Chem* 2002;**277**:38294–38304.

Hickey E, Brandon SE, Smale G, Lloyd D, Weber LA. Sequence and regulation of a gene encoding a human 89-kilodalton heat shock protein. *Mol Cell Biol* 1989;**9**: 2615–2626.

Hinault MP, Ben-Zvi A, Goloubinoff P. Chaperones and proteases: cellular fold-controlling factors of proteins in neurodegenerative diseases and aging. *J Mol Neurosci* 2006;**30**:249–265.

Hohfeld J, Jentsch S. GrpE-like regulation of the hsc70 chaperone by the anti-apoptotic protein BAG-1. *EMBO J* 1997;**16**:6209–6216.

Hohfeld J, Minami Y, Hartl FU. Hip, a novel cochaperone involved in the eukaryotic Hsc70/Hsp40 reaction cycle. *Cell* 1995;**83**:589–598.

Horwitz J. Alpha-crystallin can function as a molecular chaperone. *Proc Natl Acad Sci U S A* 1992;**89**:10449–10453.

Hsu LJ, Mallory M, Xia Y, Veinbergs I, Hashimoto M, Yoshimoto M, Thal LJ, Saitoh T, Masliah E. Expression pattern of synucleins (non-Abeta component of Alzheimer's disease amyloid precursor protein/alpha-synuclein) during murine brain development. *J Neurochem* 1998;**71**:338–344.

Hubbard SJ, Eisenmenger F, Thornton JM. Modeling studies of the change in conformation required for cleavage of limited proteolytic sites. *Protein Sci* 1994;**3**: 757–768.

Hubbard TJ, Sander C. The role of heat-shock and chaperone proteins in protein folding: possible molecular mechanisms. *Protein Eng* 1991;**4**:711–717.

Iakoucheva LM, Brown CJ, Lawson JD, Obradovic Z, Dunker AK. Intrinsic disorder in cell-signaling and cancer-associated proteins. *J Mol Biol* 2002;**323**:573–584.

Iakoucheva LM, Kimzey AL, Masselon CD, Smith RD, Dunker AK, Ackerman EJ. Aberrant mobility phenomena of the DNA repair protein XPA. *Protein Sci* 2001;**10**:1353–1362.

Ishihara N, Mizushima N. A receptor for eating mitochondria. *Dev Cell* 2009;**17**:1–2.

Iwai A, Masliah E, Yoshimoto M, Ge N, Flanagan L, de Silva HA, Kittel A, Saitoh T. The precursor protein of non-A beta component of Alzheimer's disease amyloid is a presynaptic protein of the central nervous system. *Neuron* 1995;**14**:467–475.

Jackson PK, Eldridge AG, Freed E, Furstenthal L, Hsu JY, Kaiser BK, Reimann JD. The lore of the RINGs: substrate recognition and catalysis by ubiquitin ligases. *Trends Cell Biol* 2000;**10**:429–439.

Jackson M, Lennox G, Balsitis M, Lowe J. Lewy body dysphagia. *J Neurol Neurosurg Psychiatry* 1995;**58**:756–758.

Jakes R, Spillantini MG, Goedert M. Identification of two distinct synucleins from human brain. *FEBS Lett* 1994;**345**:27–32.

Jankovic J, Kirkpatrick JB, Blomquist KA, Langlais PJ, Bird ED. Late-onset Hallervorden-Spatz disease presenting as familial parkinsonism. *Neurology* 1985;**35**:227–234.

Jaya N, Garcia V, Vierling E. Substrate binding site flexibility of the small heat shock protein molecular chaperones. *Proc Natl Acad Sci U S A* 2009;**106**:15604–15609.

Jenco JM, Rawlingson A, Daniels B, Morris AJ. Regulation of phospholipase D2: selective inhibition of mammalian phospholipase D isoenzymes by alpha- and beta-synucleins. *Biochemistry* 1998;**37**:4901–4909.

Ji H, Liu YE, Jia T, Wang M, Liu J, Xiao G, Joseph BK, Rosen C, Shi YE. Identification of a breast cancer-specific gene, BCSG1, by direct differential cDNA sequencing. *Cancer Res* 1997;**57**:759–764.

Jones DT, Ward JJ. Prediction of disordered regions in proteins from position specific score matrices. *Proteins* 2003;**53** Suppl 6: 573–578.

Jorgensen S, Skov KW, Diderichsen B. Cloning, sequence, and expression of a lipase gene from Pseudomonas cepacia: lipase production in heterologous hosts requires two Pseudomonas genes. *J Bacteriol* 1991;**173**:559–567.

Kaltashov IA, Mohimen A. Estimates of protein surface areas in solution by electrospray ionization mass spectrometry. *Anal Chem* 2005;**77**:5370–5379.

Kappe G, Franck E, Verschuure P, Boelens WC, Leunissen JA, de Jong WW. The human genome encodes 10 alpha-crystallin-related small heat shock proteins: HspB1-10. *Cell Stress Chaperones* 2003;**8**:53–61.

Karzai AW, McMacken R. A bipartite signaling mechanism involved in DnaJ-mediated activation of the Escherichia coli DnaK protein. *J Biol Chem* 1996;**271**:11236–11246.

Kim KK, Kim R, Kim SH. Crystal structure of a small heat-shock protein. *Nature* 1998;**394**:595–599.

Kim R, Lai L, Lee HH, Cheong GW, Kim KK, Wu Z, Yokota H, Marqusee S, Kim SH. On the mechanism of chaperone activity of the small heat-shock protein of Methanococcus jannaschii. *Proc Natl Acad Sci U S A* 2003;**100**:8151–8155.

Kirkin V, McEwan DG, Novak I, Dikic I. A role for ubiquitin in selective autophagy. *Mol Cell* 2009;**34**:259–269.

Klettner A. The induction of heat shock proteins as a potential strategy to treat neurodegenerative disorders. *Drug News Perspect* 2004;**17**:299–306.

Komatsu M, Waguri S, Chiba T, Murata S, Iwata J, Tanida I, Ueno T, Koike M, Uchiyama Y, Kominami E, *et al.* Loss of autophagy in the central nervous system causes neurodegeneration in mice. *Nature* 2006;**441**:880–884.

Kostenko S, Moens U. Heat shock protein 27 phosphorylation: kinases, phosphatases, functions and pathology. *Cell Mol Life Sci* 2009;**66**:3289–3307.

Krueger-Naug AM, Plumier JC, Hopkins DA, Currie RW. Hsp27 in the nervous system: expression in pathophysiology and in the aging brain. *Prog Mol Subcell Biol* 2002;**28**:235–251.

Lai BT, Chin NW, Stanek AE, Keh W, Lanks KW. Quantitation and intracellular localization of the 85K heat shock protein by using monoclonal and polyclonal antibodies. *Mol Cell Biol* 1984;**4**:2802–2810.

Lashuel HA, Hartley D, Petre BM, Walz T, Lansbury PT Jr. Neurodegenerative disease: amyloid pores from pathogenic mutations. *Nature* 2002;**418**: 291.

Latchman DS. Protection of neuronal and cardiac cells by HSP27. *Prog Mol Subcell Biol* 2002;**28**:253–265.

Laufen T, Mayer MP, Beisel C, Klostermeier D, Mogk A, Reinstein J, Bukau B. Mechanism of regulation of hsp70 chaperones by DnaJ cochaperones. *Proc Natl Acad Sci U S A* 1999;**96**:5452–5457.

Lavedan C. The synuclein family. *Genome Res* 1998;**8**:871–880.

Lavedan C, Leroy E, Dehejia A, Buchholtz S, Dutra A, Nussbaum RL, Polymeropoulos MH. Identification, localization and characterization of the human gamma-synuclein gene. *Hum Genet* 1998;**103**:106–112.

Layfield R, Fergusson J, Aitken A, Lowe J, Landon M, Mayer RJ. Neurofibrillary tangles of Alzheimer's disease brains contain 14-3-3 proteins. *Neurosci Lett* 1996;**209**:57–60.

Lee M, Hyun D, Halliwell B, Jenner P. Effect of the overexpression of wild-type or mutant alpha-synuclein on cell susceptibility to insult. *J Neurochem* 2001;**76**:998–1009.

Lee GJ, Roseman AM, Saibil HR, Vierling E. A small heat shock protein stably binds heat-denatured model substrates and can maintain a substrate in a folding-competent state. *EMBO J* 1997;**16**:659–671.

Leidhold C, Voos W. Chaperones and proteases—guardians of protein integrity in eukaryotic organelles. *Ann N Y Acad Sci* 2007;**1113**:72–86.

Leroux MR, Melki R, Gordon B, Batelier G, Candido EP. Structure-function studies on small heat shock protein oligomeric assembly and interaction with unfolded polypeptides. *J Biol Chem* 1997;**272**:24646–24656.

Levine RL, Mosoni L, Berlett BS, Stadtman ER. Methionine residues as endogenous antioxidants in proteins. *Proc Natl Acad Sci U S A* 1996;**93**:15036–15040.

Li JY, Henning Jensen P, Dahlstrom A. Differential localization of alpha-, beta- and gamma-synucleins in the rat CNS. *Neuroscience* 2002;**113**:463–478.

Li J, Qian X, Sha B. The crystal structure of the yeast Hsp40 Ydj1 complexed with its peptide substrate. *Structure* 2003;**11**:1475–1483.

Li X, Romero P, Rani M, Dunker AK, Obradovic Z. Predicting Protein Disorder for N-, C-, and Internal Regions. *Genome Inform Ser Workshop Genome Inform* 1999;**10**:30–40.

Lim KL. Ubiquitin-proteasome system dysfunction in Parkinson's disease: current evidence and controversies. *Expert Rev Proteomics* 2007;**4**:769–781.

Lindersson E, Beedholm R, Hojrup P, Moos T, Gai W, Hendil KB, Jensen PH. Proteasomal inhibition by alpha-synuclein filaments and oligomers. *J Biol Chem* 2004;**279**:12924–12934.

Linding R, Jensen LJ, Diella F, Bork P, Gibson TJ, Russell RB. Protein disorder prediction: implications for structural proteomics. *Structure* 2003a;**11**:1453–1459.

Linding R, Russell RB, Neduva V, Gibson TJ. GlobPlot: Exploring protein sequences for globularity and disorder. *Nucleic Acids Res* 2003b;**31**:3701–3708.

Lindner RA, Carver JA, Ehrnsperger M, Buchner J, Esposito G, Behlke J, Lutsch G, Kotlyarov A, Gaestel M. Mouse Hsp25, a small shock protein. The role of its C-terminal extension in oligomerization and chaperone action. *Eur J Biochem* 2000;**267**:1923–1932.

Lindquist S. The heat-shock response. *Annu Rev Biochem* 1986;**55**:1151–1191.

Lindquist S, Craig EA. The heat-shock proteins. *Annu Rev Genet* 1988;**22**:631–677.

Liu J, Rost B. NORSp: Predictions of long regions without regular secondary structure. *Nucleic Acids Res* 2003;**31**:3833–3835.

Longhi S, Uversky VN, editors. *Instrumental analysis of intrinsically disordered proteins*. Hoboken (NJ): John Wiley & Sons, Inc.; 2010.

Lorimer G. Protein folding: folding with a two-stroke motor. *Nature* 1997;**388**:720–721, 723.

Lu Z, Cyr DM. The conserved carboxyl terminus and zinc finger-like domain of the co-chaperone Ydj1 assist Hsp70 in protein folding. *J Biol Chem* 1998;**273**:5970–5978.

Lucking CB, Brice A. Alpha-synuclein and Parkinson's disease. *Cell Mol Life Sci* 2000;**57**:1894–1908.

Lundvig D, Lindersson E, Jensen PH. Pathogenic effects of alpha-synuclein aggregation. *Brain Res Mol Brain Res* 2005;**134**:3–17.

Malandrini A, Cesaretti S, Mulinari M, Palmeri S, Fabrizi GM, Villanova M, Parrotta E, Montagnani A, Montagnani M, Anichini M, *et al.* Acanthocytosis, retinitis pigmentosa, pallidal degeneration. Report of two cases without serum lipid abnormalities. *J Neurol Sci* 1996;**140**:129–131.

Manning-Bog AB, McCormack AL, Li J, Uversky VN, Fink AL, Di Monte DA. The herbicide paraquat causes up-regulation and aggregation of alpha-synuclein in mice: paraquat and alpha-synuclein. *J Biol Chem* 2002;**277**:1641–1644.

Markus G. Protein substrate conformation and proteolysis. *Proc Natl Acad Sci U S A* 1965;**54**:253–258.

Maroteaux L, Campanelli JT, Scheller RH. Synuclein: a neuron-specific protein localized to the nucleus and presynaptic nerve terminal. *J Neurosci* 1988;**8**:2804–2815.

Martinez-Yamout M, Legge GB, Zhang O, Wright PE, Dyson HJ. Solution structure of the cysteine-rich domain of the Escherichia coli chaperone protein DnaJ. *J Mol Biol* 2000;**300**:805–818.

Maruya M, Sameshima M, Nemoto T, Yahara I. Monomer arrangement in HSP90 dimer as determined by decoration with N and C-terminal region specific antibodies. *J Mol Biol* 1999;**285**:903–907.

Massol N, Lebeau MC, Renoir JM, Faber LE, Baulieu EE. Rabbit FKBP59-heat shock protein binding immunophillin (HBI) is a calmodulin binding protein. *Biochem Biophys Res Commun* 1992;**187**:1330–1335.

Matsuda N, Tanaka K. Does impairment of the ubiquitin-proteasome system or the autophagy-lysosome pathway predispose individuals to neurodegenerative disorders such as Parkinson's disease? *J Alzheimers Dis* 2010;**19**:1–9.

Mayer MP, Brehmer D, Gassler CS, Bukau B. Hsp70 chaperone machines. *Adv Protein Chem* 2001;**59**:1–44.

Mayer MP, Bukau B. Hsp70 chaperones: cellular functions and molecular mechanism. *Cell Mol Life Sci* 2005;**62**:670–684.

Mayer MP, Laufen T, Paal K, McCarty JS, Bukau B. Investigation of the interaction between DnaK and DnaJ by surface plasmon resonance spectroscopy. *J Mol Biol* 1999;**289**:1131–1144.

McCarty JS, Buchberger A, Reinstein J, Bukau B. The role of ATP in the functional cycle of the DnaK chaperone system. *J Mol Biol* 1995;**249**:126–137.

McClellan AJ, Tam S, Kaganovich D, Frydman J. Protein quality control: chaperones culling corrupt conformations. *Nat Cell Biol* 2005;**7**:736–741.

McLaughlin SH, Smith HW, Jackson SE. Stimulation of the weak ATPase activity of human hsp90 by a client protein. *J Mol Biol* 2002;**315**:787–798.

Meller N, Liu YC, Collins TL, Bonnefoy-Berard N, Baier G, Isakov N, Altman A. Direct interaction between protein kinase C theta (PKC theta) and 14-3-3 tau in T cells: 14-3-3 overexpression results in inhibition of PKC theta translocation and function. *Mol Cell Biol* 1996;**16**:5782–5791.

Melone MA, Jori FP, Peluso G. Huntington's disease: new frontiers for molecular and cell therapy. *Curr Drug Targets* 2005;**6**:43–56.

Michaud S, Morrow G, Marchand J, Tanguay RM. Drosophila small heat shock proteins: cell and organelle-specific chaperones? *Prog Mol Subcell Biol* 2002;**28**:79–101.

Mikhalyi E *Application of proteolytic enzymes to protein structure studies*. Boca Raton (FL): CRC Press; 1978.

Miyagi A, Tsunaka Y, Uchihashi T, Mayanagi K, Hirose S, Morikawa K, Ando T. Visualization of intrinsically disordered regions of proteins by high-speed atomic force microscopy. *Chemphyschem* 2008;**9**:1859–1866.

Mogk A, Schlieker C, Friedrich KL, Schonfeld HJ, Vierling E, Bukau B. Refolding of substrates bound to small Hsps relies on a disaggregation reaction mediated most efficiently by ClpB/DnaK. *J Biol Chem* 2003;**278**:31033–31042.

Moreira PI, Zhu X, Wang X, Lee HG, Nunomura A, Petersen RB, Perry G, Smith MA. Mitochondria: a therapeutic target in neurodegeneration. *Biochim Biophys Acta* 2010;**1802**:212–220.

Mukhopadhyay R, Hoh JH. AFM force measurements on microtubule-associated proteins: the projection domain exerts a long-range repulsive force. *FEBS Lett* 2001;**505**:374–378.

Murata S, Minami Y, Minami M, Chiba T, Tanaka K. CHIP is a chaperone-dependent E3 ligase that ubiquitylates unfolded protein. *EMBO Rep* 2001;**2**:1133–1138.

Nagradova N. Enzymes catalyzing protein folding and their cellular functions. *Curr Protein Pept Sci* 2007;**8**:273–282.

Nakajo S, Shioda S, Nakai Y, Nakaya K. Localization of phosphoneuroprotein 14 (PNP 14) and its mRNA expression in rat brain determined by immunocytochemistry and *in situ* hybridization. *Brain Res Mol Brain Res* 1994;**27**:81–86.

Nakajo S, Tsukada K, Kameyama H, Furuyama Y, Nakaya K. Distribution of phosphoneuroprotein 14 (PNP 14) in vertebrates: its levels as determined by enzyme immunoassay. *Brain Res* 1996;**741**:180–184.

Nakajo S, Tsukada K, Omata K, Nakamura Y, Nakaya K. A new brain-specific 14-kDa protein is a phosphoprotein. Its complete amino acid sequence and evidence for phosphorylation. *Eur J Biochem* 1993;**217**:1057–1063.

Nemoto T, Matsusaka T, Ota M, Takagi T, Collinge DB, Walther-Larsen H. Dimerization characteristics of the 94-kDa glucose-regulated protein. *J Biochem* 1996;**120**:249–256.

Ninkina NN, Alimova-Kost MV, Paterson JW, Delaney L, Cohen BB, Imreh S, Gnuchev NV, Davies AM, Buchman VL. Organization, expression and polymorphism of the human persyn gene. *Hum Mol Genet* 1998;**7**:1417–1424.

Ninkina NN, Privalova EM, Pinon LG, Davies AM, Buchman VL. Developmentally regulated expression of persyn, a member of the synuclein family, in skin. *Exp Cell Res* 1999;**246**:308–311.

Odunuga OO, Longshaw VM, Blatch GL. Hop: more than an Hsp70/Hsp90 adaptor protein. *Bioessays* 2004;**26**:1058–1068.

Oh HJ, Easton D, Murawski M, Kaneko Y, Subjeck JR. The chaperoning activity of hsp110. Identification of functional domains by use of targeted deletions. *J Biol Chem* 1999;**274**:15712–15718.

Ohtsuka K, Hata M. Molecular chaperone function of mammalian Hsp70 and Hsp40—a review. *Int J Hyperthermia* 2000;**16**:231–245.

Okazaki H, Lipkin LE, Aronson SM. Diffuse intracytoplasmic ganglionic inclusions (Lewy type) associated with progressive dementia and quadriparesis in flexion. *J Neuropathol Exp Neurol* 1961;**20**:237–244.

Oldfield CJ, Cheng Y, Cortese MS, Brown CJ, Uversky VN, Dunker AK. Comparing and combining predictors of mostly disordered proteins. *Biochemistry* 2005;**44**:1989–2000.

Ostrerova N, Petrucelli L, Farrer M, Mehta N, Choi P, Hardy J, Wolozin B. alpha-Synuclein shares physical and functional homology with 14-3-3 proteins. *J Neurosci* 1999;**19**:5782–5791.

Owens-Grillo JK, Stancato LF, Hoffmann K, Pratt WB, Krishna P. Binding of immunophilins to the 90kDa heat shock protein (hsp90) via a tetratricopeptide repeat domain is a conserved protein interaction in plants. *Biochemistry* 1996;**35**:15249–15255.

Palleros DR, Reid KL, Shi L, Welch WJ, Fink AL. ATP-induced protein-Hsp70 complex dissociation requires K+ but not ATP hydrolysis. *Nature* 1993;**365**: 664–666.

Panaretou B, Siligardi G, Meyer P, Maloney A, Sullivan JK, Singh S, Millson SH, Clarke PA, Naaby-Hansen S, Stein R, *et al.* Activation of the ATPase activity of hsp90 by the stress-regulated cochaperone aha1. *Mol Cell* 2002;**10**:1307–1318.

Patury S, Miyata Y, Gestwicki JE. Pharmacological targeting of the Hsp70 chaperone. *Curr Top Med Chem* 2009;**9**:1337–1351.

Pedersen JS, Dikov D, Flink JL, Hjuler HA, Christiansen G, Otzen DE. The changing face of glucagon fibrillation: structural polymorphism and conformational imprinting. *J Mol Biol* 2006;**355**:501–523.

Perdew GH, Hord N, Hollenback CE, Welsh MJ. Localization and characterization of the 86- and 84-kDa heat shock proteins in Hepa 1c1c7 cells. *Exp Cell Res* 1993;**209**:350–356.

Petersen K, Olesen OF, Mikkelsen JD. Developmental expression of alpha-synuclein in rat hippocampus and cerebral cortex. *Neuroscience* 1999;**91**:651–659.

Picard D. Heat-shock protein 90, a chaperone for folding and regulation. *Cell Mol Life Sci* 2002;**59**:1640–1648.

Pratt WB, Toft DO. Steroid receptor interactions with heat shock protein and immunophilin chaperones. *Endocr Rev* 1997;**18**:306–360.

Pratt WB, Toft DO. Regulation of signaling protein function and trafficking by the hsp90/hsp70-based chaperone machinery. *Exp Biol Med (Maywood)* 2003;**228**:111–133.

Prilusky J, Felder CE, Zeev-Ben-Mordehai T, Rydberg EH, Man O, Beckmann JS, Silman I, Sussman JL. FoldIndex: a simple tool to predict whether a given protein sequence is intrinsically unfolded. *Bioinformatics* 2005;**21**:3435–3438.

Privalov PL. Stability of proteins: small globular proteins. *Adv Protein Chem* 1979;**33**:167–241.

Prodromou C, Panaretou B, Chohan S, Siligardi G, O'Brien R, Ladbury JE, Roe SM, Piper PW, Pearl LH. The ATPase cycle of Hsp90 drives a molecular 'clamp' via transient dimerization of the N-terminal domains. *EMBO J* 2000;**19**:4383–4392.

Prodromou C, Roe SM, O'Brien R, Ladbury JE, Piper PW, Pearl LH. Identification and structural characterization of the ATP/ADP-binding site in the Hsp90 molecular chaperone. *Cell* 1997;**90**:65–75.

Provencher SW, Glockner J. Estimation of globular protein secondary structure from circular dichroism. *Biochemistry* 1981;**20**:33–37.

Ptitsyn O. Molten globule and protein folding. *Adv Protein Chem* 1995;**47**:83–229.

Ptitsyn OB, Uversky VN. The molten globule is a third thermodynamical state of protein molecules. *FEBS Lett* 1994;**341**:15–18.

Qureshi GA, Parvez SH, editors. *Oxidative stress and neurodegenerative disorders*. Amsterdam, Netherlands: Elsevier; 2007.

Radanyi C, Chambraud B, Baulieu EE. The ability of the immunophilin FKBP59-HBI to interact with the 90-kDa heat shock protein is encoded by its tetratricopeptide repeat domain. *Proc Natl Acad Sci U S A* 1994;**91**:11197–11201.

Radivojac P, Iakoucheva LM, Oldfield CJ, Obradovic Z, Uversky VN, Dunker AK. Intrinsic disorder and functional proteomics. *Biophys J* 2007;**92**:1439–1456.

Rambourg A, Clermont Y. Three-dimensional structure of the Golgi apparatus in mammalian cells. In: Berger EG, Roth J, editors. *The Golgi apparatus*. Basel, Switzerland: Birkhäuser; 1997. pp. 37–61.

Raviol H, Bukau B, Mayer MP. Human and yeast Hsp110 chaperones exhibit functional differences. *FEBS Lett* 2006a;**580**:168–174.

Raviol H, Sadlish H, Rodriguez F, Mayer MP, Bukau B. Chaperone network in the yeast cytosol: Hsp110 is revealed as an Hsp70 nucleotide exchange factor. *EMBO J* 2006b;**25**:2510–2518.

Receveur-Brechot V, Bourhis JM, Uversky VN, Canard B, Longhi S. Assessing protein disorder and induced folding. *Proteins* 2006;**62**:24–45.

Reddy PS, Bhagyalakshmi A. Changes in oxidative metabolism in selected tissues of the crab (Scylla serrata) in response to cadmium toxicity. *Ecotoxicol Environ Saf* 1994;**29**:255–264.

Richter K, Muschler P, Hainzl O, Buchner J. Coordinated ATP hydrolysis by the Hsp90 dimer. *J Biol Chem* 2001;**276**:33689–33696.

Richter K, Muschler P, Hainzl O, Reinstein J, Buchner J. Sti1 is a non-competitive inhibitor of the Hsp90 ATPase. Binding prevents the N-terminal dimerization reaction during the atpase cycle. *J Biol Chem* 2003;**278**:10328–10333.

Ringe D, Petsko GA. Study of protein dynamics by X-ray diffraction. *Methods Enzymol* 1986;**131**:389–433.

Romero P, Obradovic Z, Li X, Garner EC, Brown CJ, Dunker AK. Sequence complexity of disordered protein. *Proteins* 2001;**42**:38–48.

Roseman AM, Chen S, White H, Braig K, Saibil HR. The chaperonin ATPase cycle: mechanism of allosteric switching and movements of substrate-binding domains in GroEL. *Cell* 1996;**87**:241–251.

Ross OA, Braithwaite AT, Skipper LM, Kachergus J, Hulihan MM, Middleton FA, Nishioka K, Fuchs J, Gasser T, Maraganore DM, *et al.* Genomic investigation of alpha-synuclein multiplication and parkinsonism. *Ann Neurol* 2008;**63**: 743–750.

Rothman JE, Kornberg RD. Cell biology: an unfolding story of protein translocation. *Nature* 1986;**322**:209–210.

Rothman JE. Polypeptide chain binding proteins: catalysts of protein folding and related processes in cells. *Cell* 1989;**59**:591–601.

Rout MP, Aitchison JD, Suprapto A, Hjertaas K, Zhao Y, Chait BT. The yeast nuclear pore complex: composition, architecture, and transport mechanism. *J Cell Biol* 2000;**148**:635–651.

Rudiger S, Buchberger A, Bukau B. Interaction of Hsp70 chaperones with substrates. *Nat Struct Biol* 1997a;**4**:342–349.

Rudiger S, Germeroth L, Schneider-Mergener J, Bukau B. Substrate specificity of the DnaK chaperone determined by screening cellulose-bound peptide libraries. *EMBO J* 1997b;**16**:1501–1507.

Russell R, Jordan R, McMacken R. Kinetic characterization of the ATPase cycle of the DnaK molecular chaperone. *Biochemistry* 1998;**37**:596–607.

Saiki M, Honda S, Kawasaki K, Zhou D, Kaito A, Konakahara T, Morii H. Higher-order molecular packing in amyloid-like fibrils constructed with linear arrangements of hydrophobic and hydrogen-bonding side-chains. *J Mol Biol* 2005;**348**:983–998.

Scharf KD, Siddique M, Vierling E. The expanding family of Arabidopsis thaliana small heat stress proteins and a new family of proteins containing alpha-crystallin domains (Acd proteins). *Cell Stress Chaperones* 2001;**6**:225–237.

Scheufler C, Brinker A, Bourenkov G, Pegoraro S, Moroder L, Bartunik H, Hartl FU, Moarefi I. Structure of TPR domain-peptide complexes: critical elements in the assembly of the Hsp70-Hsp90 multichaperone machine. *Cell* 2000;**101**:199–210.

Schmid FX. Prolyl isomerase: enzymatic catalysis of slow protein-folding reactions. *Annu Rev Biophys Biomol Struct* 1993;**22**:123–142.

Schmid D, Baici A, Gehring H, Christen P. Kinetics of molecular chaperone action. *Science* 1994;**263**:971–973.

Schuermann JP, Jiang J, Cuellar J, Llorca O, Wang L, Gimenez LE, Jin S, Taylor AB, Demeler B, Morano KA, *et al.* Structure of the Hsp110:Hsc70 nucleotide exchange machine. *Mol Cell* 2008;**31**:232–243.

Schweers O, Schonbrunn-Hanebeck E, Marx A, Mandelkow E. Structural studies of tau protein and Alzheimer paired helical filaments show no evidence for beta-structure. *J Biol Chem* 1994;**269**:24290–24297.

Serpell LC, Sunde M, Benson MD, Tennent GA, Pepys MB, Fraser PE. The protofilament substructure of amyloid fibrils. *J Mol Biol* 2000;**300**:1033–1039.

Sha B, Lee S, Cyr DM. The crystal structure of the peptide-binding fragment from the yeast Hsp40 protein Sis1. *Structure* 2000;**8**:799–807.

Sharma SK, Christen P, Goloubinoff P. Disaggregating chaperones: an unfolding story. *Curr Protein Pept Sci* 2009;**10**:432–446.

Sharma KK, Kaur H, Kester K. Functional elements in molecular chaperone alpha-crystallin: identification of binding sites in alpha B-crystallin. *Biochem Biophys Res Commun* 1997;**239**:217–222.

Shemetov AA, Seit-Nebi AS, Gusev NB. Structure, properties, and functions of the human small heat-shock protein HSP22 (HspB8, H11, E2IG1): a critical review. *J Neurosci Res* 2008;**86**:264–269.

Shi ZZ, Zhang JW, Zheng S. What we know about ST13, a co-factor of heat shock protein, or a tumor suppressor? *J Zhejiang Univ Sci B* 2007;**8**:170–176.

Shibayama-Imazu T, Ogane K, Hasegawa Y, Nakajo S, Shioda S, Ochiai H, Nakai Y, Nakaya K. Distribution of PNP 14 (beta-synuclein) in neuroendocrine tissues: localization in Sertoli cells. *Mol Reprod Dev* 1998;**50**:163–169.

Shibayama-Imazu T, Okahashi I, Omata K, Nakajo S, Ochiai H, Nakai Y, Hama T, Nakamura Y, Nakaya K. Cell and tissue distribution and developmental change of neuron specific 14kDa protein (phosphoneuroprotein 14). *Brain Res* 1993;**622**:17–25.

Shirasu K, Lahaye T, Tan MW, Zhou F, Azevedo C, Schulze-Lefert P. A novel class of eukaryotic zinc-binding proteins is required for disease resistance signaling in barley and development in C. elegans. *Cell* 1999;**99**:355–366.

Shtilerman M, Lorimer GH, Englander SW. Chaperonin function: folding by forced unfolding. *Science* 1999;**284**:822–825.

Siegenthaler RK, Christen P. Tuning of DnaK chaperone action by nonnative protein sensor DnaJ and thermosensor GrpE. *J Biol Chem* 2006;**281**:34448–34456.

Slepenkov SV, Witt SN. Kinetic analysis of interdomain coupling in a lidless variant of the molecular chaperone DnaK: DnaK's lid inhibits transition to the low affinity state. *Biochemistry* 2002a;**41**:12224–12235.

Slepenkov SV, Witt SN. The unfolding story of the Escherichia coli Hsp70 DnaK: is DnaK a holdase or an unfoldase? *Mol Microbiol* 2002b;**45**:1197–1206.

Smith DF, Sullivan WP, Marion TN, Zaitsu K, Madden B, McCormick DJ, Toft DO. Identification of a 60-kilodalton stress-related protein, p60, which interacts with hsp90 and hsp70. *Mol Cell Biol* 1993;**13**:869–876.

Smulders R, Carver JA, Lindner RA, van Boekel MA, Bloemendal H, de Jong WW. Immobilization of the C-terminal extension of bovine alphaA-crystallin reduces chaperone-like activity. *J Biol Chem* 1996;**271**:29060–29066.

Smyth E, Syme CD, Blanch EW, Hecht L, Vasak M, Barron LD. Solution structure of native proteins with irregular folds from Raman optical activity. *Biopolymers* 2001;**58**:138–151.

Soti C, Nagy E, Giricz Z, Vigh L, Csermely P, Ferdinandy P. Heat shock proteins as emerging therapeutic targets. *Br J Pharmacol* 2005;**146**:769–780.

Spillantini MG, Crowther RA, Jakes R, Cairns NJ, Lantos PL, Goedert M. Filamentous alpha-synuclein inclusions link multiple system atrophy with Parkinson's disease and dementia with Lewy bodies. *Neurosci Lett* 1998a;**251**:205–208.

Spillantini MG, Crowther RA, Jakes R, Hasegawa M, Goedert M. alpha-Synuclein in filamentous inclusions of Lewy bodies from Parkinson's disease and dementia with lewy bodies. *Proc Natl Acad Sci U S A* 1998b;**95**:6469–6473.

Spillantini MG, Goedert M. The alpha-synucleinopathies: Parkinson's disease, dementia with Lewy bodies, and multiple system atrophy. *Ann N Y Acad Sci* 2000;**920**:16–27.

Spillantini MG, Schmidt ML, Lee VM, Trojanowski JQ, Jakes R, Goedert M. Alpha-synuclein in Lewy bodies. *Nature* 1997;**388**:839–840.

Stamler R, Kappe G, Boelens W, Slingsby C. Wrapping the alpha-crystallin domain fold in a chaperone assembly. *J Mol Biol* 2005;**353**:68–79.

Stefanis L, Kholodilov N, Rideout HJ, Burke RE, Greene LA. Synuclein-1 is selectively up-regulated in response to nerve growth factor treatment in PC12 cells. *J Neurochem* 2001;**76**:1165–1176.

Stromer T, Fischer E, Richter K, Haslbeck M, Buchner J. Analysis of the regulation of the molecular chaperone Hsp26 by temperature-induced dissociation: the N-terminal domail is important for oligomer assembly and the binding of unfolding proteins. *J Biol Chem* 2004;**279**:11222–11228.

Studer S, Narberhaus F. Chaperone activity and homo- and hetero-oligomer formation of bacterial small heat shock proteins. *J Biol Chem* 2000;**275**:37212–37218.

Sugiyama H, Hainfellner JA, Schmid-Siegel B, Budka H. Neuroaxonal dystrophy combined with diffuse Lewy body disease in a young adult. *Clin Neuropathol* 1993;**12**:147–152.

Suh WC, Lu CZ, Gross CA. Structural features required for the interaction of the Hsp70 molecular chaperone DnaK with its cochaperone DnaJ. *J Biol Chem* 1999;**274**:30534–30539.

Sullivan WP, Owen BA, Toft DO. The influence of ATP and p23 on the conformation of hsp90. *J Biol Chem* 2002;**277**:45942–45948.

Sunde M, Blake C. The structure of amyloid fibrils by electron microscopy and X-ray diffraction. *Adv Protein Chem* 1997;**50**:123–159.

Sunde M, Serpell LC, Bartlam M, Fraser PE, Pepys MB, Blake CC. Common core structure of amyloid fibrils by synchrotron X-ray diffraction. *J Mol Biol* 1997;**273**:729–739.

Swaiman KF. Hallervorden-Spatz syndrome and brain iron metabolism. *Arch Neurol* 1991;**48**:1285–1293.

Szabo A, Korszun R, Hartl FU, Flanagan J. A zinc finger-like domain of the molecular chaperone DnaJ is involved in binding to denatured protein substrates. *EMBO J* 1996;**15**:408–417.

Szabo A, Langer T, Schroder H, Flanagan J, Bukau B, Hartl FU. The ATP hydrolysis-dependent reaction cycle of the Escherichia coli Hsp70 system DnaK, DnaJ, and GrpE. *Proc Natl Acad Sci U S A* 1994;**91**:10345–10349.

Szyperski T, Pellecchia M, Wall D, Georgopoulos C, Wuthrich K. NMR structure determination of the Escherichia coli DnaJ molecular chaperone: secondary structure and backbone fold of the N-terminal region (residues 2–108) containing the highly conserved J domain. *Proc Natl Acad Sci U S A* 1994;**91**:11343–11347.

Takayama S, Krajewski S, Krajewska M, Kitada S, Zapata JM, Kochel K, Knee D, Scudiero D, Tudor G, Miller GJ, *et al.* Expression and location of Hsp70/Hsc-binding anti-apoptotic protein BAG-1 and its variants in normal tissues and tumor cell lines. *Cancer Res* 1998;**58**:3116–3131.

Takayama S, Reed JC. Molecular chaperone targeting and regulation by BAG family proteins. *Nat Cell Biol* 2001;**3**: E237–E241.

Takeda A, Mallory M, Sundsmo M, Honer W, Hansen L, Masliah E. Abnormal accumulation of NACP/alpha-synuclein in neurodegenerative disorders. *Am J Pathol* 1998;**152**:367–372.

Tamo W, Imaizumi T, Tanji K, Yoshida H, Mori F, Yoshimoto M, Takahashi H, Fukuda I, Wakabayashi K, Satoh K. Expression of alpha-synuclein, the precursor of non-amyloid beta component of Alzheimer's disease amyloid, in human cerebral blood vessels. *Neurosci Lett* 2002;**326**:5–8.

Taylor TD, Litt M, Kramer P, Pandolfo M, Angelini L, Nardocci N, Davis S, Pineda M, Hattori H, Flett PJ, *et al.* Homozygosity mapping of Hallervorden-Spatz syndrome to chromosome 20p12.3-p13. *Nat Genet* 1996;**14**:479–481.

Tetzlaff JE, Putcha P, Outeiro TF, Ivanov A, Berezovska O, Hyman BT, McLean PJ. CHIP targets toxic alpha-Synuclein oligomers for degradation. *J Biol Chem* 2008;**283**:17962–17968.

Tobe T, Nakajo S, Tanaka A, Mitoya A, Omata K, Nakaya K, Tomita M, Nakamura Y. Cloning and characterization of the cDNA encoding a novel brain-specific 14-kDa protein. *J Neurochem* 1992;**59**:1624–1629.

Tompa P. Intrinsically unstructured proteins. *Trends Biochem Sci* 2002;**27**:527–533.

Tompa P. The interplay between structure and function in intrinsically unstructured proteins. *FEBS Lett* 2005;**579**:3346–3354.

Tompa P, Csermely P. The role of structural disorder in the function of RNA and protein chaperones. *FASEB J* 2004;**18**:1169–1175.

Tompa P, Szasz C, Buday L. Structural disorder throws new light on moonlighting. *Trends Biochem Sci* 2005;**30**:484–489.

Trojanowski JQ, Goedert M, Iwatsubo T, Lee VM. Fatal attractions: abnormal protein aggregation and neuron death in Parkinson's disease and Lewy body dementia. *Cell Death Differ* 1998;**5**:832–837.

Trojanowski JQ, Lee VM. Parkinson's disease and related alpha-synucleinopathies are brain amyloidoses. *Ann N Y Acad Sci* 2003;**991**:107–110.

Tzivion G, Luo Z, Avruch J. A dimeric 14-3-3 protein is an essential cofactor for Raf kinase activity. *Nature* 1998;**394**:88–92.

Ueda K, Fukushima H, Masliah E, Xia Y, Iwai A, Yoshimoto M, Otero DA, Kondo J, Ihara Y, Saitoh T. Molecular cloning of cDNA encoding an unrecognized component of amyloid in Alzheimer disease. *Proc Natl Acad Sci U S A* 1993;**90**:11282–11286.

Uversky VN. A multiparametric approach to studies of self-organization of globular proteins. *Biochemistry (Mosc)* 1999;**64**:250–266.

Uversky VN. Natively unfolded proteins: a point where biology waits for physics. *Protein Sci* 2002a;**11**:739–756.

Uversky VN. What does it mean to be natively unfolded? *Eur J Biochem* 2002b;**269**:2–12.

Uversky VN. A protein-chameleon: conformational plasticity of alpha-synuclein, a disordered protein involved in neurodegenerative disorders. *J Biomol Struct Dyn* 2003a;**21**:211–234.

Uversky VN. Protein folding revisited: a polypeptide chain at the folding-misfolding-nonfolding cross-roads: which way to go? *Cell Mol Life Sci* 2003b;**60**:1852–1871.

Uversky VN. Neuropathology, biochemistry, and biophysics of alpha-synuclein aggregation. *J Neurochem* 2007;**103**:17–37.

Uversky VN. Alpha-synuclein misfolding and neurodegenerative diseases. *Curr Protein Pept Sci* 2008;**9**:507–540.

Uversky VN. Intrinsic disorder in proteins associated with neurodegenerative diseases. *Front Biosci* 2009;**14**:5188–5238.

Uversky VN, Eliezer D. Biophysics of Parkinson's disease: structure and aggregation of alpha-synuclein. *Curr Protein Pept Sci* 2009;**10**:483–499.

Uversky VN, Fink AL. Conformational constraints for amyloid fibrillation: the importance of being unfolded. *Biochim Biophys Acta* 2004;**1698**:131–153.

Uversky VN, Gillespie JR, Fink AL. Why are "natively unfolded" proteins unstructured under physiologic conditions? *Proteins* 2000;**41**:415–427.

Uversky VN, Li J, Fink AL. Evidence for a partially folded intermediate in alpha-synuclein fibril formation. *J Biol Chem* 2001;**276**:10737–10744.

Uversky VN, Li J, Souillac P, Millett IS, Doniach S, Jakes R, Goedert M, Fink AL. Biophysical properties of the synucleins and their propensities to fibrillate: inhibition of alpha-synuclein assembly by beta- and gamma-synucleins. *J Biol Chem* 2002a;**277**:11970–11978.

Uversky VN, Oldfield CJ, Dunker AK. Showing your ID: intrinsic disorder as an ID for recognition, regulation and cell signaling. *J Mol Recognit* 2005;**18**:343–384.

Uversky VN, Ptitsyn OB. "Partly folded" state, a new equilibrium state of protein molecules: four-state guanidinium chloride-induced unfolding of beta-lactamase at low temperature. *Biochemistry* 1994;**33**:2782–2791.

Uversky VN, Ptitsyn OB. All-or-none solvent-induced transitions between native, molten globule and unfolded states in globular proteins. *Fold Des* 1996a;**1**:117–122.

Uversky VN, Ptitsyn OB. Further evidence on the equilibrium "pre-molten globule state": four-state guanidinium chloride-induced unfolding of carbonic anhydrase B at low temperature. *J Mol Biol* 1996b;**255**:215–228.

Uversky VN, Yamin G, Souillac PO, Goers J, Glaser CB, Fink AL. Methionine oxidation inhibits fibrillation of human alpha-synuclein *in vitro*. *FEBS Lett* 2002b;**517**:239–244.

Valla J, Schneider L, Niedzielko T, Coon KD, Caselli R, Sabbagh MN, Ahern GL, Baxter L, Alexander G, Walker DG, *et al.* Impaired platelet mitochondrial activity in Alzheimer's disease and mild cognitive impairment. *Mitochondrion* 2006;**6**: 323–330.

van Montfort RL, Basha E, Friedrich KL, Slingsby C, Vierling E. Crystal structure and assembly of a eukaryotic small heat shock protein. *Nat Struct Biol* 2001b;**8**:1025–1030.

van Montfort R, Slingsby C, Vierling E. Structure and function of the small heat shock protein/alpha-crystallin family of molecular chaperones. *Adv Protein Chem* 2001a;**59**:105–156.

Velten M, Villoutreix BO, Ladjimi MM. Quaternary structure of the HSC70 cochaperone HIP. *Biochemistry* 2000;**39**:307–315.

Vila M, Vukosavic S, Jackson-Lewis V, Neystat M, Jakowec M, Przedborski S. Alpha-synuclein up-regulation in substantia nigra dopaminergic neurons following administration of the parkinsonian toxin MPTP. *J Neurochem* 2000;**74**:721–729.

Voisine C, Craig EA, Zufall N, von Ahsen O, Pfanner N, Voos W. The protein import motor of mitochondria: unfolding and trapping of preproteins are distinct and separable functions of matrix Hsp70. *Cell* 1999;**97**:565–574.

Volles MJ, Lee SJ, Rochet JC, Shtilerman MD, Ding TT, Kessler JC, Lansbury PT Jr. Vesicle permeabilization by protofibrillar alpha-synuclein: implications for the pathogenesis and treatment of Parkinson's disease. *Biochemistry* 2001;**40**:7812–7819.

Vucetic S, Xie H, Iakoucheva LM, Oldfield CJ, Dunker AK, Obradovic Z, Uversky VN. Functional anthology of intrinsic disorder. 2. Cellular components, domains, technical terms, developmental processes, and coding sequence diversities correlated with long disordered regions. *J Proteome Res* 2007;**6**:1899–1916.

Wakabayashi K, Matsumoto K, Takayama K, Yoshimoto M, Takahashi H. NACP, a presynaptic protein, immunoreactivity in Lewy bodies in Parkinson's disease. *Neurosci Lett* 1997;**239**:45–48.

Wakabayashi K, Yoshimoto M, Tsuji S, Takahashi H. Alpha-synuclein immunoreactivity in glial cytoplasmic inclusions in multiple system atrophy. *Neurosci Lett* 1998;**249**:180–182.

Wang XY, Chen X, Oh HJ, Repasky E, Kazim L, Subjeck J. Characterization of native interaction of hsp110 with hsp25 and hsc70. *FEBS Lett* 2000;**465**:98–102.

Wang X, Su B, Fujioka H, Zhu X. Dynamin-like protein 1 reduction underlies mitochondrial morphology and distribution abnormalities in fibroblasts from sporadic Alzheimer's disease patients. *Am J Pathol* 2008;**173**:470–482.

Ward JJ, McGuffin LJ, Bryson K, Buxton BF, Jones DT. The DISOPRED server for the prediction of protein disorder. *Bioinformatics* 2004a;**20**:2138–2139.

Ward JJ, Sodhi JS, McGuffin LJ, Buxton BF, Jones DT. Prediction and functional analysis of native disorder in proteins from the three kingdoms of life. *J Mol Biol* 2004b;**337**:635–645.

Waxman EA, Giasson BI. Molecular mechanisms of alpha-synuclein neurodegeneration. *Biochim Biophys Acta* 2009;**1792**:616–624.

Weaver AJ, Sullivan WP, Felts SJ, Owen BA, Toft DO. Crystal structure and activity of human p23, a heat shock protein 90 co-chaperone. *J Biol Chem* 2000;**275**:23045–23052.

Wegele H, Haslbeck M, Reinstein J, Buchner J. Sti1 is a novel activator of the Ssa proteins. *J Biol Chem* 2003;**278**:25970–25976.

Wegele H, Muller L, Buchner J. Hsp70 and Hsp90—a relay team for protein folding. *Rev Physiol Biochem Pharmacol* 2004;**151**:1–44.

Weikl T, Abelmann K, Buchner J. An unstructured C-terminal region of the Hsp90 co-chaperone p23 is important for its chaperone function. *J Mol Biol* 1999;**293**:685–691.

Weinreb PH, Zhen W, Poon AW, Conway KA, Lansbury PT Jr. NACP, a protein implicated in Alzheimer's disease and learning, is natively unfolded. *Biochemistry* 1996;**35**:13709–13715.

Welch WJ, Feramisco JR. Purification of the major mammalian heat shock proteins. *J Biol Chem* 1982;**257**:14949–14959.

Westermark P. Amyloidosis and Amyloid Proteins: Brief History and Definitions. In: Sipe JD, editor. *Amyloid proteins the beta sheet conformation and disease*. Weinheim: WILEY-VCH Verlag GmbH & Co. KGaA; 2005. pp. 3–27.

Westhof E, Altschuh D, Moras D, Bloomer AC, Mondragon A, Klug A, Van Regenmortel MH. Correlation between segmental mobility and the location of antigenic determinants in proteins. *Nature* 1984;**311**:123–126.

Wiederrecht G, Hung S, Chan HK, Marcy A, Martin M, Calaycay J, Boulton D, Sigal N, Kincaid RL, Siekierka JJ. Characterization of high molecular weight FK-506 binding activities reveals a novel FK-506-binding protein as well as a protein complex. *J Biol Chem* 1992;**267**:21753–21760.

Williams RM, Obradovic Z, Mathura V, Braun W, Garner EC, Young J, Takayama S, Brown CJ, Dunker AK. The protein non-folding problem: amino acid determinants of intrinsic order and disorder. *Pac Symp Biocomput* 2001;**6**:89–100.

Witt SN. Hsp70 molecular chaperones and Parkinson's disease. *Biopolymers* 2010;**93**:218–228.

Wittung-Stafshede P, Guidry J, Horne BE, Landry SJ. The J-domain of Hsp40 couples ATP hydrolysis to substrate capture in Hsp70. *Biochemistry* 2003;**42**:4937–4944.

Woody RW. Circular dichroism. *Methods Enzymol* 1995;**246**:34–71.

Wright PE, Dyson HJ. Intrinsically unstructured proteins: re-assessing the protein structure-function paradigm. *J Mol Biol* 1999;**293**:321–331.

Xie H, Vucetic S, Iakoucheva LM, Oldfield CJ, Dunker AK, Obradovic Z, Uversky VN. Functional anthology of intrinsic disorder. 3. Ligands, post-translational modifications, and diseases associated with intrinsically disordered proteins. *J Proteome Res* 2007a;**6**:1917–1932.

Xie H, Vucetic S, Iakoucheva LM, Oldfield CJ, Dunker AK, Uversky VN, Obradovic Z. Functional anthology of intrinsic disorder. 1. Biological processes and functions of proteins with long disordered regions. *J Proteome Res* 2007b;**6**:1882–1898.

Xu Z, Horwich AL, Sigler PB. The crystal structure of the asymmetric GroEL-GroES-(ADP)7 chaperonin complex. *Nature* 1997;**388**:741–750.

Xu Z, Sigler PB. GroEL/GroES: structure and function of a two-stroke folding machine. *J Struct Biol* 1998;**124**:129–141.

Yamin G, Glaser CB, Uversky VN, Fink AL. Certain metals trigger fibrillation of methionine-oxidized alpha-synuclein. *J Biol Chem* 2003;**278**:27630–27635.

Yamin G, Munishkina LA, Karymov MA, Lyubchenko YL, Uversky VN, Fink AL. Forcing nonamyloidogenic beta-synuclein to fibrillate. *Biochemistry* 2005;**44**:9096–9107.

Young JC, Agashe VR, Siegers K, Hartl FU. Pathways of chaperone-mediated protein folding in the cytosol. *Nat Rev Mol Cell Biol* 2004;**5**:781–791.

Young JC, Hartl FU. Polypeptide release by Hsp90 involves ATP hydrolysis and is enhanced by the co-chaperone p23. *EMBO J* 2000;**19**:5930–5940.

Zha J, Harada H, Yang E, Jockel J, Korsmeyer SJ. Serine phosphorylation of death agonist BAD in response to survival factor results in binding to 14-3-3 not BCL-X(L). *Cell* 1996;**87**:619–628.

Zhao X, Braun AP, Braun JE. Biological roles of neural J proteins. *Cell Mol Life Sci* 2008;**65**:2385–2396.

Zimmerman MC, Oberley LW, Flanagan SW. Mutant SOD1-induced neuronal toxicity is mediated by increased mitochondrial superoxide levels. *J Neurochem* 2007;**102**:609–618.

Zou J, Guo Y, Guettouche T, Smith DF, Voellmy R. Repression of heat shock transcription factor HSF1 activation by HSP90 (HSP90 complex) that forms a stress-sensitive complex with HSF1. *Cell* 1998;**94**:471–480.

2

REDOX REGULATION OF PROTEIN MISFOLDING, SYNAPTIC DAMAGE, AND NEURONAL LOSS IN NEURODEGENERATIVE DISEASES

Tomohiro Nakamura

Del E. Webb Center for Neuroscience, Aging, and Stem Cell Research, Sanford-Burnham Medical Research Institute, La Jolla, CA, USA

Stuart A. Lipton

Del E. Webb Center for Neuroscience, Aging, and Stem Cell Research, Sanford-Burnham Medical Research Institute, La Jolla, CA, USA; Departments of Neurosciences and Psychiatry, University of California, San Diego, CA, USA; Molecular Neurobiology Laboratory, The Salk Institute for Biological Studies, La Jolla, CA, USA; Departments of Molecular and Integrative Neurosciences, and Molecular and Experimental Medicine, The Scripps Research Institute, La Jolla, CA, USA

2.1 INTRODUCTION

In general, many neurodegenerative diseases are characterized by the accumulation of misfolded proteins that adversely affect neuronal connectivity and plasticity, and trigger cell death signaling pathways (Bence *et al.*, 2001;

Protein Chaperones and Protection from Neurodegenerative Diseases, First Edition.
Edited by Stephan N. Witt.

Muchowski and Wacker, 2005). For example, in Parkinson's disease (PD) the brain contains aberrant accumulations of misfolded, aggregated proteins, such as α-synuclein and synphilin-1, and in Alzheimer's disease (AD), amyloid-β (Aβ) and tau. The inclusions observed in PD are called *Lewy bodies* and are mostly found in the cytoplasm. AD brains manifest intracellular neurofibrillary tangles, which contain hyperphosphorylated tau, and extracellular plaques, which contain Aβ. These aggregates may consist of complexes of protein with nonnative secondary structures, and demonstrate poor solubility in aqueous or detergent solvent. Other disorders manifesting protein aggregation include Huntington's disease (HD, a polyQ disorder), amyotrophic lateral sclerosis (ALS), and prion disease (Ciechanover and Brundin, 2003). The aforementioned disorders are also termed *conformational diseases* because of the emergence of protein aggregation in the brain (Kopito and Ron, 2000).

An additional feature of most neurodegenerative diseases is excessive generation of reactive nitrogen species (RNS) and reactive oxygen species (ROS), which can contribute to neuronal cell injury and death (Barnham *et al.*, 2004; Beal, 2001; Emerit *et al.*, 2004; Lin and Beal, 2006; Muchowski, 2002). While many intra- and extracellular molecules may participate in neuronal injury, accumulation of nitrosative stress due to excessive generation of nitric oxide (NO) appears to be a potential factor contributing to neuronal cell damage and death (Lipton, 2006; Lipton and Rosenberg, 1994). A well-established mechanism for NO production in neurons involves a central role of the *N*-methyl-D-aspartate (NMDA)-type of glutamate receptor in the nervous system. Hyperactivation of NMDA receptors allows excessive Ca^{2+} influx through receptor-coupled ion channels, which in turn activates neuronal NO synthase (nNOS) as well as the generation of ROS (Bredt *et al.*, 1991; Garthwaite *et al.*, 1988). As our laboratory in collaboration with that of Jonathan Stamler first described, NO can mediate both protective and neurotoxic effects by reacting with cysteine residues of target proteins to form *S*-nitrosothiols (SNOs), a process termed *S-nitrosylation* because of its effects on the chemical biology of protein function. Importantly, normal mitochondrial respiration may also generate free radicals, principally ROS, and one such molecule, superoxide anion (O_2^-), reacts rapidly with free radical NO to form the very toxic product peroxynitrite ($ONOO^-$) (Beckman *et al.*, 1990; Lipton *et al.*, 1993).

Importantly, protein aggregation can result from either (i) a rare mutation in the disease-related gene encoding the protein or (ii) posttranslational changes to the protein engendered by nitrosative/oxidative stress, which may well account for the more common sporadic cases of many of these diseases (Zhang and Kaufman, 2006). Therefore, the key theme of this chapter is the hypothesis that nitrosative and oxidative stress contribute to protein misfolding in the brains of the majority of patients with neurodegenerative disorders such as AD and PD. In this chapter, we discuss specific examples showing that *S*-nitrosylation of (i) ubiquitin E3 ligases, such as parkin or (ii) endoplasmic reticulum (ER) chaperones, such as protein-disulfide isomerase (PDI), is critical for the accumulation of misfolded proteins in neurodegenerative diseases such as PD, AD, or other conditions (Chung *et al.*, 2004; Lipton *et al.*, 2005; Uehara *et al.*, 2006; Yao *et al.*,

2004). We also discuss recent findings that *S*-nitrosylation of the mitochondrial fission protein dynamin-related protein 1 (Drp1) mediates excessive mitochondrial fragmentation, bioenergetic compromise, and consequent synaptic injury in AD (Cho *et al.*, 2009).

2.2 PROTEIN MISFOLDING AND AGGREGATION IN NEURODEGENERATIVE DISEASES

In general, protein aggregates do not accumulate in unstressed, healthy neurons owing in part to the existence of cellular "quality control machineries." For example, molecular chaperones are believed to provide a defense mechanism against the toxicity of misfolded proteins because chaperones can prevent inappropriate interactions within and between polypeptides, and can promote refolding of proteins that have been misfolded because of cell stress. In accordance with these precepts, homozygous mutations in the *SIL1* gene, which encodes an ER co-chaperone, cause neurodegeneration in Marinesco–Sjögren syndrome patients and in the woozy mouse mutant (Anttonen *et al.*, 2005; Senderek *et al.*, 2005; Zhao *et al.*, 2005). In addition to the quality control of proteins provided by molecular chaperones, the ubiquitin–proteasome system (UPS) and autophagy/lysosomal degradation are involved in the clearance of abnormal or aberrant proteins. When chaperones cannot repair misfolded proteins, they may be tagged via addition of polyubiquitin chains for degradation by the proteasome. In neurodegenerative conditions, intra- or extracellular protein aggregates are thought to accumulate in the brain as a result of decrease in molecular chaperone or proteasome activities. In fact, several mutations that disturb the activity of molecular chaperones or UPS-associated enzymes can cause neurodegeneration (Cookson, 2005; Muchowski and Wacker, 2005; Zhao *et al.*, 2005). Along these lines, postmortem samples from the substantia nigra of PD patients (vs. non-PD controls) manifest a significant reduction in proteasome activity (McNaught *et al.*, 2004). Moreover, overexpression of the molecular chaperone HSP70 can prevent neurodegeneration *in vivo* in models of PD (Auluck *et al.*, 2002).

Historically, lesions that contain aggregated proteins were considered to be pathogenic. Recently, several lines of evidence have suggested that aggregates are formed through a complex multistep process by which misfolded proteins, their chaperones, and perhaps other proteins assemble into inclusion bodies. Currently, soluble (micro-)oligomers of these aberrant proteins are thought to be the most toxic forms via interference with normal cell activities, while frank macroscopic aggregates may be an attempt by the cell to wall off potentially toxic material (although such aggregates could potentially be toxic by their location or if the misfolded proteins are not sequestered by the proper chaperones) (Arrasate *et al.*, 2004; Bredt *et al.*, 1991). Additionally, at least in yeast and cell culture models, highly toxic aggregates accumulate in a perivacuolar compartment where the autophagic pathway catalyzes clearance of aggresomes. Relatively less toxic misfolded proteins are sequestered in juxtanuclear inclusions, which often contain molecular chaperones and proteasomes as part of the quality control machinery (Kaganovich *et al.*, 2008).

2.3 GENERATION OF RNS/ROS VIA NMDA RECEPTOR-MEDIATED GLUTAMATERGIC SIGNALING PATHWAYS

It is well known that the amino acid glutamate is the major excitatory neurotransmitter in the brain. Glutamate is present in high concentrations in the adult central nervous system and is released for milliseconds from nerve terminals in a Ca^{2+}-dependent manner. After the glutamate enters the synaptic cleft, it diffuses across the cleft to interact with its corresponding receptors on the postsynaptic face of an adjacent neuron. Excitatory neurotransmission is necessary for the normal development and plasticity of synapses, and for some forms of learning or memory; however, excessive activation of glutamate receptors is implicated in neuronal damage in many neurological disorders ranging from acute hypoxic-ischemic brain injury to chronic neurodegenerative diseases. It is currently thought that overstimulation of extrasynaptic NMDA receptors mediate this neuronal damage, while, in contrast, synaptic activity predominantly activates survival pathways (Hardingham *et al.*, 2002; Papadia *et al.*, 2005, 2008). Intense hyperstimulation of excitatory receptors leads to necrotic cell death, but more mild or chronic overstimulation can result in apoptotic or other forms of cell death (Ankarcrona *et al.*, 1995; Bonfoco *et al.*, 1995; Budd *et al.*, 2000).

There are two large families of glutamate receptors in the nervous system, ionotropic receptors (representing ligand-gated ion channels), and metabotropic receptors (coupled to G-proteins). Ionotropic glutamate receptors are further divided into three broad classes, NMDA receptors, α-amino-3-hydroxy-5 methyl-4-isoxazole propionic acid (AMPA) receptors, and kainate receptors, each of which is named after synthetic ligands that can selectively activate these receptors. The NMDA receptor has attracted attention for a long period of time because it has several properties that set it apart from other ionotropic glutamate receptors. One such characteristic, in contrast to most AMPA and kainate receptors, is that NMDA receptor-coupled channels are highly permeable to Ca^{2+}, thus permitting Ca^{2+} entry after ligand binding if the cell is depolarized to relieve the block of the receptor-associated ion channel by Mg^{2+} (Mayer *et al.*, 1984; Nowak *et al.*, 1984). Subsequent binding of Ca^{2+} to various intracellular molecules can lead to many significant consequences. In particular, excessive activation of NMDA receptors leads to the production of damaging free radicals (e.g., NO and ROS) and other enzymatic processes contributing to cell death (Bonfoco *et al.*, 1995; Budd *et al.*, 2000; Dawson *et al.*, 1991; Lafon-Cazal *et al.*, 1993; Lipton and Rosenberg, 1994; Lipton *et al.*, 1993).

Excessive activation of glutamate receptors is implicated in neuronal damage in many neurological disorders. John Olney coined the term *excitotoxicity* to describe this phenomenon (Olney, 1969, 1997). This form of toxicity is mediated at least in part by excessive activation of NMDA-type receptors (Chen and Lipton,

2006; Lipton, 2006; Lipton and Rosenberg, 1994), resulting in excessive Ca^{2+} influx through a receptor's associated ion channel. Increased levels of neuronal Ca^{2+}, in conjunction with the Ca^{2+}-binding protein CaM, trigger activation of nNOS and subsequent generation of NO from the amino acid L-arginine (Abu-Soud and Stuehr, 1993; Bredt *et al.*, 1991) (Figure 2.1). NO is a gaseous free radical (thus highly diffusible) and a key molecule that plays a vital role in normal signal transduction, but when in excess can lead to neuronal cell damage and death. The discrepancy of NO effects on neuronal survival can also be caused by the formation of different NO species or intermediates: NO radical ($NO^{\cdot}$), nitrosonium cation (NO^+), nitroxyl anion (NO^-, with high energy singlet

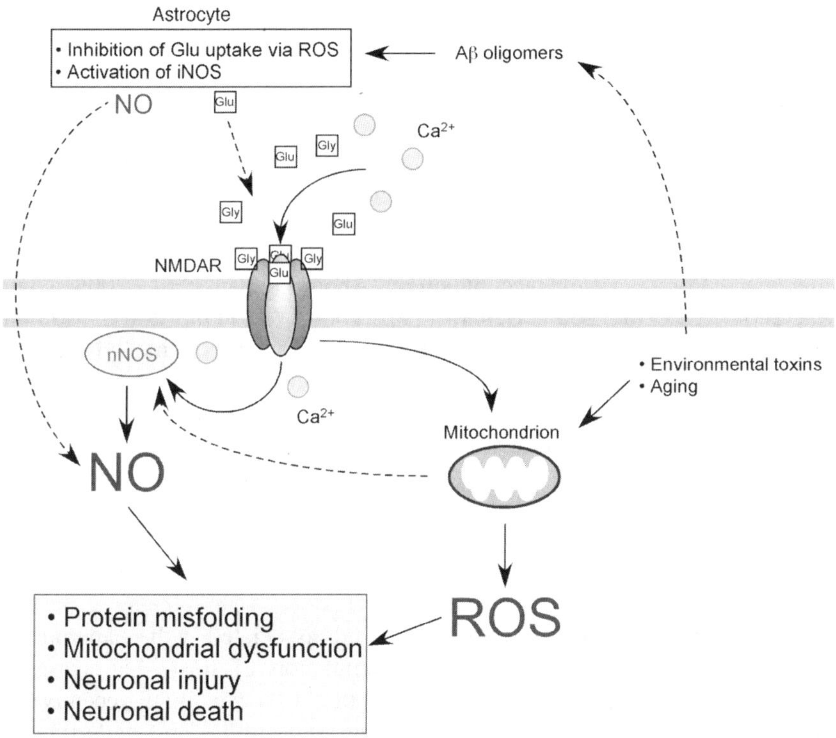

Figure 2.1 Excessive activation of the NMDA receptor (NMDAR) by glutamate (Glu) and glycine (Gly) induces ROS/NO generation and stimulates excitotoxic pathways. NMDAR hyperactivation triggers generation of NO by neuronal NO synthase (nNOS), and production and release of ROS from mitochondria. In addition, generation of mitochondrial ROS occurs upon environmental toxin stimuli or during the normal aging process. Soluble oligomers of Aβ peptide, thought to be a key mediator in AD pathogenesis, can facilitate neuronal NO production in both NMDAR-dependent and -independent manners. Excessively produced NO and ROS contribute to protein misfolding, mitochondrial dysfunction, neuronal cell injury, and death. (*A full color version of this figure appears in the color plate section.*)

and lower energy triplet forms) (Lipton *et al.*, 1993). Three subtypes of NOS have been identified; two constitutive forms of NOS—nNOS and endothelial NOS (eNOS)—take their names from the cell type in which they were first found. The name of the third subtype—inducible NOS (iNOS)—indicates that expression of the enzyme is induced by acute inflammatory stimuli. For example, activated microglia may produce neurotoxic amounts of NO via iNOS expression in various neurodegenerative diseases. All three isoforms are widely distributed in the brain. Each NOS isoform contains an oxidase domain at its amino-terminal end and a reductase domain at its carboxy-terminal end, separated by a Ca^{2+}/CaM binding site (Abu-Soud and Stuehr, 1993; Boucher *et al.*, 1999; Bredt *et al.*, 1991; Forstermann *et al.*, 1998; Groves and Wang, 2000). Constitutive and inducible NOS are also further distinguished by CaM binding: nNOS and eNOS bind CaM in a reversible Ca^{2+}-dependent manner. In contrast, iNOS binds CaM so tightly at resting intracellular Ca^{2+} concentrations that its activity does not appear to be affected by transient variations in Ca^{2+} concentration. Interestingly, to terminate iNOS-mediated NO production, microglia may redistribute iNOS to the aggresome for inactivation (Kolodziejska *et al.*, 2005).

Recent studies further pointed out the potential connection between ROS/RNS and mitochondrial dysfunction in neurodegenerative diseases, especially in PD (Beal, 2001; Betarbet *et al.*, 2000). Pesticide and other environmental toxins that inhibit mitochondrial complex I result in oxidative and nitrosative stress, and consequent aberrant protein accumulation (Abou-Sleiman *et al.*, 2006; Chung *et al.*, 2004; He *et al.*, 2003; Uehara *et al.*, 2006; Yao *et al.*, 2004). Administration to animal models of complex I inhibitors, such as 1-methyl-4-phenyl-1,2,3,6-tetrahydropyridine (MPTP), 6-hydroxydopamine, rotenone, and paraquat, which result in overproduction of ROS/RNS, reproduces many of the features of sporadic PD, such as dopaminergic neuron degeneration, upregulation and aggregation of α-synuclein, Lewy body-like intraneuronal inclusions, and behavioral impairment (Beal, 2001; Betarbet *et al.*, 2000). In addition, it has recently been proposed that mitochondrial cytochrome oxidase can produce NO in a nitrite (NO_2^-) - and pH-dependent but non-Ca^{2+}-dependent manner (Castello *et al.*, 2006).

Increased nitrosative and oxidative stress are associated with chaperone and proteasomal dysfunction, resulting in accumulation of misfolded aggregates (Isaacs *et al.*, 2006; Zhang and Kaufman, 2006). However, until recently little was known regarding the molecular and pathogenic mechanisms underlying the contribution of NO to the formation of inclusion bodies such as amyloid plaques in AD or Lewy bodies in PD.

2.4 NITROSATIVE STRESS REGULATES PROTEIN MISFOLDING AND NEURONAL CELL DEATH

Extreme nitrosative/oxidative stress can facilitate protein misfolding and aggregation—and very likely vice-versa. This relationship between ROS/RNS and protein misfolding is thought to play a role as a pathogenic trigger of neurodegenerative diseases, although the exact mechanism underlying ROS/RNS-mediated aggregate formation has remained elusive. Recent scientific advances, however, have implied that NO-related species may significantly participate in the process of protein misfolding through protein *S*-nitrosylation (and possibly nitration) under degenerative conditions.

Early investigations indicated that NO participates in cellular signaling pathways, which regulate broad aspects of brain function, including synaptic plasticity, normal development, and neuronal cell death (Bredt and Snyder, 1994; Dawson *et al.*, 1991; O'Dell *et al.*, 1991; Schuman and Madison, 1994). In general, NO exerts physiological and some pathophysiological effects via stimulation of guanylate cyclase to form cyclic guanosine-3′, 5′-monophosphate (cGMP) or through *S*-nitros(yl)ation of regulatory protein thiol groups (Garthwaite *et al.*, 1988; Isaacs *et al.*, 2006; Kandel and O'Dell, 1992; Lei *et al.*, 1992; Lipton *et al.*, 1993; Stamler *et al.*, 1992a). *S*-Nitrosylation is the covalent addition of an NO group to a critical cysteine thiol/sulfhydryl (RSH or, more properly, thiolate anion, RS^-) to form an SNO derivative (R-SNO). Such modification modulates the function of a broad spectrum of mammalian, plant, and microbial proteins. In general, a consensus motif of amino acids comprising nucleophilic residues (generally an acid and a base) surround a critical cysteine, which increases the cysteine sulfhydryl's susceptibility to *S*-nitrosylation (Hess *et al.*, 2005; Stamler *et al.*, 1997). In contrast, denitrosylating enzymes and pathways, such as those mediated by thioredoxin (TRX)/TRX reductase, PDI, and intracellular glutathione, can decrease the lifespan of protein SNOs (Benhar *et al.*, 2008; Nikitovic and Holmgren, 1996; Romero and Bizzozero, 2008). Our group first identified the physiological relevance of *S*-nitrosylation by showing that NO and related RNS exert paradoxical effects via redox-based mechanisms—NO is neuroprotective via *S*-nitrosylation of NMDA receptors (as well as other subsequently discovered targets, including caspases), and yet can also be neurodestructive by formation of peroxynitrite (or, as later discovered, reaction with additional molecules such as matrix metalloproteinase (MMP)-9 and glyceraldehyde 3-phosphate dehydrogenase (GAPDH)) (Choi *et al.*, 2000; Dimmeler *et al.*, 1997; Gu *et al.*, 2002; Hara *et al.*, 2005; Kim *et al.*, 1999; Lipton *et al.*, 1993; Mannick *et al.*, 1999; Melino *et al.*, 1997; Tenneti *et al.*, 1997). Over the past decade, accumulating evidence has suggested that *S*-nitrosylation can regulate the biological activity of a great variety of proteins, in some ways akin to phosphorylation (Chung *et al.*, 2004; Gu *et al.*, 2002; Haendeler *et al.*, 2002; Hara *et al.*, 2005; Hess *et al.*, 2005; Jaffrey *et al.*, 2001; Lipton *et al.*, 1993, 2002; Sliskovic *et al.*, 2005; Stamler, 1994; Stamler *et al.*, 1992b, 2001; Uehara *et al.*, 2006; Yao *et al.*, 2004) (Figure 2.2). Chemically, NO is often a good "leaving group," facilitating further oxidation of critical thiol to disulfide bonds among neighboring (vicinal) cysteine residues or, via reaction with ROS, to sulfenic (–SOH), sulfinic ($-SO_2H$), or sulfonic ($-SO_3H$) acid derivatization of the protein (Gu *et al.*, 2002; Stamler and Hausladen, 1998; Uehara *et al.*, 2006; Yao *et al.*, 2004).

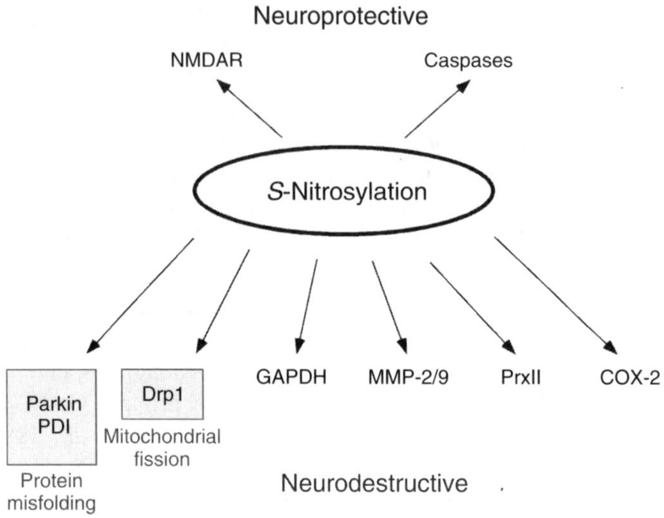

Figure 2.2 *S*-Nitrosylation of neuronal proteins regulates neuronal survival. Physiological levels of NO mediate neuroprotective effects, at least in part, by *S*-nitrosylating the NMDAR and caspases, thus inhibiting their activity. In contrast, overproduction of NO can be neurotoxic via *S*-nitrosylation of Parkin, PDI, GAPDH, MMP-2/9, PrxII, and COX-2. *S*-Nitrosylated parkin and PDI contribute to neuronal cell injury by triggering accumulation of misfolded proteins. *S*-Nitrosylation of Drp1 causes excessive mitochondrial fragmentation in neurodegenerative conditions.

Alternatively, *S*-nitrosylation may possibly produce a nitroxyl disulfide, in which the NO group is shared by close cysteine thiols (Houk *et al.*, 2003).

Analyses of mice deficient in either nNOS or iNOS confirmed that NO is an important mediator of cell injury and death after excitotoxic stimulation; NO generated from nNOS or iNOS is detrimental to neuronal survival (Huang *et al.*, 1994; Iadecola *et al.*, 1997). In addition, inhibition of NOS activity ameliorates the progression of disease pathology in animal models of PD, AD, and ALS, suggesting that excess generation of NO plays a pivotal role in the pathogenesis of several neurodegenerative diseases (Chabrier *et al.*, 1999; Hantraye *et al.*, 1996; Liberatore *et al.*, 1999; Przedborski *et al.*, 1996). Intriguingly, the levels of glutathione diminish by approximately 30% in the aged brain (Chen *et al.*, 1989), potentially assisting the accumulation of SNOs in the elderly. Although the involvement of NO in neurodegeneration has been widely accepted, the chemical relationship between nitrosative stress and accumulation of misfolded proteins has remained obscure. Recent findings, however, have shed light on molecular events underlying this relationship. Specifically, we recently mounted physiological and chemical evidence that *S*-nitrosylation modulates the (i) ubiquitin E3 ligase activity of parkin (Chung *et al.*, 2004; Lipton *et al.*, 2005; Yao *et al.*, 2004), (ii) chaperone and isomerase activities of PDI (Uehara *et al.*, 2006); and (iii) GTPase activity of Drp1, contributing to protein misfolding and neurotoxicity in models of neurodegenerative disorders.

Additionally, peroxynitrite-mediated nitration of tyrosine residue(s) may potentially contribute to dysfunctional protein folding and neuronal cell injury. For instance, nitration of α-synuclein and tau affects oligomer formation *in vitro*. Furthermore, it has been reported that nitrated α-synuclein and tau selectively accumulate in inclusion bodies in PD and neurofibrillary tangles in AD brains (Giasson *et al.*, 2000; Reynolds *et al.*, 2007, 2006; Uversky *et al.*, 2005). Collectively, these findings support the proposition that *S*-nitrosylation and possibly nitration can influence aggregate formation and neurotoxicity.

2.5 *S*-NITROSYLATION OF PARKIN

2.5.1 Parkin and the UPS

Recent studies on rare genetic forms of PD have found that mutations in the genes encoding parkin (*PARK2*), PINK1 (*PARK6*), α-synuclein (*PARK1/4*), DJ-1 (*PARK7*), ubiquitin C-terminal hydrolase L1 (UCH-L1) (*PARK5*), leucine-rich repeat kinase-2 (LRRK2) (*PARK8*), or ATP13A2 (*PARK9*) are associated with PD pathology (Bonifati *et al.*, 2003; Kitada *et al.*, 1998; Leroy *et al.*, 1998; Paisan-Ruiz *et al.*, 2004; Polymeropoulos *et al.*, 1997; Ramirez *et al.*, 2006; Valente *et al.*, 2004; Zimprich *et al.*, 2004). The discovery that mutations in these genes predispose patients to very rare familial forms of PD have allowed us to begin to understand the mechanism of protein aggregation and neuronal loss in the more common sporadic forms of PD. For instance, the identification of α-synuclein as a familial PD gene led to the recognition that one of the major constituents of Lewy bodies in sporadic PD brains is α-synuclein. In addition, identification of errors in the genes encoding parkin (a ubiquitin E3 ligase) and UCH-L1 in rare familial forms of PD has implicated possible dysfunction of the UPS in the pathogenesis of sporadic PD as well. The UPS represents an important mechanism for proteolysis in mammalian cells. Formation of polyubiquitin chains constitutes the signal for proteasomal attack and degradation. An isopeptide bond covalently attaches the C-terminus of the first ubiquitin in a polyubiquitin chain to a lysine residue in the target protein. The cascade of activating (E1), conjugating (E2), and ubiquitin-ligating (E3) type enzymes catalyzes the conjugation of the ubiquitin chain to proteins. In addition, individual E3 ubiquitin ligases play a key role in the recognition of specific substrates (Ross and Pickart, 2004).

Mutations in the parkin gene can cause autosomal recessive juvenile parkinsonism (ARJP), accounting for some cases of hereditary PD manifest in young patients with onset anywhere from the teenage years through the 40s (Cookson, 2005; Kitada *et al.*, 1998; Shimura *et al.*, 2000). Parkin is a member of a large family of E3 ubiquitin ligases that are related to one another by the presence of RING finger domains. Parkin contains a total of 35 cysteine residues, the majority of which resides within its RING domains, which coordinate a structurally important zinc atom often involved in catalysis (Marin and Ferrus, 2002). Parkin has two RING finger domains separated by an "in between RING" (IBR) domain.

This motif allows parkin to recruit substrate proteins as well as an E2 enzyme (e.g., UbcH7, UbcH8, or UbcH13). Point mutations, stop mutations, truncations, and deletions in both alleles of the parkin gene will eventually cause dysfunction in its activity and are responsible for many cases of ARJP as well as rare adult forms of PD. Parkin mutations usually do not facilitate the formation of Lewy bodies, although there is at least one exception—familial PD patients with the R275W parkin mutant manifest Lewy bodies (Farrer *et al.*, 2001). Biochemical characterization of parkin mutants show that not all parkin mutations result in loss of parkin E3 ligase activity; some of the familial-associated parkin mutants (e.g., the R275W mutant) have increased ubiquitination activity compared to the wild type (wt) (Hampe *et al.*, 2006; Matsuda *et al.*, 2006; Sriram *et al.*, 2005). Additionally, parkin can mediate the formation of nonclassical and "nondegradative" lysine 63-linked polyubiquitin chains (Lim *et al.*, 2005, 2006). Likewise, parkin can mono-ubiquitinate Eps15, HSP70, and itself possibly at multiple sites. This finding may explain how some parkin mutations induce the formation of Lewy bodies and why proteins are stabilized within the inclusions.

Several putative target substrates have been identified for parkin E3 ligase activity. One group has reported that mutant parkin failed to bind glycosylated α-synuclein for ubiquitination, leading to α-synuclein accumulation (Shimura *et al.*, 2001), but most authorities do not feel that α-synuclein is a direct substrate of parkin. Synphilin-1 (α-synuclein interacting protein), on the other hand, is considered to be a substrate for parkin ubiquitination, and it is included in Lewy body-like inclusions in cultured cells when coexpressed with α-synuclein (Chung *et al.*, 2001). Other substrates for parkin include parkin-associated endothelin receptor-like receptor (Pael-R) (Imai *et al.*, 2001), cell division control-related protein (CDCrel-1) (Zhang *et al.*, 2000), cyclin E (Staropoli *et al.*, 2003), p38 tRNA synthase (Corti *et al.*, 2003), synaptotagmin XI (Huynh *et al.*, 2003), α/β tubulin heterodimers (Ren *et al.*, 2003), as well as possibly parkin itself (auto-ubiquitination). It is generally accepted that accumulation of these substrates can lead to disastrous consequences for the survival of dopaminergic neurons in familial PD and possibly also in sporadic PD. Therefore, characterization of potential regulators that affect parkin E3 ligase activity may reveal important molecular mechanisms for the pathogenesis of PD. Heretofore, two cellular components have been shown to regulate the substrate specificity and ubiquitin E3 ligase activity of parkin. The first represents posttranslational modification of parkin through *S*-nitrosylation (see below for details) or phosphorylation (Yamamoto *et al.*, 2005), and the second, binding partners of parkin, such as carboxyl terminus of Hsc70-interacting protein (CHIP) (Imai *et al.*, 2002) and BAG5 (Kalia *et al.*, 2004). CHIP enhances the ability of parkin to inhibit cell death through upregulation of parkin-mediated ubiquitination, while BAG5-mediated inhibition of parkin E3 ligase activity facilitates neuronal cell death. In addition, several groups have recently reported that parkin-mediated mono-ubiquitination could contribute to neuronal survival via a proteasome-independent pathway (Fallon *et al.*, 2006; Hampe *et al.*, 2006; Matsuda *et al.*, 2006; Moore *et al.*, 2008). For example, parkin mono-ubiquitinates the epidermal growth factor receptor

(EGFR)-associated protein, Eps15, leading to inhibition of EGFR endocytosis (Fallon *et al.*, 2006). The resulting prolongation of EGFR signaling via the phosphoinositide-3 kinase/Akt (PKB) signaling pathway is postulated to enhance neuronal survival.

Another important molecule that links aberrant UPS activity and PD is the ubiquitin hydrolase Uch-L1, a deubiquitinating enzyme that recycles ubiquitin. Autosomal dominant mutations of Uch-L1 have been identified in two siblings with PD (Leroy *et al.*, 1998). Interestingly, a recent study suggested that a novel ubiquitin–ubiquitin ligase activity of Uch-L1 might also be important in the pathogenesis of PD (Liu *et al.*, 2002). Additional mutations in α-synuclein, DJ-1, PINK1, and LRRK2 may contribute to UPS dysfunction and subsequently lead to PD.

2.5.2 *S*-Nitrosylation of Parkin Impairs UPS in Models of PD

PD is the second most prevalent neurodegenerative disease and is characterized by the progressive loss of dopamine (DA) neurons in the substantia nigra pars compacta. Appearance of Lewy bodies that contain misfolded and ubiquitinated proteins generally accompany the loss of dopaminergic neurons in the PD brain. Such ubiquitinated inclusion bodies are the hallmark of many neurodegenerative disorders. Age-associated defects in intracellular proteolysis of misfolded or aberrant proteins might lead to accumulation and ultimately deposition of aggregates within neurons or glial cells. Although such aberrant protein accumulation had been observed in patients with genetically encoded mutant proteins, recent evidence from our laboratory suggests that nitrosative and oxidative stress are potential causal factors for protein accumulation in the much more common sporadic form of PD. As illustrated below, nitrosative/oxidative stress, commonly found during normal aging, can mimic rare genetic causes of disorders, such as PD, by promoting protein misfolding in the absence of a genetic mutation (Chung *et al.*, 2004; Lipton *et al.*, 2005; Yao *et al.*, 2004). For example, *S*-nitrosylation and further oxidation of parkin or Uch-L1 result in dysfunction of these enzymes and thus of the UPS (Choi *et al.*, 2004; Chung *et al.*, 2004, 2005; Gu *et al.*, 2005; Nishikawa *et al.*, 2003; Yao *et al.*, 2004). We and others recently discovered that nitrosative stress triggers *S*-nitrosylation of parkin (forming SNO-parkin) not only in rodent models of PD but also in the brains of human patients with PD and the related α-synucleinopathy, DLBD (diffuse Lewy body disease). SNO-parkin initially stimulates ubiquitin E3 ligase activity, resulting in enhanced ubiquitination as observed in Lewy bodies, followed by a decrease in enzyme activity, producing a futile cycle of dysfunctional UPS (Lim *et al.*, 2005; Lipton *et al.*, 2005; Yao *et al.*, 2004) (Figure 2.2). We also found that rotenone led to the generation of SNO-parkin and thus dysfunctional ubiquitin E3 ligase activity. Moreover, *S*-nitrosylation appears to compromise the neuroprotective effect of parkin (Chung *et al.*, 2004). These mechanisms involve *S*-nitrosylation of critical cysteine residues in the first RING domain of parkin

(Yao *et al.*, 2004). Nitrosative and oxidative stress can also alter the solubility of parkin via posttranslational modification of cysteine residues, which may concomitantly compromise its protective function (LaVoie *et al.*, 2007; Wang *et al.*, 2005; Wong *et al.*, 2007). Additionally, it is likely that other ubiquitin E3 ligases with RING-finger thiol motifs are *S*-nitrosylated in a similar manner to parkin to affect their enzymatic function; hence, *S*-nitrosylation of E3 ligases may be involved in a number of degenerative conditions.

The neurotransmitter DA may also impair parkin activity and contribute to neuronal demise via the modification of cysteine residue(s) (LaVoie *et al.*, 2005). DA can be oxidized to DA quinone, which can react with and inactivate proteins through covalent modification of cysteine sulfhydryl groups; peroxynitrite has been reported to promote oxidation of DA to form DA quinone (LaVoie and Hastings, 1999). DA quinone can preferentially attack cysteine residues (C268 and C323) in the RING1 and IBR domains of parkin, forming a covalent adduct that abrogates its E3 ubiquitin ligase activity (LaVoie *et al.*, 2005; Wong *et al.*, 2007). DA quinone also reduces the solubility of parkin, possibly inducing parkin misfolding after disruption of the RING-IBR-RING motif. Therefore, oxidative/nitrosative species may either directly or indirectly contribute to altered parkin activity within the brain, and subsequent loss of parkin-dependent neuroprotection results in increased cell death.

2.6 *S*-NITROSYLATION OF PDI

2.6.1 The Unfolded Protein Response (UPR) and PDI

The ER normally participates in protein processing and folding but undergoes a stress response when immature or misfolded proteins accumulate (Andrews and Johnson, 1996; Ellgaard *et al.*, 1999; Sidrauski *et al.*, 1998; Szegezdi *et al.*, 2006). ER stress stimulates two critical intracellular responses (Figure 2.3). The first represents the expression of chaperones that prevent protein aggregation via the unfolded protein response (UPR), and is implicated in protein refolding, posttranslational assembly of protein complexes, and protein degradation. This response is believed to contribute to adaptation during altered environmental conditions, promoting maintenance of cellular homeostasis. At least three ER transmembrane sensor proteins are involved in the UPR: PKR-like ER kinase (PERK), activating transcription factor 6 (ATF6), and inositol-requiring enzyme 1 (IRE1). The activation of all three proximal sensors results in the attenuation of protein synthesis via eukaryotic initiation factor-2 (eIF2) kinase and increased protein folding capacity of the ER (Kaufman, 1999; Mori, 2000; Patil and Walter, 2001; Yoshida *et al.*, 2001). The second ER stress response, termed ER-associated degradation (ERAD), specifically recognizes terminally misfolded proteins for retro-translocation across the ER membrane to the cytosol, where they can be degraded by the UPS. In addition, although severe ER stress can induce apoptosis, the ER withstands relatively mild insults via expression of stress proteins such as glucose-regulated protein (GRP) and PDI. These proteins behave as

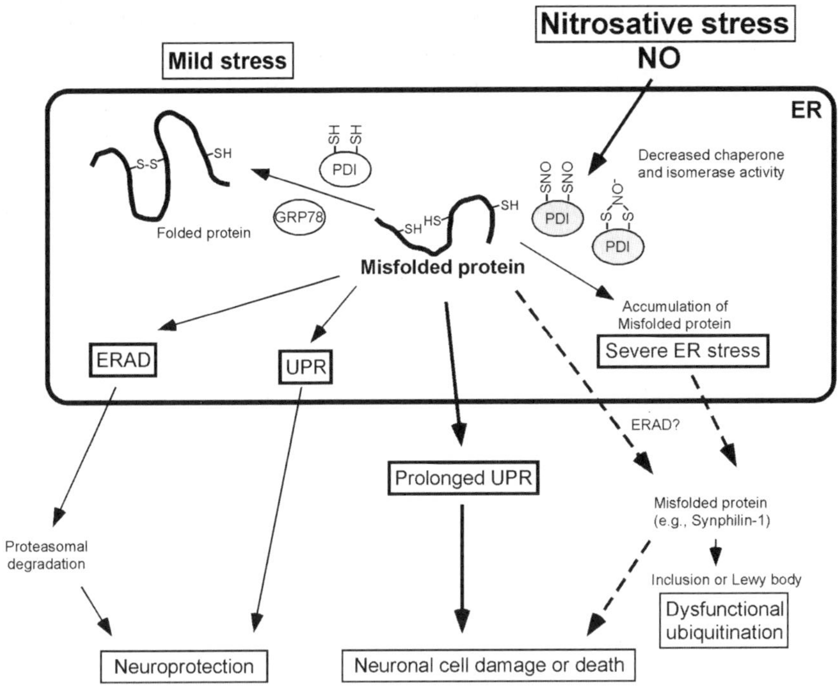

Figure 2.3 Possible mechanism of *S*-nitrosylated PDI (SNO-PDI) contributing to the accumulation of aberrant proteins and neuronal cell damage or death. ER stress is triggered when misfolded proteins accumulate within the ER lumen, inducing the unfolded protein response (UPR). The UPR is usually a transient homeostatic mechanism for cell survival, while a prolonged UPR elicits neuronal cell death. PDI modulates the activity of UPR sensors by mediating proper protein folding in the ER. Proteins that fail to attain their native folded state are eventually retro-translocated across the ER membrane to be disposed of by cytosolic proteasomes. This process, known as *ER-associated degradation*, is essential in preventing protein accumulation and aggregation in the ER. Under conditions of severe nitrosative stress, *S*-nitrosylation of neuronal PDI inhibits normal protein folding in the ER, activates ER stress, and induces a prolonged UPR, thus contributing to aberrant protein accumulation and cell damage or death. For simplicity, *S*-nitrosylation of only one (of two) thioredoxin domains of PDI is shown, resulting in the formation of SNO-PDI or possibly nitroxyl-PDI, as described in Uehara *et al.* (2006) and Forrester *et al.* (2006).

molecular chaperones that assist in the maturation, transport, and folding of secretory proteins.

During protein folding in the ER, PDI can introduce disulfide bonds into proteins (oxidation), break disulfide bonds (reduction), and catalyze thiol/disulfide exchange (isomerization), thus facilitating disulfide bond formation, rearrangement reactions, and structural stability (Lyles and Gilbert, 1991). PDI has four domains that are homologous to TRX (termed a, b, b′, and a′). Only two of the four TRX-like domains (a and a′) contain a characteristic redox-active CXXC motif, and these two-thiol/disulfide centers function as independent active sites (Edman *et al.*, 1985; Ellgaard and Ruddock, 2005; Gruber *et al.*, 2006;

Vuori *et al.*, 1992). These active-site cysteines can be found in two different redox states: oxidized (disulfide) or reduced (free sulfhydryls or thiols). During oxidation of a target protein, oxidized PDI catalyzes disulfide formation in the substrate protein, resulting in the reduction of PDI. In contrast, the reduced form of the active-site cysteines can initiate isomerization by attacking the disulfide of a substrate protein and forming a transient intermolecular disulfide bond. As a consequence, an intramolecular disulfide rearrangement occurs within the substrate itself, resulting in the generation of reduced PDI. The recently determined structure of yeast PDI revealed that the four TRX-like domains form a twisted "U" shape with the two active sites facing each other on opposite sides of the "U" (Tian *et al.*, 2006). Hydrophobic residues line the inside surface of the "U," facilitating interactions between PDI and misfolded proteins. Specifically, the b′ domain of PDI constitutes a part of the base of the "U"-shaped structure and contributes to the efficient binding of misfolded proteins (Klappa *et al.*, 1998). Several mammalian PDI homologs, such as ERp57, PDIp, and Erdj5, also localize to the ER and may manifest similar functions (Conn *et al.*, 2004; Hetz *et al.*, 2005). Increased expression of PDIp in neuronal cells under conditions mimicking PD suggests the possible contribution of PDIp to neuronal survival (Conn *et al.*, 2004). Additionally, ERdj5, an ER reductase that contains four TRX-like domains, forms a functional ERAD complex with GRP78, promoting the degradation of misfolded proteins via ERAD (Ushioda *et al.*, 2008).

2.6.2 Neuroprotective Roles of Other ER Chaperones in Neurodegenerative Conditions

In addition to PDI, the ER contains GRP78 and calnexin/calreticulin chaperones to provide proper folding and assembly for their respective target proteins, thus preventing misfolding and aggregation (Buck *et al.*, 2007; Ni and Lee, 2007). Similar to PDI, these chaperones are also upregulated by ER stress. In fact, these proteins can form large complexes that include both chaperones and co-chaperones, such as PDI, GRP78, GRP94, and SIL1, and may act simultaneously on substrate proteins (Meunier *et al.*, 2002). Among these ER chaperones, the best-characterized protein is probably GRP78/BiP. Using energy from adenosine triphosphate (ATP) hydrolysis, GRP78 facilitates the translocation of nascent proteins into the ER lumen, corrects folding of substrate proteins and degradation of misfolded proteins via ERAD. Additionally, if misfolded proteins accumulate in the ER, then GRP78 dissociates from the luminal domain of Ire1 and triggers the UPR. Recent evidence demonstrated that these physiological activities of GRP78 are particularly important for neuronal survival. For example, a specific deletion of the *grp78* gene from Purkinje cells induced a prolonged UPR and apoptotic cell death (Wang *et al.*, 2009b), whereas targeted knockout of GRP78 from prostate epithelium did not result in cell death (Fu *et al.*, 2008). Along these lines, a Purkinje cell-specific Grp78 conditional knockout mouse manifests severe motor coordination defects and cerebellar atrophy. Moreover, mutation of the *sil1* gene, encoding a co-chaperone of GRP78, causes Marinesco–Sjögren syndrome, a rare form of neurodegenerative disease associated with cerebellar ataxia and progressive myopathy (Anttonen *et al.*, 2005; Senderek *et al.*, 2005). Taken together, this evidence suggests that GRP78 contributes to the maintenance of a proper ER environment and may thereby serve a neuroprotective function.

ER membrane-bound calnexin and its luminal homolog, calreticulin, recognize monoglucosylated N-linked glycans on newly synthesized proteins and facilitate both protein folding and ERAD. In fact, essentially all glycoproteins interact transiently with one or both of these chaperone proteins during maturation (Ellgaard *et al.*, 1999). GRP58, a member of the PDI family, also often associates with calnexin/calreticulin and participates in the folding steps via introduction of disulfide bonds to substrate proteins. In addition to its chaperone activity, calreticulin can bind to Ca^{2+} in the ER lumen with high capacity and thus serve as a Ca^{2+} storage/buffering protein. In the nervous system, expression of mouse calreticulin is high in the embryonic stage compared to that in adult brains, suggesting that calreticulin may play an important role during central nervous system development (Zhang *et al.*, 2007). Indeed, calreticulin deficiency leads to exencephaly due to a defect in neural tube closure (Rauch *et al.*, 2000). Additionally, dysfunction in the calnexin/calreticulin system may also contribute to age-associated neurodegenerative diseases, such as AD, because overexpression of calnexin (or GRP78) can suppress the production of Aβ peptide in cultured cell lines (Hoshino *et al.*, 2007).

In many neurodegenerative disorders and cerebral ischemia, the accumulation of immature and denatured proteins results in ER dysfunction (Atkin *et al.*, 2006; Conn *et al.*, 2004; Hu *et al.*, 2000; Rao and Bredesen, 2004), but upregulation of PDI and other ER chaperones (e.g., GRP78) represent an adaptive response promoting protein refolding and may offer neuronal cell protection (Conn *et al.*, 2004; Hetz *et al.*, 2005; Ko *et al.*, 2002; Tanaka *et al.*, 2000). In a recent study, we reported that the *S*-nitrosylation of PDI (to form SNO-PDI) disrupts its neuroprotective role and GRP probably undergoes a similar fate (Uehara *et al.*, 2006).

2.6.3 *S*-Nitrosylation of PDI Mediates Protein Misfolding and Neurotoxicity in Cell Models of PD or AD

Disturbance of Ca^{2+} homeostasis within the ER plays a critical role in the accumulation of misfolded proteins and ER stress because the function of several ER chaperones requires high concentrations of Ca^{2+}. In addition, it is generally accepted that excessive generation of NO can contribute to activation of the ER stress pathway, at least in some cell types (Gotoh *et al.*, 2002; Oyadomari *et al.*, 2001). Molecular mechanisms by which NO induces protein misfolding and ER stress, however, have remained enigmatic until recently. The ER normally manifests a relatively positive redox potential in contrast to the highly reducing environment of the cytosol and mitochondria. This redox environment can influence the stability of protein *S*-nitrosylation and oxidation reactions (Forrester *et al.*, 2006). *S*-Nitrosylation can enhance the activity of the ER Ca^{2+} channel-ryanodine receptor (Xu *et al.*, 1998), which may provide a clue to how

NO disrupts Ca^{2+} homeostasis in the ER and activates the cell death pathway. Interestingly, we have recently reported that excessive NO can also lead to *S*-nitrosylation of the active-site thiol groups of PDI, and this reaction inhibits both its isomerase and chaperone activities (Uehara *et al.*, 2006). Mitochondrial complex I insult by rotenone can also result in *S*-nitrosylation of PDI in cell culture models. Moreover, we found that PDI is *S*-nitrosylated in the brains of virtually all cases of sporadic AD and PD examined. Under pathological conditions, it is possible that both cysteine sulfhydryl groups in the TRX-like domains of PDI form SNOs. Unlike the formation of a single SNO, which is commonly seen after de-nitrosylation reactions catalyzed by PDI (Sliskovic *et al.*, 2005), dual nitrosylation may be relatively more stable and prevent subsequent disulfide formation on PDI. Therefore, we speculate that these pathological *S*-nitrosylation reactions on PDI are more easily detected during neurodegenerative conditions. In addition, it is possible that vicinal (nearby) cysteine thiols reacting with NO can form nitroxyl disulfide (Houk *et al.*, 2003), and such reaction may potentially occur in the catalytic side of PDI to inhibit enzymatic activity. In order to determine the consequences of *S*-nitrosylated PDI (SNO-PDI) formation in neurons, we exposed cultured cerebrocortical neurons to neurotoxic concentrations of NMDA, thus inducing excessive Ca^{2+} influx and consequent NO production from nNOS. Under these conditions, we found that PDI was *S*-nitrosylated in an NOS-dependent manner. SNO-PDI formation led to the accumulation of polyubiquitinated/misfolded proteins and activation of the UPR. Moreover, *S*-nitrosylation abrogated the inhibitory effect of PDI on aggregation of proteins observed in Lewy body inclusions (Chung *et al.*, 2001; Uehara *et al.*, 2006). *S*-Nitrosylation of PDI also prevented its attenuation of neuronal cell death triggered by ER stress, misfolded proteins, or proteasome inhibition (Figure 2.3). Further evidence suggested that SNO-PDI may in effect transport NO to the extracellular space, where it could conceivably exert additional adverse effects (Sliskovic *et al.*, 2005). Another form of cysteine modification, *S*-glutathionylation, inhibits PDI activity and therefore may also contribute to protein misfolding in degenerative conditions (Townsend *et al.*, 2009). In addition, NO can possibly mediate cell death or injury via *S*-nitrosylation or nitration reactions on other TRX-like proteins, such as TRX itself and glutaredoxin (Aracena-Parks *et al.*, 2006; Haendeler *et al.*, 2002; Tao *et al.*, 2006).

In addition to PDI, *S*-nitrosylation is likely to affect critical thiol groups on other chaperones, such as HSP90 in the cytoplasm (Martinez-Ruiz *et al.*, 2005) and possibly GRP78/GRP94 in the ER. Normally, HSP90 stabilizes misfolded proteins and modulates the activity of cell signaling proteins including NOS and calcineurin (Muchowski and Wacker, 2005). In AD brains, the levels of HSP90 are increased in both the cytosolic and membranous fractions, where HSP90 is thought to maintain tau and Aβ in a soluble conformation, thereby averting their aggregation (Dou *et al.*, 2003; Kakimura *et al.*, 2002). Martinez-Ruiz *et al.*, (2005) recently demonstrated that *S*-nitrosylation of HSP90 can occur in endothelial cells, and this modification abolishes its ATPase activity, which is required for its function as a molecular chaperone. These studies imply that *S*-nitrosylation of HSP90 in neurons of AD brains may contribute to the accumulation of tau and Aβ aggregates.

The UPS is apparently impaired in the aging brain. In addition, inclusion bodies similar to those found in neurodegenerative disorders can appear in brains of normal aged individuals or those with subclinical manifestations of disease (Gray *et al.*, 2003). These findings suggest that the activity of the UPS and molecular chaperones may decline in an age-dependent manner (Paz Gavilan *et al.*, 2006). Given that we have not found detectable quantities of SNO-parkin and SNO-PDI in normal aged brain (Chung *et al.*, 2004; Uehara *et al.*, 2006; Yao *et al.*, 2004), we speculate that *S*-nitrosylation of these and similar proteins may represent a key event that contributes to susceptibility of the aging brain to neurodegenerative conditions.

2.6.4 PDI Activity in ALS and Prion Disease

Recently, PDI has been implicated in the pathophysiology of familial ALS (Atkin *et al.*, 2006). Mutations in Cu/Zn superoxide dismutase (SOD1) are known to be involved in motor neuron death in some forms of familial ALS. SOD1 is an intracellular homodimeric metalloprotein that forms a stable intra-subunit disulfide bond. Biochemical evidence suggests that the disulfide-reduced monomer of mutant SOD1 (mtSOD1) forms inclusion bodies (Arnesano *et al.*, 2004; Doucette *et al.*, 2004; Furukawa and O'Halloran, 2005; Rakhit *et al.*, 2004; Tiwari and Hayward, 2003), and aggregates of misfolded mtSOD1 are commonly associated with the disease, as seen at postmortem examination. In addition, although wtSOD1 is found predominantly in the cytoplasm, mtSOD1 forms monomers or insoluble high molecular mass multimers within the ER (Kikuchi *et al.*, 2006). Intriguingly, PDI colocalizes and binds to intracellular aggregates of mtSOD1. Upregulation of the UPR has also been observed in mtSOD1 mice. Recent studies have shown that inhibition of PDI activity with bacitracin can increase aggregation of mtSOD1 in neuronal cells, and that regulation of endogenous PDI activity by reticulons protects against neurodegeneration (Atkin *et al.*, 2006; Yang *et al.*, 2009). In contrast, overexpression of PDI decreases mtSOD aggregation and mtSOD1-induced neuronal cell death. These findings suggest that ER stress contributes to the pathophysiology of familial ALS, and increased PDI activity may decrease mtSOD1 aggregation and promote neuronal survival (Atkin *et al.*, 2006; Walker *et al.*, 2010; Yang *et al.*, 2009). Moreover, SNO levels have also been found to be abnormal in the spinal cords of mtSOD1 transgenic mice (Schonhoff *et al.*, 2006), including elevated SNO-PDI (Walker *et al.*, 2010). Sporadic ALS patients also manifest increased SNO-PDI levels, suggesting that *S*-nitrosylation of PDI may play a role in the pathogenesis of sporadic ALS (Walker *et al.*, 2010). It will be informative to determine whether SNO-PDI is causally involved in protein aggregation and motor neuron injury in ALS in the absence of SOD1 mutation. Chapter 10 also covers aspects of SOD1.

In addition, transmissible spongiform encephalopathies (TSEs), also known as *prion diseases*, are transmissible neurodegenerative disorders and include

Creutzfeldt–Jacob disease, bovine spongiform encephalopathy, and scrapie. Cerebral accumulation of misfolded prion protein (PrP) and extensive neuronal apoptosis represent pathological hallmarks of these prion diseases. Recent reports have suggested that a prolonged UPR due to PrP misfolding in the ER may contribute to neuronal dysfunction (Hetz *et al.*, 2003, 2005; Yoo *et al.*, 2002). This ER stress response is mainly associated with upregulation of GRP58, an ER chaperone with PDI-like activity, suggesting that this chaperone may play an important role in the cellular response to prion infection (Hetz *et al.*, 2005). In fact, *in vitro* studies on GRP58, either overexpressing (via transfection) or downregulating (via RNAi), demonstrated that this ER chaperone protects cells against PrP misfolding and toxicity. Collectively, these studies raise the possibility that SNO-PDI and *S*-nitrosylation of other chaperone molecules may represent potential therapeutic targets to prevent protein aggregation in several neurodegenerative diseases.

2.7 *S*-NITROSYLATION OF Drp1

2.7.1 Dysregulation of Mitochondrial Dynamics in AD and Other Neurodegenerative Diseases

Dysfunction in mitochondria represents another hallmark of neurodegenerative diseases. For example, patients with early-stage AD regularly exhibit declining mitochondrial energy metabolism and ATP production, which may subsequently cause synaptic loss and neuronal damage (Liang *et al.*, 2008; Parker *et al.*, 1994; Reddy, 2007). Interestingly, neurons in AD and other neurodegenerative brains often display abnormal mitochondrial morphology (Baloyannis, 2006; Wang *et al.*, 2009c).

Normally, mitochondria are known to continuously undergo fission and fusion (known as *mitochondrial dynamics*) to generate smaller organelles or elongated, tubular structures, respectively. This normal mitochondrial fission and fusion can facilitate formation of new mitochondria (biogenesis), repair of defective mitochondrial DNA through mixing, and redistribution of mitochondria to sites requiring high-energy production (Chen and Chan, 2006; Frederick and Shaw, 2007; Knott *et al.*, 2008). Conversely, an imbalance in fission or fusion initiates malfunctions in mitochondrial morphology and bioenergetics, and may thus contribute to neuronal injury during neurodegeneration (Barsoum *et al.*, 2006; Bossy-Wetzel *et al.*, 2003; Knott *et al.*, 2008). Interestingly, dysfunction in mitochondrial dynamics can result from rare genetic mutations in fission- or fusion-related genes, as occurs in Charcot–Marie–Tooth (CMT) disease and autosomal dominant optic atrophy (ADOA) (Delettre *et al.*, 2000; Zuchner *et al.*, 2004).

Neurons are particularly vulnerable to mitochondrial defects because they require high levels of energy for their survival and specialized function. Mitochondrial biogenesis is required in regions that demand high concentrations of ATP, especially the synapse. The distribution of mitochondria at the nerve terminal can control synaptic transmission and structure (Chen and Chan, 2006; Li *et al.*, 2004, 2008).

In healthy neurons, the fission/fusion machinery proteins maintain mitochondrial integrity and ensure their presence at critical locations. These proteins include Drp1 and Fis1, acting as fission proteins, and Mitofusin (Mfn) and Opa1, operating as fusion proteins (Youle and Karbowski, 2005). In both familial and sporadic neurodegenerative conditions, abnormal mitochondria regularly appear in the brain as a result of dysfunction in the fission/fusion machinery. Genetic mutations in Mfn2 can cause CMT disease, a hereditary peripheral neuropathy that affects both motor and sensory neurons (Kijima *et al.*, 2005; Zuchner *et al.*, 2004). Mutations in Opa1 cause ADOA, characterized by the loss of retinal ganglion cells and the optic nerve, representing their axons (Delettre *et al.*, 2000). Recently, Waterham *et al.* described a heterozygous, dominant-negative mutation of Drp1 in a patient whose symptoms were broadly similar to those of CMT neuropathy and ADOA (Waterham *et al.*, 2007). Drp1 includes four distinct structural domains: an N-terminal GTPase domain, a dynamin-like middle domain, an insert B domain, and a C-terminal GED domain. The mutation (Ala395 to Asp) was found in the middle domain of Drp1. This case study further suggested that a defect in mitochondrial fission may have more severe consequences than those of fusion defects, since the Drp1 mutation caused a much earlier onset (prenatal) and fatal outcome. In addition, it is apparent that the balance between fission and fusion is critical for normal function of mitochondria and determination of phenotype in disease. These fission/fusion proteins are widely expressed in human tissues, clearly supporting the notion that neurons are particularly sensitive to mitochondrial dysfunction.

Emerging evidence suggests that mitochondrial dysfunction plays a prominent role in the pathogenesis of AD (Wang *et al.*, 2009d). An estimated 26 million people globally have AD, which is thought to be the most common form of dementia. In AD brains, synaptic damage and neuronal loss in the hippocampus and cerebral cortex mainly accounts for the cognitive decline. Analyses of autopsy and biopsy samples revealed that mitochondria isolated from the AD brains exhibit diminished respiratory capacity (Parker *et al.*, 1994), and that AD neurons contain a number of mitochondria with fractured cristae (Hirai *et al.*, 2001). In addition, electron microscopic studies have described an increase in mitochondrial fragmentation in human AD brains (Baloyannis, 2006; Wang *et al.*, 2009c). In cell-based experiments, Aβ production resulted in the appearance of fragmented and abnormally distributed mitochondria (Barsoum *et al.*, 2006; Wang *et al.*, 2008), suggesting that Aβ (possibly in the form of soluble oligomers) may trigger excessive mitochondrial fission in AD patients. Pathological forms of tau may also contribute to mitochondrial fragmentation in AD brains since expression of caspase-cleaved tau induced mitochondrial fission in a calcineurin-dependent manner (Quintanilla *et al.*, 2009).

Similarly, dysfunction in mitochondrial integrity is associated with PD (Büeler, 2009). For instance, the Parkinsonian neurotoxins, rotenone and 1-methyl-4-phenylpyridinium (MPP^+), which inhibit complex I of the mitochondrial electron transport chain, can induce excessive mitochondrial fragmentation and cell death (Barsoum *et al.*, 2006). In addition, multiple groups recently observed that a deficiency in familial PD-related proteins, such as parkin and PINK1, led to the appearance of mitochondrial pathology (Exner *et al.*, 2007; Greene *et al.*, 2003; Lutz *et al.*, 2009; Poole *et al.*, 2008). Exogenous expression of mitochondrial fusion proteins, Mfn2 and OPA1, or dominant negative Drp1 rescued the altered mitochondrial morphology, suggesting that parkin or PINK1 deficiency promoted mitochondrial fragmentation (Lutz *et al.*, 2009). Interestingly, Drp1 seems to activate autophagy/mitophagy pathways for morphologic remodeling of mitochondria in PINK1-deficient neuroblastoma cells (Dagda *et al.*, 2009). Taken together, dysregulation of mitochondrial dynamics may contribute to a common pathway leading to the pathogenesis of several neurodegenerative diseases, including AD and PD.

2.7.2 *S*-Nitrosylation of Drp1 Mediates Mitochondrial Fission and Neurotoxicity in Cell Models of AD

In addition to the rare genetic mutations seen in the genes encoding mitochondrial fission and fusion proteins, recent studies have demonstrated that posttranslational modification of these molecules can contribute to altered mitochondria dynamics. For example, phosphorylation, ubiquitination, sumoylation, and proteolytic cleavage of Drp1 regulate mitochondrial fission by affecting Drp1 activity, at least in cell culture systems (Breckenridge *et al.*, 2008; Chang and Blackstone, 2007; Cribbs and Strack, 2007; Karbowski *et al.*, 2007; Nakamura *et al.*, 2006; Taguchi *et al.*, 2007; Wasiak *et al.*, 2007; Yonashiro *et al.*, 2006). Excessive activation of mitochondrial fission or fusion proteins by posttranslational modification was posited to contribute to neurodegeneration by compromising mitochondrial function. Interestingly, along these lines, we recently reported that excessive NO can also lead to *S*-nitrosylation of Drp1 at Cys644 (Cho *et al.*, 2009). Cys644 resides within the GTPase effector domain (GED) of Drp1, which influences both GTPase activity and oligomer formation of Drp1 (Low and Lowe, 2006; Pitts *et al.*, 2004; Ramachandran *et al.*, 2007; Zhu *et al.*, 2004). *S*-Nitrosylation of Drp1 (forming SNO-Drp1) induces formation of Drp1 dimers, which function as building blocks for tetramers and higher order structures of Drp1, and activates Drp1 GTPase activity; however, substitution of Cys644 for Ala (C644A) abrogated these effects of NO.

We further demonstrated that exposure to oligomeric Aβ peptide results in the formation of SNO-Drp1 in cell culture models. Moreover, we and others have observed that Drp1 is *S*-nitrosylated in the brains of virtually all cases of sporadic AD (Cho *et al.*, 2009; Wang *et al.*, 2009c). In order to determine the consequences of *S*-nitrosylated Drp1 in neurons, we exposed cultured cerebrocortical neurons to the physiological NO donor, *S*-nitrosocysteine (SNOC), or to Aβ oligomers and found that both induced SNO-Drp1 formation and led to the accumulation of fragmented mitochondria. Moreover, mutation of a specific cysteine residue in Drp1 (C644A) prevented these effects of SNOC or Aβ on mitochondrial fragmentation, consistent with the notion that SNO-Drp1 triggered excessive mitochondrial fission or fragmentation. Finally, in response to Aβ, SNO-Drp1 induced mitochondrial fragmentation caused synaptic damage, an early characteristic feature of AD, and eventually apoptotic neuronal cell death. Importantly, blockade of Drp1 nitrosylation (using the Drp1(C644A) mutant) prevented Aβ-mediated synaptic loss and neuronal cell death, suggesting that SNO-Drp1 may represent a potential therapeutic target to protect neurons and their synapses in AD.

In addition to AD, SNO-Drp1 may affect the pathogenesis of other neurodegenerative disorders, such as HD, as we have recently observed nitrosylation of Drp1 in HD brains as well as in AD brains. An expanded CAG repeat in the huntingtin (*htt*) gene is the cause of HD. A potential mechanism for neurodegeneration in HD that is triggered by mutant Htt (mtHtt) involves mitochondrial dysfunction. Evidence for this includes reduced activity of respiratory complexes, decreased mitochondrial membrane potential, and changes in mitochondrial ultrastructure (Gu *et al.*, 1996; Panov *et al.*, 2002). Interestingly, expression of mtHtt induces mitochondrial fragmentation in Hela cells (Wang *et al.*, 2009a) and exposure to the mitochondrial complex II inhibitor 3-nitropropionic acid (3-NP), which reproduces many of the pathological features of HD, leads to mitochondria fission (Liot *et al.*, 2009). Furthermore, accumulating evidence suggests that excitotoxic pathways, which increase NO, may contribute to the pathophysiology of HD (Bossy-Wetzel *et al.*, 2008). Hence, it is tempting to postulate that SNO-Drp1 may be involved in the mitochondrial fragmentation and neuronal injury observed in HD.

2.8 CONCLUSIONS

Excessive nitrosative and oxidative stress triggered by overactivation of NMDA receptors may result in malfunction of molecular chaperones and the UPS, thus contributing to abnormal protein accumulation and neuronal damage in sporadic forms of neurodegenerative diseases. Our elucidation of an NO-mediated pathway to dysfunction of parkin and PDI by *S*-nitrosylation provides a mechanistic link between free radical production, abnormal protein accumulation, and neuronal cell injury in neurodegenerative disorders such as PD. In addition, nitrosylation reactions can lead to mitochondrial dysfunction, thus contributing to synaptic damage and neuronal loss in sporadic forms of neurodegenerative diseases. For example, SNO-Drp1-mediated excessive fragmentation of mitochondria can contribute to Aβ toxicity. Elucidation of these new pathways may lead to the development of additional new therapeutic approaches to prevent aberrant protein misfolding by targeted disruption or prevention of nitrosylation of specific proteins such as parkin and PDI.

ACKNOWLEDGMENTS

This work was supported in part by NIH grants P01 HD29587, R01 EY05477, R01 EY09024, the American Parkinson's Disease Association, San Diego Chapter, and an Ellison Senior Scholars Award in Aging (to S.A.L.).

REFERENCES

Abou-Sleiman PM, Muqit MM, Wood NW. Expanding insights of mitochondrial dysfunction in Parkinson's disease. *Nat Rev Neurosci* 2006;**7**:207–219.

Abu-Soud HM, Stuehr DJ. Nitric oxide synthases reveal a role for calmodulin in controlling electron transfer. *Proc Natl Acad Sci U S A* 1993;**90**:10769–10772.

Andrews DW, Johnson AE. The translocon: more than a hole in the ER membrane? *Trends Biochem Sci* 1996;**21**:365–369.

Ankarcrona M, Dypbukt JM, Bonfoco E, Zhivotovsky B, Orrenius S, Lipton SA, Nicotera P. Glutamate-induced neuronal death: a succession of necrosis or apoptosis depending on mitochondrial function. *Neuron* 1995;**15**:961–973.

Anttonen AK, Mahjneh I, Hamalainen RH, Lagier-Tourenne C, Kopra O, Waris L, Anttonen M, Joensuu T, Kalimo H, Paetau A, *et al.* The gene disrupted in Marinesco–Sjogren syndrome encodes SIL1, an HSPA5 cochaperone. *Nat Genet* 2005;**37**:1309–1311.

Aracena-Parks P, Goonasekera SA, Gilman CP, Dirksen RT, Hidalgo C, Hamilton SL. Identification of cysteines involved in *S*-nitrosylation, s-glutathionylation, and oxidation to disulfides in ryanodine receptor type 1. *J Biol Chem* 2006;**281**:40354–40368.

Arnesano F, Banci L, Bertini I, Martinelli M, Furukawa Y, O'Halloran TV. The unusually stable quaternary structure of human Cu,Zn-superoxide dismutase 1 is controlled by both metal occupancy and disulfide status. *J Biol Chem* 2004;**279**:47998–48003.

Arrasate M, Mitra S, Schweitzer ES, Segal MR, Finkbeiner S. Inclusion body formation reduces levels of mutant huntingtin and the risk of neuronal death. *Nature* 2004;**431**:805–810.

Atkin JD, Farg MA, Turner BJ, Tomas D, Lysaght JA, Nunan J, Rembach A, Nagley P, Beart PM, Cheema SS, *et al*. Induction of the unfolded protein response in familial amyotrophic lateral sclerosis and association of protein-disulfide isomerase with superoxide dismutase 1. *J Biol Chem* 2006;**281**:30152–30165.

Auluck PK, Chan HY, Trojanowski JQ, Lee VM, Bonini NM. Chaperone suppression of alpha-synuclein toxicity in a Drosophila model for Parkinson's disease. *Science* 2002;**295**:865–868.

Baloyannis SJ. Mitochondrial alterations in Alzheimer's disease. *J Alzheimers Dis* 2006;**9**:119–126.

Barnham KJ, Masters CL, Bush AI. Neurodegenerative diseases and oxidative stress. *Nat Rev Drug Discov* 2004;**3**:205–214.

Barsoum MJ, Yuan H, Gerencser AA, Liot G, Kushnareva Y, Graber S, Kovacs I, Lee WD, Waggoner J, Cui J, *et al.* Nitric oxide-induced mitochondrial fission is regulated by dynamin-related GTPases in neurons. *EMBO J* 2006;**25**:3900–3911.

Beal MF. Experimental models of Parkinson's disease. *Nat Rev Neurosci* 2001;**2**: 325–334.

Beckman JS, Beckman TW, Chen J, Marshall PA, Freeman BA. Apparent hydroxyl radical production by peroxynitrite: implications for endothelial injury from nitric oxide and superoxide. *Proc Natl Acad Sci U S A* 1990;**87**:1620–1624.

Bence NF, Sampat RM, Kopito RR. Impairment of the ubiquitin-proteasome system by protein aggregation. *Science* 2001;**292**:1552–1555.

Benhar M, Forrester MT, Hess DT, Stamler JS. Regulated protein denitrosylation by cytosolic and mitochondrial thioredoxins. *Science* 2008;**320**:1050–1054.

Betarbet R, Sherer TB, MacKenzie G, Garcia-Osuna M, Panov AV, Greenamyre JT. Chronic systemic pesticide exposure reproduces features of Parkinson's disease. *Nat Neurosci* 2000;**3**:1301–1306.

Bonfoco E, Krainc D, Ankarcrona M, Nicotera P, Lipton SA. Apoptosis and necrosis: two distinct events induced, respectively, by mild and intense insults with N-methyl-D-aspartate or nitric oxide/superoxide in cortical cell cultures. *Proc Natl Acad Sci U S A* 1995;**92**:7162–7166.

Bonifati V, Rizzu P, van Baren MJ, Schaap O, Breedveld GJ, Krieger E, Dekker MC, Squitieri F, Ibanez P, Joosse M, *et al.* Mutations in the DJ-1 gene associated with autosomal recessive early-onset parkinsonism. *Science* 2003;**299**:256–259.

Bossy-Wetzel E, Barsoum MJ, Godzik A, Schwarzenbacher R, Lipton SA. Mitochondrial fission in apoptosis, neurodegeneration and aging. *Curr Opin Cell Biol* 2003;**15**:706–716.

Bossy-Wetzel E, Petrilli A, Knott AB. Mutant huntingtin and mitochondrial dysfunction. *Trends Neurosci* 2008;**31**:609–616.

Boucher JL, Moali C, Tenu JP. Nitric oxide biosynthesis, nitric oxide synthase inhibitors and arginase competition for L-arginine utilization. *Cell Mol Life Sci* 1999;**55**:1015–1028.

Breckenridge DG, Kang BH, Kokel D, Mitani S, Staehelin LA, Xue D. Caenorhabditis elegans drp-1 and fis-2 regulate distinct cell-death execution pathways downstream of ced-3 and independent of ced-9. *Mol Cell* 2008;**31**:586–597.

Bredt DS, Hwang PM, Glatt CE, Lowenstein C, Reed RR, Snyder SH. Cloned and expressed nitric oxide synthase structurally resembles cytochrome P-450 reductase. *Nature* 1991;**351**:714–718.

Bredt DS, Snyder SH. Nitric oxide: a physiologic messenger molecule. *Annu Rev Biochem* 1994;**63**:175–195.

Buck TM, Wright CM, Brodsky JL. The activities and function of molecular chaperones in the endoplasmic reticulum. *Semin Cell Dev Biol* 2007;**18**:751–761.

Budd SL, Tenneti L, Lishnak T, Lipton SA. Mitochondrial and extramitochondrial apoptotic signaling pathways in cerebrocortical neurons. *Proc Natl Acad Sci U S A* 2000;**97**:6161–6166.

Büeler, H. (2009). Mitochondrial dynamics, cell death, and the pathogenesis of Parkinson's disease. Apoptosis Feb 4, PMID: 20131004.

Castello PR, David PS, McClure T, Crook Z, Poyton RO. Mitochondrial cytochrome oxidase produces nitric oxide under hypoxic conditions: implications for oxygen sensing and hypoxic signaling in eukaryotes. *Cell Metab* 2006;**3**:277–287.

Chabrier PE, Demerle-Pallardy C, Auguet M. Nitric oxide synthases: targets for therapeutic strategies in neurological diseases. *Cell Mol Life Sci* 1999;**55**:1029–1035.

Chang CR, Blackstone C. Cyclic AMP-dependent protein kinase phosphorylation of Drp1 regulates its GTPase activity and mitochondrial morphology. *J Biol Chem* 2007;**282**:21583–21587.

Chen H, Chan DC. Critical dependence of neurons on mitochondrial dynamics. *Curr Opin Cell Biol* 2006;**18**:453–459.

Chen HS, Lipton SA. The chemical biology of clinically tolerated NMDA receptor antagonists. *J Neurochem* 2006;**97**:1611–1626.

Chen TS, Richie JP, Lang CA Jr. The effect of aging on glutathione and cysteine levels in different regions of the mouse brain. *Proc Soc Exp Biol Med* 1989;**190**:399–402.

Cho DH, Nakamura T, Fang J, Cieplak P, Godzik A, Gu Z, Lipton SA. *S*-nitrosylation of Drp1 mediates beta-amyloid-related mitochondrial fission and neuronal injury. *Science* 2009;**324**:102–105.

Choi J, Levey AI, Weintraub ST, Rees HD, Gearing M, Chin LS, Li L. Oxidative modifications and down-regulation of ubiquitin carboxyl-terminal hydrolase L1 associated with idiopathic Parkinson's and Alzheimer's diseases. *J Biol Chem* 2004;**279**:13256–13264.

Choi YB, Tenneti L, Le DA, Ortiz J, Bai G, Chen HS, Lipton SA. Molecular basis of NMDA receptor-coupled ion channel modulation by *S*-nitrosylation. *Nat Neurosci* 2000;**3**:15–21.

Chung KK, Dawson TM, Dawson VL. Nitric oxide, *S*-nitrosylation and neurodegeneration. *Cell Mol Biol (Noisy-le-grand)* 2005;**51**:247–254.

Chung KK, Thomas B, Li X, Pletnikova O, Troncoso JC, Marsh L, Dawson VL, Dawson TM. *S*-nitrosylation of parkin regulates ubiquitination and compromises parkin's protective function. *Science* 2004;**304**:1328–1331.

Chung KK, Zhang Y, Lim KL, Tanaka Y, Huang H, Gao J, Ross CA, Dawson VL, Dawson TM. Parkin ubiquitinates the alpha-synuclein-interacting protein, synphilin-1: implications for Lewy-body formation in Parkinson disease. *Nat Med* 2001;**7**:1144–1150.

Ciechanover A, Brundin P. The ubiquitin proteasome system in neurodegenerative diseases: sometimes the chicken, sometimes the egg. *Neuron* 2003;**40**:427–446.

Conn KJ, Gao W, McKee A, Lan MS, Ullman MD, Eisenhauer PB, Fine RE, Wells JM. Identification of the protein disulfide isomerase family member PDIp in experimental Parkinson's disease and Lewy body pathology. *Brain Res* 2004;**1022**:164–172.

Cookson MR. The biochemistry of Parkinson's disease. *Annu Rev Biochem* 2005;**74**:29–52.

Corti O, Hampe C, Koutnikova H, Darios F, Jacquier S, Prigent A, Robinson JC, Pradier L, Ruberg M, Mirande M, *et al.* The p38 subunit of the aminoacyl-tRNA synthetase complex is a Parkin substrate: linking protein biosynthesis and neurodegeneration. *Hum Mol Genet* 2003;**12**:1427–1437.

Cribbs JT, Strack S. Reversible phosphorylation of Drp1 by cyclic AMP-dependent protein kinase and calcineurin regulates mitochondrial fission and cell death. *EMBO Rep* 2007;**8**:939–944.

Dagda RK, Cherra SJ, Kulich SM, Tandon A, Park D, Chu CT 3rd Loss of PINK1 function promotes mitophagy through effects on oxidative stress and mitochondrial fission. *J Biol Chem* 2009;**284**:13843–13855.

Dawson VL, Dawson TM, London ED, Bredt DS, Snyder SH. Nitric oxide mediates glutamate neurotoxicity in primary cortical cultures. *Proc Natl Acad Sci U S A* 1991;**88**:6368–6371.

Delettre C, Lenaers G, Griffoin JM, Gigarel N, Lorenzo C, Belenguer P, Pelloquin L, Grosgeorge J, Turc-Carel C, Perret E, *et al.* Nuclear gene OPA1, encoding a mitochondrial dynamin-related protein, is mutated in dominant optic atrophy. *Nat Genet* 2000;**26**:207–210.

Dimmeler S, Haendeler J, Nehls M, Zeiher AM. Suppression of apoptosis by nitric oxide via inhibition of interleukin-1beta-converting enzyme (ICE)-like and cysteine protease protein (CPP)-32-like proteases. *J Exp Med* 1997;**185**:601–607.

Dou F, Netzer WJ, Tanemura K, Li F, Hartl FU, Takashima A, Gouras GK, Greengard P, Xu H. Chaperones increase association of tau protein with microtubules. *Proc Natl Acad Sci U S A* 2003;**100**:721–726.

Doucette PA, Whitson LJ, Cao X, Schirf V, Demeler B, Valentine JS, Hansen JC, Hart PJ. Dissociation of human copper-zinc superoxide dismutase dimers using chaotrope and reductant. Insights into the molecular basis for dimer stability. *J Biol Chem* 2004;**279**:54558–54566.

Edman JC, Ellis L, Blacher RW, Roth RA, Rutter WJ. Sequence of protein disulphide isomerase and implications of its relationship to thioredoxin. *Nature* 1985;**317**:267–270.

Ellgaard L, Molinari M, Helenius A. Setting the standards: quality control in the secretory pathway. *Science* 1999;**286**:1882–1888.

Ellgaard L, Ruddock LW. The human protein disulphide isomerase family: substrate interactions and functional properties. *EMBO Rep* 2005;**6**:28–32.

Emerit J, Edeas M, Bricaire F. Neurodegenerative diseases and oxidative stress. *Biomed Pharmacother* 2004;**58**:39–46.

Exner N, Treske B, Paquet D, Holmstrom K, Schiesling C, Gispert S, Carballo-Carbajal I, Berg D, Hoepken HH, Gasser T, *et al.* Loss-of-function of human PINK1 results in mitochondrial pathology and can be rescued by parkin. *J Neurosci* 2007;**27**:12413–12418.

Fallon L, Belanger CM, Corera AT, Kontogiannea M, Regan-Klapisz E, Moreau F, Voortman J, Haber M, Rouleau G, Thorarinsdottir T, *et al.*. A regulated interaction with the UIM protein Eps15 implicates parkin in EGF receptor trafficking and PI(3)K-Akt signalling. *Nat Cell Biol* 2006;**8**:834–842.

Farrer M, Chan P, Chen R, Tan L, Lincoln S, Hernandez D, Forno L, Gwinn-Hardy K, Petrucelli L, Hussey J, *et al.*. Lewy bodies and parkinsonism in families with parkin mutations. *Ann Neurol* 2001;**50**:293–300.

Forrester MT, Benhar M, Stamler JS. Nitrosative stress in the ER: a new role for *S*-nitrosylation in neurodegenerative diseases. *ACS Chem Biol* 2006;**1**:355–358.

Forstermann U, Boissel JP, Kleinert H. Expressional control of the 'constitutive' isoforms of nitric oxide synthase (NOS I and NOS III). *Faseb J* 1998;**12**:773–790.

Frederick RL, Shaw JM. Moving mitochondria: establishing distribution of an essential organelle. *Traffic* 2007;**8**:1668–1675.

Fu Y, Wey S, Wang M, Ye R, Liao CP, Roy-Burman P, Lee AS. Pten null prostate tumorigenesis and AKT activation are blocked by targeted knockout of ER chaperone GRP78/BiP in prostate epithelium. *Proc Natl Acad Sci U S A* 2008;**105**:19444–19449.

Furukawa Y, O'Halloran TV. Amyotrophic lateral sclerosis mutations have the greatest destabilizing effect on the apo- and reduced form of SOD1, leading to unfolding and oxidative aggregation. *J Biol Chem* 2005;**280**:17266–17274.

Garthwaite J, Charles SL, Chess-Williams R. Endothelium-derived relaxing factor release on activation of NMDA receptors suggests role as intercellular messenger in the brain. *Nature* 1988;**336**:385–388.

Giasson BI, Duda JE, Murray IV, Chen Q, Souza JM, Hurtig HI, Ischiropoulos H, Trojanowski JQ, Lee VM. Oxidative damage linked to neurodegeneration by selective alpha-synuclein nitration in synucleinopathy lesions. *Science* 2000;**290**:985–989.

Gotoh T, Oyadomari S, Mori K, Mori M. Nitric oxide-induced apoptosis in RAW 264.7 macrophages is mediated by endoplasmic reticulum stress pathway involving ATF6 and CHOP. *J Biol Chem* 2002;**277**:12343–12350.

Gray DA, Tsirigotis M, Woulfe J. Ubiquitin, proteasomes, and the aging brain. *Sci Aging Knowledge Environ* 2003;**2003**: RE6.

Greene JC, Whitworth AJ, Kuo I, Andrews LA, Feany MB, Pallanck LJ. Mitochondrial pathology and apoptotic muscle degeneration in Drosophila parkin mutants. *Proc Natl Acad Sci U S A* 2003;**100**:4078–4083.

Groves JT, Wang CC. Nitric oxide synthase: models and mechanisms. *Curr Opin Chem Biol* 2000;**4**:687–695.

Gruber CW, Cemazar M, Heras B, Martin JL, Craik DJ. Protein disulfide isomerase: the structure of oxidative folding. *Trends Biochem Sci* 2006;**31**:455–464.

Gu M, Gash MT, Mann VM, Javoy-Agid F, Cooper JM, Schapira AH. Mitochondrial defect in Huntington's disease caudate nucleus. *Ann Neurol* 1996;**39**:385–389.

Gu Z, Kaul M, Yan B, Kridel SJ, Cui J, Strongin A, Smith JW, Liddington RC, Lipton SA. *S*-nitrosylation of matrix metalloproteinases: signaling pathway to neuronal cell death. *Science* 2002;**297**:1186–1190.

Gu Z, Nakamura T, Yao D, Shi ZQ, Lipton SA. Nitrosative and oxidative stress links dysfunctional ubiquitination to Parkinson's disease. *Cell Death Differ* 2005;**12**:1202–1204.

Haendeler J, Hoffmann J, Tischler V, Berk BC, Zeiher AM, Dimmeler S. Redox regulatory and anti-apoptotic functions of thioredoxin depend on *S*-nitrosylation at cysteine 69. *Nat Cell Biol* 2002;**4**:743–749.

Hampe C, Ardila-Osorio H, Fournier M, Brice A, Corti O. Biochemical analysis of Parkinson's disease-causing variants of Parkin, an E3 ubiquitin-protein ligase with monoubiquitylation capacity. *Hum Mol Genet* 2006;**15**:2059–2075.

Hantraye P, Brouillet E, Ferrante R, Palfi S, Dolan R, Matthews RT, Beal MF. Inhibition of neuronal nitric oxide synthase prevents MPTP-induced parkinsonism in baboons. *Nat Med* 1996;**2**:1017–1021.

Hara MR, Agrawal N, Kim SF, Cascio MB, Fujimuro M, Ozeki Y, Takahashi M, Cheah JH, Tankou SK, Hester LD, *et al.* S-nitrosylated GAPDH initiates apoptotic cell death by nuclear translocation following Siah1 binding. *Nat Cell Biol* 2005;**7**:665–674.

Hardingham GE, Fukunaga Y, Bading H. Extrasynaptic NMDARs oppose synaptic NMDARs by triggering CREB shut-off and cell death pathways. *Nat Neurosci* 2002;**5**:405–414.

He Y, Imam SZ, Dong Z, Jankovic J, Ali SF, Appel SH, Le W. Role of nitric oxide in rotenone-induced nigro-striatal injury. *J Neurochem* 2003;**86**:1338–1345.

Hess DT, Matsumoto A, Kim SO, Marshall HE, Stamler JS. Protein *S*-nitrosylation: purview and parameters. *Nat Rev Mol Cell Biol* 2005;**6**:150–166.

Hetz C, Russelakis-Carneiro M, Maundrell K, Castilla J, Soto C. Caspase-12 and endoplasmic reticulum stress mediate neurotoxicity of pathological prion protein. *EMBO J* 2003;**22**:5435–5445.

Hetz C, Russelakis-Carneiro M, Walchli S, Carboni S, Vial-Knecht E, Maundrell K, Castilla J, Soto C. The disulfide isomerase Grp58 is a protective factor against prion neurotoxicity. *J Neurosci* 2005;**25**:2793–2802.

Hirai K, Aliev G, Nunomura A, Fujioka H, Russell RL, Atwood CS, Johnson AB, Kress Y, Vinters HV, Tabaton M, *et al.* Mitochondrial abnormalities in Alzheimer's disease. *J Neurosci* 2001;**21**:3017–3023.

Hoshino T, Nakaya T, Araki W, Suzuki K, Suzuki T, Mizushima T. Endoplasmic reticulum chaperones inhibit the production of amyloid-beta peptides. *Biochem J* 2007;**402**:581–589.

Houk KN, Hietbrink BN, Bartberger MD, McCarren PR, Choi BY, Voyksner RD, Stamler JS, Toone EJ. Nitroxyl disulfides, novel intermediates in transnitrosation reactions. *J Am Chem Soc* 2003;**125**:6972–6976.

Hu BR, Martone ME, Jones YZ, Liu CL. Protein aggregation after transient cerebral ischemia. *J Neurosci* 2000;**20**:3191–3199.

Huang Z, Huang PL, Panahian N, Dalkara T, Fishman MC, Moskowitz MA. Effects of cerebral ischemia in mice deficient in neuronal nitric oxide synthase. *Science* 1994;**265**:1883–1885.

Huynh DP, Scoles DR, Nguyen D, Pulst SM. The autosomal recessive juvenile Parkinson disease gene product, parkin, interacts with and ubiquitinates synaptotagmin XI. *Hum Mol Genet* 2003;**12**:2587–2597.

Iadecola C, Zhang F, Casey R, Nagayama M, Ross ME. Delayed reduction of ischemic brain injury and neurological deficits in mice lacking the inducible nitric oxide synthase gene. *J Neurosci* 1997;**17**:9157–9164.

Imai Y, Soda M, Hatakeyama S, Akagi T, Hashikawa T, Nakayama KI, Takahashi R. CHIP is associated with Parkin, a gene responsible for familial Parkinson's disease, and enhances its ubiquitin ligase activity. *Mol Cell* 2002;**10**:55–67.

Imai Y, Soda M, Inoue H, Hattori N, Mizuno Y, Takahashi R. An unfolded putative transmembrane polypeptide, which can lead to endoplasmic reticulum stress, is a substrate of Parkin. *Cell* 2001;**105**:891–902.

Isaacs AM, Senn DB, Yuan M, Shine JP, Yankner BA. Acceleration of amyloid beta-peptide aggregation by physiological concentrations of calcium. *J Biol Chem* 2006;**281**:27916–27923.

Jaffrey SR, Erdjument-Bromage H, Ferris CD, Tempst P, Snyder SH. Protein *S*-nitrosylation: a physiological signal for neuronal nitric oxide. *Nat Cell Biol* 2001;**3**:193–197.

Kaganovich D, Kopito R, Frydman J. Misfolded proteins partition between two distinct quality control compartments. *Nature* 2008;**454**:1088–1095.

Kakimura J, Kitamura Y, Takata K, Umeki M, Suzuki S, Shibagaki K, Taniguchi T, Nomura Y, Gebicke-Haerter PJ, Smith MA, *et al.* Microglial activation and amyloid-beta clearance induced by exogenous heat-shock proteins. *Faseb J* 2002;**16**:601–603.

Kalia SK, Lee S, Smith PD, Liu L, Crocker SJ, Thorarinsdottir TE, Glover JR, Fon EA, Park DS, Lozano AM. BAG5 inhibits parkin and enhances dopaminergic neuron degeneration. *Neuron* 2004;**44**:931–945.

Kandel ER, O'Dell TJ. Are adult learning mechanisms also used for development?. *Science* 1992;**258**:243–245.

Karbowski M, Neutzner A, Youle RJ. The mitochondrial E3 ubiquitin ligase MARCH5 is required for Drp1 dependent mitochondrial division. *J Cell Biol* 2007;**178**:71–84.

Kaufman RJ. Stress signaling from the lumen of the endoplasmic reticulum: coordination of gene transcriptional and translational controls. *Genes Dev* 1999;**13**:1211–1233.

Kijima K, Numakura C, Izumino H, Umetsu K, Nezu A, Shiiki T, Ogawa M, Ishizaki Y, Kitamura T, Shozawa Y, *et al.* Mitochondrial GTPase mitofusin 2 mutation in Charcot-Marie-Tooth neuropathy type 2A. *Hum Genet* 2005;**116**:23–27.

Kikuchi H, Almer G, Yamashita S, Guegan C, Nagai M, Xu Z, Sosunov AA, McKhann GM, Przedborski S II. Spinal cord endoplasmic reticulum stress associated with a microsomal accumulation of mutant superoxide dismutase-1 in an ALS model. *Proc Natl Acad Sci U S A* 2006;**103**:6025–6030.

Kim WK, Choi YB, Rayudu PV, Das P, Asaad W, Arnelle DR, Stamler JS, Lipton SA. Attenuation of NMDA receptor activity and neurotoxicity by nitroxyl anion, NO. *Neuron* 1999;**24**:461–469.

Kitada T, Asakawa S, Hattori N, Matsumine H, Yamamura Y, Minoshima S, Yokochi M, Mizuno Y, Shimizu N. Mutations in the parkin gene cause autosomal recessive juvenile parkinsonism. *Nature* 1998;**392**:605–608.

Klappa P, Ruddock LW, Darby NJ, Freedman RB. The b' domain provides the principal peptide-binding site of protein disulfide isomerase but all domains contribute to binding of misfolded proteins. *EMBO J* 1998;**17**:927–935.

Knott AB, Perkins G, Schwarzenbacher R, Bossy-Wetzel E. Mitochondrial fragmentation in neurodegeneration. *Nat Rev Neurosci* 2008;**9**:505–518.

Ko HS, Uehara T, Nomura Y. Role of ubiquilin associated with protein-disulfide isomerase in the endoplasmic reticulum in stress-induced apoptotic cell death. *J Biol Chem* 2002;**277**:35386–35392.

Kolodziejska KE, Burns AR, Moore RH, Stenoien DL, Eissa NT. Regulation of inducible nitric oxide synthase by aggresome formation. *Proc Natl Acad Sci U S A* 2005;**102**:4854–4859.

Kopito RR, Ron D. Conformational disease. *Nat Cell Biol* 2000;**2**: E207–E209.

Lafon-Cazal M, Pietri S, Culcasi M, Bockaert J. NMDA-dependent superoxide production and neurotoxicity. *Nature* 1993;**364**:535–537.

LaVoie MJ, Cortese GP, Ostaszewski BL, Schlossmacher MG. The effects of oxidative stress on parkin and other E3 ligases. *J Neurochem* 2007;**103**:2354–2368.

LaVoie MJ, Hastings TG. Peroxynitrite- and nitrite-induced oxidation of dopamine: implications for nitric oxide in dopaminergic cell loss. *J Neurochem* 1999;**73**:2546–2554.

LaVoie MJ, Ostaszewski BL, Weihofen A, Schlossmacher MG, Selkoe DJ. Dopamine covalently modifies and functionally inactivates parkin. *Nat Med* 2005;**11**:1214–1221.

Lei SZ, Pan ZH, Aggarwal SK, Chen HS, Hartman J, Sucher NJ, Lipton SA. Effect of nitric oxide production on the redox modulatory site of the NMDA receptor-channel complex. *Neuron* 1992;**8**:1087–1099.

Leroy E, Boyer R, Auburger G, Leube B, Ulm G, Mezey E, Harta G, Brownstein MJ, Jonnalagada S, Chernova T, *et al.* The ubiquitin pathway in Parkinson's disease. *Nature* 1998;**395**:451–452.

Li H, Chen Y, Jones AF, Sanger RH, Collis LP, Flannery R, McNay EC, Yu T, Schwarzenbacher R, Bossy B, *et al.* Bcl-xL induces Drp1-dependent synapse formation in cultured hippocampal neurons. *Proc Natl Acad Sci U S A* 2008;**105**:2169–2174.

Li Z, Okamoto K, Hayashi Y, Sheng M. The importance of dendritic mitochondria in the morphogenesis and plasticity of spines and synapses. *Cell* 2004;**119**:873–887.

Liang WS, Reiman EM, Valla J, Dunckley T, Beach TG, Grover A, Niedzielko TL, Schneider LE, Mastroeni D, Caselli R, *et al.* Alzheimer's disease is associated with reduced expression of energy metabolism genes in posterior cingulate neurons. *Proc Natl Acad Sci U S A* 2008;**105**:4441–4446.

Liberatore GT, Jackson-Lewis V, Vukosavic S, Mandir AS, Vila M, McAuliffe WG, Dawson VL, Dawson TM, Przedborski S. Inducible nitric oxide synthase stimulates dopaminergic neurodegeneration in the MPTP model of Parkinson disease. *Nat Med* 1999;**5**:1403–1409.

Lim KL, Chew KC, Tan JM, Wang C, Chung KK, Zhang Y, Tanaka Y, Smith W, Engelender S, Ross CA, *et al.* Parkin mediates nonclassical, proteasomal-independent ubiquitination of synphilin-1: implications for Lewy body formation. *J Neurosci* 2005;**25**:2002–2009.

Lim KL, Dawson VL, Dawson TM. Parkin-mediated lysine 63-linked polyubiquitination: a link to protein inclusions formation in Parkinson's and other conformational diseases? *Neurobiol Aging* 2006;**27**:524–529.

Lin MT, Beal MF. Mitochondrial dysfunction and oxidative stress in neurodegenerative diseases. *Nature* 2006;**443**:787–795.

Liot G, Bossy B, Lubitz S, Kushnareva Y, Sejbuk N, Bossy-Wetzel E. Complex II inhibition by 3-NP causes mitochondrial fragmentation and neuronal cell death via an NMDA- and ROS-dependent pathway. *Cell Death Differ* 2009;**16**:899–909.

Lipton SA. Paradigm shift in neuroprotection by NMDA receptor blockade: memantine and beyond. *Nat Rev Drug Discov* 2006;**5**:160–170.

Lipton SA, Choi YB, Pan ZH, Lei SZ, Chen HS, Sucher NJ, Loscalzo J, Singel DJ, Stamler JS. A redox-based mechanism for the neuroprotective and neurodestructive effects of nitric oxide and related nitroso-compounds. *Nature* 1993;**364**:626–632.

Lipton SA, Choi YB, Takahashi H, Zhang D, Li W, Godzik A, Bankston LA. Cysteine regulation of protein function–as exemplified by NMDA-receptor modulation. *Trends Neurosci* 2002;**25**:474–480.

Lipton SA, Nakamura T, Yao D, Shi ZQ, Uehara T, Gu Z. Comment on "*S*-nitrosylation of parkin regulates ubiquitination and compromises parkin's protective function". *Science* 2005;**308**: 1870; author reply 1870.

Lipton SA, Rosenberg PA. Excitatory amino acids as a final common pathway for neurologic disorders. *N Engl J Med* 1994;**330**:613–622.

Liu Y, Fallon L, Lashuel HA, Liu Z, Lansbury PT Jr. The UCH-L1 gene encodes two opposing enzymatic activities that affect alpha-synuclein degradation and Parkinson's disease susceptibility. *Cell* 2002;**111**:209–218.

Low HH, Lowe J. A bacterial dynamin-like protein. *Nature* 2006;**444**:766–769.

Lutz AK, Exner N, Fett ME, Schlehe JS, Kloos K, Laemmermann K, Brunner B, Kurz-Drexler A, Vogel F, Reichert AS, *et al.* Loss of parkin or PINK1 function increases DRP1-dependent mitochondrial fragmentation. *J Biol Chem* 2009;**284**:22938–22951.

Lyles MM, Gilbert HF. Catalysis of the oxidative folding of ribonuclease A by protein disulfide isomerase: dependence of the rate on the composition of the redox buffer. *Biochemistry* 1991;**30**:613–619.

Mannick JB, Hausladen A, Liu L, Hess DT, Zeng M, Miao QX, Kane LS, Gow AJ, Stamler JS. Fas-induced caspase denitrosylation. *Science* 1999;**284**:651–654.

Marin I, Ferrus A. Comparative genomics of the RBR family, including the Parkinson's disease-related gene parkin and the genes of the ariadne subfamily. *Mol Biol Evol* 2002;**19**:2039–2050.

Martinez-Ruiz A, Villanueva L, Gonzalez de Orduna C, Lopez-Ferrer D, Higueras MA, Tarin C, Rodriguez-Crespo I, Vazquez J, Lamas S. *S*-nitrosylation of Hsp90 promotes the inhibition of its ATPase and endothelial nitric oxide synthase regulatory activities. *Proc Natl Acad Sci U S A* 2005;**102**:8525–8530.

Matsuda N, Kitami T, Suzuki T, Mizuno Y, Hattori N, Tanaka K. Diverse effects of pathogenic mutations of Parkin that catalyze multiple monoubiquitylation *in vitro*. *J Biol Chem* 2006;**281**:3204–3209.

Mayer ML, Westbrook GL, Guthrie PB. Voltage-dependent block by Mg2+ of NMDA responses in spinal cord neurones. *Nature* 1984;**309**:261–263.

McNaught KS, Perl DP, Brownell AL, Olanow CW. Systemic exposure to proteasome inhibitors causes a progressive model of Parkinson's disease. *Ann Neurol* 2004;**56**:149–162.

Melino G, Bernassola F, Knight RA, Corasaniti MT, Nistico G, Finazzi-Agro A. *S*-nitrosylation regulates apoptosis. *Nature* 1997;**388**:432–433.

Meunier L, Usherwood YK, Chung KT, Hendershot LM. A subset of chaperones and folding enzymes form multiprotein complexes in endoplasmic reticulum to bind nascent proteins. *Mol Biol Cell* 2002;**13**:4456–4469.

Moore DJ, West AB, Dikeman DA, Dawson VL, Dawson TM. Parkin mediates the degradation-independent ubiquitination of Hsp70. *J Neurochem* 2008;**105**:1806–1819.

Mori K. Tripartite management of unfolded proteins in the endoplasmic reticulum. *Cell* 2000;**101**:451–454.

Muchowski PJ. Protein misfolding, amyloid formation, and neurodegeneration: a critical role for molecular chaperones? *Neuron* 2002;**35**:9–12.

Muchowski PJ, Wacker JL. Modulation of neurodegeneration by molecular chaperones. *Nat Rev Neurosci* 2005;**6**:11–22.

Nakamura N, Kimura Y, Tokuda M, Honda S, Hirose S. MARCH-V is a novel mitofusin 2- and Drp1-binding protein able to change mitochondrial morphology. *EMBO Rep* 2006;**7**:1019–1022.

Ni M, Lee AS. ER chaperones in mammalian development and human diseases. *FEBS Lett* 2007;**581**:3641–3651.

Nikitovic D, Holmgren A. S-nitrosoglutathione is cleaved by the thioredoxin system with liberation of glutathione and redox regulating nitric oxide. *J Biol Chem* 1996;**271**:19180–19185.

Nishikawa K, Li H, Kawamura R, Osaka H, Wang YL, Hara Y, Hirokawa T, Manago Y, Amano T, Noda M, *et al.* Alterations of structure and hydrolase activity of parkinsonism-associated human ubiquitin carboxyl-terminal hydrolase L1 variants. *Biochem Biophys Res Commun* 2003;**304**:176–183.

Nowak L, Bregestovski P, Ascher P, Herbet A, Prochiantz A. Magnesium gates glutamate-activated channels in mouse central neurons. *Nature* 1984;**307**:462–465.

O'Dell TJ, Hawkins RD, Kandel ER, Arancio O. Tests of the roles of two diffusible substances in long-term potentiation: evidence for nitric oxide as a possible early retrograde messenger. *Proc Natl Acad Sci U S A* 1991;**88**:11285–11289.

Olney JW. Brain lesions, obesity, and other disturbances in mice treated with monosodium glutamate. *Science* 1969;**164**:719–721.

Olney JW, Wozniak DF, Farber NB. Excitotoxic neurodegeneration in Alzheimer disease. New hypothesis and new therapeutic strategies. *Arch Neurol* 1997;**54**:1234–1240.

Oyadomari S, Takeda K, Takiguchi M, Gotoh T, Matsumoto M, Wada I, Akira S, Araki E, Mori M. Nitric oxide-induced apoptosis in pancreatic beta cells is mediated by the endoplasmic reticulum stress pathway. *Proc Natl Acad Sci U S A* 2001;**98**:10845–10850.

Paisan-Ruiz C, Jain S, Evans EW, Gilks WP, Simon J, van der Brug M, Lopez de Munain A, Aparicio S, Gil AM, Khan N, *et al.* Cloning of the gene containing mutations that cause PARK8-linked Parkinson's disease. *Neuron* 2004;**44**:595–600.

Panov AV, Gutekunst CA, Leavitt BR, Hayden MR, Burke JR, Strittmatter WJ, Greenamyre JT. Early mitochondrial calcium defects in Huntington's disease are a direct effect of polyglutamines. *Nat Neurosci* 2002;**5**:731–736.

Papadia S, Soriano FX, Leveille F, Martel MA, Dakin KA, Hansen HH, Kaindl A, Sifringer M, Fowler J, Stefovska V, *et al.* Synaptic NMDA receptor activity boosts intrinsic antioxidant defenses. *Nat Neurosci* 2008;**11**:476–487.

Papadia S, Stevenson P, Hardingham NR, Bading H, Hardingham GE. Nuclear Ca2+ and the cAMP response element-binding protein family mediate a late phase of activity-dependent neuroprotection. *J Neurosci* 2005;**25**:4279–4287.

Parker WD, Parks J, Filley CM, Kleinschmidt-DeMasters BK Jr. Electron transport chain defects in Alzheimer's disease brain. *Neurology* 1994;**44**:1090–1096.

Patil C, Walter P. Intracellular signaling from the endoplasmic reticulum to the nucleus: the unfolded protein response in yeast and mammals. *Curr Opin Cell Biol* 2001;**13**:349–355.

Paz Gavilan M, Vela J, Castano A, Ramos B, del Rio JC, Vitorica J, Ruano D. Cellular environment facilitates protein accumulation in aged rat hippocampus. *Neurobiol Aging* 2006;**27**:973–982.

Pitts KR, McNiven MA, Yoon Y. Mitochondria-specific function of the dynamin family protein DLP1 is mediated by its C-terminal domains. *J Biol Chem* 2004;**279**:50286–50294.

Polymeropoulos MH, Lavedan C, Leroy E, Ide SE, Dehejia A, Dutra A, Pike B, Root H, Rubenstein J, Boyer R, *et al.* Mutation in the alpha-synuclein gene identified in families with Parkinson's disease. *Science* 1997;**276**:2045–2047.

Poole AC, Thomas RE, Andrews LA, McBride HM, Whitworth AJ, Pallanck LJ. The PINK1/Parkin pathway regulates mitochondrial morphology. *Proc Natl Acad Sci U S A* 2008;**105**:1638–1643.

Przedborski S, Jackson-Lewis V, Yokoyama R, Shibata T, Dawson VL, Dawson TM. Role of neuronal nitric oxide in 1-methyl-4-phenyl-1,2,3,6-tetrahydropyridine (MPTP)-induced dopaminergic neurotoxicity. *Proc Natl Acad Sci U S A* 1996;**93**:4565–4571.

Quintanilla RA, Matthews-Roberson TA, Dolan PJ, Johnson GV. Caspase-cleaved tau expression results in mitochondrial dysfunction in cortical neurons. Implications for the pathogenesis of Alzheimer disease. *J Biol Chem* 2009;**284**:18754–18766.

Rakhit R, Crow JP, Lepock JR, Kondejewski LH, Cashman NR, Chakrabartty A. Monomeric Cu,Zn-superoxide dismutase is a common misfolding intermediate in the oxidation models of sporadic and familial amyotrophic lateral sclerosis. *J Biol Chem* 2004;**279**:15499–15504.

Ramachandran R, Surka M, Chappie JS, Fowler DM, Foss TR, Song BD, Schmid SL. The dynamin middle domain is critical for tetramerization and higher-order self-assembly. *EMBO J* 2007;**26**:559–566.

Ramirez A, Heimbach A, Grundemann J, Stiller B, Hampshire D, Cid LP, Goebel I, Mubaidin AF, Wriekat AL, Roeper J, *et al.* Hereditary parkinsonism with dementia is caused by mutations in ATP13A2, encoding a lysosomal type 5P-type ATPase. *Nat Genet* 2006;**38**:1184–1191.

Rao RV, Bredesen DE. Misfolded proteins, endoplasmic reticulum stress and neurodegeneration. *Curr Opin Cell Biol* 2004;**16**:653–662.

Rauch F, Prud'homme J, Arabian A, Dedhar S, St-Arnaud R. Heart, brain, and body wall defects in mice lacking calreticulin. *Exp Cell Res* 2000;**256**:105–111.

Reddy PH. Mitochondrial dysfunction in aging and Alzheimer's disease: strategies to protect neurons. *Antioxid Redox Signal* 2007;**9**:1647–1658.

Ren Y, Zhao J, Feng J. Parkin binds to alpha/beta tubulin and increases their ubiquitination and degradation. *J Neurosci* 2003;**23**:3316–3324.

Reynolds MR, Berry RW, Binder LI. Nitration in neurodegeneration: deciphering the "Hows" "nYs". *Biochemistry* 2007;**46**:7325–7336.

Reynolds MR, Reyes JF, Fu Y, Bigio EH, Guillozet-Bongaarts AL, Berry RW, Binder LI. Tau nitration occurs at tyrosine 29 in the fibrillar lesions of Alzheimer's disease and other tauopathies. *J Neurosci* 2006;**26**:10636–10645.

Romero JM, Bizzozero OA. Intracellular glutathione mediates the denitrosylation of protein nitrosothiols in the rat spinal cord. *J Neurosci Res* 2009;**87**:701–709.

Ross CA, Pickart CM. The ubiquitin-proteasome pathway in Parkinson's disease and other neurodegenerative diseases. *Trends Cell Biol* 2004;**14**:703–711.

Schonhoff CM, Matsuoka M, Tummala H, Johnson MA, Estevez AG, Wu R, Kamaid A, Ricart KC, Hashimoto Y, Gaston B, *et al.* S-nitrosothiol depletion in amyotrophic lateral sclerosis. *Proc Natl Acad Sci U S A* 2006;**103**:2404–2409.

Schuman EM, Madison DV. Locally distributed synaptic potentiation in the hippocampus. *Science* 1994;**263**:532–536.

Senderek J, Krieger M, Stendel C, Bergmann C, Moser M, Breitbach-Faller N, Rudnik-Schoneborn S, Blaschek A, Wolf NI, Harting I, *et al.* Mutations in SIL1 cause Marinesco-Sjogren syndrome, a cerebellar ataxia with cataract and myopathy. *Nat Genet* 2005;**37**:1312–1314.

Shimura H, Hattori N, Kubo S, Mizuno Y, Asakawa S, Minoshima S, Shimizu N, Iwai K, Chiba T, Tanaka K, *et al.* Familial Parkinson disease gene product, parkin, is a ubiquitin-protein ligase. *Nat Genet* 2000;**25**:302–305.

Shimura H, Schlossmacher MG, Hattori N, Frosch MP, Trockenbacher A, Schneider R, Mizuno Y, Kosik KS, Selkoe DJ. Ubiquitination of a new form of alpha-synuclein by parkin from human brain: implications for Parkinson's disease. *Science* 2001;**293**:263–269.

Sidrauski C, Chapman R, Walter P. The unfolded protein response: an intracellular signalling pathway with many surprising features. *Trends Cell Biol* 1998;**8**:245–249.

Sliskovic I, Raturi A, Mutus B. Characterization of the S-denitrosation activity of protein disulfide isomerase. *J Biol Chem* 2005;**280**:8733–8741.

Sriram SR, Li X, Ko HS, Chung KK, Wong E, Lim KL, Dawson VL, Dawson TM. Familial-associated mutations differentially disrupt the solubility, localization, binding and ubiquitination properties of parkin. *Hum Mol Genet* 2005;**14**:2571–2586.

Stamler JS. Redox signaling: nitrosylation and related target interactions of nitric oxide. *Cell* 1994;**78**:931–936.

Stamler JS, Hausladen A. Oxidative modifications in nitrosative stress. *Nat Struct Biol* 1998;**5**:247–249.

Stamler JS, Lamas S, Fang FC. Nitrosylation. The prototypic redox-based signaling mechanism. *Cell* 2001;**106**:675–683.

Stamler JS, Simon DI, Osborne JA, Mullins ME, Jaraki O, Michel T, Singel DJ, Loscalzo J. *S*-nitrosylation of proteins with nitric oxide: synthesis and characterization of biologically active compounds. *Proc Natl Acad Sci U S A* 1992a;**89**:444–448.

Stamler JS, Singel DJ, Loscalzo J. Biochemistry of nitric oxide and its redox-activated forms. *Science* 1992b;**258**:1898–1902.

Stamler JS, Toone EJ, Lipton SA, Sucher NJ. (S)NO signals: translocation, regulation, and a consensus motif. *Neuron* 1997;**18**:691–696.

Staropoli JF, McDermott C, Martinat C, Schulman B, Demireva E, Abeliovich A. Parkin is a component of an SCF-like ubiquitin ligase complex and protects postmitotic neurons from kainate excitotoxicity. *Neuron* 2003;**37**:735–749.

Szegezdi E, Logue SE, Gorman AM, Samali A. Mediators of endoplasmic reticulum stress-induced apoptosis. *EMBO Rep* 2006;**7**:880–885.

Taguchi N, Ishihara N, Jofuku A, Oka T, Mihara K. Mitotic phosphorylation of dynamin-related GTPase Drp1 participates in mitochondrial fission. *J Biol Chem* 2007;**282**:11521–11529.

Tanaka S, Uehara T, Nomura Y. Up-regulation of protein-disulfide isomerase in response to hypoxia/brain ischemia and its protective effect against apoptotic cell death. *J Biol Chem* 2000;**275**:10388–10393.

Tao L, Jiao X, Gao E, Lau WB, Yuan Y, Lopez B, Christopher T, RamachandraRao SP, Williams W, Southan G, *et al.* Nitrative inactivation of thioredoxin-1 and its role in postischemic myocardial apoptosis. *Circulation* 2006;**114**:1395–1402.

Tenneti L, D'Emilia DM, Lipton SA. Suppression of neuronal apoptosis by *S*-nitrosylation of caspases. *Neurosci Lett* 1997;**236**:139–142.

Tian G, Xiang S, Noiva R, Lennarz WJ, Schindelin H. The crystal structure of yeast protein disulfide isomerase suggests cooperativity between its active sites. *Cell* 2006;**124**:61–73.

Tiwari A, Hayward LJ. Familial amyotrophic lateral sclerosis mutants of copper/zinc superoxide dismutase are susceptible to disulfide reduction. *J Biol Chem* 2003;**278**:5984–5992.

Townsend DM, Manevich Y, He L, Xiong Y, Bowers RR, Hutchens S, Tew KD Jr. Nitrosative stress-induced s-glutathionylation of protein disulfide isomerase leads to activation of the unfolded protein response. *Cancer Res* 2009;**69**:7626–7634.

Uehara T, Nakamura T, Yao D, Shi ZQ, Gu Z, Ma Y, Masliah E, Nomura Y, Lipton SA. S-Nitrosylated protein-disulphide isomerase links protein misfolding to neurodegeneration. *Nature* 2006;**441**:513–517.

Ushioda R, Hoseki J, Araki K, Jansen G, Thomas DY, Nagata K. ERdj5 is required as a disulfide reductase for degradation of misfolded proteins in the ER. *Science* 2008;**321**:569–572.

Uversky VN, Yamin G, Munishkina LA, Karymov MA, Millett IS, Doniach S, Lyubchenko YL, Fink AL. Effects of nitration on the structure and aggregation of alpha-synuclein. *Brain Res Mol Brain Res* 2005;**134**:84–102.

Valente EM, Abou-Sleiman PM, Caputo V, Muqit MM, Harvey K, Gispert S, Ali Z, Del Turco D, Bentivoglio AR, Healy DG, *et al.* Hereditary early-onset Parkinson's disease caused by mutations in PINK1. *Science* 2004;**304**:1158–1160.

Vuori K, Pihlajaniemi T, Myllyla R, Kivirikko KI. Site-directed mutagenesis of human protein disulphide isomerase: effect on the assembly, activity and endoplasmic reticulum retention of human prolyl 4-hydroxylase in Spodoptera frugiperda insect cells. *Embo J* 1992;**11**:4213–4217.

Walker AK, Farg MA, Bye CR, McLean CA, Horne MK, Atkin JD. Protein disulphide isomerase protects against protein aggregation and is S-nitrosylated in amyotrophic lateral sclerosis. *Brain* 2010;**133**:(Pt 1): 105–116.

Wang C, Ko HS, Thomas B, Tsang F, Chew KC, Tay SP, Ho MW, Lim TM, Soong TW, Pletnikova O, *et al.* Stress-induced alterations in parkin solubility promote parkin aggregation and compromise parkin's protective function. *Hum Mol Genet* 2005;**14**:3885–3897.

Wang H, Lim PJ, Karbowski M, Monteiro MJ. Effects of overexpression of huntingtin proteins on mitochondrial integrity. *Hum Mol Genet* 2009a;**18**:737–752.

Wang M, Ye R, Barron E, Baumeister P, Mao C, Luo S, Fu Y, Luo B, Dubeau L, Hinton DR, *et al.* Essential role of the unfolded protein response regulator GRP78/BiP in protection from neuronal apoptosis. *Cell Death Differ* 2009b;**17**:488–498.

Wang X, Su B, Lee HG, Li X, Perry G, Smith MA, Zhu X. Impaired balance of mitochondrial fission and fusion in Alzheimer's disease. *J Neurosci* 2009c;**29**:9090–9103.

Wang X, Su B, Zheng L, Perry G, Smith MA, Zhu X. The role of abnormal mitochondrial dynamics in the pathogenesis of Alzheimer's disease. *J Neurochem* 2009d;**109**:(Suppl 1): 153–159.

Wang X, Su B, Siedlak SL, Moreira PI, Fujioka H, Wang Y, Casadesus G, Zhu X. Amyloid-beta overproduction causes abnormal mitochondrial dynamics via differential modulation of mitochondrial fission/fusion proteins. *Proc Natl Acad Sci U S A* 2008;**105**:19318–19323.

Wasiak S, Zunino R, McBride HM. Bax/Bak promote sumoylation of DRP1 and its stable association with mitochondria during apoptotic cell death. *J Cell Biol* 2007;**177**:439–450.

Waterham HR, Koster J, van Roermund CW, Mooyer PA, Wanders RJ, Leonard JV. A lethal defect of mitochondrial and peroxisomal fission. *N Engl J Med* 2007;**356**:1736–1741.

Wong ES, Tan JM, Wang C, Zhang Z, Tay SP, Zaiden N, Ko HS, Dawson VL, Dawson TM, Lim KL. Relative sensitivity of parkin and other cysteine-containing enzymes to stress-induced solubility alterations. *J Biol Chem* 2007;**282**:12310–12318.

Xu L, Eu JP, Meissner G, Stamler JS. Activation of the cardiac calcium release channel (ryanodine receptor) by poly-*S*-nitrosylation. *Science* 1998;**279**:234–237.

Yamamoto A, Friedlein A, Imai Y, Takahashi R, Kahle PJ, Haass C. Parkin phosphorylation and modulation of its E3 ubiquitin ligase activity. *J Biol Chem* 2005;**280**:3390–3399.

Yang YS, Harel NY, Strittmatter SM. Reticulon-4A (Nogo-A) redistributes protein disulfide isomerase to protect mice from SOD1-dependent amyotrophic lateral sclerosis. *J Neurosci* 2009;**29**:13850–13859.

Yao D, Gu Z, Nakamura T, Shi ZQ, Ma Y, Gaston B, Palmer LA, Rockenstein EM, Zhang Z, Masliah E, *et al.* Nitrosative stress linked to sporadic Parkinson's disease: *S*-nitrosylation of parkin regulates its E3 ubiquitin ligase activity. *Proc Natl Acad Sci U S A* 2004;**101**:10810–10814.

Yonashiro R, Ishido S, Kyo S, Fukuda T, Goto E, Matsuki Y, Ohmura-Hoshino M, Sada K, Hotta H, Yamamura H, *et al.* A novel mitochondrial ubiquitin ligase plays a critical role in mitochondrial dynamics. *EMBO J* 2006;**25**:3618–3626.

Yoo BC, Krapfenbauer K, Cairns N, Belay G, Bajo M, Lubec G. Overexpressed protein disulfide isomerase in brains of patients with sporadic Creutzfeldt-Jakob disease. *Neurosci Lett* 2002;**334**:196–200.

Yoshida H, Matsui T, Yamamoto A, Okada T, Mori K. XBP1 mRNA is induced by ATF6 and spliced by IRE1 in response to ER stress to produce a highly active transcription factor. *Cell* 2001;**107**:881–891.

Youle RJ, Karbowski M. Mitochondrial fission in apoptosis. *Nat Rev Mol Cell Biol* 2005;**6**:657–663.

Zhang Y, Gao J, Chung KK, Huang H, Dawson VL, Dawson TM. Parkin functions as an E2-dependent ubiquitin- protein ligase and promotes the degradation of the synaptic vesicle-associated protein, CDCrel-1. *Proc Natl Acad Sci U S A* 2000;**97**:13354–13359.

Zhang K, Kaufman RJ. The unfolded protein response: a stress signaling pathway critical for health and disease. *Neurology* 2006;**66**: S102–S109.

Zhang X, Szabo E, Michalak M, Opas M. Endoplasmic reticulum stress during the embryonic development of the central nervous system in the mouse. *Int J Dev Neurosci* 2007;**25**:455–463.

Zhao L, Longo-Guess C, Harris BS, Lee JW, Ackerman SL. Protein accumulation and neurodegeneration in the woozy mutant mouse is caused by disruption of SIL1, a cochaperone of BiP. *Nat Genet* 2005;**37**:974–979.

Zhu PP, Patterson A, Stadler J, Seeburg DP, Sheng M, Blackstone C. Intra- and intermolecular domain interactions of the C-terminal GTPase effector domain of the multimeric dynamin-like GTPase Drp1. *J Biol Chem* 2004;**279**:35967–35974.

Zimprich A, Biskup S, Leitner P, Lichtner P, Farrer M, Lincoln S, Kachergus J, Hulihan M, Uitti RJ, Calne DB, *et al.* Mutations in LRRK2 cause autosomal-dominant parkinsonism with pleomorphic pathology. *Neuron* 2004;**44**:601–607.

Zuchner S, Mersiyanova IV, Muglia M, Bissar-Tadmouri N, Rochelle J, Dadali EL, Zappia M, Nelis E, Patitucci A, Senderek J, *et al.* Mutations in the mitochondrial GTPase mitofusin 2 cause Charcot-Marie-Tooth neuropathy type 2A. *Nat Genet* 2004;**36**:449–451.

3

CHAPERONE-MEDIATED AUTOPHAGY AND PARKINSON'S DISEASE

Marta Martinez-Vicente

Vall d'Hebron Research Insititue, Barcelona, Spain

Ester Wong

Department of Developmental and Molecular Biology, Bronx, NY, USA

3.1 PROTEIN HOMEOSTASIS AND NEURODEGENERATION

Maintaining a balanced proteome is critical to the well-being of all types of cells. Despite the fidelity of gene transcription, it has been estimated that as many as one-third of all nascent proteins in eukaryotes might not attain their correctly folded conformations (Schubert *et al.*, 2000). Protein misfolding could also arise from various posttranslational modifications, such as oxidation, glycation, and nitrosylation, and/or disease-causing genetic mutations (Goldberg, 2003). These faulty proteins often organize into complex structures (oligomers, protofibrils, and small aggregates) that are toxic to the cells. For self-preservation, cells have evolved an intricate cellular protein quality control mechanism to combat the constant threat of protein misfolding. This encompasses surveillance machineries that identify and repair altered or nonfunctional proteins and also rapidly destroy those that are beyond repair in order to maintain intracellular protein homeostasis. The essential effectors of protein quality control in almost all cells

Protein Chaperones and Protection from Neurodegenerative Diseases, First Edition.
Edited by Stephan N. Witt.

are the molecular chaperones and the proteolytic systems. The chaperones act as "quality control officers" that participate in correct folding and refolding of proteins and also map out the fate of a protein by triaging the decision to repair or to destroy (Figure 3.1). The elimination of unwanted proteins is mediated by two major intracellular proteolytic systems—the ubiquitin–proteasome system (UPS) and the lysosomes (autophagy) (Figure 3.1). Proper functioning of these quality control systems has proved to be essential to guarantee normal cellular function, particularly in postmitotic cells such as neurons where unwanted products that escape the surveillance mechanisms will accumulate in the cytosol and remain there throughout the life span of the organism. Intracellular buildup of damaged or abnormal proteins leads to loss of cellular functions and eventually cell death (Kopito, 2000). Defects in both chaperones and proteolytic systems have been identified in different neurodegenerative disorders and have been proposed to contribute to their pathogenesis. A common outcome of such aberrant protein mishandling is the formation of insoluble protein inclusions (Figure 3.1). In this chapter, we have focused on how impairments of the various proteolytic systems could precipitate protein aggregation and lead to neuronal demise in neurodegeneration—using Parkinson's disease (PD) as the main model. A special emphasis is placed on the role of a selective form of lysosomal degradation—chaperone-mediated autophagy (CMA)—in the neurodegenerative process, as growing evidence supports a close relationship between malfunctioning of this autophagic pathway and the pathogenesis of PD.

3.2 FAILURE OF CELLULAR QUALITY CONTROL IN THE PD BRAIN

PD is the most prevalent neurodegenerative movement disorder (Dorsey *et al.*, 2007). Clinically, the disease is characterized by motor deficits that include bradykinesia, postural instability, rigidity, and tremor. The major pathological feature of PD is the selective loss of dopaminergic neurons in the *substantia nigra pars compacta* (SNpc) of the midbrain (Savitt *et al.*, 2006). Most cases of PD occur in a sporadic manner where the etiology remains poorly understood. Numerous hypotheses for sporadic PD include combinations of aging process, genetic propensity, and environmental exposures leading to oxidative stress, mitochondrial dysfunction, microglial activation, and excitotoxicity (Gandhi and Wood, 2005; Savitt *et al.*, 2006). However, familial forms of PD attributed to mutations in more than eight PD-associated genes have also been reported (Savitt *et al.*, 2006). In particular, the functional characterization of these PD-linked genes points to an intimate link between aberration in protein homeostasis and PD pathogenesis. This is supported by the observation that intraneuronal protein inclusions known as *Lewy bodies* (LBs) are often present in the surviving neurons of the *substantia nigra* (SN) as well as in other affected regions of the brain in PD patients (Braak *et al.*, 2003). Notably, α-synuclein, a presynaptic terminal-enriched protein that has high propensity to oligomerize and aggregate, was found to be the major component of LBs (Spillantini *et al.*, 1997).

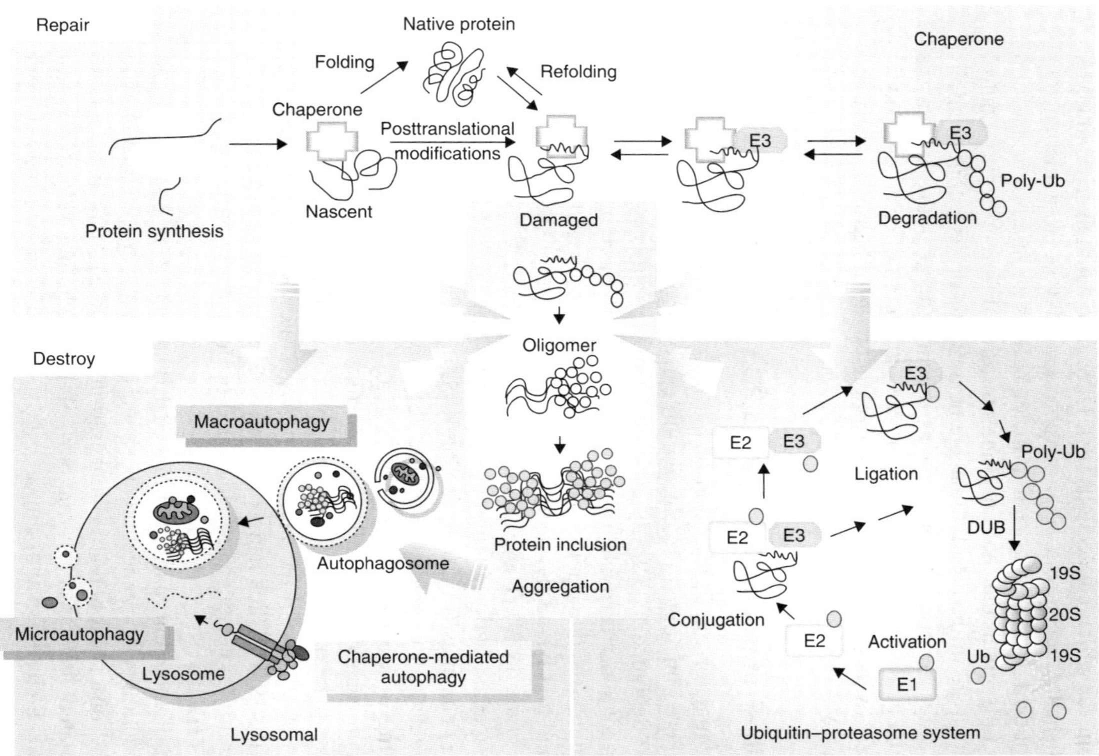

Figure 3.1 Intersections of the different quality control systems in the cells. The cellular surveillance network comprises the chaperones, the ubiquitin–proteasome system (UPS) and the lysosomal (autophagy) system. Molecular chaperones assist in both initial protein folding and subsequent maintenance of protein conformation under changing environmental conditions. Many factors (e.g., oxidation) can lead to protein unfolding or misfolding. Initially, cells may attempt to refold the protein but if this fails, the molecular chaperones recruit both the UPS and lysosomal pathways for destruction of the damaged proteins. Proteins destined for UPS degradation are tagged with a chain of ubiquitin (Ub, normally K48-linked) via repeated actions of ubiquitin-activating (E1), -conjugating (E2), and -ligating (E3) enzymes. Monomeric ubiquitins are regenerated by the hydrolytic activity of deubiquitinating enzyme (DUB). Fine balance between the three systems contributes to protein homeostasis. Aggregation is ensued when aberrations occur in any of the surveillance systems. (*A full color version of this figure appears in the color plate section.*)

Disease-causing α-synuclein mutations (A30P, A53T, E46K) have been shown by many groups to increase the accumulation of the protein and its propensity to aggregate (Kruger *et al.*, 1998; Singleton *et al.*, 2003; Zarranz *et al.*, 2004). Importantly, α-synuclein has become an interesting protein for the study of alterations of the proteolytic systems in PD and other synucleopathies as it is the first pathogenic protein for which connections with UPS, macroautophagy, and CMA have been established (Cuervo *et al.*, 2004; Webb *et al.*, 2003). How misfolded or aggregated α-synuclein evades degradation inside cells is an interesting question to explore, especially given that not only one but three different quality control systems also participate in the cellular management of this protein. Since various degradative pathways are found to be interdependent in the cellular milieu, the interaction of pathogenic α-synuclein proteins with all the implicated proteolytic systems, followed by a more in-depth description of the relationship between α-synuclein and CMA will be discussed.

3.2.1 Ubiquitin–Proteasome System and PD

The UPS is essential for the degradation of the majority of short-lived proteins and subsets of altered misfolded proteins within cells (Figure 3.1). The selection of proteins to be degraded by UPS is a highly regulated process. The proteins destined for degradation are covalently tagged with a polymeric ubiquitin chain in which the terminal residue (G76) of one ubiquitin molecule is linked through an isopeptide bond to a lysine (K) residue (mostly K48) within another ubiquitin molecule (Chau *et al.*, 1989). The ligation reaction is elaborate and requires the sequential and repeated action of ubiquitin-activating (E1), -conjugating (E2), and -ligating (E3) enzymes (Figure 3.1). The K48-linked polyubiquitinated substrate is then targeted for degradation by the 26S proteasome—a large protease complex consisting of a barrel-shaped 20S proteolytic core with two 19S regulatory caps, one on each side of the barrel's opening (Pickart and Cohen, 2004). The components of the 19S cap play important roles in substrate recognition, unfolding, and translocation of the substrate proteins into the lumen of the barrel core (Braun *et al.*, 1999; Glickman *et al.*, 1998; Navon and Goldberg, 2001). Monomeric ubiquitin molecules are regenerated in the process by the deubiquitylation enzymes (DUBs).

The presence of intraneuronal ubiquitinated LB in PD is suggestive of possible UPS dysfunction. The connections between UPS impairment and PD were reinforced by the observation that administration of proteasomal inhibitors to rodents recapitulated the cardinal symptoms of PD including selective nigral cell loss, LB-like inclusions, and clinical signs of bradykinesia, rigidity, and tremor (McNaught *et al.*, 2004). Depletion of the level of 26S proteasome in the SN of mice also showed extensive degeneration in the nigrastriatal pathway accompanied by α-synuclein positive inclusions (Bedford *et al.*, 2008). However, a more direct association of UPS with PD pathogenesis came from the identification of parkin and ubiquitin carboxyl-terminal hydrolase L1 (UCH-L1) as familial PD-linked genes (Kitada *et al.*, 1998; Leroy *et al.*, 1998). Both the gene products perform specific functions in the proteasomal degradation pathway.

Parkin functions as an E3 ligase, which catalyzes the addition of ubiquitin chains to target proteins for destruction by the proteasome. Mutations in parkin are common and underscore the cause of almost half of early onset autosomal recessive parkinsonism. Mutations in UCH-L1, an ubiquitin C-terminal hydrolase L1, which normally functions to cleave polyubiquitin chains into monomeric ubiquitin have been identified in some PD patients.

Disruption of UPS is thought to promote the toxic accumulation of proteins that are detrimental to neuronal survival, including α-synuclein (Fornai *et al.*, 2003; McNaught *et al.*, 2002, 2004; Rideout *et al.*, 2005). Furthermore, pathologically misfolded or altered proteins often assume complex conformations that prevent their entry into the narrow catalytic pore of the barrel-shaped proteasome complex (Bennett *et al.*, 2005). It is also common that the interaction of these altered proteins with the proteasome core causes steric occlusions as seen for aggregated α-synuclein, which preferentially interacts with the 19S cap (Snyder *et al.*, 2003), thereby preventing efficient proteasomal clearance of other proteins (Bennett *et al.*, 2005; Stefanis *et al.*, 2001; Verhoef *et al.*, 2002) (Figure 3.1). Conceivably, this proteasomal impairment further drives the aggregation process and exacerbates the toxicity of the misfolded proteins (Figure 3.1). The "bulk removal mechanism" of the macroautophagic pathway makes it better suited to handle aggregates and inclusions (Figure 3.2).

Interestingly, some of the causative gene products of PD participate in both UPS and lysosomal pathways. For example, parkin, besides mediating K48-linked polyubiquitin chain conjugation on substrates, which typically targets protein for UPS degradation, also assembles K63-linked ubiquitination of substrate (Lim *et al.*, 2005; Olzmann *et al.*, 2007; Tan *et al.*, 2008a) that promotes substrate aggregation and preferentially targets it to autophagy for removal (Tan *et al.*, 2008a, 2008b). This highlights a potential way of modulating the degradative fate of a protein substrate via tagging it with different ubiquitin chains (see Section 3.4.1). Besides facilitating autophagic degradation of aggregated proteins, parkin also partners with PINK1, a serine/threonine protein kinase linked to familial PD and contributes to the selective clearance of dysfunctional mitochondria by autophagy (mitophagy) (Dagda *et al.*, 2009; Narendra *et al.*, 2008; Vives-Bauza *et al.*, 2010). UCH-L1, an enzyme with both deubiquitinating and ligase activities, has recently been found to interact with components of CMA and affects the activity of this pathway (Kabuta *et al.*, 2008). Hence, these findings highlight that the recently described proteasomal-independent functions of parkin, PINK1, and UCH-L1 should be taken into account when investigating their roles in PD pathogenesis.

3.3 DIFFERENT FLAVORS OF AUTOPHAGY: MECHANISMS AND FUNCTIONS

Autophagy is the delivery of intracellular components—soluble proteins, macromolecules, and organelles—to lysosomes for degradation. Increasing roles of autophagy in human diseases and physiology have continued to unravel over the past decade (Mizushima *et al.*, 2008). It is now recognized that autophagy is not only involved in cellular housekeeping but also performs key functions in immune surveillance of pathogens, cellular response to stressors, development, and as a critical component of the intracellular quality control systems to manage abnormal or altered components (Cuervo, 2006; He and Klionsky, 2009; Klionsky, 2007; Mizushima *et al.*, 2008; Rubinsztein *et al.*, 2007). Three types of autophagy have been described in mammalian cells: *macroautophagy, microautophagy*, and *CMA*. Although all three pathways mediate lysosomal degradation of intracellular components, they differ in (i) the mechanisms by which substrates are delivered to the lysosomes; (ii) the nature of the substrates degraded; and (iii) their regulation and activation conditions (Figure 3.2). Both macro- and microautophagy are able to engulf cargo in both selective and nonselective manners, whereas CMA only functions to degrade soluble proteins in a selective way.

3.3.1 Macroautophagy

Macroautophagy is the best characterized form of autophagy and is frequently referred to as *autophagy*. It is considered an inducible form of autophagy, even

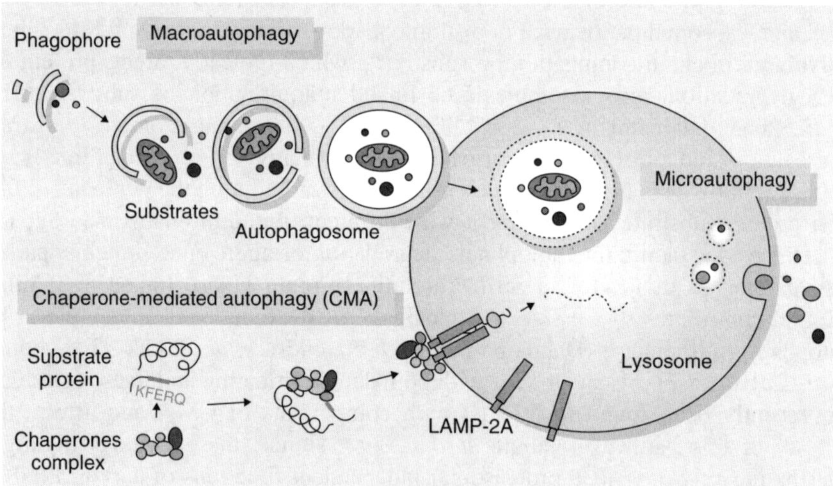

Figure 3.2 Types of autophagy in mammalian cells. Three different types of autophagy occur in mammalian cells. In macroautophagy, intracellular components (including protein aggregates) are first enclosed in a double membrane vesicle or autophagosome, which then fuse directly with lysosomes to facilitate breakdown of substrates by hydrolases. In microautophagy, whole regions of the cytosol are sequestered by lysosomes through invaginations or tubulations of the lysosomal membrane. Soluble cytosolic proteins can also be degraded by chaperone-mediated autophagy. In this pathway, proteins with a targeting motif are directed to lysosomes by a chaperone complex. Upon binding to the membrane receptor, the substrates are translocated across the lysosomal membrane into the lumen for degradation.

though most cell types have considerable basal macroautophagic activity, which is essential for the maintenance of cellular homeostasis (Cuervo, 2006; Hara *et al.*, 2006; Komatsu *et al.*, 2006). Notably, macroautophagy is activated in the early stage of nutrient deprivation, providing amino acids and other essential basic elements as an adaptative response to such stress condition (He and Klionsky, 2009). In addition, this pathway also participates in other cellular functions described above (Klionsky, 2007; Mizushima *et al.*, 2008; Rubinsztein *et al.*, 2007).

In this pathway, a portion of the cytoplasm including proteins and entire organelles is surrounded by a double membrane formed *de novo*, which seals upon itself to form a vesicle called *autophagic vacuole* or *autophagosome* (Mizushima *et al.*, 2002; Wang and Klionsky, 2003) (Figure 3.2). The engulfed cytosolic contents are subsequently degraded by lysosomal hydrolytic enzymes upon fusion of the autophagosome with mature lysosomes. The macroautophagic process can be divided into different steps: nucleation, cargo recognition and selection, vesicle formation, fusion with lysosomes, and degradation of the cargo.

Nucleation occurs in response to cellular stress, which results in the formation of a limiting membrane, also known as a *phagophore*. Genetic screens in yeast have led to the identification of a family of genes, the autophagy-related genes or *ATG*s, which code for proteins involved in different steps of macroautophagy (He and Klionsky, 2009; Mizushima *et al.*, 2002; Noda *et al.*, 2002). Most of these *ATG* genes identified in yeast are also conserved in mammals and other organisms, thus revealing that macroautophagy is a highly conserved process in eukaryotes. In yeast, the recruitment of the Atg proteins for the formation and elongation of the *phagophore* occurs at the pre-autophagosome structure or PAS (He and Klionsky, 2009; Mizushima *et al.*, 2002; Noda *et al.*, 2002). The elongation step consists of two ubiquitin-like cascades, the protein–protein Atg5–Atg12 system and the protein–lipid conjugation LC3-phosphatidyl ethanolamine system (Klionsky and Emr, 2000; Ohsumi and Mizushima, 2004). Once the autophagosome is sealed, most of the Atg proteins dissociate from the vesicle prior to its fusion with lysosomes (Figure 3.2). No analogous PAS structure is found in mammalian cells where autophagosomes are hypothesized to emerge from different sites. Most recently, a discrete region in the endoplasmic reticulum (ER) (the omegasome) is suggested to serve as the nucleation site for the formation of autophagosomes in mammalian cells (Axe *et al.*, 2008).

Macroautophagy is regulated by two signaling pathways—the class I and III phosphatidylinositol-3-kinase (PI3K) pathways. The class I PI3K complex is found to negatively regulate macroautophagy through activation of Akt/PKB, which activates mTOR (mammalian target of rapamycin) kinase—the major negative regulator of macroautophagy (Kanazawa *et al.*, 2004). Chemical inhibition of mTOR is often used to upregulate macroautophagy and has been explored for therapeutic application in the treatment of proteinopathies (Ravikumar *et al.*, 2004; Sarkar and Rubinsztain, 2008). The class III PI3K complex regulates macroautophagy by interacting with beclin-1, which is essential for the nucleation of the limiting membrane (Kihara *et al.*, 2001; Zhong *et al.*, 2009). Although cellular levels of beclin-1 have been shown to correlate with autophagic activity

(Pickford *et al.*, 2008), its assembly into specific multiprotein complexes is likely to be more indicative of autophagy activation.

Macroautophagy has been originally described as a nonselective degradation pathway, but recent studies have demonstrated some level of selectivity in cargo recognition and degradation. According to these observations, selective macroautophagic degradation is possible for organelles such as mitochondria (mitophagy) (Lemasters *et al.*, 2002), peroxisomes (pexophagy) (Sakai *et al.*, 1998; Yokota, 2003), ER (reticulophagy) (Yorimitsu and Klionsky, 2005), ribosomes (ribophagy) (Kraft *et al.*, 2008), and lipid droplets (lipophagy) (Singh *et al.*, 2009). Additionally, selective macroautophagic degradation of ubiquitinated protein aggregates (aggrephagy) has also been observed in mammalians cells (Kim *et al.*, 2008; Pankiv *et al.*, 2007). To date, the mechanism of selective autophagy is still not well understood. However, we and others have found evidence of the involvement of ubiquitination in this process (Kim *et al.*, 2008; Kirkin *et al.*, 2009; Komatsu *et al.*, 2007b; Pankiv *et al.*, 2007; Tan *et al.*, 2008a, b). Analogous to the UPS, where ubiquitinated proteins are recognized by ubiquitin-binding receptors for delivery to proteasome, autophagic clearance of protein aggregates also requires ubiquitin-binding receptors such as p62 and NBR1. The simultaneous binding of p62 and/or NBR1 proteins to the ubiquitinated aggregates and the autophagosome-associated proteins LC3 (light-chain 3 of microtubule-associated protein 1) and/or GABARAP (γ-aminobutyric acid type A ($GABA_A$) receptor-associated protein) helps to mediate docking of ubiquitinated protein aggregates to the autophagosome, thereby facilitating their selective degradation (Figure 3.3). Cargo recognition by p62 is not only limited to protein aggregates but it also includes organelles. In a similar fashion, p62 also acts as the adaptor receptor for selectively targeting peroxisomes coated with ubiquitins for autophagic removal (Kim *et al.*, 2008). Recently, NIX, a proapoptotic protein localizes in the outer membrane of mitochondria, has been shown to bind to GABARAP and was proposed to target mitochondria to macroautophagy for clearance in the erythrocyte (Sandoval *et al.*, 2008).

3.3.2 Microautophagy

Microautophagy involves the direct invagination of the lysosomal membrane to form finger-like protrusions to sequestrate portions of the cytosol containing cytosolic proteins and organelles (mitochondria, peroxisomes, and the nucleus) for degradation in lysosomes (Mortimore *et al.*, 1988; Sakai *et al.*, 1998) (Figure 3.2). Although microautophagy is the least characterized form of autophagy, it is generally accepted that microautophagy is constitutively active and participates in the basal turnover of both whole organelles and proteins (Dubouloz *et al.*, 2005; Roberts *et al.*, 2003; Yokota, 2003). Additionally, there are numerous descriptions in yeast of selective microautophagic degradation of various organelles, namely microautophagy of peroxisomes (micropexophagy) (Dunn *et al.*, 2005; Farré and Subramani, 2004), of mitochondria (micromitophagy) (Kissová *et al.*, 2007), and of the nucleus (piecemeal microautophagy

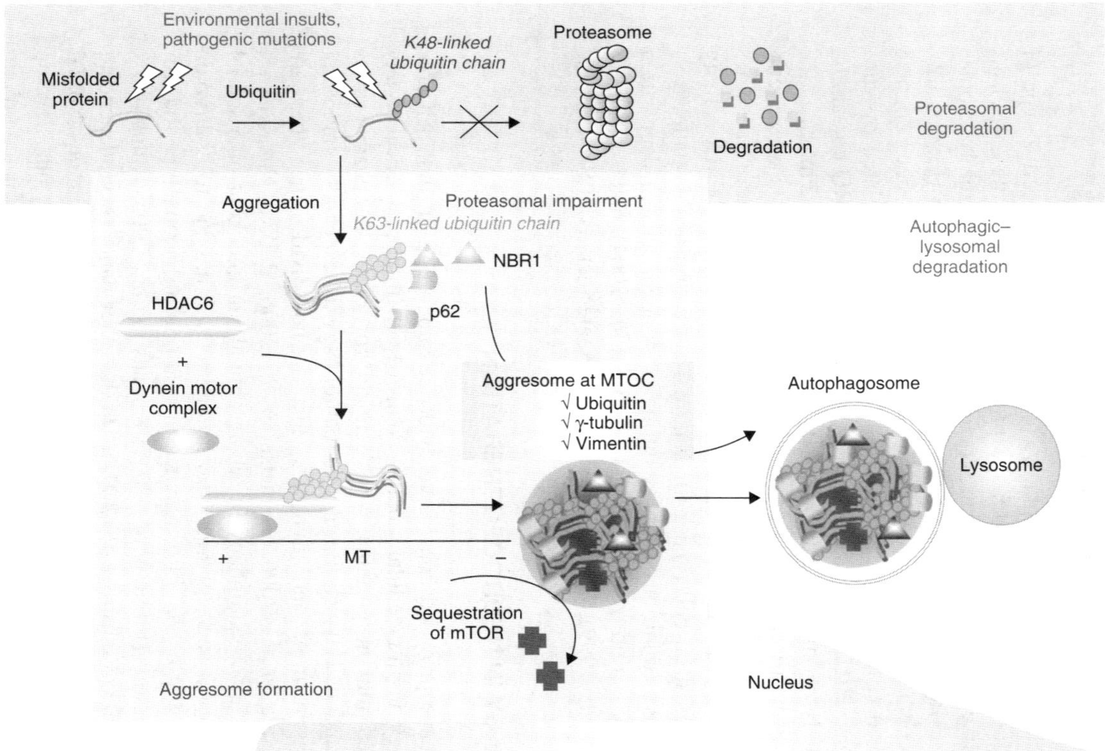

Figure 3.3 Aggresome biogenesis and clearance. Under normal cellular conditions, proteins destined for degradation by the proteasome are tagged with a chain of K48-linked ubiquitin. Under conditions of proteasome impairment, the cell switches to K63-linked ubiquitination to divert the protein load originally targeted for proteasomal degradation away from the otherwise overloaded machinery. Alternatively, K48-linked polyubiquitinated substrates could be remodeled under conditions of stress. K63-linked polyubiquitin, through its interactions with HDAC6 and p62, helps sequestering the diverted protein load into an aggresome. At the same time, K63-linked polyubiquitin also acts as a cargo recognition signal, presumably in partnership with p62 and NBR1, for the autophagy machinery, thereby facilitating the clearance of aggresomes *via* lysosomal-mediated degradation.

of the nucleus (PMN)/micronucleophagy) (Krick *et al.*, 2008; Park *et al.*, 2009). Studies from these selective forms of microautophagy have led to the understanding that much of the core ATG machinery seems to be required for microautophagic processes.

3.3.3 Chaperone-Mediated Autophagy

In contrast to macro- and microautophagy, CMA is a highly selective type of autophagy in which only soluble cytosolic proteins that contain the consensus motif (KFERQ-like) in their amino acid sequences are targeted to the lysosomes for degradation (Majeski and Dice, 2004; Massey *et al.*, 2006a) (Figure 3.2). This motif is recognized by the cytosolic chaperone Hsc70 (heat shock cognate protein of 70 kDa) (Chiang *et al.*, 1989), whose activity is modulated by various co-chaperones. Hsc70 and its co-chaperones then target KFERQ-containing substrates to the lysosomal membrane. Once at the lysosomal membrane, the substrate binds to the receptor for this pathway, the lysosome-associated membrane protein type-2A (LAMP-2A), and gets internalized into the lysosomes (Cuervo and Dice, 1996). Similar to macroautophagy, some basal CMA activity is detectable in almost all cell types, but maximal activation of this pathway is attained under stress conditions such as during prolonged nutrient deprivation (Cuervo *et al.*, 1995), mild oxidative stress (Kiffin *et al.*, 2004), and exposure to toxins (Cuervo *et al.*, 1999). CMA is reviewed in greater details in Section 3.5.

3.4 AUTOPHAGY IN NEURODEGENERATION

Recent studies have provided conclusive evidence supporting an indispensable role of autophagy in intracellular quality control even in the absence of any disease-associated mutant proteins. The demand for basal autophagy varies among tissues and cell types. Basal macroautophagy is especially important in cells, such as neurons and myocytes, that do not divide once they are differentiated (Hara *et al.*, 2006; Komatsu *et al.*, 2006, 2007a; Nakai *et al.*, 2007). Ablation of macroautophagy in whole animal via knockout of essential macroautophagy genes such as *atg5, atg7*, and *beclin-1* has led to embryonic lethality (in the case of *beclin-1*) or resulted in death during neonatal period (for *atg5* and *atg7* ablation) (Fimia *et al.*, 2007; Komatsu *et al.*, 2005; Kuma *et al.*, 2004; Qu *et al.*, 2003; Takashi *et al.*, 2007). On the other hand, two independent studies have shown that mice with neural tissue-specific suppression of basal macroautophagy survive the initial neonatal starvation period but develop progressive motor deficits accompanied by accumulation of ubiquitinated protein inclusions in their neurons (Hara *et al.*, 2006; Komatsu *et al.*, 2006). All these findings emphasize that constitutive macroautophagy is essential for neuronal survival.

The importance of autophagy is even more evident in disease state. Besides UPS, cells also rely heavily on the autophagic system to remove pathogenic disease-linked proteins. Compensatory activation of the autophagic system in general has been well documented in many protein conformational disorders (Bendiske and Bahr, 2003; Cataldo *et al.*, 1995; Kegel *et al.*, 2000). Expression of mutant huntingtin, the protein found in protein inclusions associated with Huntington's disease (HD), results in upregulation of macroautophagy along with the general stimulation of endosomal–lysosomal activity (Kegel *et al.*, 2000). Enhanced macroautophagy and lysosomal protease levels were also observed in different models of Alzheimer's disease (AD) (Bendiske and Bahr, 2003; Butler *et al.*, 2005; Cataldo *et al.*, 1995). Interventions aimed at enhancing the macroautophagic activity have been successful in facilitating the removal of some of these misbehaving proteins (i.e., mutant peripheral myelin protein, mutant proteins with expanded polyglutamine tracts, and α-synuclein) in different cellular models (Fortun *et al.*, 2003; Iwata *et al.*, 2005; Ravikumar *et al.*, 2002; Webb *et al.*, 2003). Furthermore, pharmacological upregulation of macroautophagy with analogs of rapamycin, an inhibitor of the negative regulator of autophagy, mTOR, reduced huntingtin aggregates, decreased toxicity, and improved disease-associated symptoms in fly and animal models for HD (Ravikumar *et al.*, 2004). These promising studies support the idea that the compensatory upregulation of macroautophagy may slow down disease progression and explain the extended disease duration characteristic of many age-related neurodegenerative disorders.

3.4.1 Macroautophagy and Protein Inclusions

When the proteasome becomes compromised in its function, it is difficult to imagine that the cell will continue to burden the proteasomal machinery under such conditions with an endless stream of cargo proteins to be degraded. Notably, many pathogenic proteins are susceptible to degradation by both the proteasomal and the autophagic pathways (Cuervo *et al.*, 2004; Webb *et al.*, 2003). Although what controls the choice of degradation system for a protein is not fully understood, factors such as the intrinsic propensity of the protein to aggregate seem crucial in this decision. Interestingly, protein inclusions associated with several neurodegenerative diseases bear striking resemblance to aggresomes (Mishra *et al.*, 2003; Olanow *et al.*, 2004). As aggresomes are seemingly inert structures, their formation could be considered as a proactive way by which the cell deals with its nondisposable proteins, resulting in a protective response by the cell in times of proteolytic stress. Consistent with this, several studies supporting a neuroprotective role for inclusion formation have emerged recently (Arrasate *et al.*, 2004; Bowman *et al.*, 2005). These studies collectively suggest that the accumulation of disease-associated proteins in a diffuse nonaggregated form is more toxic than when they are sequestered into inclusion bodies. We have recently proposed that cells may utilize ubiquitination of proteins to help divert proteins originally destined for proteasomal degradation away from the system particularly under conditions of cellular stress (Tan *et al.*, 2008a) (Figure 3.3). The diverted ubiquitin-enriched proteins are then sequestered into aggresomes to be acted upon by the autophagy system. In this way, the cell could preserve its proteasome function over prolonged periods of proteolytic stress and recover thereafter. Importantly, our recent work identified K63-linked polyubiquitin, a proteasome-independent ubiquitin signal, as a novel cargo selection signal for macroautophagy-mediated clearance of aggresomes (Tan *et al.*, 2008a, b). We found that the ubiquitin-positive, aggresome-like inclusions that formed in cultured cells ectopically expressing either K63 ubiquitin mutant or the heterodimeric Ubc13/Uev1a E2 pair (which promote endogenous K63-linked ubiquitination) are rapidly cleared when autophagy is induced. Furthermore, efficient recruitment of lysosomal structures to K63 polyubiquitinated proteins during aggresome formation has also been observed. An important clue pointing to the participation of protein ubiquitination in the clearance of aggresomes by autophagy is that these structures are frequently enriched with p62. As discussed earlier, p62 mediates clearance of ubiquitinated aggresome inclusions by binding to the ubiquitin in the aggresomes and to the autophagosome marker LC3 at the same time (Bjorkoy *et al.*, 2005; Pankiv *et al.*, 2007). Studies have shown that p62 appears to preferentially bind long K63-linked ubiquitin chains (Kirkin *et al.*, 2009; Tan *et al.*, 2008b). Interestingly, another ubiquitin receptor, the NBR1 protein has recently been identified to serve similar function as that of p62 (Kirkin *et al.*, 2009). Besides p62, the microtubule-associated deacetylase HDAC6 has also been proposed to connect aggresomes with the autophagy machinery (Figure 3.3). Like p62, HDAC6 interacts with polyubiquitinated misfolded proteins but binds additionally to dynein, a minus end-directed microtubule motor that transports cargo from the cell periphery toward its center (Ravikumar *et al.*, 2005). Dynein mutations have previously been demonstrated to impair autophagic clearance of aggregation-prone proteins (Ravikumar *et al.*, 2005). By virtue of its dual interaction with dynein and ubiquitinated proteins, HDAC6 not only helps to transport aggregates to the microtubule organizing center (MTOC) but also appears to facilitate the delivery of the autophagy apparatus to aggresomes (Iwata *et al.*, 2005; Olzmann *et al.*, 2007). Indeed, cells deficient in HDAC6 cannot form aggresomes properly and fail to clear misfolded proteins from the cytoplasm (Kawaguchi *et al.*, 2003). Most recently, we have also identified a novel role of HDAC6 in ubiquitin selective quality control autophagy where it functions to coordinate actin remodeling to facilitate fusion between autophagosome and lysosome (Lee *et al.*, 2010). Additionally, parkin-mediated K63 polyubiquitination of misfolded DJ-1 has been demonstrated to couple the protein to the dynein motor complex via the HDAC6 adaptor, thereby promoting its sequestration into aggresomes (Olzmann *et al.*, 2007). Thus, by virtue of its association with HDAC6 and also p62, K63-linked polyubiquitin would have the ability to reroute these protein cargoes toward the MTOC to facilitate the formation of an aggresome (Figure 3.3). Although our results favor the participation of K63-linked polyubiquitination under these circumstances, it is possible that monoubiquitination might also have a role (Rott *et al.*, 2008). Taking all these findings together, we envisage a model of inclusion biogenesis and clearance whereby the nature of ubiquitin linkage plays a deterministic role in deciding the fate of a misfolded protein. Alternatively, it has been suggested that macroautophagy may not directly clear aggregates but instead it clears aggregate precursors (Rubinsztein, 2006). The removal of such precursors may shift the equilibrium away from aggregate formation, thereby reducing the size and the number of protein inclusions. Nevertheless, evidence supporting the role of autophagy in clearance of protein inclusions is unequivocal, and it is tempting to think that autophagy could clear proteins in all conformations, particularly under conditions of proteasome impairment.

3.4.2 Autophagic Failure in Neurodegenerative Diseases

It is evident that upregulation of autophagy prevents intracellular accumulation of protein aggregates. How then do we account for the persistence of intraneuronal protein inclusions and cell toxicity in the various pathophysiological states? The answer to this question is that dysfunction in autophagic system often time also accompanies the disease states.

The enormous amounts of autophagosomes seen in the affected neurons of some neurodegenerative disorders such as AD and HD (Cataldo *et al.*, 1995; Kegel *et al.*, 2000; Yu *et al.*, 2005) may be interpreted as an increase in the stimulation of autophagy in response to the higher load of altered proteins and damaged organelles. However, an expansion of the autophagosome compartment is not always an indication of increased level of autophagy as cells could display higher number of autophagosomes not only when macroautophagy is upregulated (more synthesis of autophagosomes) but also when clearance of autophagosomes is blocked (less fusion/degradation of autophagosomes by lysosomes). In fact, emerging evidence indicates that a failure of the autophagic process in the advanced states of the disease is responsible for the accumulation of autophagosomes (Martinez-Vicente *et al.*, 2010; Yu *et al.*, 2005). During early disease state, autophagy is often induced to protect neurons against toxic proteins, stress, or damaged organelles that may cause cell toxicity and apoptosis. However, overloading of this system or impairing it at different steps because of other aggravating factors as the disease progresses could lead to the failure of macroautophagy and result in accumulation of autophagic vacuoles. At this point, an inadequate level of autophagic activity also contributes toward the degeneration process and neuronal cell death (Figure 3.4) (Boland and Nixon, 2006).

Among the different factors that could account for the dysfunction of the autophagic systems under these circumstances, aging has been proposed as one of the most aggravating factors. The activities of both CMA and macroautophagy have been shown to decline with age (Cuervo *et al.*, 2005). Fusion between autophagosomes and lysosomes decreases in aging cells and deficiency in the ability of lysosomes to degrade the autophagosome cargo as a result of reduced activities of some lysosomal proteases has also been reported (Brunk and Terman, 2002; Butler *et al.*, 2005). In addition, a lower level of beclin-1, a protein that plays a key role in the activation of macroautophagy, has also been seen in the brains of aged individuals (Pickford *et al.*, 2008). These

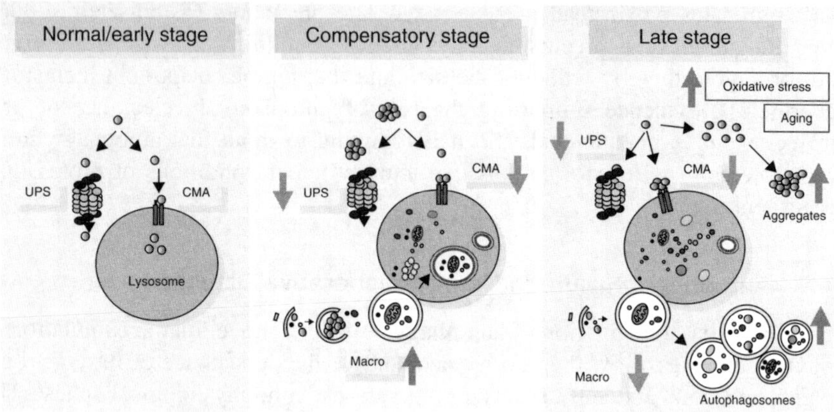

Figure 3.4 Autophagy in the progression of neurodegenerative disorders. During the normal/early stage, most soluble altered proteins are targeted for degradation via the UPS or CMA. However with time, some altered proteins are sequestered into aggregates that cannot be degraded by these two pathways. At the same time, these toxic proteins may also inhibit the activities of UPS and CMA, leading to compensatory upregulation of macroautophagy to facilitate the removal of these toxic proteins (compensatory stage). In most neurodegenerative disorders, this compensatory stage is followed by a late/failure stage, in which the cell survivability is compromised. In this late stage, further blockage of UPS and CMA accompanied by a decrease in macroautophagic activity accentuate the accumulation of toxic products and aggregates. In addition, other aggravating factors such as increased oxidative stress and aging also contribute to the general failure of the proteolytic systems.

age-related impairments in autophagy may contribute directly to the accumulation of altered proteins. Furthermore, the decreased turnover rates of intracellular components such as mitochondria may explain their accumulation in aging neurons and the consequent increase in oxidative stress (Brunk and Terman, 2002). As the malfunction worsens, destabilization of autophagosome and autophagolysosome membranes triggers the release of undigested products and hydrolytic enzymes into the cytoplasm (Boland and Nixon, 2006). Often in parallel, the pathogenic proteins also contribute toward impairment of the autophagic pathway as exemplified by the inhibitory effect of pathogenic α-synuclein on CMA (Cuervo *et al.*, 2004; Martinez-Vicente *et al.*, 2008). In some instances, failure of autophagy is linked to the dysfunction of proteins that play a direct role in autophagy. This is exemplified by parkin and PINK1 in PD, both of which participate in mitophagy (Dagda *et al.*, 2009; Narendra *et al.*, 2008; Vives-Bauza *et al.*, 2010).

The particular sequence of alterations in the intracellular proteolytic systems in various neurodegenerative disorders could vary depending on the cell type and the intrinsic properties of the pathogenic protein. Accordingly, therapeutic interventions based on manipulations of the proteolytic pathways should be customized depending on the disease stage. For example, upregulation of macroautophagy could be beneficial to prolong the duration of the compensation stage, but it may have detrimental effects during the failure stage when the clearance of autophagosomes is already compromised. Better characterization of the molecular defects behind the failure of each of the different proteolytic systems in different pathophysiological conditions is essential for the development of more targeted restorative intervention. The specific failures in autophagy associated with selected neurodegenerative diseases have been described below.

3.4.2.1 Parkinson's Disease (PD). Failure of macroautophagy in PD (impairment of CMA in PD is described in Section 3.7) has recently been confirmed through the newly identified roles of parkin and PINK1 in maintaining mitochondrial homeostasis by mediating trafficking and autophagic clearance of dysfunctional mitochondria (Dagda *et al.*, 2009; Narendra *et al.*, 2008; Vives-Bauza *et al.*, 2010). Decrements in the activities of parkin and PINK1 due either to familial mutations in PD, environmental stress, or to haploinsufficiency in PD could lead to deregulation of mitochondrial turnover. This persistence of damaged mitochondria in the cytosol may contribute to oxidative stress, and dopamine (DA)-producing neurons are particularly susceptible to such stress. Parkin is also a multifunctional ubiquitin ligase capable of performing both proteasomal-linked K48 polyubiquitination and macroautophagy-linked K63 polyubiquitination (Section 3.4.1). In this aspect, parkin may serve an important role in the triage of misfolded proteins between proteasomal and lysosomal degradation. A loss of parkin function could potentially cripple the ability of the neurons to adequately handle misfolded protein stress through the manipulation of both degradative systems. Additionally, LRRK2 has also been implicated in macroautophagy regulation. This leucine-rich repeat kinase protein is present in the membrane microdomains of some of the vesicles of the endosomal–autophagic system and has been shown to be involved in actin cytoskeleton mediated vesicle trafficking (Alegre-Abarrategui *et al.*, 2009). Malfunctioning of LRRK2 in PD patients could affect the recruitment of this protein to specific membrane microdomains and affect the activity of the endosomal–autophagic pathway.

The presence of mature LBs in neurons may arise from gross autophagy system dysfunction or, alternately, from an inability of certain types of LBs to recruit the autophagy machinery. We have recently demonstrated that the composition of an aggresome influences its clearance by macroautophagy (Wong *et al.*, 2008). For example, whereas aggresomes generated in cells expressing α-synuclein and synphilin-1 are amenable to clearance by macroautophagy, those produced in cells expressing AIMP2/p38 (also a component of LB) are apparently resistant to autophagic clearance (Wong *et al.*, 2008). In fact, AIMP2/p38 is a *bona fide* substrate of parkin and has been reported to accumulate in the brains of PD patients carrying parkin mutations (Corti *et al.*, 2003; Wang *et al.*, 2005). Such impairments in the clearance of intracellular inclusions could initiate the neurodegeneration process as apparent in mice deficient in macroautophagy (Hara *et al.*, 2006; Komatsu *et al.*, 2006).

3.4.2.2 Alzheimer's Disease (AD). It is now known that the massive expansion of the endosomal–lysosomal compartments and accumulation of autophagosomes in the affected brain regions of AD represent abnormalities due to a defect in fusion between autophagosomes and lysosomes (Boland *et al.*, 2008; Nixon, 2005, 2007). Accumulation of these autophagosomes can act as a source of cytotoxic products. The presence of the amyloid precursor protein (APP) in the accumulating autophagosomes along with the APP secretases responsible for its cleavage into the pathogenic peptide αβ1–42 converts autophagosomes into an internal source for this pathogenic product. Conceivably, this contributes toward β-amyloid deposition in AD brains (Yu *et al.*, 2004). Furthermore, the accumulated autophagic compartments can become "leaky" as a result of the buildup of pathogenic products and release lysosomal hydrolases into the cytosol that could lead to cell death (Kaasik *et al.*, 2005).

A reduced ability to activate macroautophagy is also partly responsible for the autophagic deficit in AD. Beclin-1 is found decreased in the affected brain regions of AD patients. Remarkably, the overexpression of beclin-1 in AD animal models resulted in a decrease in both intracellular and extracellular amyloid pathology leading to restored normal functions (Pickford *et al.*, 2008).

Neurofibrillary tangles that arise from the aggregation of mutant tau protein also associate with AD and other tauopathies (Pickart and Cohen, 2004). We have recently identified tau as a substrate of CMA and uncovered that mutant tau is degraded by CMA in an atypical way (Wang *et al.*, 2009). Upon binding to the CMA translocation machinery, the partially membrane-inserted tau protein undergoes regulated cleavage by cathepsin L, a lysosomal luminal protease, resulting in the generation of smaller pathogenic tau fragments at the lysosomal membrane. This favors the oligomerization of tau at or near the surface of lysosomes, which not only results in membrane disruption and lysosomal leakage but also further seeds the aggregation of tau in the cytosol by acting as a nucleating center when released from the lysosome. Therefore, the inability of mutant tau to undergo complete CMA contributes to the generation of amyloidogenic fragments that organize into aggregates better handled by macroautophagy. Overloading of the macroautophagic pathway may lead to its failure and further exacerbate the pathology.

3.4.2.3 Huntington's Disease (HD). HD is caused by an expansion of the polyglutamine tract in the N-terminal region of the protein huntingtin (htt) that converts it into a pathogenic aggregate-prone protein (Rubinsztein, 2002). When mutant htt is not efficiently removed, it causes toxicity to cells and eventually leads to cell death and onset of the disease. While wild-type and mutant htt in the nucleus are normally degraded by the proteasome (Iwata *et al.*, 2009), the removal of toxic forms of cytosolic mutant htt occurs preferentially via autophagy (Ravikumar *et al.*, 2004; Shibata *et al.*, 2006). Interestingly, autophagic removal of htt is partly modulated through htt acetylation (Jeong *et al.*, 2009). Furthermore, htt protein inclusions are found to sequester mTOR, the negative regulator of macroautophagy, thereby leading to activation of macroautophagy to facilitate its own removal (Ravikumar *et al.*, 2004).

The molecular mechanisms that contribute to HD pathogenesis are still being elucidated. Recent studies have proposed a role for huntingtin in the normal function of the autophagic system. This came from the observations that huntingtin is associated with the ER (Atwal *et al.*, 2007), late endosomes, and autophagosomes (Kim *et al.*, 2002), as well as partnering with Rab5 to facilitate the formation of autophagosome (Ravikumar *et al.*, 2008). However, the possible function of huntingtin in macroautophagy and the step(s) of the autophagic process affected by mutant huntingtin are still poorly understood. Recently, our group has found that the previously reported massive expansion of the autophagic compartments in the neuronal cells of HD does not result in the expected increase in proteolysis. Instead, the turnover of cytosolic components is significantly impaired in these cells. Our study shows that the autophagic deficiency is not due to a defect in fusion between autophagosomes and lysosomes nor a reduced proteolytic activity, but rather originates from a problem with cargo recognition, leading to inefficient cargo loading into autophagosomes in HD cells (Martinez-Vicente *et al.*, 2010). Analysis of these autophagosomes has revealed a marked decrease in their cargo content, that is, they appear as "empty" autophagosomes in EM micrographs. Consequently, we observed large amounts of lipid droplets being present and altered mitochondria, which are both cargo of autophagy, in the cytosol of HD cells.

3.5 NUTS AND BOLTS OF CHAPERONE-MEDIATED AUTOPHAGY

A key to understand how aberrations in CMA contribute to the pathogenesis of PD is an appreciation of the different unique features and steps involved in this pathway (Figure 3.5). CMA is distinguished from the other lysosomal degradative pathways by its selective mode of degradation. Only proteins carrying a particular pentapeptide motif in their amino acid sequences (described in greater details in Section 3.5.1) are targeted for degradation by CMA. This selective form of autophagy originated from the observation that prolonged starvation in animals or the removal of serum growth factors from cultured cells for more than 8–10 h accelerated the degradation of only a subset of cytosolic proteins in lysosomes (Wing *et al.*, 1991). The signature CMA targeting motifs in these proteins are recognized by the cytosolic chaperone involved in this pathway to preferentially target them to lysosome for removal. This selectivity of CMA provides cells with an advantage under conditions where specific degradation is preferred over nondiscriminative bulk removal.

The other trademark of CMA is the direct translocation of cargo across the lysosomal membrane into the lumen in a molecule-by-molecule fashion. This process is mediated by an integral lysosomal membrane receptor, LAMP-2A, and requires substrate unfolding for the substrate protein to transverse the lysosomal membrane (Salvador *et al.*, 2000). The requirement for unfolding and the

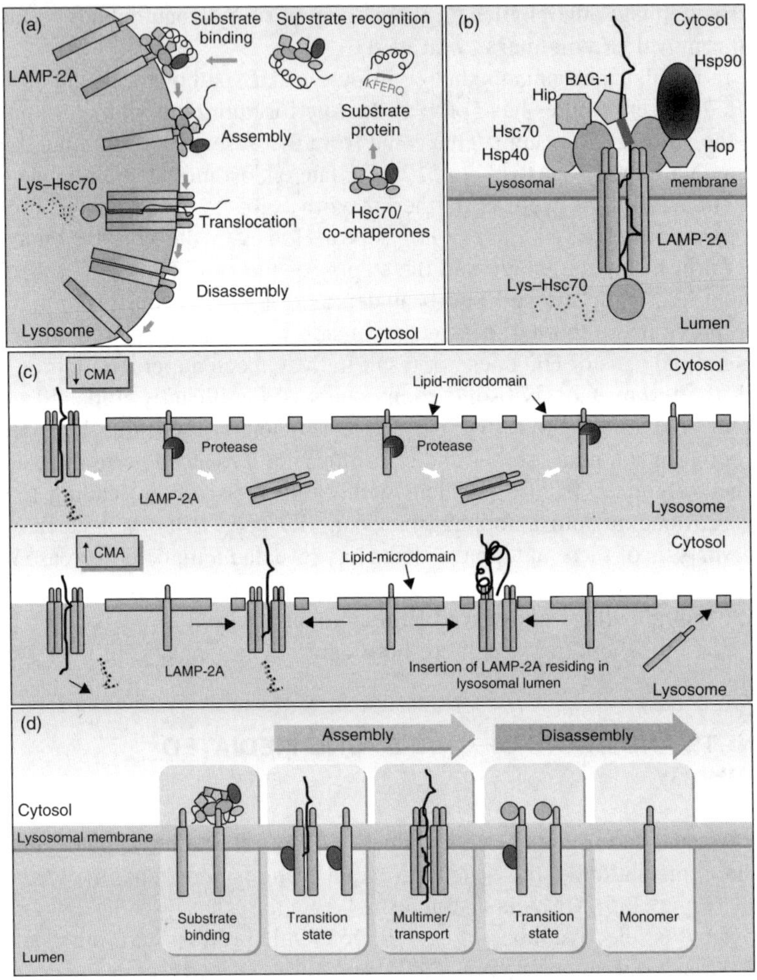

Figure 3.5 Components of chaperone-mediated autophagy (CMA). (a) Overview of the steps involved in CMA. Cytosolic proteins bearing KFERQ-like motif are recognized by Hsc70 and co-chaperones in the cytosol. The substrate–Hsc70–co-chaperone complex is then delivered to the surface of the lysosome where it binds to the cytosolic tail of LAMP-2A. Substrate binding initiates multimerization of LAMP-2A into an active translocation complex. Substrate unfolding occurs by unknown mechanisms prior to translocation into the lysosomal lumen assisted by a luminal Hsc70 (Lys–Hsc70). Once inside lysosomes, the substrate is degraded rapidly. In the absence of substrate, Hsc70 promotes the disassembly of LAMP-2A from the translocation complex into monomeric forms again. (b) Schematic model of the substrate–Hsc70–co-chaperone complex at the lysosomal membrane. (c) Dynamics of LAMP-2A is partly regulated by its association with discrete lipid microdomain in the lysosomal membrane. Under basal condition, a percentage of LAMP-2A is continuously turned over in the microdomain where the proteases responsible for its cleavage reside. Activation of CMA is associated with mobilization of the LAMP-2A out of the microdomain to prevent its degradation, resulting in a net increase in LAMP-2A levels at the lysosomal membrane. Increase in LAMP-2A levels are also achieved via lysosomal membrane insertion of a pool of full size LAMP-2A residing in lysosomal lumen. (d) LAMP-2A undergoes cycles of assembly and disassembly upon demand for substrate binding and uptake.

presence of a selective receptor for the CMA substrates at the lysosomal membrane make this a saturable form of autophagy.

Current understanding of the steps pertinent to CMA activity—substrate recognition and binding, formation of active lysosomal translocon for this pathway, and substrate unfolding and translocation—are described in the following sections.

3.5.1 Chaperoning of CMA Substrates

CMA substrates are earmarked by the presence of a pentapeptide targeting motif biochemically related to KFERQ in their amino acid sequence. The CMA tag is degenerate and allows a series of amino acid combinations guided by the parameters that a Q should be flanked on either side by four amino acids consisting of a basic (K, R), an acidic (D, E), a bulky hydrophobic (F, I, L, V), and a repeated basic or bulky hydrophobic amino acid (K, R, F, I, L, V) (Dice, 1990). KFERQ-like motifs are present in 30% of cytosolic soluble proteins, among which α-synuclein, the gene product of PARK1 and PARK4 implicated in PD (Table 3.1) is one of them. In fact, our sequence analysis revealed that CMA consensus motifs are found in all of the PD-linked gene products identified to date (Table 3.1), thus implying a greater than expected role of CMA in PD pathogenesis.

The KFERQ-like motif in proteins is recognized by the constitutively expressed Hsc70 in the cytosol (Chiang *et al.*, 1989) (Figure 3.5a), which also binds a series of co-chaperones to form the substrate–Hsc70–co-chaperone complex (Agarraberes and Dice, 2001) (Figure 3.5a, b). The binding of Hsc70 to the substrates is an adenosine triphosphate (ATP)-dependent process in which substrate protein binding is favored by hydrolytic conversion of ATP into adenosine diphosphate (ADP). The functions of the co-chaperones that bind to Hsc70 have not been clearly elucidated but by analogy to other process they may modulate the ATPase activity of Hsc70. Heat shock protein of 40 kDa (Hsp40) stimulates the ATPase activity of Hsc70 leading to increased binding and release of substrate proteins by Hsc70. Hip (Hsc70 interacting protein) blocks the binding of new ATP molecules, hence stabilizing the Hsc70, Hsp40, and protein substrate complex, whereas Hop (Hsc70–Hsp90 organizer protein) and BAG-1 (Bcl2-associated athanogene-1) promote substrate release by favoring ATP binding. It is possible that some of the co-chaperones may act directly on the substrate. In other systems, the heat shock protein of 90 kDa (Hsp90) is able to bind unfolded regions of a substrate protein to prevent it from aggregating during the cycles of binding/release from Hsc70. Blockage of the substrate binding site in Hsc70, depletion of nucleotides, or disruption of the chaperone–co-chaperone interactions would all block the delivery of substrate proteins into lysosomes (Agarraberes and Dice, 2001).

Many substrates of CMA can also be substrates for other degradative systems. For example, evidence has shown that α-synuclein is degraded by the proteasome, CMA, and macroautophagy (Cuervo *et al.*, 2004; Vogiatzi *et al.*, 2008;

Table 3.1 CMA targeting motifs in proven and potential CMA substrate proteins

Protein	Sequence	Properties	Location	Proven
I. Others				
Aldolase B	QKKEL	Q++−Δ	N-terminus	Yes
	QFREL	QΔ + −Δ	N-terminus	Yes
	IKLDQ	Δ + Δ−Q	N-terminus	Yes
Annexin I	EFLKQ	−ΔΔ+Q	N-terminus	Yes
Annexin II	QKVFD	Q+ΔΔ−	C-terminus	Yes
Annexin IV	QELLR	Q−ΔΔ+	N-terminus	Yes
Annexin VI	QEFIK	Q−ΔΔ+	C-terminus	Yes
Aspartate aminotransferase	RKVEQ	+ + Δ−Q	N-terminus	Yes
c-fos	NLLKE	NΔΔ + −	Middle	Yes
GAPDH	NRVVD	N+ΔΔ−	C-terminus	Yes
Glutathione transferase	NKKFE	N+ + Δ−	C-terminus	Yes
Hemoglobin (β-chain)	QRFFE	Q+ΔΔ−	N-terminus	Yes
Hsc70	QRDKV	Q+ − +Δ	Middle	Yes
	QKILD	Q+ΔΔ−	C-terminus	—
IκB	VKELQ	Δ + −ΔQ	C-terminus	Yes
α-2-microglobulin	VDKLN	Δ − +Δ N	N-terminus	Yes
	RIKEN	+Δ + − N	C-terminus	—
Pax-2	DVVRQ	+ΔΔ+Q	N-terminus	Yes
	QRIVE	Q+ΔΔ−	N-terminus	—
	QLLRE	QΔΔ + −	Middle	—
26S proteasome (C8)	IEKLQ	Δ − +ΔQ	Middle	Yes
19S proteasome (PA28)	QEKVF	Q − +ΔΔ	Middle	Yes
Pyruvate kinase	QDLKF	+Δ − +Q	Middle	Yes
RNase A	KFERQ	Δ + +−Q	N-terminus	Yes
Tau	QVEVK	QΔ − Δ+	C-terminus	Yes
	KDRVQ	+ − +ΔQ	C-terminus	—
II. Parkinson's Disease Related				
α-synuclein (*PARK1, PARK4*)	VKKDQ	Δ + +−Q	C-terminus	Yes
Parkin (*PARK2*)	QLKEV	QΔ + −Δ	N-terminus	Putative
UCH-L1 (*PARK5*)	SVLKQ	⊘ΔΔ+Q	Middle	Putative
PINK1 (*PARK6*)	NLKLD	NΔ + Δ−	C-terminus	Putative
DJ-1 (*PARK7*)	ILKEQ	ΔΔ + −Q	Middle	Putative
LRRK2/Dardarin (*PARK8*)	KEVFQ	+ − ΔΔQ	N-terminus	Putative
	QDVRK	Q−Δ + +	N-terminus	—
	QIRDK	QΔ + −+	Middle	—
	KIRDQ	+Δ + −Q	Middle	—
	QLVRE	QΔΔ + −	Middle	—
	LEKFQ	Δ − +ΔQ	Middle	—
	DRLLQ	− + ΔΔQ	C-terminus	—
	QKLIE	Q+ΔΔ−	C-terminus	—
Nurr1 (*NR4A2*)	KRtVQ	+ + ⊘ΔQ	Middle	Putative
	QRIFy	Q+ΔΔ⊘	C-terminus	—
Synphilin-1	QKVVE	Q+ΔΔ−	N-terminus	Putative
	QEKIL	Q- + ΔΔ	N-terminus	

Δ = hydrophobic; + = positively charged; − = negatively charged; Q= potential site for phosphorylation to substitute as negatively charged residue.

Webb *et al.*, 2003). Naturally, the question is what decides the degradation fate of a protein with multiple proteolytic options? Molecular size may be a determinant for siphoning substrates from the proteasome and CMA toward macroautophagy since unfolding is coupled to proteolysis in the former two systems. Hence, aggregated proteins are handled more efficiently by macroautophagy. However, what determines the flagging of a soluble substrate for CMA removal over proteasomal degradation and whether the two pathways work synergistically or competitively for the same substrates is unclear. For CMA proteolysis, uncovering of the KFERQ-like motifs that often time may be buried in the core of the protein substrate is required for recognition by Hsc70 (Cuervo *et al.*, 2004; Kiffin *et al.*, 2004). Therefore, protein modifications that either enhance or prevent the accessibility of the CMA tag will directly impact substrate CMA. Although ubiquitination is not required for CMA of most substrates (Cuervo *et al.*, 1998), it is highly possible that ubiquitination may modulate CMA of specific proteins by modifying the accessibility of the CMA tag. Whether or not ubiquitination affects CMA activity is unknown. Furthermore, posttranslational modifications such as phosphorylation—when the negative residue is missing—or acetylation—when a positive residue is lacking—could give rise to a functional noncanonical CMA tag out of an amino acid stretch missing just one of the residues of the CMA motif.

3.5.2 Binding of CMA Substrates to Lysosomal Receptor

Once bound to the molecular chaperone complex, the CMA substrate is delivered to the lysosome membrane (Figure 3.5a) where it binds to the cytosolic tail of a LAMP-2A, the receptor responsible for CMA (Cuervo and Dice, 1996). LAMP-2A is one of the three isoforms derived from the alternative splicing of the *lamp-2* gene. The other two LAMP-2 isoforms are LAMP-2B and LAMP-2C. All three isoforms have an identical luminal region but differ in their transmembrane regions and cytosolic tails. Only LAMP-2A acts as a receptor for CMA (Cuervo and Dice, 1996). The substrate–receptor interaction does not require the KFERQ-like motif but involves a stretch of four positively charged amino acids in the cytosolic tail of LAMP-2A (Cuervo and Dice, 2000b). The specific amino acids in the substrate protein responsible for binding to LAMP-2A remain unknown.

The levels of LAMP-2A at the lysosomal membrane, the ability of this receptor to multimerize in an active translocon complex, and the need for substrate unfolding prior to translocation into the lysosomal lumen are the rate-determining steps in CMA (Cuervo and Dice, 2000b). These requirements make CMA a saturable pathway. In fact, blockage of LAMP-2A with antibodies against its cytosolic tail prevented substrate binding and decreased CMA activity, whereas overexpression of LAMP-2A increases the capability of the system (Cuervo and Dice, 1996). More recently, we have also shown that siRNA knockdown of LAMP-2A level in cultured cells significantly reduced CMA activity (Massey *et al.*, 2006b). In the case of PD, the pathogenic mutations in α-synuclein have caused the mutant proteins to bind with higher affinity to LAMP-2A but mutant α-synucleins are unable to translocate efficiently, hence saturating the pathway by precluding the association to the membrane and translocation of other CMA cargo.

3.5.3 Assembly and Disassembly of the "LAMP-2A Gateway" to CMA

The levels and multimer organization of LAMP-2A at the lysosomal membrane are highly dynamic and influence CMA activity. When CMA is less active, most of the LAMP-2A molecules in the lysosomal membrane are found associated with specific lipid microdomains at this membrane (Figure 3.5c) where LAMP-2A undergoes continuously turnover by a pair of lysosomal proteases, cathepsin A, and a yet unidentified metalloprotease (Cuervo *et al.*, 2003; Kaushik *et al.*, 2006). Induction of CMA is associated with an increase in the levels of LAMP-2A at the lysosomal membrane. This is achieved through the mobilization of LAMP-2A out of the lipid microdomains to prevent its degradation. Besides this, the level of LAMP-2A at the lysosomal membrane is also augmented through transcriptional upregulation of the *lamp-2* gene and by the recruitment of a population of full length LAMP-2A residing in the lumen of lysosomes to the lysosomal surface (Cuervo and Dice, 2000b).

Dynamics in the multimeric organization of LAMP-2A is also observed at the lysosomal membrane. We have found that rather than a stable translocon, LAMP-2A undergoes continuous cycles of assembly and disassembly into translocation complex active for CMA in response to the availability of substrate proteins (Bandyopadhyay *et al.*, 2008). CMA substrates only bind to monomeric form of LAMP-2A and this binding initiates the organization of LAMP-2A into a 700-kDa multimeric complex required for substrate translocation (Bandyopadhyay *et al.*, 2008). Both chaperones Hsc70 and Hsc90 assist in this dynamic association/disassociation process but are not part of the translocation complex. Besides participating in delivery and presentation of substrates to LAMP-2A, Hsc70 also serves a second function in CMA facilitating the disassembly of LAMP-2A from the translocating complex. This dual function of Hsc70 is regulated by the presence of substrate proteins. When substrate is available, Hsc70 favors their binding to LAMP-2A thereby driving the multimerization of LAMP-2A into active translocons. In the absence of substrate, Hsc70 facilitates dissociation of LAMP-2A into monomers. Hsc90 aids in this dynamic process by stabilizing LAMP-2A, whereas in transitionary complexes between assembly and disassembly (Figure 3.5d). Interestingly, stabilization of LAMP-2A is attributed to a lysosome-resident population of Hsc90 that binds to the luminal side of the lysosomal membrane. Although the mechanism is not fully understood, one possibility is that Hsp90 stabilizes the LAMP-2A complexes by binding to exposed surfaces on LAMP-2A that become accessible to protease attack because of conformational changes for multimerization thereby shielding it from degradation. Indeed, treatment with different Hsp90 inhibitors remarkably reduces the amount of LAMP-2A at the lysosomal membrane (Bandyopadhyay *et al.*, 2008). It is conceivable that the persistent binding of pathogenic α-synuclein to LAMP-2A would interfere with this dynamic formation of LAMP-2A translocon.

3.5.4 Substrate Unfolding and Translocation

Substrate translocation is coupled to substrate unfolding and this requires the combined actions of two forms of Hsc70, one at the cytosolic side of the lysosomal membrane and a second more acidic Hsc70 isoform present inside the lysosome (Lys–Hsc70 in Figure 3.5a) (Salvador *et al.*, 2000). Although the exact mechanism remains unknown, the cytosolic Hsc70 is probably involved in the unfolding of substrate, which occurs after binding but before translocation (Salvador *et al.*, 2000). The lysosomal Hsc70 may aid substrate translocation by binding to the substrate and actively pulling it inside the lysosome lumen or serve to lock the portion of protein, which is already translocated, in order to prevent any translocation back to the cytosol (Figure 3.5a). The levels of Lys–Hsc70 also change with the activation of CMA where starvation and oxidative stress have shown to increase the amount of Lys–Hsc70 per lysosome (Kiffin *et al.*, 2004). In fact, the subset of lysosomes able to perform CMA in cells is often identified by the presence of Lys–Hsc70 (Cuervo *et al.*, 1997).

3.6 PHYSIOLOGICAL ROLES OF CMA

The best characterized function of CMA is to provide an alternative source of amino acids during periods of prolonged nutrient deprivation (Cuervo *et al.*, 1995). At the early phase of starvation, macroautophagy is first activated and reaches its maximal activity around 6 h of nutrient deprivation. During this period, cells take advantage of the high capacity bulk degradation mediated by macroautophagy to supply amino acids and other essential components it needs for cell survival (Mizushima, 2005). CMA is activated when starvation persists beyond 6–8 h, reaches its maximal activity around 24 h, and remains active up to three days into starvation. This switch from nondiscriminative mass removal to a more targeted proteolysis allows the cells to selectively degrade proteins of lesser importance for cellular maintenance while preserving those proteins and structures critical for cell survival in these conditions (Figure 3.6).

As with other types of autophagy, CMA plays an essential role in cellular quality control (Figure 3.6). Like the UPS, CMA is also responsible for the removal of damaged oxidized proteins and its activity has been shown to be upregulated by the presence of oxidized, misfolded, or truncated proteins (Finn and Dice, 2005; Kiffin *et al.*, 2004). During mild oxidative stress, oxidized proteins can be detected in the lumen of CMA active lysosomes. Furthermore, oxidized CMA substrates bind and internalize into isolated CMA active lysosomes at higher rates than their unmodified counterparts. Selectivity may also be the reason behind the activation of CMA during oxidative stress (Finn and Dice, 2005; Kiffin *et al.*, 2004) where CMA would allow removal of oxidized or misfolded proteins without compromising the normal functional proteins in the vicinity. Proteins modified by mild oxidative stress may become better substrates for CMA because of the fact that they are partially unfolded, hence facilitating their recognition by the chaperone and at the same time increasing the ease of unfolding required for

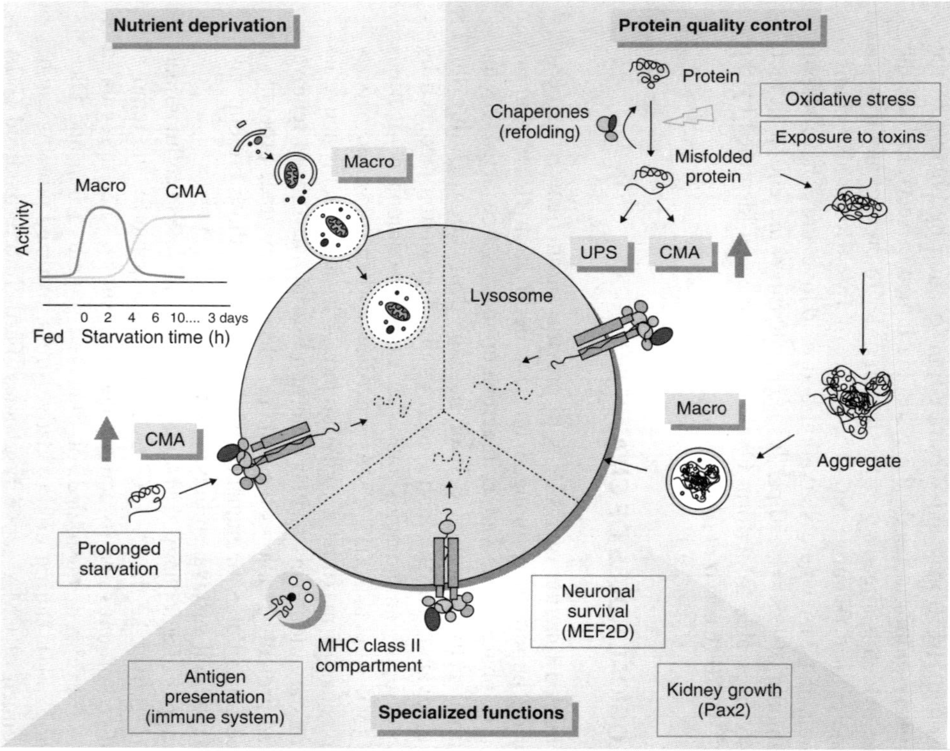

Figure 3.6 Physiological role of CMA. CMA performs two general functions shared with other autophagic pathways. Together with macroautophagy, CMA is activated under nutrient deprivation conditions to provide an alternative source of energy. CMA also participates in cellular quality control. This pathway is shown to be activated in the presence of oxidized, misfolded, or truncated proteins in order to eliminate damaged or toxic proteins. However, when CMA and UPS are overwhelmed and when misfolded proteins are organized into higher structures like oligomers, fibrils, and aggregates, macroautophagy would be the main pathway for their elimination. In addition, CMA also carries out specialized functions in particular cell types.

translocating across the CMA receptor. In this aspect, both UPS and CMA can only degrade modified proteins while they are still soluble since unraveling is required in order to reach the proteolytic core or the luminal proteases, respectively. The activities of UPS and CMA could become impaired in the process of removing these oxidized misfolded proteins (Bennett *et al.*, 2005; Cuervo *et al.*, 2004; Snyder *et al.*, 2003; Stefanis *et al.*, 2001), accentuating the intracellular accumulation of damaged proteins. In contrast, when oxidative damage of proteins becomes more severe leading to overt aggregation, macroautophagy will be the main pathway for the elimination of such insoluble structures. Despite the compensatory activities between CMA and macroautophagy, both the pathways are not redundant as the blockage of either CMA or macroautophagy in cultured cells causes the cells to be sensitive to different types of cellular stressors (Kaushik *et al.*, 2008; Massey *et al.*, 2006b).

In some specialized cell types, CMA also serves specific functions (Figure 3.6). For example, CMA may regulate neuronal survival by regulating the level of myocyte enhancer factor 2D (MEF2D) in neurons (Yang *et al.*, 2009). As described in more detail in the following section, UCH-L1, one of the genetic factors implicated in familial PD has been demonstrated to bind directly to LAMP-2A (Kabuta *et al.*, 2008). On the basis of the restricted expression of UCH-L1, found abundantly only in nerve cells, it is tempting to suggest that CMA may participate in other neuronal functions. In antigen presenting cells, CMA is involved in the presentation of endogenous peptides by major histocompatibility complex class II (MHC-II) (Zhou *et al.*, 2005), thereby influencing autoimmunity. CMA also plays a role in modulating kidney growth by controlling the degradation of specific proteins such as the transcriptional factor Pax2 in renal tubular cells (Franch *et al.*, 2001).

3.7 IMPAIRMENT OF CMA IN PD

In addition to the UPS, α-synuclein is also selectively degraded in lysosomes via CMA (Cuervo *et al.*, 2004; Martinez-Vicente *et al.*, 2008). α-Synuclein contains the CMA targeting motif (Table 3.1) that is needed by Hsc70 for recognition, which delivers it to the LAMP-2A receptor for elimination via CMA. Indeed, α-synuclein has been shown experimentally to interact with the chaperones and LAMP-2A receptor—the essential components of CMA. Disruption of the CMA motif by amino acid mutations prevented the binding of α-synuclein to LAMP-2A and increased the half-life of α-synuclein, clearly indicating that CMA is responsible for the degradation of a percentage of this pathogenic protein that needs to be tightly regulated (Cuervo *et al.*, 2004). Furthermore, it is possible to reproduce the direct translocation of wild-type α-synuclein across the lysosomal membrane in a well-established *in vitro* system using isolated lysosomes (Cuervo *et al.*, 2004; Martinez-Vicente *et al.*, 2008). In the *in vitro* lysosomal transport assay, wild-type α-synuclein was able to inhibit the lysosomal binding and uptake of established CMA substrates like glyceraldehyde 3-phosphate dehydrogenase

(GAPDH) and RNase A in a competitive manner and vice versa. Notably, α-synuclein with the disrupted CMA motif exhibited markedly reduced association with and translocation across the lysosomal membrane and was unable to interfere with the lysosomal binding of GAPDH. Therefore, α-synuclein fulfills all the criteria of a bona fide CMA substrate.

A role of CMA dysfunction in the pathogenesis of PD came from the observation that α-synuclein proteins carrying familial PD mutations (e.g., A30P, A53T) failed to be degraded by CMA (Cuervo *et al.*, 2004). The pathogenic mutant α-synuclein proteins are still being recognized by Hsc70 and delivered to the lysosomes since none of the familial PD mutations described to date affect the CMA targeting motif. In fact, both A30P and A53T mutant α-synuclein bind to the LAMP-2A receptor with abnormally high affinity at the lysosomal membrane compared to wild-type α-synuclein, but are poorly internalized into the lysosomes. Therefore, these pathogenic α-synuclein proteins act like uptake blockers of CMA, inhibiting not only their own degradation but also the degradation of other CMA substrates (Cuervo *et al.*, 2004) (Figure 3.7). Consequentially, the increase in cytosolic α-synuclein levels could favor its aggregation and thus increase the concentration of aggregated protein in the cell.

Besides this link of CMA with familial forms of PD, which only account for a small percentage of the PD cases, decrements in CMA have also been

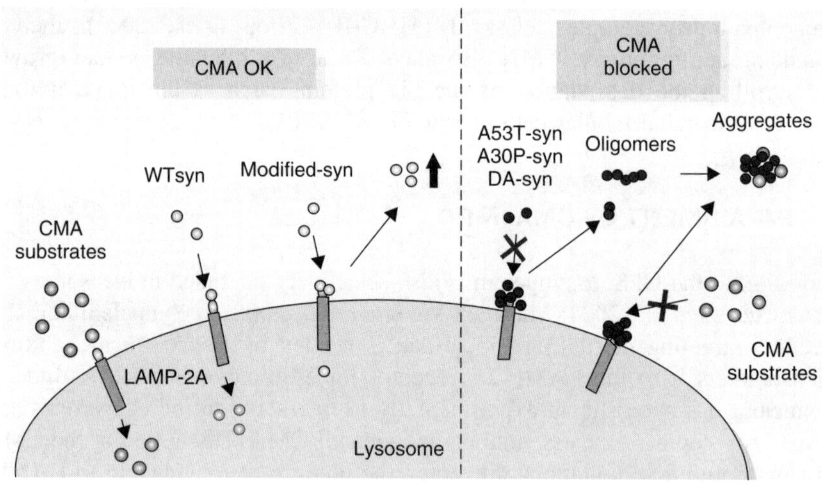

Figure 3.7 Altered turnover of α-synuclein via CMA in Parkinson's disease. A percentage of intracellular α-synuclein is normally degraded via CMA in lysosomes. Pathogenic mutations in α-synuclein and particular posttranslational modifications (see text) alter the lysosomal translocation of α-synuclein by CMA. The pathogenic α-synucleins bind to the lysosomal membrane with unusually high affinity and the consequences of impaired CMA of α-synucleins are two folds: on one hand, decreased degradation of α-synuclein leads to its accumulation in the cytosol and promotes its aggregation; on the other hand, the blockage of CMA by the altered α-synucleins prevents degradation of other CMA substrates, which could then accumulate in protein inclusions characteristics of the disease state.

implicated in idiopathic forms of PD, which represent the majority of the PD cases. PD related environmental stress, a major contributing factor to sporadic PD, was also found to impact CMA. We have found that certain posttranslational modifications of α-synuclein affect its ability to be degraded by CMA (Martinez-Vicente *et al.*, 2008). Phosphorylation and nitration of α-synuclein result in reduced lysosomal uptake. However, because these modified forms of α-synuclein bind loosely to the CMA receptor, they can be easily displaced by other CMA substrate proteins; hence they do not affect CMA (Martinez-Vicente *et al.*, 2008). In contrast, DA modification of α-synuclein impairs CMA-mediated degradation by a mechanism similar to that of familial mutant forms of α-synuclein. DA-modified α-synuclein binds with high affinity to the CMA translocation complex but does not translocate into the lumen of lysosomes. Like the familial mutant α-synuclein proteins, this inefficient translocation also precludes the degradation of other CMA substrates (Figure 3.7). In support for some cell type-dependent effect, CMA inhibition following L-DOPA treatment is more prominent in ventral midbrain cultures containing dopaminergic neurons than in non-DA producing cortical neurons. Importantly, α-synuclein appears to be the primary mediator of DA-induced blockage of CMA as ventral midbrain neurons derived from α-synuclein null mice are relatively spared from the inhibitory effects of DA on CMA activity (Martinez-Vicente *et al.*, 2008). Various studies have indicated that protofibrillar, oligomeric forms of α-synuclein are toxic and may actually be more potent neuronal killers than fibrillar aggregated α-synuclein species. In fact, α-synuclein protofibrils can form pore-like structures capable of permeabilizing membranes and vesicles and causing serious cellular injuries (Lashuel *et al.*, 2002). Most importantly, DA-modified α-synuclein proteins are more prone to form α-synuclein protofibrils than fibrils (Conway *et al.*, 2001), which may explain why DA-producing neurons are especially susceptible to degeneration in PD. Hence, an accumulation of this modified form of α-synuclein as a result of CMA impairment could considerably increase cell toxicity. Although the specific mechanisms by which mutant and DA-modifiedα-synuclein block CMA remain to be elucidated, we have found that both forms of α-synuclein are able to seed the formation of high molecular weight oligomeric structures at the lysosomal membrane (Martinez-Vicente *et al.*, 2008). Direct steric blockage of the translocation complex by this α-synuclein oligomers, changes in the lateral mobility of membrane proteins, or changes in essential lysosomal properties caused by oligomers formed at the lysosomal membrane are all possible mechanisms that require future investigation.

A novel link between CMA and PD has recently been uncovered when another PD-linked gene product, UCH-L1, was found to physically interact with LAMP-2A as well as Hsc70 and Hsp90, the principal chaperones involved in CMA (Kabuta *et al.*, 2008). The levels of these interactions were abnormally increased by the PD-associated I93M mutation in UCH-L1. Similar enhanced bindings to LAMP-2A, Hsc70, and Hsp90 were also observed for carbonyl-modified UCH-L1 (Kabuta *et al.*, 2008). Particularly, the study reports an accumulation of α-synuclein attributed to blockage of CMA degradation when I93M mutant UCH-L1 is overexpressed in cells. UCH-L1 is a highly abundant protein expressed mainly in the neurons. Interestingly, although most of UCH-L1 exists in the soluble form, a subset of UCH-L1 is farnesylated and associated with cellular membranes, including ER (Liu *et al.*, 2008). The amount of membrane-associated UCH-L1 is shown to correlate with the intracellular level of α-synuclein and neurotoxicity. This membrane-associated UCH-L1 appears to negatively regulate the lysosome degradation of α-synuclein. However, it remains to be elucidated whether such decrease is related to a decline in the macroautophagic or CMA degradation of α-synuclein. In fact, the significance of wild-type UCH-L1 interacting with LAMP-2A in normal function of CMA is unknown. This can be partly resolved by confirming whether UCH-L1 is a true substrate of CMA. We have detected a noncanonical CMA motif in the amino acid sequence of UCH-L1 (Table 3.1). Besides α-synuclein and UCH-L1, we have also identified putative CMA targeting motifs in the rest of the familial PD gene products identified to date including parkin, PINK1, DJ-1, LRRK2, and Nurr1 (Table 3.1). Two CMA motifs are also detected in synphilin-1, an α-synuclein interacting protein. This may indicate a greater involvement of CMA in the pathogenesis of PD and the potential pathogenic effects of the PD relevant mutations in these proteins on CMA activity await evaluation.

Recently, neuronal survival factor MEF2D has been shown to be a substrate of CMA. Deregulation of CMA of MEF2D has also been suggested to contribute to PD pathogenesis (Yang *et al.*, 2009). This transcriptional factor is required for neuronal survival and is shuttled from the nucleus to the cytoplasm, where it interacts with Hsc70 to be delivered to lysosomes for degradation. Interestingly, high levels of α-synuclein like those observed in PD patients with triplication of α-synuclein gene or by the overexpression of α-synuclein in animal and cellular models, cause disruption of this process. This leads to accumulation of MEF2D in the cytoplasm that subsequently results in neuronal death. These findings highlight the importance of CMA in neuronal survival.

3.8 CELLULAR CONSEQUENCES OF CMA BLOCKAGE AND THERAPEUTIC IMPLICATIONS

Efforts to elucidate the cellular consequences of the blockage of CMA have confirmed the critical role of CMA in the cellular response to stress. RNA interference against LAMP-2A decreases CMA activity in fibroblast cells and renders these cells more susceptible to specific stressors (Massey *et al.*, 2006b). Under normal conditions, CMA-deficient cells behave like control cells and show similar levels of viability and sensitivity to stressors such as heat shock and starvation. However, CMA-deficient cells are considerably more sensitive to oxidative stressors like H_2O_2, paraquat, and cadmium as well as exposure to UV light (Massey *et al.*, 2006b). Dysfunction of CMA is highly detrimental to the midbrain DA neurons shown to be affected in PD patients as this subset of neurons are particularly vulnerable to oxidative stress because of the oxidative nature of the DA metabolism. This is corroborated by our findings that downregulation of LAMP-2A expression in DA neurons accentuates neuronal cell death when exposed to oxidative stressors (Martinez-Vicente *et al.*, 2008). Interestingly, an upregulation of macroautophagy is observed in cells with reduced CMA activity as a compensatory mechanism (Massey *et al.*, 2006b). Unfortunately, the bulk removal nature of macroautophagy cannot totally replace CMA under stress conditions, particularly when selectivity is needed.

The severity of the cellular consequences of impaired CMA activity is also modulated by the activity of other proteolytic systems (Figure 3.4). As discussed earlier, macroautophagy is constitutively upregulated in cells when CMA is compromised, which may account for the ability of CMA-defective cells to cope basally and against stressors such as starvation (Massey *et al.*, 2006b). On the other hand, CMA is also found to be maximally activated in cells deficient in macroautophagy (Kaushik *et al.*, 2008). Chronic and acute inhibition of UPS also led to the induction of macroautophagy (Ding *et al.*, 2003; Iwata *et al.*, 2005; Pandey *et al.*, 2007). This clearly indicates that the different intracellular proteolytic systems are interdependent and cross talk exists between them. Hence, understanding the mechanisms that underlie the interpathway communication may prove to be useful to modulate the cellular management of excess protein load and toxic proteins. Elucidating the molecular determinants that efficiently divert a protein substrate from one degradative pathway to another or upregulating the activity of one pathway over the other would allow us to better manipulate the various proteolytic systems to slow down disease progression. For example, it is well characterized that CMA is induced by oxidative stress but the signaling mechanism responsible for this activation is still unknown. Therefore, deciphering the signaling complexes involved would help in identification of activators, enhancers, or inhibitors that we could use to more appropriately control this pathway. In addition, we have shown that cells utilize specific ubiquitin topology to promote aggregation of pathogenic proteins in the transitory entity, aggresome, for more efficient removal by macroautophagy. Understanding the molecular players involved in such ubiquitin topology remodeling should permit the ability to preferentially enhance this stress-coping mechanism to prevent overloading and blockage of UPS and CMA by pathogenic proteins such as α-synuclein in PD in the future. However, as a cautionary note, the types of therapeutic interventions based on the manipulation of the proteolytic pathways should be customized depending on the stages in the disease. Enhancing aggregation when clearance of autophagic vacuoles is already compromised, especially in the advance stages, would contribute to further clogging in the affected neurons (Martinez-Vicente and Cuervo, 2007) (Figure 3.4).

Modulating the aggravating effects of aging on the proteolytic systems is another possible point of intervention. In the case of CMA, age-related decline in the activity of this pathway has been attributed mainly to a decrease in LAMP-2A levels at the lysosomal membrane (Cuervo and Dice, 2000a; Dice, 1982).

Recent work from our laboratory has demonstrated that preserving the LAMP-2A levels in mice until later in life help maintain high levels of CMA activity in these aged animals (Zhang and Cuervo, 2008). This results in lower cellular content of oxidized and aggregated proteins in the transgenic LAMP-2A animals and a concomitant improvement in the functioning of other quality control systems such as macroautophagy (Zhang and Cuervo, 2008). Hence, restoring or improving CMA activity through interventions able to stabilize LAMP-2A at the lysosomal membrane until advanced ages may prove beneficial in the prevention or treatment of PD and other neurodegenerative diseases.

3.9 CONCLUDING REMARKS

Over the last decade, protein misfolding, aggregation, and dysfunction in the intracellular surveillance systems have taken central stage in the pathogenesis of PD and a myriad of other degenerative diseases. Better characterization of the molecular players in PD pathogenesis has uncovered multiple evidence of involvement of these gene products in both the proteasomal and autophagy–lysosomal degradation systems. Conceivably, this may reflect the economic sense of the cells to utilize the same protein in markedly different pathways, which will offer one additional layer of control over different pathways, simultaneously emphasizing the intimate connections between the various intracellular quality control systems. As such, "traditional" roles of these PD causative proteins in precipitating aberrations in the proteolytic systems need to be reevaluated. Remarkably, the cell could tolerate these pathogenic events over an extended period of time as seen in patients with aggregation-prone α-synuclein mutations. This reflects the amazing capability of the intracellular protein quality control systems to deal with misfolded/aggregated proteins. Preserving or harnessing the functionality of the various cellular degradative pathways until late in life may thus represent feasible approaches in the treatment of PD as well as other neurodegenerative diseases.

ACKNOWLEDGMENT

We are particularly indebted to our wonderful mentor, Professor Ana Maria Cuervo, for critically reviewing this chapter. Ester Wong is grateful for the fellowship support from Hereditary Disease Foundation.

REFERENCES

Agarraberes FA, Dice JF. A molecular chaperone complex at the lysosomal membrane is required for protein translocation. *J Cell Sci* 2001;**114**:2491–2499.

Alegre-Abarrategui J, *et al.* LRRK2 regulates autophagic activity and localizes to specific membrane microdomains in a novel human genomic reporter cellular model. *Hum Mol Genet* 2009;**18**:4022–4034.

Arrasate M, Mitra S, Schweitzer ES, Segal MR, Finkbeiner S. Inclusion body formation reduces levels of mutant huntingtin and the risk of neuronal death. *Nature* 2004;**431**:805–810.

Atwal RS, *et al.* Huntingtin has a membrane association signal that can modulate huntingtin aggregation, nuclear entry and toxicity. *Hum Mol Genet* 2007;**16**:2600–2615.

Axe EL, *et al.* Autophagosome formation from membrane compartments enriched in phosphatidylinositol 3-phosphate and dynamically connected to the endoplasmic reticulum. *J Cell Biol* 2008;**182**:685–701.

Bandyopadhyay U, *et al.* The chaperone-mediated autophagy receptor organizes in dynamic protein complexes at the lysosomal membrane. *Mol Cell Biol* 2008;**28**:5747–5763.

Bedford L, *et al.* Depletion of 26S proteasomes in mouse brain neurons causes neurodegeneration and Lewy-like inclusions resembling human pale bodies. *J Neurosci* 2008;**28**:8189–8198.

Bendiske J, Bahr B. Lysosomal activation is a compensatory response against protein accumulation and associated synaptopathogenesis-an approach for slowing Alzheimer's disease? *J Neuropathol Exp Neurol* 2003;**62**:451–463.

Bennett EJ, Bence N, Jayakumar R, Kopito R. Global impairment of the ubiquitin-proteasome system by nuclear or cytoplasmic protein aggregates precedes inclusion body formation. *Mol Cell* 2005;**4**:351–365.

Bjorkoy G, *et al.* p62/SQSTM1 forms protein aggregates degraded by autophagy and has a protective effect on huntingtin-induced cell death. *J Cell Biol* 2005;**171**:603–614.

Boland B, Nixon R. Neuronal macroautophagy: from development to degeneration. *Mol Aspects Med* 2006;**27**:503–519.

Boland B, *et al.* Autophagy induction and autophagosome clearance in neurons: relationship to autophagic pathology in Alzheimer's disease. *J Neurosci* 2008;**28**:6926–6937.

Bowman AB, Yoo SY, Dantuma NP, Zoghbi HY. Neuronal dysfunction in a polyglutamine disease model occurs in the absence of ubiquitin-proteasome system impairment and inversely correlates with the degree of nuclear inclusion formation. *Hum Mol Genet* 2005;**14**:679–691.

Braak H, Del Tredici K, Rub U, de Vos RA, Jansen Steur EN, Braak E. Staging of brain pathology related to sporadic Parkinson's disease. *Neurobiol Aging* 2003;**24**:197–211.

Braun BC, Glickman M, Kraft R, Dahlmann B, Kloetzel PM, Finley D, Schmidt M. The base of the proteasome regulatory particle exhibits chaperone-like activity. *Nat Cell Biol* 1999;**1**:221–226.

Brunk UT, Terman A. The mitochondrial-lysosomal axis theory of aging: accumulation of damaged mitochondria as a result of imperfect autophagocytosis. *Eur J Biochem* 2002;**269**:1996–2002.

Butler D, Brown Q, Chin D, Batey L, Karim S, Mutneja M, Karanian D, Bahr B. Cellular responses to protein accumulation involve autophagy and lysosomal enzymes activation. *Rejuvenation Res* 2005;**8**:227–237.

Cataldo A, Barnett JL, Berman SA, Li J, Quarless S, Bursztajn S, Lippa C, Nixon RA. Gene expression and cellular content of cathepsin D in human neocortex: evidence for early upregulation of the endosomal-lysosomal system in pyramidal neurons in Alzheimer's disease. *Neuron* 1995;**14**:1–20.

Chau V, Tobias JW, Bachmair A, Marriott D, Ecker DJ, Gonda DK, Varshavsky A. A multiubiquitin chain is confined to specific lysine in a targeted short-lived protein. *Science* 1989;**243**:1576–1583.

Chiang H, Terlecky S, Plant C, Dice J. A role for a 70 kDa heat shock protein in lysosomal degradation of intracellular protein. *Science* 1989;**246**:382–385.

Conway KA, Rochet JC, Bieganski RM, Lansbury PT Jr. Kinetic stabilization of the alpha-synuclein protofibril by a dopamine-alpha-synuclein adduct. *Science* 2001;**294**:1346–1349.

Corti O, *et al.* The p38 subunit of the aminoacyl-tRNA synthetase complex is a Parkin substrate: linking protein biosynthesis and neurodegeneration. *Hum Mol Genet* 2003;**12**:1427–1437.

Cuervo AM. Autophagy in neurons: it is not all about food. *Trends Mol Med* 2006;**12**:461–464.

Cuervo AM, Bergamini E, Brunk UT, Droge W, Ffrench M, Terman A. Autophagy and aging: the importance of maintaining "clean" cells. *Autophagy* 2005;**1**:131–140.

Cuervo AM, Dice JF. A receptor for the selective uptake and degradation of proteins by lysosomes. *Science* 1996;**273**:501–503.

Cuervo AM, Dice JF. Age-related decline in chaperone-mediated autophagy. *J Biol Chem* 2000a;**275**:31505–31513.

Cuervo AM, Dice JF. Regulation of lamp2a levels in the lysosomal membrane. *Traffic* 2000b;**1**:570–583.

Cuervo AM, Hildebrand H, Bomhard EM, Dice JF. Direct lysosomal uptake of alpha 2-microglobulin contributes to chemically induced nephropathy. *Kidney Int* 1999;**55**:529–545.

Cuervo AM, Knecht E, Terlecky SR, Dice JF. Activation of a selective pathway of lysosomal proteolysis in rat liver by prolonged starvation. *Am J Physiol* 1995;**269**:1200–1208.

Cuervo AM, Stefanis L, Fredenburg R, Lansbury PT, Sulzer D. Impaired degradation of mutant alpha-synuclein by chaperone-mediated autophagy. *Science* 2004;**305**:1292–1295.

Cuervo AM, *et al.* A population of rat liver lysosomes responsible for the selective uptake and degradation of cytosolic proteins. *J Biol Chem* 1997;**272**:5606–5615.

Cuervo AM, *et al.* IkB is a substrate for a selective pathway of lysosomal proteolysis. *Mol Biol Cell* 1998;**9**:1995–2010.

Cuervo AM, *et al.* Cathepsin A regulates chaperone-mediated autophagy through cleavage of the lysosomal receptor. *EMBO J* 2003;**22**:12–19.

Dagda RK, *et al.* Loss of PINK1 function promotes mitophagy through effects on oxidative stress and mitochondrial fission. *J Biol Chem* 2009;**284**:13843–13855.

Dice J. Altered degradation of proteins microinjected into senescent human fibroblasts. *J Biol Chem* 1982;**257**:14624–14627.

Dice J. Peptide sequences that target cytosolic proteins for lysosomal proteolysis. *Trends Biochem Sci* 1990;**15**:305–309.

Ding Q, *et al.* Characterization of chronic low-level proteasome inhibition on neural homeostasis. *J Neurochem* 2003;**86**:489–497.

Dorsey ER, *et al.* Projected number of people with Parkinson disease in the most populous nations, 2005 through 2030. *Neurology* 2007;**68**:384–386.

Dubouloz F, Deloche O, Wanke V, Cameroni E, De Virgilio C. The TOR and EGO protein complexes orchestrate microautophagy in yeast. *Mol Cell* 2005;**19**:15–26.

Dunn WA, *et al.* Pexophagy: the selective autophagy of peroxisomes. *Autophagy* 2005;**1**:75–83.

Farré JC, Subramani S. Peroxisome turnover by micropexophagy: an autophagy related process. *Trends Cell Biol* 2004;**14**:515–523.

Fimia GM, *et al.* Ambra1 regulates autophagy and development of the nervous system. *Nature* 2007;**447**:1121–1125.

Finn PF, Dice JF. Ketone bodies stimulate chaperone-mediated autophagy. *J Biol Chem* 2005;**280**:25864–25870.

Fornai F, *et al.* Fine structure and biochemical mechanisms underlying nigrostriatal inclusions and cell death after proteasome inhibition. *J Neurosci* 2003;**23**:8955–8966.

Fortun J, Dunn WA Jr, Joy S, Li J, Notterpek L. Emerging role for autophagy in the removal of aggresomes in Schwann cells. *J Neurosci* 2003;**23**:10672–10680.

Franch H, *et al.* A mechanism regulating proteolysis of specific proteins during renal tubular cell growth. *J Biol Chem* 2001;**276**:19126–19131.

Gandhi S, Wood NW. Molecular pathogenesis of Parkinson's disease. *Hum Mol Genet* 2005;**14** Spec No. 2: 2749–2755.

Glickman MH, Rubin DM, Coux O, Wefes I, Pfeifer G, Cjeka Z, Baumeister W, Fried VA, Finley D. A subcomplex of the proteasome regulatory particle required for ubiquitin conjugate degradation and related to the COP9-signalosome and eIF3. *Cell* 1998;**94**:615–623.

Goldberg AL. Protein degradation and protection against misfolded or damaged proteins. *Nature* 2003;**426**:895–899.

Hara T, *et al.* Suppression of basal autophagy in neural cells causes neurodegenerative disease in mice. *Nature* 2006;**441**:885–889.

He C, Klionsky DJ. Regulation mechanisms and signaling pathways of autophagy. *Annu Rev Genet* 2009;**43**:67–93.

Iwata A, *et al.* Intra-nuclear degradation of polyglutamine aggregates by the ubiquitin-proteasome system. *J Biol Chem* 2009;**280**:40282–40292.

Iwata A, Riley BE, Johnston JA, Kopito RR. HDAC6 and microtubules are required for autophagic degradation of aggregated huntingtin. *J Biol Chem* 2005;**280**:40282–40292.

Jeong H, *et al.* Acetylation targets mutant huntingtin to autophagosomes for degradation. *Cell* 2009;**137**:60–72.

Kaasik A, Rikk T, Piirsoo A, Zharkovsky T, Zharkovsky A. Up-regulation of lysosomal cathepsin L and autophagy during neuronal death induced by reduced serum and potassium. *Eur J Neurosci* 2005;**22**:1023–1031.

Kabuta T, *et al.* Abberant interaction between Parkinson's disease-associated mutant UCH-L1 and the lysosomal receptor for chaperone-mediated autophagy. *J Biol Chem* 2008;**283**:23731–23738.

Kanazawa T, *et al.* Amino acids and insulin control autophagic proteolysis through different signaling pathways in relation to mTOR in isolated rat hepatocytes. *J Biol Chem* 2004;**279**:8452–8459.

Kaushik S, *et al.* Lysosome membrane lipid microdomains: novel regulators of chaperone-mediated autophagy. *EMBO J* 2006;**25**:3921–3933.

Kaushik S, *et al.* Constitutive activation of chaperone-mediated autophagy in cells with impaired macroautophagy. *Mol Biol Cell* 2008;**19**:2179–2192.

Kawaguchi Y, Kovacs JJ, McLaurin A, Vance JM, Ito A, Yao TP. The deacetylase HDAC6 regulates aggresome formation and cell viability in response to misfolded protein stress. *Cell* 2003;**115**:727–738.

Kegel K, Kim M, Sapp E, McIntyre C, Castano J, Aronin N, Difiglia M. Huntingtin expression stimulates endosomal-lysosomal activity, endosome tubulation, and autophagy. *J Neurosci* 2000;**20**:7268–7278.

Kiffin R, Christian C, Knecht E, Cuervo AM. Activation of chaperone-mediated autophagy during oxidative stress. *Mol Biol Cell* 2004;**15**:4829–4840.

Kihara A, Kabeya Y, Ohsumi Y, Yoshimori T. Beclin-phosphatidylinositol 3-kinase complex functions at the trans-Golgi network. *EMBO Rep* 2001;**2**:330–335.

Kim J, *et al.* Convergence of multiple autophagy and cytoplasm to vacuole targeting components to a perivacuolar membrane compartment prior to de novo vesicle formation. *J Biol Chem* 2002;**277**:763–773.

Kim PK, *et al.* Ubiquitin signals autophagic degradation of cytosolic proteins and peroxisomes. *Proc Natl Acad Sci USA* 2008;**105**:20567–20574.

Kirkin V, *et al.* A role for NBR1 in autophagosomal degradation of ubiquitinated substrates. *Mol Cell* 2009;**33**:505–516.

Kissová I, *et al.* Selective and non-selective autophagic degradation of mitochondria in yeast. *Autophagy* 2007;**3**:329–336.

Kitada T, *et al.* Mutations in the parkin gene cause autosomal recessive juvenile parkinsonism. *Nature* 1998;**392**:605–608.

Klionsky DJ. Autophagy: from phenomenology to molecular understanding in less than a decade. *Nat Rev Mol Cell Biol* 2007;**8**:931–937.

Klionsky DJ, Emr SD. Autophagy as a regulated pathway of cellular degradation. *Science* 2000;**290**:1717–1721.

Komatsu M, *et al.* Impairment of starvation-induced and constitutive autophagy in Atg7-deficient mice. *J Cell Biol* 2005;**169**:425–434.

Komatsu M, *et al.* Loss of autophagy in the central nervous system causes neurodegeneration in mice. *Nature* 2006;**441**:880–884.

Komatsu M, *et al.* Essential role for autophagy protein Atg7 in the maintenance of axonal homeostasis and the prevention of axonal degeneration. *Proc Natl Acad Sci USA* 2007a;**104**:14489–14494.

Komatsu M, *et al.* Homeostatic levels of p62 control cytoplasmic inclusion body formation in autophagy-deficient mice. *Cell* 2007b;**131**:1149–1163.

Kopito R. Aggresomes, inclusion bodies and protein aggregation. *Trends Cell Biol* 2000;**10**:523–530.

Kraft C, *et al.* Mature ribosomes are selectively degraded upon starvation by an autophagy pathway requiring the Ubp3p/Bre5p ubiquitin protease. *Nat Cell Biol* 2008;**10**:602–610.

Krick R, *et al.* Piecemeal microautophagy of the nucleus requires the core macroautophagy genes. *Mol Biol Cell* 2008;**19**:4492–4505.

Kruger R, *et al.* Ala30Pro mutation in the gene encoding alpha-synuclein in Parkinson's disease. *Nat Genet* 1998;**18**:106–108.

Kuma A, *et al.* The role of autophagy during the early neonatal starvation period. *Nature* 2004;**432**:1032–1036.

Lashuel HA, Petre BM, Wall J, Simon M, Nowak RJ, Walz T, Lansbury PT Jr. Alpha-synuclein, especially the Parkinson's disease-associated mutants, forms pore-like annular and tubular protofibrils. *J Mol Biol* 2002;**322**:1089–1102.

Lee JY, *et al.* HDAC6 controls autophagosome maturation essential for ubiquitin-selective quality-control autophagy. *EMBO J* 2010;**29**:969–980.

Lemasters JJ, *et al.* Role of mitochondrial inner membrane permeabilization in necrotic cell death, apoptosis, and autophagy. *Antioxid Redox Signal* 2002;**4**:769–781.

Leroy E, *et al.* The ubiquitin pathway in Parkinson's disease. *Nature* 1998;**395**:451–452.

Lim KL, Chew KC, Tan JM, Wang C, Chung KK, Zhang Y, Tanaka Y, Smith W, Engelender S, Ross CA, Dawson VL, Dawson TM. Parkin mediates nonclassical, proteasomal-independent ubiquitination of synphilin-1: implications for Lewy body formation. *J Neurosci* 2005;**25**:2002–2009.

Liu Z, *et al.* Membrane-associated farnesylated UCH-L1 promotes α-synuclein neurotoxicity and is a therapeutic target for Parkinson's disease. *Proc Natl Acad Sci USA* 2008;**106**:4635–4640.

Majeski A, Dice J. Mechanisms of chaperone-mediated autophagy. *Int J Biochem Cell Biol* 2004;**36**:2435–2444.

Martinez-Vicente M, Cuervo AM. Autophagy and neurodegeneration: when the cleaning crew goes on strike. *Lancet Neurol* 2007;**6**:352–361.

Martinez-Vicente M, *et al.* Dopamine-modified alpha-synuclein blocks chaperone-mediated autophagy. *J Clin Invest* 2008;**118**:777–788.

Massey A, Zhang C, Cuervo A. Chaperone-mediated autophagy in aging and disease. *Curr Top Dev Biol* 2006a;**73**:205–235.

Massey AC, *et al.* Consequences of the selective blockage of chaperone-mediated autophagy. *Proc Natl Acad Sci USA* 2006b;**103**:5805–5810.

McNaught KS, *et al.* Impairment of the ubiquitin-proteasome system causes dopaminergic cell death and inclusion body formation in ventral mesencephalic cultures. *J Neurochem* 2002;**81**:301–306.

McNaught KS, Perl DP, Brownell AL, Olanow CW. Systemic exposure to proteasome inhibitors causes a progressive model of Parkinson's disease. *Ann Neurol* 2004;**56**:149–162.

Mishra RS, Bose S, Gu Y, Li R, Singh N. Aggresome formation by mutant prion proteins: the unfolding role of proteasomes in familial prion disorders. *J Alzheimer's Dis* 2003;**5**:15–23.

Mizushima N. The pleiotropic role of autophagy: from protein metabolism to bactericide. *Cell Death Differ* 2005;**12**:1535–1541.

Mizushima N, Levine B, Cuervo AM, Klionsky DJ. Autophagy fights disease through cellular self-digestion. *Nature* 2008;**451**:1069–1074.

Mizushima N, Ohsumi Y, Yoshimori T. Autophagosome formation in mammalian cells. *Cell Struct Funct* 2002;**27**:421–429.

Mortimore G, *et al.* Regulation of microautophagy and basal protein turnover in rat liver. Effects of short-term starvation. *J Biol Chem* 1988;**263**:2506–2512.

Nakai A, *et al.* The role of autophagy in cardiomyocytes in the basal state and in response to hemodynamic stress. *Nat Med* 2007;**13**:619–624.

Narendra D, Tanaka A, Suen DF, Youle RJ. Parkin is recruited selectively to impaired mitochondria and promotes their autophagy. *J Cell Biol* 2008;**183**:795–803.

Navon A, Goldberg AL. Proteins are unfolded on the surface of the ATPase ring before transport into the proteasome. *Mol Cell* 2001;**8**:1339–1349.

Nixon RA. Endosome function and dysfunction in Alzheimer's disease and other neurodegenerative diseases. *Neurobiol Aging* 2005;**26**:373–382.

Nixon RA. Autophagy, amyloidogenesis and Alzheimer disease. *J Cell Sci* 2007;**120**:4081–4091.

Noda T, Suzuki K, Ohsumi Y. Yeast autophagosomes: de novo formation of a membrane structure. *Trends Cell Biol* 2002;**12**:231–235.

Ohsumi Y, Mizushima N. Two ubiquitin-like conjugation systems essential for autophagy. *Semin Cell Dev Biol* 2004;**15**:231–236.

Olanow CW, Perl DP, DeMartino GN, McNaught KS. Lewy-body formation is an aggresome-related process: a hypothesis. *Lancet Neurol* 2004;**3**:496–503.

Olzmann JA, Li L, Chudaev MV, Chen J, Perez FA, Palmiter RD, Chin LS. Parkin-mediated K63-linked polyubiquitination targets misfolded DJ-1 to aggresomes via binding to HDAC6. *J Cell Biol* 2007;**178**:1025–1038.

Pandey UB, *et al.* HDAC6 rescues neurodegeneration and provides an essential link between autophagy and the UPS. *Nature* 2007;**447**:860–864.

Pankiv S, *et al.* p62/SQSTM1 binds directly to Atg8/LC3 to facilitate degradation of ubiquitinated protein aggregates by autophagy. *J Biol Chem* 2007;**282**:24131–24145.

Park YE, *et al.* Autophagic degradation of nuclear components in mammalian cells. *Autophagy* 2009;**5**:795–804.

Pickart CM, Cohen RE. Proteasomes and their kin: proteases in the machine age. *Nat Rev Mol Cell Biol* 2004;**5**:177–187.

Pickford F, *et al.* The autophagy-related protein beclin 1 shows reduced expression in early Alzheimer disease and regulates amyloid beta accumulation in mice. *J Clin Invest* 2008;**118**:2190–2199.

Qu X, *et al.* Promotion of tumorigenesis by heterozygous disruption of the beclin-1 autophagy gene. *J Clin Invest* 2003;**112**:1809–1820.

Ravikumar B, Duden R, Rubinsztein D. Aggregate-prone proteins with polyglutamine and polyalanine expansions are degraded by autophagy. *Hum Mol Genet* 2002;**11**:1107–1117.

Ravikumar B, *et al.* Inhibition of mTOR induces autophagy and reduces toxicity of polyglutamine expansions in fly and mouse models of Huntington disease. *Nat Genet* 2004;**36**:585–595.

Ravikumar B, *et al.* Dynein mutations impair autophagic clearance of aggregate-prone proteins. *Nat Genet* 2005;**37**:771–776.

Ravikumar B, *et al.* Rab5 modulates aggregation and toxicity of mutant huntingtin through macroautophagy in cell and fly models of Huntington disease. *J Cell Sci* 2008;**121**:1649–1660.

Rideout HJ, Lang-Rollin IC, Savalle M, Stefanis L. Dopaminergic neurons in rat ventral midbrain cultures undergo selective apoptosis and form inclusions, but do not up-regulate iHSP70, following proteasomal inhibition. *J Neurochem* 2005;**93**:1304–1313.

Roberts P, Moshitch-Moshkovitz S, Kvam E, O'Toole E, Winey M, Goldfarb D. Piecemeal microautophagy of nucleus in *Saccharomyces cerevisiae*. *Mol Biol Cell* 2003;**14**:129–141.

Rott R, *et al.* Monoubiquitylation of alpha-synuclein by seven in absentia homolog (SIAH) promotes its aggregation in dopaminergic cells. *J Biol Chem* 2008;**283**:3316–3328.

Rubinsztein DC. Lessons from animal models of Huntington's disease. *Trends Genet* 2002;**18**:202–209.

Rubinsztein D. The roles of intracellular protein-degradation pathways in neurodegeneration. *Nature* 2006;**443**:780–786.

Rubinsztein DC, Gestwicki JE, Murphy LO, Klionsky DJ. Potential therapeutic applications of autophagy. *Nat Rev Drug Discov* 2007;**6**:304–312.

Sakai Y, *et al.* Peroxisome degradation by microautophagy in *Pichia pastoris*. Identification of specific steps and morphological intermediates. *J Cell Biol* 1998;**141**:625–636.

Salvador N, Aguado C, Horst M, Knecht E. Import of a cytosolic protein into lysosomes by chaperone-mediated autophagy depends on its folding state. *J Biol Chem* 2000;**275**:27447–27456.

Sandoval H, *et al.* Essential role for Nix in autophagic maturation of erythroid cells. *Nature* 2008;**454**:232–236.

Sarkar S, Rubinsztain DC. Small molecule enhancers of autophagy for neurodegenerative diseases. *Mol Biosyst* 2008;**4**:895–901.

Savitt JM, Dawson VL, Dawson TM. Diagnosis and treatment of Parkinson disease: molecules to medicine. *J Clin Invest* 2006;**116**:1744–1754.

Schubert U, Anton LC, Gibbs J, Norbury CC, Yewdell JW, Bennink JR. Rapid degradation of a large fraction of newly synthesized proteins by proteasomes. *Nature* 2000;**404**:770–774.

Shibata M, *et al.* Regulation of intracellular accumulation of mutant Huntingtin by Beclin 1. *J Biol Chem* 2006;**281**:14474–14485.

Singh R, *et al.* Autophagy regulates lipid metabolism. *Nature* 2009;**458**:1131–1135.

Singleton AB, *et al.* alpha-Synuclein locus triplication causes Parkinson's disease. *Science* 2003;**302**: 841.

Snyder H, Mensah K, Theisler C, Lee J, Matouschek A, Wolozin B. Aggregated and monomeric alpha-synuclein bind to the S 6′ proteasomal protein and inhibit proteasomal function. *J Biol Chem* 2003;**278**:11753–11759.

Spillantini MG, Schmidt ML, Lee VM, Trojanowski JQ, Jakes R, Goedert M. Alpha-synuclein in Lewy bodies. *Nature* 1997;**388**:839–840.

Stefanis L, Larsen KE, Rideout HJ, Sulzer D, Greene LA. Expression of A53T mutant but not wild-type alpha-synuclein in PC12 cells induces alterations of the ubiquitin-dependent degradation system, loss of dopamine release, and autophagic cell death. *J Neurosci* 2001;**21**:2549–2560.

Takashi Y, *et al.* Bif-1 interacts with Beclin 1 through UVRAG and regulates autophagy and tumorigenesis. *Nat Cell Biol* 2007;**9**:1142–1151.

Tan JM, Wong ES, Dawson VL, Dawson TM, Lim KL. Lysine 63-linked polyubiquitin potentially partners with p62 to promote the clearance of protein inclusions by autophagy. *Autophagy* 2008a;**4**:251–253.

Tan JM, Wong ES, Kirkpatrick DS, Pletnikova O, Ko HS, Tay SP, Ho MW, Troncoso J, Gygi SP, Lee MK, Dawson VL, Dawson TM, Lim KL. Lysine 63-linked ubiquitination promotes the formation and autophagic clearance of protein inclusions associated with neurodegenerative diseases. *Hum Mol Genet* 2008b;**17**:431–439.

Verhoef LG, Lindsten K, Masucci MG, Dantuma NP. Aggregate formation inhibits proteasomal degradation of polyglutamine proteins. *Hum Mol Genet* 2002;**11**:2689–2700.

Vives-Bauza C, *et al.* PINK1-dependent recruitment of Parkin to mitochondria in mitophagy. *Proc Natl Acad Sci USA* 2010;**107**:378–383.

Vogiatzi T, *et al.* Wild type alpha-synuclein is degraded by chaperone-mediated autophagy and macroautophagy in neuronal cells. *J Biol Chem* 2008;**283**:23542–23556.

Wang CW, Klionsky DJ. The molecular mechanism of autophagy. *Mol Med* 2003;**9**:65–76.

Wang Y, Martinez-Vicente M, Krüger U, Kaushik S, Wong E, Mandelkow EM, Cuervo AM, Mandelkow E. Tau fragmentation, aggregation and clearance: the dual role of lysosomal processing. *Hum Mol Genet* 2009;**18**:4153–4170.

Wang C, *et al.* Stress-induced alterations in parkin solubility promote parkin aggregation and compromise parkin's protective function. *Hum Mol Genet* 2005;**14**:3885–3897.

Webb JL, Ravikumar B, Atkins J, Skepper J, Rubinsztein D. Alpha-synuclein is degraded by both autophagy and the proteasome. *J Biol Chem* 2003;**278**:25009–25013.

Wing S, *et al.* Proteins containing peptide sequences related to KFERQ are selectively depleted in liver ans heart, but not skeletal muscle, of fasted rats. *Biochem J* 1991;**275**:165–169.

Wong ES, Tan JM, Soong WE, Hussein K, Nukina N, Dawson VL, Dawson TM, Cuervo AM, Lim KL. Autophagy-mediated clearance of aggresomes is not a universal phenomenon. *Hum Mol Genet* 2008;**17**:2570–2582.

Yang Q, *et al.* Regulation of neuronal survival factor MEF2D by chaperone-mediated autophagy. *Science* 2009;**323**:124–127.

Yokota S. Degradation of normal and proliferated peroxisomes in rat hepatocytes: regulation of peroxisomes quantity in cells. *Microsc Res Tech* 2003;**61**:151–160.

Yorimitsu T, Klionsky DJ. Autophagy: molecular machinery for self-eating. *Cell Death Differ* 2005;**12**:1542–1552.

Yu WH, *et al.* Autophagic vacuoles are enriched in amyloid precursor protein-secretase activities: implications for beta-amyloid peptide over-production and localization in Alzheimer's disease. *Int J Biochem Cell Biol* 2004;**36**:2531–2540.

Yu WH, *et al.* Macroautophagy-a novel β-amyloid peptide-generating pathway activated in Alzheimer's disease. *J Cell Biol* 2005;**171**:87–98.

Zarranz JJ, *et al.* The new mutation, E46K, of alpha-synuclein causes Parkinson and Lewy body dementia. *Ann Neurol* 2004;**55**:164–173.

Zhang C, Cuervo AM. Restoration of chaperone-mediated autophagy in aging liver improves cellular maintenance and hepatic function. *Nat Med* 2008;**14**:959–965.

Zhong Y, *et al.* Distinct regulation of autophagic activity by Atg14L and Rubicon associated with Beclin 1-phosphatidylinositol-3-kinase complex. *Nat Cell Biol* 2009;**11**:468–476.

Zhou D, *et al.* Lamp-2a facilitates MHC class II presentation of cytoplasmic antigens. *Immunity* 2005;**22**:571–581.

4

CHAPERONE AND ANTI-CHAPERONE PROPERTIES OF SYNUCLEIN: IMPLICATIONS FOR DEVELOPMENT, AGING, AND NEURODEGENERATIVE DISEASE

Makoto Hashimoto, Kazuanri Sekiyama, Akio Sekigawa, and Masayo Fujita

Laboratory for Chemistry and Metabolism, Tokyo Metropolitan Institute for Neuroscience, Tokyo, Japan

4.1 INTRODUCTION

Neurodegenerative diseases such as Alzheimer's disease (AD) and Parkinson's disease (PD) are characterized by various histopathological features, including intracellular and extracellular amyloid deposits, axonal swelling, synaptic loss,

Protein Chaperones and Protection from Neurodegenerative Diseases, First Edition.
Edited by Stephan N. Witt.

glial activation, and neuronal cell death. The precise mechanism of neurodegeneration is unclear, but mounting evidence has shown that protein aggregation plays a central role in the pathogenesis (Hashimoto and Masliah, 1999). The current prevailing view is that gene mutations and/or environmental factors cause disease-specific amyloidogenic proteins such as amyloid-β (Aβ) in AD and α-synuclein (α-syn) in PD to misfold and aggregate to form neurotoxic protofibrils, leading to neurodegeneration (Hashimoto and Masliah, 1999).

It is well established that chaperones play a protective role against neurodegeneration. In support of this concept, overexpression of the molecular chaperone Hsp70 results in amelioration of various models of neurodegenerative diseases, including PD (Witt, 2010). When chaperones cannot repair the misfolded proteins, they stimulate degradation of these proteins through the ubiquitin–proteasome system (UPS) or the autophagy–lysosomal pathway (Dice, 2007; Murata *et al.*, 2009). Mutations that disturb the activity of molecular chaperones or UPS-associated enzymes can cause neurodegeneration (Olanow and McNaught, 2006), and analysis of postmortem brain samples from PD patients shows a significant decrease in proteasome activity in the substantia nigra compared to non-PD brains (McNaught and Jenner, 2001). Thus, these results show that chaperones play an essential role in the defense against neurotoxic protein aggregates in neurodegenerative diseases.

Such a conventional view suggests that disease-specific amyloidogenic proteins may be simply substrates of chaperones, and that the aggregation of amyloidogenic proteins depends on the activity of the chaperones (Figure 4.1a). In young brains, chaperones are active enough to cope with misfolding of the amyloidogenic proteins. During the course of aging, the activities of chaperones decrease and misfolded proteins gradually accumulate, resulting in aggregation and formation of amyloid fibrils. In this view, the roles of chaperones are more important than those of amyloidogenic proteins. Thus, aggregation of amyloidogenic proteins is a result, rather than a cause, of decreased chaperone function in aging and neurodegenerative diseases. This scheme is simple and clear, and has been established on the basis of a plethora of data, but it does not provide insight beyond the idea that amyloidogenic proteins, as well as other aggregation-prone proteins, are passive substrates of chaperones.

4.2 CHAPERONE AND ANTI-CHAPERONE ACTIVITIES OF AMYLOIDOGENIC PROTEINS

Contrary to the conventional view, the main goal of this chapter is to apply the concept of chaperone and anti-chaperone to various actions of disease-specific amyloidogenic proteins. Indeed, this concept was primarily described for endoplasmic reticulum (ER) chaperones, including protein disulfide isomerase (PDI) and binding immunoglobulin protein (BiP) (Puig and Gilbert, 1994a,b). Since these ER chaperone proteins at low concentrations stimulate protein aggregation, it was interpreted that retaining misfolded and aggregated proteins in ER through

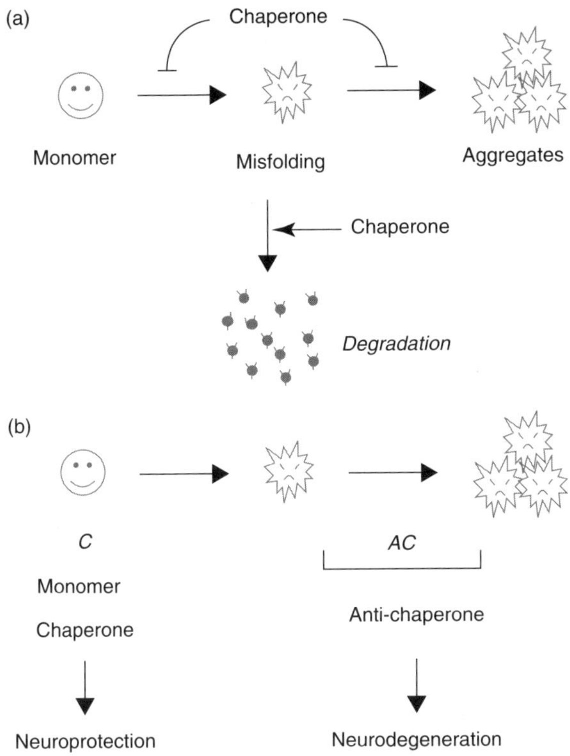

Figure 4.1 Neurodegenerative disease-specific amyloidogenic proteins are chaperones and anti-chaperones. (a) Chaperones such as HSPs inhibit misfolding and aggregation of amyloidogenic proteins and stimulate degradation of these proteins. Disease-specific amyloidogenic proteins are passive substrates of chaperones, and failure of chaperone regulation of the amyloidogenic proteins due to causes such as aging, gene mutation, and environmental factors may lead to neurodegeneration. (b) The positive actions of disease-specific amyloidogenic proteins are illustrated. These proteins are chaperones and anti-chaperones themselves, and may participate in regulation of neuroprotection and neurodegeneration.

anti-chaperone actions was beneficial under certain circumstances, such as adenosine triphosphate (ATP) depletion and redox imbalance of cells. Similar chaperone and anti-chaperone activities were later described for *Escherichia coli* trigger factor, a well-characterized peptidyl-prolyl cis-trans isomerase (Huang *et al.*, 2002). Thus, it is important to note that anti-chaperone activities of some proteins might have evolutionarily developed in terms of physiological context.

In this chapter, we suggest that a number of biological functions of synuclein proteins can be attributed to chaperone and anti-chaperone activities. Considering that the precise mechanism of chaperone and anti-chaperone actions of synuclein proteins is still obscure at the moment, the concept of chaperone and anti-chaperone refers to inhibitory and stimulatory activities on protein aggregation, respectively. Accordingly, anti-chaperones are used to represent all types of aggregates rather than specific populations of toxic aggregates. Although aggregation of α-syn may play a central role in the pathogenesis of synucleinopathies, including PD and dementia with Lewy bodies (DLB) (Trojanowski *et al.*, 1998), this molecule is protective against neurodegeneration caused by deletion of cysteine-string protein-α (CSP-α), the presynaptic co-chaperone to Hsc70, in the premature stage (Chandra *et al.*, 2005). Thus, α-syn is not simply a chaperone substrate but a chaperone itself. A recent study suggested that similar to α-syn, β-synuclein (β-syn) and γ-synuclein (γ-syn) may also have chaperone and anti-chaperone activities. β-syn protects against neurotoxicity caused by α-syn through chaperone-like activity (Hashimoto *et al.*, 2001; Uversky *et al.*, 2002); two missense mutations of β-syn have been identified in familial and sporadic DLB (Ohtake *et al.*, 2004), and wild-type β-syn can be induced to form fibril structures *in vitro* (Yamin *et al.*, 2005). The chaperone activity of γ-syn may be important for cancer progression, whereas transgenic (tg) mice overexpressing γ-syn exhibit neurodegeneration phenotypes (Ninkina *et al.*, 2009). Collectively, these results suggest that synuclein family members are not just passive substrates of chaperones, but behave as chaperones and anti-chaperones (Figure 4.1b). Owing to these dual activities, synuclein proteins may be actively involved in both neuroprotection and neurodegeneration.

We further suggest that the dual chaperone and anti-chaperone activities may be applicable not only to synuclein proteins but also to other amyloidogenic proteins that are causative for various neurodegenerative diseases, including AD, tauopathies, Huntington's disease (HD), and Prion disease. Moreover, a similar concept could be used for small HSPs (sHSPs), including HSP27 and α-crystallins, in which missense mutations have been identified in several protein conformation diseases, such as peripheral neuropathies and cataract. Finally, we discuss the biological role of the dual chaperone and anti-chaperone activities of amyloidogenic proteins in terms of the regulation of aging. Conversion between the chaperone and anti-chaperone of amyloidogenic proteins may be linked to many aging associated pathologies, including mitochondrial dysfunction, decreased activity of protein degradation, and glial activation. Furthermore, the conversion of amyloidogenic proteins and aging are either delayed or stimulated by various common mechanisms, including signaling pathways. On the basis of these observations, we speculate that amyloidogenic proteins may be actively involved in the regulation of aging and neurodegenerative disease.

4.2.1 Role of α-Syn as an Anti-chaperone in Neurodegenerative Disease

4.2.1.1 Synuclein Family of Peptides. α-Syn is a presynaptic molecule of ~140 amino acids, which was identified independently from AD brain, mouse brain, and song bird (reviewed in Hashimoto and Masliah (1999)). α-Syn belongs to the synuclein family of peptides, which contains two other members, β- and γ-syn (Figure 4.2). β-Syn was originally discovered as a phosphoprotein PNP14 in bovine brain, while γ-syn was reported independently as a breast cancer-specific

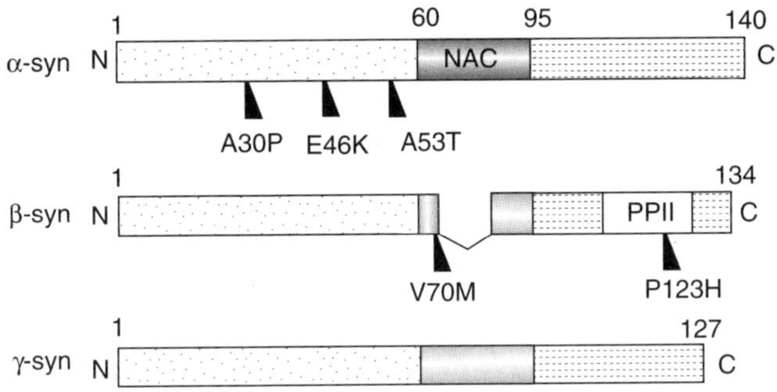

Figure 4.2 Schematic of the structure of synuclein proteins. Synucleins are composed of N-terminal basic regions with a repeating motif (KTKEGV) and divergent C-terminal acidic regions. The central region of α-syn includes a strongly hydrophobic NAC domain which has been purified from Alzheimer's brain, whereas the majority of the corresponding region in β-syn is naturally deleted and the same region in γ-syn is less hydrophobic. β-Syn is characterized by the presence of a PPII sequence in the C-terminal region. For α-syn, two mutations (A53T and A30P) have been found in familial PD, while another mutation (E46K) was later identified in familial DLB. For β-syn, two mutations (V70M and P123H) have been reported in sporadic and familial DLB, respectively.

gene product, persyn and synoretin (Hashimoto and Masliah, 1999). Compared to γ-syn, β-syn is more homologous with α-syn. Furthermore, both α-syn and β-syn are abundantly expressed in the central nervous system, whereas γ-syn is mainly expressed in the peripheral tissues. Structurally, the synuclein proteins share a highly homologous N-terminal region with a unique amino acid repeating motif (KTKEGV) and a less conserved C-terminal acidic region (Figure 4.2). The most striking feature in these proteins is that α-syn, but not β- or γ-syn, possesses the hydrophobic *n*on-β-*a*myloid *c*omponent of the AD amyloid (NAC) domain, which has previously been purified from homogenates of AD brain. The NAC sequence is essential for β-sheet formation and formation of amyloid fibrils, but it remains unclear whether α-syn plays a causative or secondary role in the pathogenesis of typical AD.

4.2.1.2 α-Syn Mutations Are Linked to Lewy Body Disease. The involvement of α-synuclein in the pathogenesis of neurodegenerative disease was demonstrated when two separate missense mutations (A53T, A30P) were discovered in the *α-syn* gene in kindreds with early-onset familial PD (Kruger *et al.*, 1998; Polymeropoulos *et al.*, 1997). Another missense mutation (E46K) was later linked to familial DLB (Zarranz *et al.*, 2004). In addition to the amino acid substitutions, multiplications of wild-type α-syn were shown to cause PD in rare families (Singleton *et al.*, 2003), indicating that increased expression of wild-type α-syn is sufficient to cause parkinsonism. Following the discovery of *α-syn* gene mutations, numerous histopathological studies have detected α-syn immunoreactivities in Lewy bodies and dystrophic Lewy neuritis in both familial and sporadic cases of PD and DLB (Figure 4.3) (Trojanowski *et al.*, 1998). Subsequently,

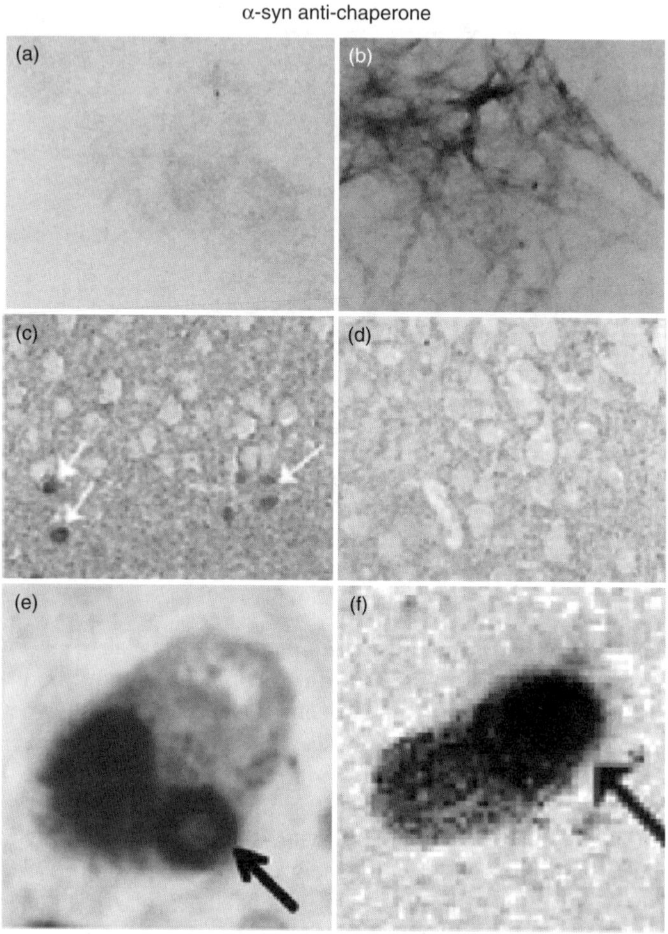

Figure 4.3 Anti-chaperone actions of α-syn in various experimental models. (a, b) Ultrastructural analysis of α-syn aggregates. Incubation of α-syn recombinant protein (a) at pH 6.9 at 65°C results in fibril formation (b). *Source:* reprinted from Hashimoto *et al.* (1998) with permission. (c, d) Analysis of α-syn tg-mice brains showing somatodendritic accumulation of α-syn in α-syn mice (c), but not in non-tg mice (d). *Source:* reprinted from Masliah *et al.* (2000) with permission. (e, f) Analysis of DLB brains showing Lewy bodies with typical α-syn immunoreactivity (e) and double labeling for α-syn and cytochrome-*c* (f). (e) — *Source:* reprinted from Wei *et al.* (2007) with permission. (f) — *Source*: reprinted from Hashimoto *et al.* (1999b) with permission. (*A full color version of this figure appears in the color plate section.*)

immunoreactivities for α-syn in Lewy bodies has been found in other Lewy body diseases, including familial AD with amyloid precursor protein (APP) or presenilin-1 mutations, Lewy body variant of AD, and neurodegeneration with brain iron accumulation type I (NBIA 1, formerly known as *Hallervorden–Spatz syndrome*) (Galvin *et al.*, 2000; Lippa *et al.*, 1998). Moreover, immunoreactivity for α-syn has been observed in glial cell inclusions, the hallmark of multiple system atrophy (Wakabayashi *et al.*, 1998). Consequently, these diseases have been categorized as α-synucleinopathies and constitute a major class of neurodegenerative disease, together with tauopathies and TDP-43 proteinopathies.

4.2.1.3 Biochemical Analysis of α-Syn Aggregation. Studies performed *in vitro* have characterized the aggregation properties of α-syn. It has been shown that α-syn in solution does not have a stable structure, except for some residual helical structure in the N-terminus of the protein (Weinreb *et al.*, 1996). Owing to the coexistence of a "natively unfolded" structure and the hydrophobic NAC domain, α-syn molecules may tend to self-associate and form amyloid-like fibrils (Figure 4.3). Indeed, aggregation of α-syn is a nucleation-dependent process similar to that for other amyloidogenic proteins such as Aβ and prion proteins (PrPs) (Wood *et al.*, 1999). Aggregation of α-syn can be induced by conditions such as an increased time lag, high temperature, low pH, and oxidative stress (Hashimoto *et al.*, 1998). Formation of α-syn aggregates is also stimulated by molecules such as dopamine (Conway *et al.*, 2001), Aβ peptides (Yoshimoto *et al.*, 1995), and polyunsaturated fatty acids (Sharon *et al.*, 2003). Importantly, α-syn mutants have a greater propensity for self-association and aggregation compared to wild-type α-syn (El-Agnaf *et al.*, 1998; Greenbaum *et al.*, 2005). The A30P mutant also exhibits impaired phospholipid binding compared to wild type α-syn and other mutants (Perrin *et al.*, 2000).

It has gradually become clear that neurotoxicity in the pathogenesis of neurodegenerative disorders is due to intermediates of protein aggregates, such as oligomers and protofibrils, rather than to mature fibrils (Lansbury, 1999). In this context, two pathogenic mutations (A53T and A30P) relative to wild-type α-syn have been shown to promote formation of annular protofibrils and result in increased formation of pore-like structures in membranes, as observed for other amyloidogenic proteins such as Aβ and amylin (Lashuel *et al.*, 2002). Thus, membrane disruption via formation of protofibrils could be a possible neurotoxic mechanism of α-syn.

4.2.1.4 Tg Mice Model of Synucleinopathies. Tg mice and drosophila overexpressing wild-type or mutant forms of α-syn have been produced to examine the pathogenic role of α-syn *in vivo*, and these rodent and fly models recapitulate a PD-like pathology (Feany and Bender, 2000; Masliah *et al.*, 2000). Masliah and colleagues produced the first α-syn tg mouse model using the platelet-derived growth factor-β promoter (Masliah *et al.*, 2000). Histological analyses showed formation of granular inclusions of α-syn in neuronal cell bodies and neuritis, which were also immunoreactive with ubiquitin in different brain regions (Figure 4.3). Furthermore, these tg mice showed impaired locomotor performance on the rotarod, in addition to a reduction of tyrosine hydrolase-positive terminals in the striatum, which are features reminiscent of PD. Subsequently, van der Putten *et al.* generated Thy1 promoter-driven tg mice that expressed wild-type or mutant (A53T and A30P) α-syn (van der Putten *et al.*, 2000). In these mice, α-syn accumulated as granular deposits in somatodendritic regions that were biochemically detected in detergent-insoluble fractions. Moreover, these mice displayed rapid deterioration of rotarod performance and a severe locomotor phenotype, which could be ascribed to degeneration of neuromuscular junctions. Many other groups have generated α-syn tg mice using different promoters and these models have been useful for investigation of the pathogenesis of synucleinopathies. However, current α-syn tg mouse models are not perfect replicas of PD since neither neuronal loss nor Lewy body formation is evident, and further work is required to improve these models.

4.2.2 Role of α-syn as a Presynaptic Chaperone

4.2.2.1 Chaperone Activities of α-Syn In Vitro. The native unfolded structure of α-syn in solution suggests that α-syn is a chaperone-like protein (Weinreb *et al.*, 1996), and this concept has been greatly advanced since Wolozin and colleagues first described the similarity of α-syn with 14-3-3 proteins (Ostrerova *et al.*, 1999). Curiously, the N-terminal portion of α-syn (residues 1–61) shares 40% amino acid homology with the 14-3-3 proteins, which are ubiquitously expressed in the brain and have been shown to associate with protein kinase C (PKC), BCL2-antagonist of cell death (BAD), extracellular signal-regulated kinases (ERK), and Raf-1 (Broadie *et al.*, 1997). Similarly, cytoplasmic molecules, including 14-3-3, PKC, BAD, ERK, and the microtubule-associated protein tau, but not Raf-1 (Ostrerova *et al.*, 1999). The interaction of α-syn with these signaling molecules led to the proposal that synucleins may function as molecular chaperones, and the chaperone activity of α-syn has subsequently been shown *in vitro* (Kim *et al.*, 2000; Souza *et al.*, 2000). Thus, α-syn protects glutathione *S*-transferase and aldolase from heat-induced precipitation, and inhibits dithiothreitol-induced precipitation ofα-lactalbumin and bovine serum albumin. Consistent with the chaperone activity of α-syn *in vitro*, overexpression of α-syn protects neuronal cells against apoptotic stimuli, including hydrogen peroxide and 1-methyl-4-phenyl-1,2,3,6-tetrahydropyridine (MPTP) (da Costa *et al.*, 2000).

4.2.2.2 Chaperone Activities of α-Syn In Vivo. *In vivo*, the chaperone activity of α-syn was demonstrated in cross-breeding experiments of CSP-α knockout (KO) mice with α-syn tg mice or α-syn KO mice (Chandra *et al.*, 2005). CSP-α is thought to act as a presynaptic chaperone that recruits a co-chaperone, Hsc70, to misfolded proteins via an interaction with the J domain of CSP-α (reviewed in (Evans *et al.*, 2003)). In this way, CSP-α maintains the conformational integrity of proteins involved in presynaptic functions such as exocytosis and vesicle trafficking. CSP-α homologs have a conserved domain structure, and the presence of the J domain indicates that CSP-α is a molecular chaperone, since other J-domain-containing proteins of the HSP40 chaperone family bind to misfolded proteins and recruit a 70-kDa heat shock cognate protein to regulate protein folding. Indeed, CSP-α binds to Hsc70 and stimulates its ATPase activity, thereby preventing aggregation of a denatured protein. This chaperone activity is augmented by a small glutamine-rich tetratricopeptide repeat-containing protein, which serves as a co-chaperone binding partner for both CSP-α and HSC70. This suggests that the trimeric complex acts as a presynaptic chaperone machine.

CSP-α KO mice are characterized by impaired neurotransmission, presynaptic neurodegeneration, and premature death (Fernandez-Chacon *et al.*, 2004). Surprisingly, cross-breeding of CSP-α KO mice with mice overexpressing α-syn mitigated the neuropathology and shortened life span (Chandra *et al.*, 2005). The abnormalities in SNARE proteins in CSP-α KO mice indicate that CSP-α is part of a SNARE-repair chaperone, and wild-type but not A30P mutant α-syn partly rescued this impairment. Conversely, the reverse effects on neuropathology and life span were observed in the double (CSP-α/α-syn) KO mice (Chandra *et al.*, 2005). Collectively, these results suggest that the function of CSP-α in preventing neurodegeneration during the developmental stage in mice is complementary to that of α-syn.

The overlapping functions of CSP-α and α-syn suggest that a chaperone function of α-syn might be involved in the regulation of various processes in exocytosis, such as neurotransmitter synthesis (e.g., gamma-amino butyric acid (GABA)) and vesicle filling, vesicle docking, Ca^{2+}-entry via voltage-dependent Ca^{2+}-channels, and vesicle fusion through interaction with a member of the SNARE complex. Consistent with this, it has been proposed that α-syn may be involved in the regulation of the size of distinct pools of synaptic vesicles in mature neurons (Murphy *et al.*, 2000). Thus, the aggregation of α-syn may result in failure to maintain the fidelity of synaptic membrane traffic, leading to synaptic dysfunction and neurodegeneration.

4.2.3 β-Syn as a Neuroprotective Chaperone

4.2.3.1 β-Syn Inhibits α-Syn Aggregation In Vitro. Despite the strong homology between α- and β-syn, the majority of the corresponding NAC region in β-syn is naturally deleted, suggesting that β-syn may be less prone to aggregate compared to α-syn. In addition, α- and β-syn are colocalized in the presynapse and the expression level of β-syn is comparable to or higher than that of α-syn in many regions in the brain. These results suggest that β-syn could be a negative regulator of α-syn aggregation, and *in vitro* studies have shown that aggregation of α-syn is dose dependently inhibited by β-syn (Hashimoto *et al.*, 2001; Uversky *et al.*, 2002). Subsequently, it was shown under similar cell-free conditions that β-syn inhibits formation of α-syn protofibrils, the most pathogenic α-syn aggregates (Park and Lansbury, 2003). These results suggest that β-syn acts as a chaperone to inhibit aggregation of α-syn. Consistent with this idea, the effect of β-syn on the suppression of heat-induced aggregation of proteins such as aldolase, alcohol dehydrogenase, and citrate synthase is significantly stronger than those of α- and γ-syn (Lee *et al.*, 2004).

4.2.3.2 Role of Polyproline II (PPII) Helices in the Chaperone Activity of β-Syn. The mechanisms through which β-syn acts as a chaperone are elusive, but emerging evidence suggests that the polyproline II (PPII) helix structure might play a key role. (The polyproline helix, referred to as PPII, is a left-handed helix of all trans-proline residues.) Proteins are generally characterized by specific three-dimensional structures that include the α-helix, β-sheet, and β-turn. In addition to these elements, the PPII helix is increasingly recognized as a dominant conformation in the unfolded state of peptides (Creamer and Campbell, 2002). Furthermore, the PPII helix domain is important for protein–protein interactions and has been implicated in diverse biological activities including signal transduction, transcription, cell motility, and immune response (Rath *et al.*, 2005).

Using a combination of high resolution heteronuclear nuclear magnetic resonance (NMR) and other sophisticated techniques, Bertoncini *et al.* characterized the extended conformations in the native state of β-syn (Bertoncini *et al.*, 2007). Despite the lack of defined secondary structure, β-syn adopted transient PPII helix conformations clustered at the C-terminal region. Indeed, the region 105–115 is characterized by the presence of the motif EPLXEPLXEPE, which resembles the proline-rich sequences of proteins implicated in vesicular endocytosis and trafficking. Therefore, the extended conformation and flexibility of the PPII structure of β-syn might provide a chaperone function through interaction with target proteins. The role of PPII helix structures in chaperone-like activities in other amyloidogenic proteins, including tau, APP/Aβ*s* and PrP, is discussed further below.

4.2.3.3 β-Syn Ameliorates α-Syn Neuropathology. The hypothesis that β-syn inhibits aggregation of α-syn through a chaperone-like function has been examined in tg mice. Platelet derived growth factor (PDGF) promoter-driven α-syn tg mice exhibited motor deficits and had Lewy body-like inclusions immunoreactive for α-syn and ubiquitin (Masliah *et al.*, 2000), whereas mice overexpressing β-syn showed no evidence of neurodegeneration (Hashimoto *et al.*, 2001). To determine whether reestablishing a normal balance between α- and β-syn prevents the α-syn pathology, α-syn tg mice were crossed with β-syn tg mice, and bigenic mice overexpressing both α- and β-syn were analyzed (Hashimoto *et al.*, 2001). The results showed significantly decreased formation of Lewy bodies and amelioration of motor function deficits in the bigenic mice compared to those in α-syn tg mice, suggesting that β-syn is protective against the neuropathology caused by α-syn (Figure 4.4).

The mechanism by which β-syn ameliorates α-syn neuropathology is unclear, but it is reasonable to speculate that chaperone functions of β-syn may underlie the prevention of neurotoxicity caused by protofibrils of α-syn. In support of this notion, a chaperone-like interaction of β-syn with α-syn was observed by coimmunoprecipitation in brain homogenates, and similar results were observed using extracts from cell cultures (Hashimoto *et al.*, 2001). The efficiency of binding of α- to β-syn was low, but this is compatible with the concept that the binding

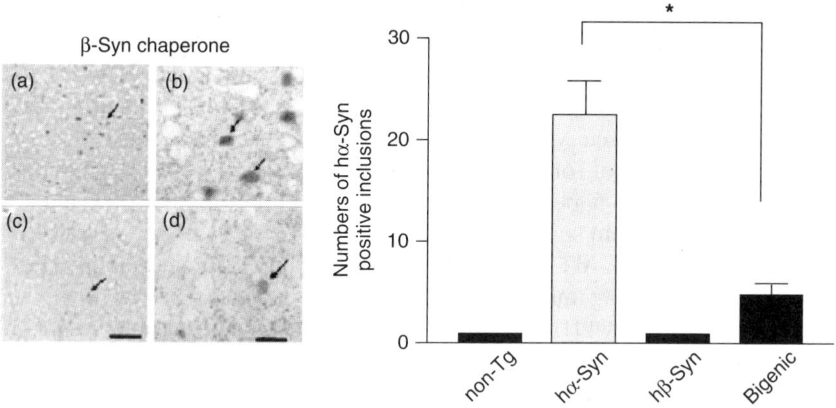

Figure 4.4 β-Syn chaperone ameliorates α-syn pathology in tg mice. Neuropathological analysis of hα-synuclein inclusion formation in tg mice. Sections were obtained from the temporal cortex of 12-month-old mice and labeled with anti-human (h) α-syn. (a, b) Light microscopy showed that anti-hα-syn immunoreactive inclusions were abundant in hα-syn tg mice, whereas (c, d) bigenic mice showed a reduced number of anti-hα-synuclein immunoreactive inclusions. Statistical analysis is shown in the right panel. *Source*: Reprinted from Hashimoto *et al.* (2001) with permission.

of chaperones to target proteins is generally transient and of low affinity. Also of relevance to the chaperone function of β-syn, it has been shown that β-syn has protective effects on the proteasome pathology caused by α-syn. Aggregated α-syn binds to S6', a component of the 19S subunit in the 26S proteasome, and inhibits 26S proteasomal degradation activity (Ghee *et al.*, 2000). β-Syn alone has little effect on proteasomal activity, but coincubation with α-syn antagonized the proteasomal inhibition of aggregated α-syn (Snyder *et al.*, 2005). Thus, β-syn may protect many intracellular organelles from the toxic protofibrils of α-syn.

LaSpada and colleagues have also performed cross-breeding experiments using prion promoter-driven α-syn A53T mice and β-syn mice (Fan *et al.*, 2006). Consistent with the earlier study, overexpression of β-syn was shown to retard the progression of impaired motor performance, reduce α-syn aggregation, and extend survival in double tg mice. In this study, α-syn protein expression was markedly decreased in the cortex of bigenic mice compared to α-syn single tg mice. Since the reduced α-syn protein level was not accompanied by a decrease in α-syn mRNA, a posttranslational mechanism was proposed to account for the decrease in α-syn in the bigenic mice. Further studies are required to determine the detailed mechanism.

4.2.3.4 Upregulation of Akt Activity by β-Syn. Since β-syn has strong chaperone activity, it has been assumed that the neuroprotective actions of β-syn are attributable in part to the stabilizing effects of β-syn on signaling molecules. In this context, the PI3K pathway is a probable candidate because this pathway may also play a protective role in various neurodegenerative disorders, including PD.

It has been well documented that HSPs such as HSP90 and HSP27 coordinately regulate Akt, the central molecule in the PI3K signaling pathway (Brazil *et al.*, 2002), and the chaperone activity of β-syn suggests that it might also regulate Akt activity. In support of this hypothesis, transfection of β-syn into neuroblastoma cells resulted in increased Akt activity and conferred protection of the cells against the neurotoxic effects of rotenone (Hashimoto *et al.*, 2004a). The effects ofbreak β-syn on Akt activity appeared to occur through a direct interaction and were independent of upstream signaling molecules in the PI3K pathway, such as PDK-1 and PI3K. Increased Akt phosphorylation was found in brain homogenates of β-syn overexpressing tg mice, in which Akt coprecipitated preferentially with β-syn but not with α-syn (Hashimoto *et al.*, 2004a). It is also of note that mRNA levels for β-syn are decreased in selectively affected brain regions, such as the superior temporal cortex and substantia nigra, in postmortem DLB brains compared to controls (Rockenstein *et al.*, 2001). This suggests that decreased β-syn expression could underlie progression of the disease.

Collectively, β-syn may be an endogenous chaperone-like molecule that acts in a protective manner against α-syn neuropathology. This indicates that β-syn could have therapeutic potential. To evaluate this possibility, a lentiviral vector expressing human β-syn (lenti-β-syn) was tested in an α-syn tg mouse model (Hashimoto *et al.*, 2004b). Intracerebral injection of lenti-β-syn reduced the formation of α-syn inclusions and ameliorated the neurodegenerative changes in α-syn tg mice. Co-immunoprecipitation and immunoblot experiments showed that the mechanisms of β-syn neuroprotection involved binding to α-syn and Akt, which resulted in decreased aggregation and decreased accumulation of α-syn in the synaptic membrane. Thus, these results suggest that *in vivo* transfer of genes encoding β-syn might provide a novel approach to the development of therapy for PD and related diseases.

4.2.4 β-Syn and Neurodegenerative Disease: β-Syn as an Anti-chaperone

4.2.4.1 Accumulation of Wild-Type β-Syn in Synucleinopathies. Several investigations have suggested that not only α-syn but also β- and γ-syn are involved in the pathogenesis of neurodegenerative disease. Galvin and colleagues showed that although Lewy bodies were not immunopositive for β- or γ-syn, these proteins were abnormally concentrated in dystrophic neurites in the hippocampus region in PD and DLB brains (Galvin *et al.*, 1999). In a similar context, β- and γ-syn immunoreactivities were detected in spheroids but not in Lewy body-like or glial inclusions in neurodegeneration with brain iron accumulation, type 1 (NBIA 1) or Hallervorden–Spatz syndrome (Galvin *et al.*, 2000). Furthermore, β- and γ-syn, but not α-syn, accumulate in axonal spheroid bodies in gracile axonal dystrophy mice, which are characterized by natural truncation of the *UCHL-1* (park5) gene (Wang *et al.*, 2004). Finally, γ-syn tg mice are characterized by the presence of γ-syn-positive spheroids and dystrophic neuritis, most prominently in the spinal cord, leading to the loss of spinal motor neurons (Ninkina *et al.*, 2009).

Taken together, these results suggest that β- and γ-syn are involved in aspects of neuropathology, including neuritic accumulation and axonal degeneration, rather than somatodendritic pathology.

4.2.4.2 *Identification of β-Syn Missense Mutations in DLBD.* In 2004, Ohtake *et al.* identified two missense mutations of β-syn in unrelated diffuse Lewy body disease (DLBD) cases (Ohtake *et al.*, 2004). A valine to methionine substitution at position 70 (V70M) was found in a sporadic DLB case in Japan, and a P123H mutation was identified in familial DLBD cases in Seattle. These amino acid changes occurred at conserved residues in highly conserved regions of β-syn, and did not seem to be single nucleotide polymorphisms. Furthermore, based on cosegregation analysis of the Seattle pedigree, the P123H mutation was shown to be a dominant trait with incomplete penetration.

There are at least two possibilities for the mechanism of the incomplete dominant mode of inheritance of the P123H β-syn mutation. One is that mutant β-syn may interact with and sequester wild-type β-syn, leading to a decrease in the chaperone activity of wild-type β-syn. That is, the β-syn mutant may have dominant negative effects on wild-type β-syn. The alternate but not mutually exclusive possibility is that β-syn mutants might have obtained a toxic gain of function. Consistent with this, we observed that the mutant proteins (P123H and V70M) were prone to self-aggregation and stimulation of β-syn aggregation under cell-free conditions (Wei *et al.*, 2007). Furthermore, wild-type β-syn aggregates form amyloid fibrils in the presence of specific metal ions, glycosaminoglycans, or macromolecular crowding agents (Yamin *et al.*, 2005). Collectively, these results suggest that mutations or environmental factors might underlie aggregation of β-syn, which then leads to enhanced aggregation of other proteins, including α-syn. Histopathological analyses of the Seattle case (P123H mutation) showed α-syn aggregation in Lewy body pathology, but failed to detect β-syn immunoreactivity (Ohtake *et al.*, 2004). These results support the dominant negative mechanism over a toxic gain of function, but further studies may detect evidence for β-syn accumulation in neuritic pathology.

4.2.4.3 *Pathogenic Effects of Mutant β-Syn in Neuroblastoma Cells.* To investigate the pathogenic role of the DLB-linked β-syn mutations (P123H and V70M), these mutants were stably transfected into B103 neuroblastoma cells (Wei *et al.*, 2007). The cells overexpressing mutant β-syn were abundant in various cytoplasmic membranous inclusions resembling the histopathology of lysosomal storage disease (Figure 4.5). In these cells, the inclusion bodies were immunopositive for mutant β-syn and for lysosomal markers, such as cathepsin B and lysosome-associated membrane protein (LAMP-2). Notably, formation of these lysosomal inclusions was strongly stimulated by the coexpression of α-syn. Collectively these results demonstrate that overexpression of mutant (P123H and V70M) β-syn in neuroblastoma cells results in an enhanced lysosomal pathology.

Since a recent study suggested that gangliosides are involved in the pathogenesis of synucleinopathies, we wanted to determine whether the changes

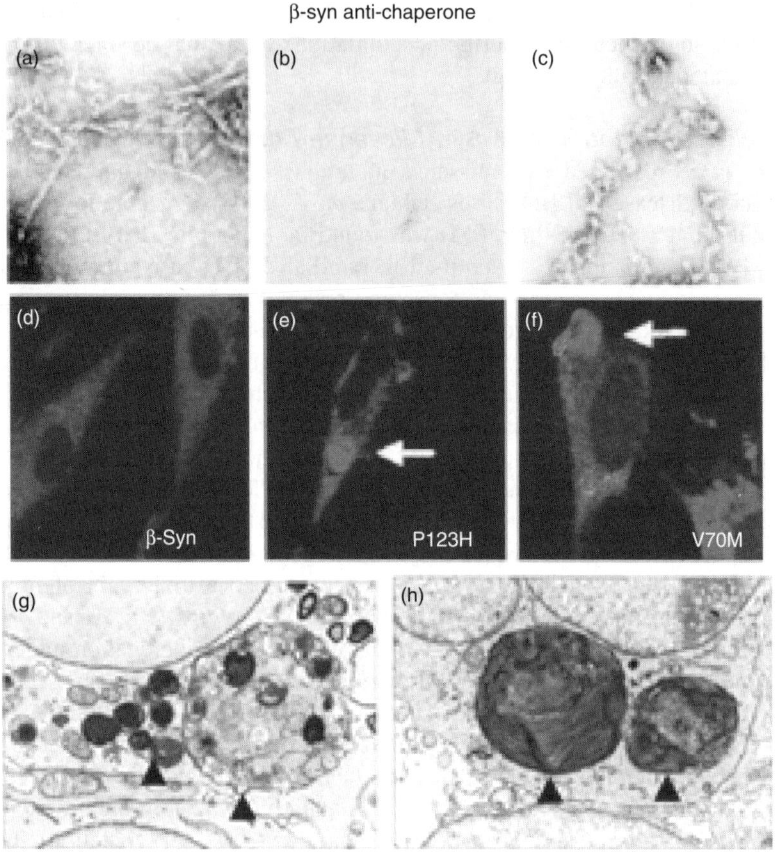

Figure 4.5 DLB-linked β-syn mutations are pathogenic. (a–c) Recombinant proteins of α-syn, β-syn, and DLB-linked P123H β-syn were incubated under acidic conditions for four days, and protein preparations were analyzed by electron microscopy. α-Syn exhibited straight filaments of (a), while the mutant β-syn P123H formed twisted and granular fibrils with diameters of approximately 10 nm (c). No visible aggregates were formed by β-syn (b). (d–h) B103 rat neuroblastoma cells overexpressing β-syn (d), P123H β-syn (e, g, h), or V70M β-syn (f) were analyzed by immunofluorescence (d–f), and electron microscopy (g, h). Immuno-staining with anti-β-syn revealed that P123H β-syn and V70M β-syn, but not β-syn, accumulated in cytoplasmic inclusions (e, f). Ultrastructurally, some large electron-dense inclusions composed of various types of giant autophagosomes were apparent (g and h). The bars represent 20 μm (a–f) and 2 μm (g, h). *Source:* Reprinted from Wei *et al.* (2007) with permission. (*A full color version of this figure appears in the color plate section.*)

in endogenous gangliosides might affect the lysosomal pathology in our cellular model of DLB. To elucidate the mechanism that drives ganglioside-mediated protection of synucleopathies, neuroblastoma cells overexpressing α-syn/P123Hβ-syn were treated with D-threo-1-phenyl-2-decanoylamino -3-morpholino-1-propanol (PDMP), an inhibitor of glycosyl ceramide synthase. The PDMP-treated cells developed lysosomal disease characterized by reduced lysosomal activity, enhanced lysosomal membrane permeability, and decreased expression of lysosomal membrane proteins. PDMP-mediated inhibition of the autophagy–lysosomal pathway resulted in accumulation of synuclein proteins and cellular cytotoxicity. Ganglioside treatment reversed this phenotype, which suggests that gangliosides may protect against the lysosomal pathology of synucleopathies (Wei *et al.*, 2009).

On the basis of this background, it is likely that mutant β-syn tg mice could recapitulate some aspects of DLB pathology, particularly neuritic pathology, as revealed by axonal degeneration and synaptic swelling. Since many lines of α-syn tg mice exhibit somatodendritic accumulation of α-syn, it will be of considerable interest to determine whether combined effects of these different pathological features induced by two synuclein proteins occur in bigenic mice. Thus, a more accelerated phenotype of the DLB mice model may be obtained by cross-breeding of α-syn tg mice with mutant β-syn tg mice.

4.2.5 γ-Syn as a Chaperone and Anti-chaperone

Compared to the NAC region that is critical for α-syn fibrillation, the corresponding NAC sequence in γ-syn exhibits increased α-helical propensity. Therefore, γ-syn may be situated between α- and β-syn in terms of aggregation properties. This possibility is supported by *in vitro* biochemical analyses showing that γ-syn possesses a weak propensity to form amyloid fibrils (Marsh *et al.*, 2006) but suppresses protein aggregation (Souza *et al.*, 2000). Such chaperone and anti-chaperone properties of γ-syn might be involved in various biological processes *in vivo*, including development of cancer, neurodegeneration, and retinal dystrophy (RD).

4.2.5.1 Chaperone Action of γ-Syn in Cancer. γ-Syn has been extensively studied in cancers such as breast cancer and ovarian cancer (Hashimoto and Masliah, 1999). Since the expression level of γ-syn is well correlated with the presence of metastatic lesions, γ-syn may be related to stimulation of tumor invasiveness and metastasis. Mechanistically, the chaperone action of γ-syn may play a crucial role in the proliferation and survival of hormone-responsive cancer. Tumor cells often have elevated levels of HSPs, raising the possibility that such cells have a selective prosurvival advantage that contributes to tumorigenesis (Mosser and Morimoto, 2004). A similar idea may be applicable to the role of the chaperone activity of γ-syn in cancer, since γ-syn participates in the HSP–estrogen receptor-α (ER-α) complex, enhances the high-affinity ligand-binding capacity of ER-α, and stimulates ligand-dependent activation of ER-α. The γ-syn-mediated stimulation of ER-α transcriptional activity is consistent with its stimulation of mammary tumorigenesis in response to estrogen. These data indicate that γ-syn is a chaperone protein in the HSP-based multiprotein chaperone complex for stimulation of ligand-dependent ER-α signaling, and thus γ-syn may stimulate hormone-responsive mammary tumorigenesis (Liu *et al.*, 2007).

γ-Syn also has chaperone-like interactions with other signaling molecules in cancer, and these interactions might be related to resistance against chemotherapy. For instance, γ-syn may associate with ERK1/2 and JNK, resulting in altered activities of these molecules (Pan *et al.*, 2002). Overexpression of γ-syn in ovarian cell lines resulted in the activation of ERK1/2 and downregulation of JNK1, and resistance of these cells to therapeutic agents was eliminated by a MEK1/2 inhibitor. Similarly, activation of JNK and caspase-3 by therapeutic agents was downregulated in cells overexpressing γ-syn, suggesting blockage of the apoptotic pathway by the chaperone activity of γ-syn.

4.2.5.2 Role of γ-Syn in Neurodegeneration and Retinal Dystrophy. Anti-chaperone activity of γ-syn has been demonstrated in primary cultured sensory neurons (Buchman *et al.*, 1998) and mouse brain (Ninkina *et al.*, 2009). Microinjection of an expression plasmid of γ-syn into cultured sensory neurons resulted in dysregulation of neurofilament network integrity (Buchman *et al.*, 1998). Furthermore, tg mice overexpressing γ-syn under the control of the Thy-1 promoter develop severe age- and transgene dose-dependent neuropathologies and motor deficits. Histopathologically, γ-syn accumulated in neuronal cell bodies and processes, and astrogliosis was evident in motor neurons in the spinal cord. The motor neurons were further characterized by downregulation of HSP27 and disintegration of the neurofilament network, as well as cell death, suggesting that γ-syn might be involved in neuropathophysiological changes and the death of susceptible neurons. Thus, the loss of chaperone activity of γ-syn may have led to disintegration of the neurofilament network. However, it is also possible that a gain of toxic function might be at play in γ-syn-induced neurodegeneration in tg mice.

γ-Syn has also been shown to be involved in the pathogenesis of RD, an age-associated degenerative disease of the eye comprising a group of heterogeneous retinal disorders, including macular degeneration and retinitis pigmentosa. Expression of rhodopsin carrying a P23H mutation causes accumulation of γ-syn in intracellular inclusion bodies in the perinuclear area of photoreceptor cells. Since β-syn, γ-syn, and HSP70, but not α-syn, protect cultured ocular cells from mutant rhodopsin accumulation, these molecules might play protective roles through chaperone functions, and loss of these activities may underlie the pathogenesis (Surgucheva *et al.*, 2005). Simultaneously, it is possible that a gain of toxic function might gradually develop during the later stage of pathogenesis.

4.3 CHAPERONE AND ANTI-CHAPERONE ACTIVITIES OF OTHER AMYLOIDOGENIC PROTEINS

Beyond the synucleins, dual chaperone and anti-chaperone activities may also characterize other amyloidogenic proteins associated with neurodegenerative diseases. There are several reports showing chaperone-like activities of amyloidogenic proteins in AD, prion disease, and polyglutamine (polyQ) disease. As for β-syn, some of these amyloidogenic proteins are characterized by the presence of PPII domains in their disordered structures.

4.3.1 APP Family of Peptides as Transmembrane Chaperones

APP belongs to a conserved gene family that also includes mammalian amyloid precursor-like protein 1 (APLP1) and amyloid precursor-like protein 2 (APLP2), drosophila APPL, and *C. elegans* APL-1 (Walsh *et al.*, 2007). These peptides are type I integral membrane glycoproteins composed of large extracellular domains and short cytoplasmic tails. APP and APLP1 and 2 have conventionally been thought to exist and act as monomers, but accumulating biochemical and structural data suggest that APP family members may exist as dimers or even more complex oligomers (Scheuermann *et al.*, 2001). APP peptides are critically involved in synaptic processes important for memory function in normal adult brain, whereby homo- and heterodimerization of these peptides promotes intercellular adhesion in synaptically enriched membrane compartments (Soba *et al.*, 2005). Moreover, APLP2 associates with the major histocompatibility complex class I molecule, Kd, and regulates endocytosis of this molecule through a chaperone-like function (Tuli *et al.*, 2009). Collectively, these results suggest that APP family peptides could be transmembrane chaperones and that interactions of these peptides through chaperone-like activities might underlie their biological activities in the nervous system. Furthermore, APP peptides are involved in a variety of essential physiological functions in the nervous system, including neuronal cell growth and differentiation, adhesion, and synaptic plasticity (Saitoh *et al.*, 1989). Therefore, it is interesting to reevaluate whether these numerous but apparently nonspecific actions of APP family members are attributable to chaperone activities of these molecules.

Because the Aβ domain is unique to APP and is absent in both APLP1 and APLP2, it is generally believed that neither APLP1 nor APLP2 is actively involved in the pathogenesis of AD. However, several observations suggest that this view may be incorrect. A pathological role of APLP1 and APLP2 in AD is supported by the observation that both proteins are present in typical AD-associated neuropathological lesions, including dystrophic neurites, a subset of senile plaques, and reactive astrocytes in AD brain (Crain *et al.*, 1996). Therefore, it is possible that structural changes of APLP1 and APLP2 and loss of chaperone activities of these molecules might be involved in the pathogenesis of AD. Evidence is also accumulating that formation of heteromeric complexes of APP with other members through their chaperone-like activities have particularly important implications for understanding cellular regulation and mechanisms of Aβ production (Figure 4.6a). The coexpression of APP with either APLP1 or APLP2 results in decreased generation of Aβ 1-42 in cell cultures, supporting this concept. Moreover, chaperone-like interactions may also be important in the interactions of various Aβ species. Although Aβ is characterized by a disordered, random-coil structure, a recent biophysical study has shown that the monomeric form of synthetic Aβ (1-28) predominantly adopts a PPII helix conformation in an

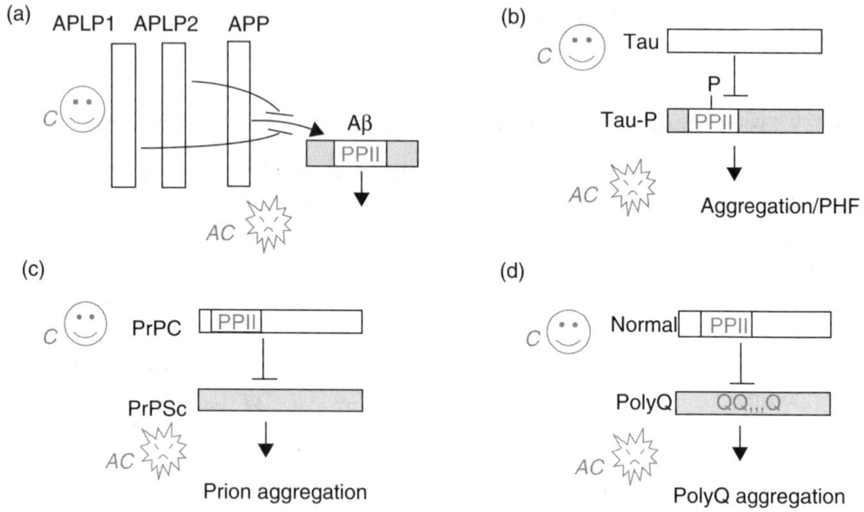

Figure 4.6 Chaperone and anti-chaperone actions of amyloidogenic proteins associated with various neurodegenerative diseases. (a) APP family members APLP1 and APLP2 may interact with APP through chaperone-like actions, thereby negatively regulating Aβ production from APP. (b) Tau may associate with various enzymes through chaperone-like activity, and phosphorylation of tau may result in a loss of chaperone activity, leading to aggregation. (c) The N-terminal domain of PrP has chaperone-like activity, and alteration of this domain may contribute to conversion of PrP^{C} to PrP^{Sc}. (d) Aggregation of the elongated polyQ protein may be inhibited by the normal gene product through a chaperone-like interaction. Similar to β-syn, the presence of a PPII domain and/or PPII-like activities have been demonstrated in various amyloidogenic proteins and related proteins.

acidic solution (Baumketner and Shea, 2006). This result suggests that aberrant PPII activities of Aβ may play an important role in the pathogenesis of AD.

4.3.2 Chaperone-like Activities of Other Amyloidogenic Proteins

In AD brain, aberrant polymerization of the microtubule-associated protein tau is the main component of neurofibrillary tangles. Tau also accumulates in various structures during the pathogenesis of various tauopathies, which include frontotemporal dementia and parkinsonism linked to chromosome-17, Pick's disease, and corticobasal degeneration (Lee and Trojanowski, 1999). Tau associates with enzymes such as glyceraldehyde 3-phosphate dehydrogenase (GAPDH) and lactate dehydrogenases (LDH) through chaperone-like activities (Tian *et al.*, 2004), and the PPII activity of tau is enhanced through phosphorylation by kinases such as GSK-3 and CDK5. Therefore, aberrant phosphorylation of the PPII domain of tau may result in loss of chaperone activity and/or gain of an aggregative property of this molecule (Figure 4.6b).

Transmissible spongiform encephalopathies, including bovine spongiform encephalopathy and Creutzfeldt–Jakob disease, are characterized by conversion of a cell surface host protein PrP^C to a protease-resistant isoform, PrP^{Sc} (Prusiner, 1998). The mechanism of conversion from PrP^C to PrP^{Sc} is poorly understood, but a role for the N-terminal region of PrP is increasingly likely, particularly because of its potential to form a PPII helix structure (Gill *et al.*, 2000). Thus, the N-terminal domain of PrP may have chaperone-like activity and aberrant regulation of this chaperone domain might contribute to conversion of PrP^C to PrP^{Sc} (Figure 4.6c).

Finally, in the majority of polyQ diseases, such as HD, one chromosome provides a normal gene with a nonamyloidogenic product, whereas the other encodes a gene containing an elongated CAG repeat that gives a disease-causing polyQ protein that is amyloidogenic and causative of neurodegeneration. Speculatively, the normal gene product may interact with the elongated polyQ protein and negatively regulate its aggregation (Figure 4.6d), and it is of note that the normal gene product may possess PPII activity (Chellgren *et al.*, 2006). Previous studies have shown that the backbone structure of monomeric polyQ is completely disordered, but a recent computational analysis demonstrated that the glutamine residues in polyQ tracts have a significant propensity to adopt a PPII helical conformation. NMR spectroscopy and circular dichroism have also revealed that monomeric polyQ with up to 15 residues has the propensity to adopt a PPII conformation, while other structures such as α-helices and β-sheets are predominant in longer polyQ proteins (Chellgren *et al.*, 2006). Since the PPII helical conformation is not thought to be a precursor to polyQ aggregation, the normal gene product might act as a chaperone but not as an anti-chaperone. Consistent with this idea, the normal gene product may be incorporated into polyQ aggregates, leading to the formation of heterologous polyQ fibrils (Busch *et al.*, 2003).

Many disease-specific amyloidogenic proteins and their family members are characterized by chaperone and anti-chaperone activities associated with a PPII domain. Future studies may reveal that structural changes in this domain play critical roles in loss of chaperone activity and/or gain of toxic aggregate formation in the pathogenesis of neurodegenerative disease.

4.4 CHAPERONE AND ANTI-CHAPERONE ACTIVITIES OF sHSPs

In the sections above, we showed that synucleins and other amyloidogenic proteins exhibit dual chaperone and anti-chaperone activities. In this section, we examine the possibility that sHSPs may also be endowed with similar chaperone and anti-chaperone activities.

4.4.1 sHSPs as Inhibitors of Protein Aggregation

There are 10 sHSPs in humans and mice (HSPB1–10), of which HSP27 (HSPB1), αA-crystallin (HSPB4), and αB-crystallin (HSPB5) are the best known (Kappe *et al.*, 2003). The monomeric molecular masses of sHSPs are around 16–43 kDa, and the proteins assemble into large oligomeric complexes that range in size from 200 to 800 kDa (Kappe *et al.*, 2003). Several reports have shown that αA- and αB-crystallins interact with other members of the α-crystallin family and that these proteins can be copurified from cell or tissue extracts, suggesting that both homo- and heterooligomerization commonly occur in mammalian cells (Narberhaus, 2002). The oligomerization results from multiple inter- and intramolecular interactions through the α-crystallin domains of sHSPs, which are characterized by sandwich structures of two β sheets (Narberhaus, 2002). The N-terminal region that precedes the α-crystallin domain is variable in length and amino acid sequence, thereby contributing to structural diversity among different sHSPs and playing a regulatory role in stabilization of multimers. The α-crystallin domain is followed by the C-terminal extension, which is involved in protein solubility and shares no sequence homology among family members. Phosphorylation may also play a critical role in structural remodeling and deoligomerization of sHSP into dimmers (Narberhaus, 2002). Phosphorylation is induced by mitogen-activated protein (MAP) kinase-activated kinase 2, which is situated downstream of p38 and essential in many protective activities of sHSPs against various stresses. Regardless of the subunit composition, the diverse assemblies of sHSPs efficiently protect other proteins from aggregation (Narberhaus, 2002).

Similar to high molecular mass HSPs such as HSP70, expression of sHSPs are induced in response to a variety of stresses, including heat shock, oxidative stress, osmo-stress, and ischemia, although some sHSPs are constitutively expressed under physiological conditions (Hartl, 1996). HSP70 is mainly involved in ATP-dependent protein folding, but sHSPs stabilize proteins and prevent their aggregation in an ATP-independent manner. Furthermore, the activity of HSP70 is enhanced by the J-domain protein co-chaperone belonging to the HSP40 family, whereas sHSPs do not require association with co-chaperones. Given the relatively close molecular masses and the lack of requirements for ATP and co-chaperones, the protective actions of sHSPs appear to be similar to those of synucleins.

4.4.2 Missense Mutations of sHSPs in Protein Conformation Diseases

A number of mutations of sHSP family members have been identified in protein conformation diseases, including cataract, myopathies, and distal neuropathies, highlighting the essential role of these proteins in such disorders. Given the dominant inheritance of the disease, it is likely that altered structures of sHSPs due to gene mutations might underlie the disease onset. A point mutation in the α-crystallin domain of αA-crystallin (R116C) is responsible for a dominant congenital cataract disease in humans (Litt *et al.*, 1998), and a similar mutation in αB-crystallin (R120G) was found in a French family with an autosomal dominant desmin-related myopathy (Vicart *et al.*, 1998). sHSP mutations have also been linked to distal hereditary motor neuropathies, which are genetically heterogeneous diseases of the peripheral nervous system that cause nerve degeneration and distal limb muscle atrophy. A K141N point mutation in HSP22 (HSPB8) has been found in two families with distal hereditary motor neuropathy and a second mutation (K141E) was found in two other pedigrees (Irobi *et al.*, 2004). K141 is in the α-crystallin domain and is equivalent to R116 in αA-crystallin and R120 in αB-crystallin, pointing to the importance of this residue for maintenance of correct protein folding. Indeed, biophysical studies have shown that the mutations of R116C in αA crystallin and R120G in αB crystallin significantly reduce chaperone-like activity compared to wild-type α-crystallins (Kumar *et al.*, 1999). Subsequently, S135F, R127W, T151I, and P182L mutations in HSP27 (HSPB1) have been discovered in families with distal hereditary motor neuropathy, while individuals with Charcot–Marie–Tooth disease, the most common inherited motor and sensory neuropathy and a genetically and clinically heterogeneous syndrome, exhibit a K141N mutation in HSP22, and S135F and R136W mutations in HSP27 (Evgrafov *et al.*, 2004).

4.4.3 Pathogenic Mechanism of sHSPs: Loss of Chaperone or Gain of Anti-chaperone Function?

The molecular mechanisms through which sHSPs are involved in the pathogenesis of protein conformation diseases are unclear. Many reports have suggested that loss of chaperone activities of these molecules could play a causative role in the pathogenesis of these disorders, but it is also possible that sHSP misfolding results in a toxic gain of function.

Cataracts develop because of factors such as aging, exposure to ultraviolet light or radiation, trauma, or secondary effects of diabetes and hypertension, all of which may ultimately cause aggregation of crystallins (reviewed in (Harding and Dilley, 1976)). The crystallins are highly stable proteins that are organized in a supramolecular β-sheet structure within the lens. In mammals, there are three classes of crystallins, α, β*s* and γ, and each has distinct subunits. αA- and αB-crystallins are abundantly present in the lens, in which their likely role is as a chaperone to all crystallin proteins including themselves, thus preventing protein aggregation and precipitation.

Loss of chaperone activity due to the αA-crystallin R116C mutation or a loss of capacity of this mutant to polymerize with αB-crystallin may be responsible for the precipitation of αB-crystallin and the resulting increase in lens opacity. In sporadic cases, α-crystallin taken from the lens nucleus shows an age-dependent decrease in chaperone function (Derham and Harding, 1997). Furthermore, high molecular mass aggregates and α-crystallin found within the nucleus from cataract lenses may have reduced chaperone function. Therefore, posttranslational modifications known to occur during aging, such as glycation, carbamylation, oxidation, phosphorylation, and truncation, may cause a decrease in chaperone function. However, it is unclear whether loss of the chaperone activity of α-crystallin is sufficient to explain all the aspects of the pathogenesis of the disease. Indeed, formation of amyloid deposits in the eye lens may disturb the short range order of the crystallins and thus lead to lens opacity and cataract. In this context,

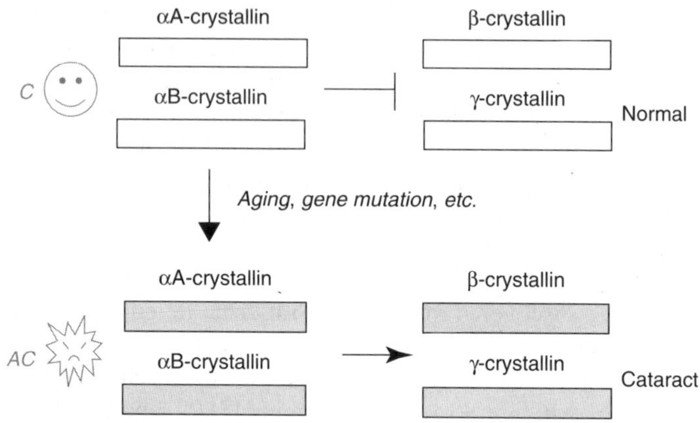

Figure 4.7 Chaperone and anti-chaperone actions of α-crystallins. α-Crystallins including αA- and αB-crystallin are molecular chaperones in mammalian lens that prevent aggregation of themselves and β- and γ-crystallins. Aging and causes such as gene mutations and diabetes induce misfolding and aggregation of α-crystallins, which stimulates aggregation of β- and γ-crystallins. Studies performed *in vitro* have shown that all the crystallin family of peptides are capable of forming amyloid fibrils.

all three classes of wild-type crystallin proteins, including the molecular chaperone α-crystallins, have been shown to be capable of forming amyloid fibrils under unfolding conditions *in vitro* (Meehan *et al.*, 2004). Thus, the conversion of crystallins into toxic fibrils might contribute to the development of cataract (Figure 4.7).

Regarding HSP27, overexpression of the HSP27 R148G mutant in cell cultures leads to the formation of cytoplasmic aggregates that accumulate as amorphous perinuclear structures (Chavez Zobel *et al.*, 2005). Since wild-type HSP27 is involved in the organization of the neurofilament network, which is important for maintaining the axonal cytoskeleton and transport, it is possible that mutant HSP27 fails to control filament–filament interactions and neurofilament assembly, leading to formation of cytoplasmic aggregates. Similarly, overexpression of the HSP22 K141N mutant in COS cells dramatically increases the number of cytoplasmic and perinuclear aggregates (Irobi *et al.*, 2004). Furthermore, the αB-crystallin R120G mutant causes loss of a specific chaperone function of αB-crystallin at the level of the intermediate filament, resulting in the aggregation of desmin, and this has been suggested to be a cause of the αB-crystallin (R120G)-mediated myopathy (Chavez Zobel *et al.*, 2003). In majority of the cases, loss of chaperone function of sHSPs has been proposed to contribute to the disintegrity of the intermediate filament network, leading to aggresome formation. However, an alternative interpretation is that the gain of anti-chaperone activity by sHSPs might contribute to aggresome formation, since this process is triggered by various aggregation-prone amyloidogenic proteins (Kopito, 2000). In support of this hypothesis, it was recently shown that the αB-crystallin R120G mutant directly promotes desmin filament aggregation, with desmin networks being most vulnerable when they were being made *de novo* compared to those that were already established (Perng *et al.*, 2004).

4.4.4 Protective Role of sHSPs Against Neurodegenerative Disease

The sHSP family of proteins has been implicated in a number of protective activities in the brain. These include resistance to apoptosis, cytoskeleton modulation, prevention of amyloid formation, regulation of signaling during the developmental processes, protection against oxidative stress, thermotolerance, eye lens transparency, and nucleic acid preservation (Narberhaus, 2002). A potential link between sHSPs and membrane function has also received recent attentions (Narberhaus, 2002). Therefore, it is likely that sHSPs may play a critical role in anti-apoptosis and cytoprotection, and that structural alteration of these molecules might be involved in the pathogenesis of neurodegenerative disease.

A protective role of sHSPs has been shown in cellular and animal models of various neurodegenerative diseases, including polyQ disease. For example, overexpression of HSP27 resulted in protection against polyQ-mediated toxicity in cellular models of HD, whereby HSP27 suppressed the increased levels of reactive oxygen species (ROS) caused by mutant huntingtin (Wyttenbach *et al.*, 2002). Consistent with this, decreased expression of HSP27 was observed in both neuronal and non-neuronal cell cultures under conditions of cytotoxicity induced by overexpression of polyQ proteins derived from spinocerebellar ataxia type 3 (Wen *et al.*, 2003). Similar to HSP27, HSP22 prevents protein aggregation in CCL39 Chinese hamster lung fibroblasts overexpressing the polyQ protein Huntingtin43Q (Carra *et al.*, 2005). In this model cellular system, neither HSP27 nor αB-crystallin had protective effects against polyQ-mediated cell toxicity. In this regard, the specificity of HSP22 could be accounted for by a demonstration that HSP22, but not other sHSPs, interacts with Bag3, a protein that may facilitate the disposal of doomed proteins by stimulating macroautophagy (Carra *et al.*, 2008). Finally, the increased HSP27 level in a tg model of Machado-Joseph disease (MJD) and in the brains from patients suggests the upregulation of expression of HSP27 in response to increasing toxicity during the course of neurodegeneration (Chang *et al.*, 2005).

Increased expression and accumulation of sHSPs have been described in many other neurodegenerative diseases. Upregulation of HSP27 is required for the survival of injured sensory and motor neurons (Benn *et al.*, 2002), while augmented αB-crystallin discriminates between neurons in various neurodegenerative diseases, including Creutzfeldt–Jakob disease (Kato *et al.*, 1992). In PD, McLean and coworkers investigated the immunoreactivities of various chaperones in the substantia nigra of PD brains and found that both Lewy bodies and Lewy neuritis were highly immunopositive for HSP27, HSP70, and the HSP40 family members, HDJ-1 and HDJ-2, and to lesser extents for HSP110, HSP60, and HSP90 (McLean *et al.*, 2002). In AD, immunoreactivities of sHSPs, including HSP20, HSP27, and HSPB2, have been extensively detected in diffuse and mature senile plaques (Wilhelmus *et al.*, 2006). In contrast, immunoreactivities of sHSPs in neurofibrillary tangles are less clear, despite the expression levels of HSP27 and αB-crystallin increasing with the severity of AD-specific morphological changes and the duration of dementia (Wilhelmus *et al.*, 2006). Since biochemical data have shown that HSP27 is cross-linked with highly ubiquitinated tau by transglutaminase in the AD brain (Nemes *et al.*, 2004), one possibility is that the epitope of sHSP in the neurofibrillary tangles might be extensively modified such that it escapes detection by immunological procedures. Curiously, both HSP27 and αB-crystallin are also found in a large number of proliferating astrocytes in AD (Shinohara *et al.*, 1993), with the highest expression of HSP27 exhibited by degenerative astrocytes in areas rich in senile plaques, indicating a role of sHSPs in gliosis in AD.

Finally, it is of note that sHSPs are involved in the pathogenesis of Alexander's disease, a leukodystrophy-like neurodegenerative disease characterized by the abundance of "Rosenthal fibers" in astrocytes (Iwaki *et al.*, 1993). Rosenthal fibers are composed of protein aggregates of which the major component is glial fibrillary acidic protein (GFAP). These fibers have been shown to contain sHSPs, including αB-crystallin and HSP27, which presumably reflects the protective role of these molecules against toxic aggregates of GFAP.

Taken together, increased levels and accumulation of sHSPs are observed in a variety of neuropathological regions. During the course of neurodegeneration, sHSPs may be upregulated as chaperones to cope with increasing protofibrils and oxidative stress, until a later stage when most of the sHSPs are consumed by aggregated proteins and/or sequestered in pathological inclusions. An alternative interpretation of misfolding and aggregation of sHSPs is that these proteins may actively participate in disease progression in cooperation with the disease-specific amyloidogenic proteins.

4.5 POSITIVE FEEDBACK IN THE CONVERSION OF AMYLOIDOGENIC PROTEINS FROM CHAPERONE TO ANTI-CHAPERONE (C → AC) IN VARIOUS NEUROPATHOLOGIES

In the sections above, we have discussed how synucleins, disease-specific amyloidogenic proteins, and sHSPs may all have chaperone and anti-chaperone activities. This raises the question of the biological role of the chaperone and anti-chaperone duality. In this section, we examine the possibility that C → AC conversion of amyloidogenic proteins may be linked to aging-associated pathologies such as mitochondrial dysfunction, decreased activity of protein degradation including the UPS and autophagy–lysosome pathway, and glial activation.

4.5.1 C → AC Conversion of Amyloidogenic Proteins and Mitochondrial Dysfunction

According to the mitochondrial theory of aging, a variant of the free radical theory of aging, accumulation of damage to mitochondria and mitochondrial DNA (mtDNA) may lead to the aging of humans and animals (Balaban *et al.*, 2005). mtDNA is continually exposed to a high steady-state level of ROS and free radicals in the matrix of mitochondria, since it has no protection by histones or DNA-binding proteins. Elevation of oxidative stress and associated oxidative damage gradually occurs in the mitochondria of neurons in aging, and the enhanced oxidative stress may stimulate protein aggregation. Thus, oxidative stress produced by iron and hydrogen peroxide has been shown to induce amyloid-like aggregates of α-syn *in vitro* (Hashimoto *et al.*, 1999a) and in cell cultures (Ostrerova-Golts *et al.*, 2000). Indeed, oxidative stress is thought to contribute to PD because dopamine is a strong free radical generator and the principle neurotransmitter in the substantia nigra (Jenner and Olanow, 1998). In addition, iron, which also stimulates free radical production, accumulates in the substantia nigra with age (Jenner and Olanow, 1998). Thus, it is reasonable to speculate that oxidative stress caused by mitochondrial dysfunction may stimulate aggregation of α-syn and presumably of other amyloidogenic proteins. In turn, amyloidogenic proteins may interfere with mitochondrial functions. For example, α-syn may be imported into mitochondria, in which it accumulates in PD brains, and impair respiratory complex I activity (Vila *et al.*, 2008). Accumulation of α-syn oligomers and protofibrils in the mitochondrial membrane might result in the release of cytochrome C, with subsequent activation of the apoptosis cascade, leading to neuronal cell death (Figure 4.3e) (Hashimoto *et al.*, 1999b). Collectively, these studies suggest that C → AC conversion of amyloidogenic proteins and mitochondrial dysfunction are linked and form a vicious cycle that either triggers or perpetuates neurodegeneration.

4.5.2 C → AC Conversion of Amyloidogenic Proteins and Decreased Protein Degradation

An age-associated decline in the activities of the protein degradation system is an apparent risk factor for protein aggregation. Ubiquitination of endogenous proteins is one of the key regulatory steps of protein degradation, followed by regulation of proteasome activity. Many amyloidogenic proteins, such as α-syn and tau, are subjected to degradation in the proteasome. Proteasomal activity decreases during the aging process in various model systems of neurodegenerative diseases and in human brains (Friguet *et al.*, 2000) and this may lead to decreased degradation of amyloidogenic proteins. Aggregated α-syn may in turn interfere with the proteasome, since α-syn has been shown to bind to S6′, a component of the 19S subunit in the 26S proteasome (Ghee *et al.*, 2000). Aggregated α-syn, but not its monomeric form, has also been shown to inhibit the activity of the 26S proteasome under cell-free conditions (Snyder *et al.*, 2003).

The autophagy–lysosome pathway is also involved in the degradation of amyloidogenic proteins. Degradation of cytoplasmic components is achieved through distinct autophagic pathways, including macroautophagy and chaperone-mediated autophagy (Cuervo *et al.*, 2004; Webb *et al.*, 2003). Macroautophagy involves sequestration of cytosolic regions into autophagosomes that deliver their contents to late endosomal and lysosomal compartments for degradation. In contrast, chaperone-mediated autophagy selectively targets proteins containing a KFERQ peptide motif to lysosomes. A recent study suggests that amyloidogenic proteins such as α-syn and Huntingtin are susceptible to degradation through macroautophagy and chaperone-mediated autophagy (see Chapter 3) (Cuervo *et al.*, 2004; Webb *et al.*, 2003; Wei *et al.*, 2007). Thus, a decrease in the activity of autophagy and the lysosomal processing capacity during aging of brains may decrease degradation and cause accumulation of amyloidogenic proteins. The accumulation of proteins such as Aβ and α-syn may in turn cause dysfunction of the autophagy–lysosome pathway. Amyloid fibrils have been shown to destabilize lysosomal membranes and stimulate lysosomal membrane permeabilization (LMP), resulting in leakage of lysosomal enzymes into the cytoplasm and leading to cell death (Wei *et al.*, 2009). Amyloidogenic proteins might also directly interfere with the autophagy pathway, since it has been shown that Huntingtin associates with and dysregulates the activity of mammalian target of rapamycin (mTOR), a regulatory molecule in the autophagy–lysosome pathway (Ravikumar *et al.*, 2004).

4.5.3 C → AC Conversion of Amyloidogenic Proteins and Glial Activation

There is a consistent age-dependent upregulation of astrocytic and microglial markers in brains derived from experimental animal models and humans, which could be interpreted as age-dependent upregulation of neuroinflammation (Lucin and Wyss-Coray, 2009). Astrocytes are indispensable to neurons for their glutamate-buffering effects and regulation of inflammation, and aberrant activation of astrocytes may lead to impaired synaptic plasticity and neuronal damage. Furthermore, an enhanced microglial cell-driven inflammatory response results in the release of inflammatory mediators, including an array of neurotoxic cytokines and chemokines. Thus, it is possible that an abnormally upregulated inflammatory response may exacerbate neurodegeneration and stimulate C → AC conversion of amyloidogenic proteins.

Amyloidogenic proteins may also stimulate microglial activation. In AD brains, senile plaques surrounded by numerous microglia reflect direct activation of microglia by Aβ. Indeed, *in vitro* studies have shown that exposure of fibrillar Aβ to microglial cells leads to time- and dose-dependent activation of signaling molecules, including Lyn, Syk, and FAK, with generation of superoxide radicals (McDonald *et al.*, 1997). Similarly, aggregated α-syn has been shown to stimulate microglial activation in cell cultures (Zhang *et al.*, 2005). Of relevance to these findings, a recent study showed that α-syn may be released from neuronal cells into the culture medium (Lee, 2008). These results suggest that α-syn and other cytoplasmic amyloidogenic proteins may be released from degenerating neurons, leading to stimulation of microglia.

4.5.4 C → AC Conversion of Amyloidogenic Proteins in Mediation of Aging-Associated Phenomena

Collectively, the results described above suggest that C → AC conversion of amyloidogenic proteins may be centrally situated in the neurodegeneration process and may act as a mediator of various aging-associated pathological phenomena through positive feedback loops (Figure 4.8a). A particular pathology that is triggered by gene mutations or environmental factors may stimulate aggregation of amyloidogenic proteins. The protein aggregates then stimulate not only the primary pathology but also other pathologies. All of the pathologies in turn stimulate aggregation of amyloidogenic proteins. Such a positive feedback process is repeated until all the pathologies are greatly amplified and amyloidogenic proteins form mature amyloid fibrils. In this scheme, the primary pathology can be relatively flexible, which is consistent with hereditary cases of neurodegenerative diseases being caused by mutations of various genes encoding proteins with different cellular functions. Whatever the primary pathology, once C → AC conversion is triggered, the central core of neurodegeneration is similar, solid and progressive. If this view is correct, then it could explain why neurodegeneration is resistant to many radical treatments.

4.6 C → AC CONVERSION OF AMYLOIDOGENIC PROTEINS: A ROLE OF PROGRAMMED AGING?

The hypothesis that C → AC conversion of amyloidogenic proteins may play an important role in the regulation of aging is supported by two pieces of

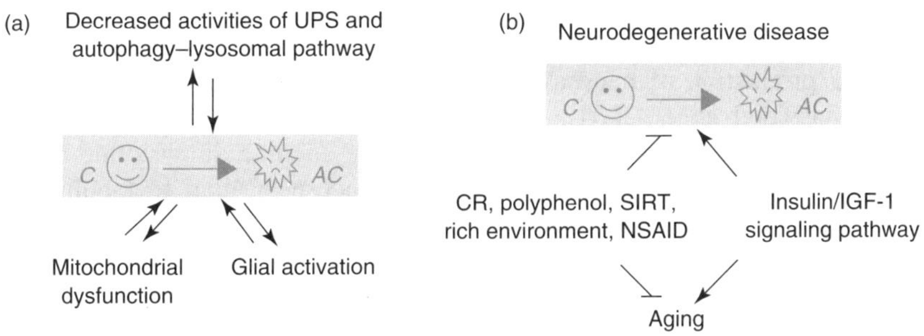

Figure 4.8 Relationship between C → AC changes of amyloidogenic proteins and aging. (a) A positive feedback loop between C → AC conversion of amyloidogenic proteins and aging-associated pathologies such as decreased activities of the UPS and autophagy–lysosomal pathway, mitochondrial dysfunction, and glial activation. (b) Both C → AC conversion of amyloidogenic proteins and aging are suppressed by anti-aging strategies such as CR, polyphenols, SIRTs, an improved environment, and treatment with NSAIDs, and are stimulated by common signaling pathways including the insulin/IGF-1 pathway.

circumstantial evidence. First, several different strategies for prophylaxis of neurodegenerative diseases as described below, such as caloric restriction (CR), uptake of dietary polyphenol, quality of environments, anti-inflammation, and hormonal replacements, appear to be effective for extension of longevity. Second, aging and C → AC conversion are regulated by common signaling pathways, including the insulin/insulin growth factor-1 (IGF-1) pathway.

4.6.1 Anti-aging and Suppression of C → AC Conversion of Amyloidogenic Proteins Are Synonymous

The first line of evidence supporting a role for C → AC conversion of amyloidogenic proteins in regulation of aging is that suppression of these changes is similar to anti-aging. Despite extensive efforts to exploit radical treatment of neurodegenerative diseases, most are ineffective for the advanced stage of the disease. Therefore, attention has turned to prophylaxis, and various protective treatments have been identified on the basis of epidemiological data and results in animal models. These strategies include CR, uptake of dietary polyphenols, improved quality of environment, anti-inflammatory drugs, and hormonal replacement, and most are associated with extended longevity in aging research.

CR prolongs the life span in a variety of organisms (Canto and Auwerx, 2009). Furthermore, epidemiological reports have shown that high caloric diets increase the risk of AD, suggesting that dietary interventions in adult life might slow the disease progression of AD (Pasinetti and Eberstein, 2008). Consistent with this, CR substantially decreases accumulation of Aβ in senile plaques and also decreases astrocytic activation, as assessed by GFAP immunoreactivity in AD-tg mice (Halagappa *et al.*, 2007). CR and lower insulin levels may slow age-dependent processes and extend life span through various mechanisms. For example, CR may reduce oxidative stress (Youngman *et al.*, 1992) while increasing production of neurotrophic factors and cytoprotective protein chaperones (Maswood *et al.*, 2004). CR regimens and CR mimetics have recently been shown to trigger sirtuin 1, an nicotinamide adenine dinucleotide (NAD)-dependent histone deacetylase that regulates glucose or lipid metabolism and insulin signaling through its deacetylase activity. Sirtuin 1 has been associated with longevity in a variety of organisms from bacteria to mouse (Canto and Auwerx, 2009), and this may account for the effectiveness of CR in ameliorating the neuropathology of AD and extending longevity.

Dietary intake of antioxidants may be another important factor that confers anti-aging and protection against neurodegenerative disease. In this context, it is well known that resveratrol, a red wine polyphenol, may extend longevity in the French population. Resveratrol protects against neurological disorders such as strokes, ischemia, and neurodegenerative disease (Anekonda, 2006). In a mouse model of HD, resveratrol-induced sirtuin 1 was found to protect neurons against polyQ toxicity (Anekonda, 2006). In slow Wallerian degeneration mice, stimulation with resveratrol was found to protect against degeneration of neurons from axotomy, suggesting that resveratrol has therapeutic value for neuronal degeneration (Araki *et al.*, 2004).

The influence of the quality of the environment may also be important for longevity of aging and onset of neurodegenerative disease. Environmental enrichment can delay the onset of symptoms in R6 HD mice (van Dellen *et al.*, 2000), and exposure of tg mice coexpressing familial AD-linked APP and PS1 variants to an "enriched environment" resulted in pronounced reductions in cerebral Aβ levels and amyloid deposits, compared to animals raised under "standard housing" conditions (Lazarov *et al.*, 2005). In this experimental system, enrichment of environment led to reductions in steady state levels of cerebral Aβ peptides and amyloid deposition. Lifelong exercise and physical training are also known to extend mean life span and to ameliorate cognitive function in older adults and AD patients (Mattson *et al.*, 2004).

Finally, both aging and onset of neurodegenerative disease are regulated by the interaction of the nervous system with other biological systems, such as the immune system. Epidemiological studies have shown that long-term users of non-steroidal anti-inflammatory drugs (NSAIDs) such as ibuprofen and indomethacin have a significantly lower frequency of neurodegenerative diseases (McGeer and McGeer, 1998). On the basis of these findings, studies are in progress to determine whether anti-inflammatory drugs are effective therapy for neurodegenerative diseases. On the other hand, aging may be characterized by chronic low-grade inflammation, the so called "inflammaging," which manifests as a two- to four-fold increase in the production of proinflammatory cytokines and acute phase proteins (Giunta *et al.*, 2008). Therefore, appropriate control of inflammation by anti-inflammatory genes may result in increased longevity and successful aging. Despite the epidemiological data, a large-scale clinical trial using NSAIDs (naproxen and rofecoxib) failed to demonstrate the prevention of cognitive decline in mild-to-moderate AD (Thal, 2003). Further studies are required to clarify the discrepancy between the epidemiological data and clinical trials.

In conclusion, protection against neurodegenerative disease and anti-aging are almost synonymous (Figure 4.8b). Therefore, enhancement of prolongevity signaling may be a promising approach against aging-associated neurodegenerative diseases. Conversely, a strategy of prophylaxis against neurodegenerative diseases may lead to a prolonged life span.

4.6.2 Regulation of C → AC Conversion of Amyloidogenic Proteins and Aging by Common Signaling

The second line of evidence to support a role of C → AC conversion of amyloidogenic proteins in the regulation of aging is that both aging and the chaperone/anti-chaperone conversion may be regulated by common signaling pathways. In contrast to the classical notion that aging is a random and passive phenomenon, evidence is accumulating to suggest that aging is actively regulated in a manner similar to other biological processes. Among many signals, the insulin/IGF-1 pathway has been most extensively characterized (reviewed in (Kenyon, 2005)). This pathway has primarily been linked to the life span of the nematode worm *Caenorhabditis elegans* (*C. elegans*), in which mutations in daf-2, an ortholog of the insulin/IGF-1 receptor, have been shown to double the life span. This life span extension caused by daf-2 mutations required the activity of daf-16, an FOXO family transcription factor. Furthermore, the daf-2 receptor was shown to activate a conserved PI-3 kinase signaling pathway that may also affect life span (Kenyon, 2005). Essentially similar results have been observed in other biological system, including yeast, drosophila, and mouse (Kenyon, 2005). Finally, the FOXO3A genotype has been strongly associated with human longevity, indicating that aging may at least partly be regulated by the insulin/IGF-1 pathway.

Notably, it was recently shown that aggregation-mediated Aβ(1–42) toxicity was reduced in *C. elegans* when aging was slowed by decreased insulin/IGF-1-like signaling (Cohen *et al.*, 2006). Under these conditions, the downstream transcription factors, heat shock factor 1 and DAF-16, regulate the opposing disaggregation and aggregation activities to promote cellular survival in response to constitutive toxic protein aggregation. More studies are needed to elucidate the role of the insulin/IGF-1 pathway in the development of AD, but the results raise an intriguing possibility that the aging process and aggregation-mediated proteotoxicity are regulated by a common signal pathway (Figure 4.8b). Earlier work had also shown a role for altered signal transduction pathways in the pathogenesis of AD and other neurodegenerative diseases. For example, several signal pathways, including PKC and casein kinase II, are upregulated in AD brains compared to controls (Iimoto *et al.*, 1990). Such classic and important observations should now be reevaluated from a new perspective on the relationship between aging and protein aggregation.

4.7 CONCLUSION AND PERSPECTIVE

In this review, we have proposed that amyloidogenic proteins are dual proteins with chaperone and anti-chaperone activities, and that C → AC conversion of these proteins plays a central role in both aging and neurodegenerative disease. This new perspective has the potential to become a unified theory for neurodegenerative disease because the chaperone and anti-chaperone dual activities are observed not only for synuclein proteins but also for other amyloidogenic proteins and sHSPs in which mutations are causative for neurodegenerative and protein conformation diseases.

As long as amyloidogenic proteins behave as normal chaperones, they are useful protectors of neurons. However, once they convert to anti-chaperones, they become killers of neurons. Such "Jekyll and Hyde" dual activities of these proteins might serve an efficacious switch from neuroprotection to neurodegeneration during the course of aging. Thus, it is intriguing to speculate that C → AC conversion of amyloidogenic proteins may have evolved to actively regulate aging of neurons. This hypothesis is in line with recent progress in aging research, given the emerging evidence that aging is at least partly a programmed phenomenon and not a passive and random process. However, since most studies of programmed aging have been performed in primitive organisms such as

C. elegans and drosophila, further work is required to extend this theory to mammalian brains. The views presented here raise the possibility that C → AC conversion of amyloidogenic proteins may have evolved to regulate programmed aging in mammalian brains.

We previously proposed the concept that C → AC conversion of amyloidogenic proteins might produce a diversity of toxic protein aggregates that are beneficial for preconditioning of the brain to cope with forthcoming stresses during aging (Fujita *et al.*, 2006). This previous hypothesis and the current view are not mutually exclusive, but are similar and complementary in postulating an active role for C → AC conversion in the regulation of aging of the brain. Given the apparent importance of chaperone actions in the pathophysiology of the aging process, construction of a new theory on chaperone actions may be helpful for the understanding of the relationships among chaperones, aging, and neurodegenerative disease. This novel theoretical concept and the accumulating experimental data on chaperone biology may open a new avenue of therapy for neurodegenerative disease and anti-aging.

REFERENCES

Anekonda TS. Resveratrol–a boon for treating Alzheimer's disease? *Brain Res Rev* 2006;**52**:316–326.

Araki T, Sasaki Y, Milbrandt J. Increased nuclear NAD biosynthesis and SIRT1 activation prevent axonal degeneration. *Science (New York, NY)* 2004;**305**:1010–1013.

Balaban RS, Nemoto S, Finkel T. Mitochondria, oxidants, and aging. *Cell* 2005;**120**:483–495.

Baumketner A, Shea JE. Folding landscapes of the Alzheimer amyloid-beta(12–28) peptide. *J Mol Biol* 2006;**362**:567–579.

Benn SC, Perrelet D, Kato AC, Scholz J, Decosterd I, Mannion RJ, Bakowska JC, Woolf CJ. Hsp27 upregulation and phosphorylation is required for injured sensory and motor neuron survival. *Neuron* 2002;**36**:45–56.

Bertoncini CW, Rasia RM, Lamberto GR, Binolfi A, Zweckstetter M, Griesinger C, Fernandez CO. Structural characterization of the intrinsically unfolded protein beta-synuclein, a natural negative regulator of alpha-synuclein aggregation. *J Mol Biol* 2007;**372**:708–722.

Brazil DP, Park J, Hemmings BA. PKB binding proteins. Getting in on the Akt. *Cell* 2002;**111**:293–303.

Broadie K, Rushton E, Skoulakis EM, Davis RL. Leonardo, a Drosophila 14-3-3 protein involved in learning, regulates presynaptic function. *Neuron* 1997;**19**:391–402.

Buchman VL, Adu J, Pinon LG, Ninkina NN, Davies AM. Persyn, a member of the synuclein family, influences neurofilament network integrity. *Nat Neurosci* 1998;**1**:101–103.

Busch A, Engemann S, Lurz R, Okazawa H, Lehrach H, Wanker EE. Mutant huntingtin promotes the fibrillogenesis of wild-type huntingtin: a potential mechanism for loss of huntingtin function in Huntington's disease. *J Biol Chem* 2003;**278**:41452–41461.

Canto C, Auwerx J. Caloric restriction, SIRT1 and longevity. *Trends Endocrinol Metab* 2009;**20**:325–331.

Carra S, Seguin SJ, Lambert H, Landry J. HspB8 chaperone activity toward poly(Q)-containing proteins depends on its association with Bag3, a stimulator of macroautophagy. *J Biol Chem* 2008;**283**:1437–1444.

Carra S, Sivilotti M, Chavez Zobel AT, Lambert H, Landry J. HspB8, a small heat shock protein mutated in human neuromuscular disorders, has *in vivo* chaperone activity in cultured cells. *Hum Mol Genet* 2005;**14**:1659–1669.

Chandra S, Gallardo G, Fernandez-Chacon R, Schluter OM, Sudhof TC. Alpha-synuclein cooperates with CSPalpha in preventing neurodegeneration. *Cell* 2005;**123**:383–396.

Chang WH, Cemal CK, Hsu YH, Kuo CL, Nukina N, Chang MH, Hu HT, Li C, Hsieh M. Dynamic expression of Hsp27 in the presence of mutant ataxin-3. *Biochem Biophys Res Commun* 2005;**336**:258–267.

Chavez Zobel AT, Lambert H, Theriault JR, Landry J. Structural instability caused by a mutation at a conserved arginine in the alpha-crystallin domain of Chinese hamster heat shock protein 27. *Cell Stress Chaperones* 2005;**10**:157–166.

Chavez Zobel AT, Loranger A, Marceau N, Theriault JR, Lambert H, Landry J. Distinct chaperone mechanisms can delay the formation of aggresomes by the myopathy-causing R120G alphaB-crystallin mutant. *Hum Mol Genet* 2003;**12**:1609–1620.

Chellgren BW, Miller AF, Creamer TP. Evidence for polyproline II helical structure in short polyglutamine tracts. *J Mol Biol* 2006;**361**:362–371.

Cohen E, Bieschke J, Perciavalle RM, Kelly JW, Dillin A. Opposing activities protect against age-onset proteotoxicity. *Science (New York, NY)* 2006;**313**:1604–1610.

Conway KA, Rochet JC, Bieganski RM, Lansbury PT Jr. Kinetic stabilization of the alpha-synuclein protofibril by a dopamine-alpha-synuclein adduct. *Science (New York, NY)* 2001;**294**:1346–1349.

da Costa CA, Ancolio K, Checler F. Wild-type but not Parkinson's disease-related ala-53–> Thr mutant alpha -synuclein protects neuronal cells from apoptotic stimuli. *J Biol Chem* 2000;**275**:24065–24069.

Crain BJ, Hu W, Sze CI, Slunt HH, Koo EH, Price DL, Thinakaran G, Sisodia SS. Expression and distribution of amyloid precursor protein-like protein-2 in Alzheimer's disease and in normal brain. *Am J Pathol* 1996;**149**:1087–1095.

Creamer TP, Campbell MN. Determinants of the polyproline II helix from modeling studies. *Adv Protein Chem* 2002;**62**:263–282.

Cuervo AM, Stefanis L, Fredenburg R, Lansbury PT, Sulzer D. Impaired degradation of mutant alpha-synuclein by chaperone-mediated autophagy. *Science (New York, NY)* 2004;**305**:1292–1295.

van Dellen A, Blakemore C, Deacon R, York D, Hannan AJ. Delaying the onset of Huntington's in mice. *Nature* 2000;**404**:721–722.

van der Putten H, Wiederhold KH, Probst A, Barbieri S, Mistl C, Danner S, Kauffmann S, Hofele K, Spooren WP, Ruegg MA, *et al.* Neuropathology in mice expressing human alpha-synuclein. *J Neurosci* 2000;**20**:6021–6029.

Derham BK, Harding JJ. Effect of aging on the chaperone-like function of human alpha-crystallin assessed by three methods. *Biochem J* 1997;**328**:(Pt 3): 763–768.

Dice JF. Chaperone-mediated autophagy. *Autophagy* 2007;**3**:295–299.

El-Agnaf OM, Jakes R, Curran MD, Wallace A. Effects of the mutations Ala30 to Pro and Ala53 to Thr on the physical and morphological properties of alpha-synuclein protein implicated in Parkinson's disease. *FEBS Lett* 1998;**440**:67–70.

Evans GJ, Morgan A, Burgoyne RD. Tying everything together: the multiple roles of cysteine string protein (CSP) in regulated exocytosis. *Traffic (Copenhagen, Denmark)* 2003;**4**:653–659.

Evgrafov OV, Mersiyanova I, Irobi J, Van Den Bosch L, Dierick I, Leung CL, Schagina O, Verpoorten N, Van Impe K, Fedotov V, *et al.* Mutant small heat-shock protein 27 causes axonal Charcot-Marie-Tooth disease and distal hereditary motor neuropathy. *Nat Genet* 2004;**36**:602–606.

Fan Y, Limprasert P, Murray IV, Smith AC, Lee VM, Trojanowski JQ, Sopher BL, La Spada AR. Beta-synuclein modulates alpha-synuclein neurotoxicity by reducing alpha-synuclein protein expression. *Hum Mol Genet* 2006;**15**:3002–3011.

Feany MB, Bender WW. A Drosophila model of Parkinson's disease. *Nature* 2000;**404**:394–398.

Fernandez-Chacon R, Wolfel M, Nishimune H, Tabares L, Schmitz F, Castellano-Munoz M, Rosenmund C, Montesinos ML, Sanes JR, Schneggenburger R, *et al.* The synaptic vesicle protein CSP alpha prevents presynaptic degeneration. *Neuron* 2004;**42**:237–251.

Friguet B, Bulteau AL, Chondrogianni N, Conconi M, Petropoulos I. Protein degradation by the proteasome and its implications in aging. *Ann N Y Acad Sci* 2000;**908**:143–154.

Fujita M, Wei J, Nakai M, Masliah E, Hashimoto M. Chaperone and anti-chaperone: two-faced synuclein as stimulator of synaptic evolution. *Neuropathology* 2006;**26**:383–392.

Galvin JE, Giasson B, Hurtig HI, Lee VM, Trojanowski JQ. Neurodegeneration with brain iron accumulation, type 1 is characterized by alpha-, beta-, and gamma-synuclein neuropathology. *Am J Pathol* 2000;**157**:361–368.

Galvin JE, Uryu K, Lee VM, Trojanowski JQ. Axon pathology in Parkinson's disease and Lewy body dementia hippocampus contains alpha-, beta-, and gamma-synuclein. *Proc Natl Acad Sci U S A* 1999;**96**:13450–13455.

Ghee M, Fournier A, Mallet J. Rat alpha-synuclein interacts with Tat binding protein 1, a component of the 26S proteasomal complex. *J Neurochem* 2000;**75**:2221–2224.

Gill AC, Ritchie MA, Hunt LG, Steane SE, Davies KG, Bocking SP, Rhie AG, Bennett AD, Hope J. Post-translational hydroxylation at the N-terminus of the prion protein reveals presence of PPII structure *in vivo*. *EMBO J* 2000;**19**:5324–5331.

Giunta B, Fernandez F, Nikolic WV, Obregon D, Rrapo E, Town T, Tan J. Inflammaging as a prodrome to Alzheimer's disease. *J Neuroinflammation* 2008;**5**: 51.

Greenbaum EA, Graves CL, Mishizen-Eberz AJ, Lupoli MA, Lynch DR, Englander SW, Axelsen PH, Giasson BI. The E46K mutation in alpha-synuclein increases amyloid fibril formation. *J Biol Chem* 2005;**280**:7800–7807.

Halagappa VK, Guo Z, Pearson M, Matsuoka Y, Cutler RG, Laferla FM, Mattson MP. Intermittent fasting and caloric restriction ameliorate age-related behavioral deficits in the triple-transgenic mouse model of Alzheimer's disease. *Neurobiol Dis* 2007;**26**:212–220.

Harding JJ, Dilley KJ. Structural proteins of the mammalian lens: a review with emphasis on changes in development, aging and cataract. *Exp Eye Res* 1976;**22**:1–73.

Hartl FU. Molecular chaperones in cellular protein folding. *Nature* 1996;**381**:571–579.

Hashimoto M, Bar-On P, Ho G, Takenouchi T, Rockenstein E, Crews L, Masliah E. Beta-synuclein regulates Akt activity in neuronal cells. A possible mechanism for neuroprotection in Parkinson's disease. *J Biol Chem* 2004a;**279**:23622–23629.

Hashimoto M, Rockenstein E, Mante M, Crews L, Bar-On P, Gage FH, Marr R, Masliah E. An antiaggregation gene therapy strategy for Lewy body disease utilizing beta-synuclein lentivirus in a transgenic model. *Gene Ther* 2004b;**11**:1713–1723.

Hashimoto M, Hsu LJ, Sisk A, Xia Y, Takeda A, Sundsmo M, Masliah E. Human recombinant NACP/alpha-synuclein is aggregated and fibrillated *in vitro*: relevance for Lewy body disease. *Brain Res* 1998;**799**:301–306.

Hashimoto M, Hsu LJ, Xia Y, Takeda A, Sisk A, Sundsmo M, Masliah E. Oxidative stress induces amyloid-like aggregate formation of NACP/alpha-synuclein *in vitro*. *Neuroreport* 1999a;**10**:717–721.

Hashimoto M, Takeda A, Hsu LJ, Takenouchi T, Masliah E. Role of cytochrome c as a stimulator of alpha-synuclein aggregation in Lewy body disease. *J Biol Chem* 1999b;**274**:28849–28852.

Hashimoto M, Masliah E. Alpha-synuclein in Lewy body disease and Alzheimer's disease. *Brain Pathol* 1999;**9**:707–720.

Hashimoto M, Rockenstein E, Mante M, Mallory M, Masliah E. beta-Synuclein inhibits alpha-synuclein aggregation: a possible role as an anti-parkinsonian factor. *Neuron* 2001;**32**:213–223.

Huang GC, Chen JJ, Liu CP, Zhou JM. Chaperone and antichaperone activities of trigger factor. *Eur J Biochem* 2002;**269**:4516–4523.

Iimoto DS, Masliah E, DeTeresa R, Terry RD, Saitoh T. Aberrant casein kinase II in Alzheimer's disease. *Brain Res* 1990;**507**:273–280.

Irobi J, Van Impe K, Seeman P, Jordanova A, Dierick I, Verpoorten N, Michalik A, De Vriendt E, Jacobs A, Van Gerwen V, *et al.* Hot-spot residue in small heat-shock protein 22 causes distal motor neuropathy. *Nat Genet* 2004;**36**:597–601.

Iwaki T, Iwaki A, Tateishi J, Sakaki Y, Goldman JE. Alpha B-crystallin and 27-kd heat shock protein are regulated by stress conditions in the central nervous system and accumulate in Rosenthal fibers. *Am J Pathol* 1993;**143**:487–495.

Jenner P, Olanow CW. Understanding cell death in Parkinson's disease. *Ann Neurol* 1998;**44**: S72–S84.

Kappe G, Franck E, Verschuure P, Boelens WC, Leunissen JA, de Jong WW. The human genome encodes 10 alpha-crystallin-related small heat shock proteins: HspB1-10. *Cell Stress Chaperones* 2003;**8**:53–61.

Kato S, Hirano A, Umahara T, Llena JF, Herz F, Ohama E. Ultrastructural and immunohistochemical studies on ballooned cortical neurons in Creutzfeldt-Jakob disease: expression of alpha B-crystallin, ubiquitin and stress-response protein 27. *Acta Neuropathol* 1992;**84**:443–448.

Kenyon C. The plasticity of aging: insights from long-lived mutants. *Cell* 2005;**120**: 449–460.

Kim TD, Paik SR, Yang CH, Kim J. Structural changes in alpha-synuclein affect its chaperone-like activity *in vitro*. *Protein Sci* 2000;**9**:2489–2496.

Kopito RR. Aggresomes, inclusion bodies and protein aggregation. *Trends Cell Biol* 2000;**10**:524–530.

Kruger R, Kuhn W, Muller T, Woitalla D, Graeber M, Kosel S, Przuntek H, Epplen JT, Schols L, Riess O. Ala30Pro mutation in the gene encoding alpha-synuclein in Parkinson's disease. *Nat Genet* 1998;**18**:106–108.

Kumar LV, Ramakrishna T, Rao CM. Structural and functional consequences of the mutation of a conserved arginine residue in alpha A and alpha B crystallins. *J Biol Chem* 1999;**274**:24137–24141.

Lansbury PT Jr. Evolution of amyloid: what normal protein folding may tell us about fibrillogenesis and disease. *Proc Natl Acad Sci U S A* 1999;**96**:3342–3344.

Lashuel HA, Hartley D, Petre BM, Walz T, Lansbury PT Jr. Neurodegenerative disease: amyloid pores from pathogenic mutations. *Nature* 2002;**418**: 291.

Lazarov O, Robinson J, Tang YP, Hairston IS, Korade-Mirnics Z, Lee VM, Hersh LB, Sapolsky RM, Mirnics K, Sisodia SS. Environmental enrichment reduces Abeta levels and amyloid deposition in transgenic mice. *Cell* 2005;**120**:701–713.

Lee SJ. Origins and effects of extracellular alpha-synuclein: implications in Parkinson's disease. *J Mol Neurosci* 2008;**34**:17–22.

Lee D, Paik SR, Choi KY. Beta-synuclein exhibits chaperone activity more efficiently than alpha-synuclein. *FEBS Lett* 2004;**576**:256–260.

Lee VM, Trojanowski JQ. Neurodegenerative tauopathies: human disease and transgenic mouse models. *Neuron* 1999;**24**:507–510.

Lippa CF, Fujiwara H, Mann DM, Giasson B, Baba M, Schmidt ML, Nee LE, O'Connell B, Pollen DA, St George-Hyslop P, *et al.* Lewy bodies contain altered alpha-synuclein in brains of many familial Alzheimer's disease patients with mutations in presenilin and amyloid precursor protein genes. *Am J Pathol* 1998;**153**:1365–1370.

Litt M, Kramer P, LaMorticella DM, Murphey W, Lovrien EW, Weleber RG. Autosomal dominant congenital cataract associated with a missense mutation in the human alpha crystallin gene CRYAA. *Hum Mol Genet* 1998;**7**:471–474.

Liu YE, Pu W, Jiang Y, Shi D, Dackour R, Shi YE. Chaperoning of estrogen receptor and induction of mammary gland proliferation by neuronal protein synuclein gamma. *Oncogene* 2007;**26**:2115–2125.

Lucin KM, Wyss-Coray T. Immune activation in brain aging and neurodegeneration: too much or too little? *Neuron* 2009;**64**:110–122.

Marsh JA, Singh VK, Jia Z, Forman-Kay JD. Sensitivity of secondary structure propensities to sequence differences between alpha- and gamma-synuclein: implications for fibrillation. *Protein Sci* 2006;**15**:2795–2804.

Masliah E, Rockenstein E, Veinbergs I, Mallory M, Hashimoto M, Takeda A, Sagara Y, Sisk A, Mucke L. Dopaminergic loss and inclusion body formation in alpha-synuclein mice: implications for neurodegenerative disorders. *Science (New York, NY)* 2000;**287**:1265–1269.

Maswood N, Young J, Tilmont E, Zhang Z, Gash DM, Gerhardt GA, Grondin R, Roth GS, Mattison J, Lane MA, *et al.* Caloric restriction increases neurotrophic factor levels and attenuates neurochemical and behavioral deficits in a primate model of Parkinson's disease. *Proc Natl Acad Sci U S A* 2004;**101**:18171–18176.

Mattson MP, Duan W, Wan R, Guo Z. Prophylactic activation of neuroprotective stress response pathways by dietary and behavioral manipulations. *NeuroRx* 2004;**1**:111–116.

McDonald DR, Brunden KR, Landreth GE. Amyloid fibrils activate tyrosine kinase-dependent signaling and superoxide production in microglia. *J Neurosci* 1997;**17**:2284–2294.

McGeer PL, McGeer EG. Glial cell reactions in neurodegenerative diseases: pathophysiology and therapeutic interventions. *Alzheimer Dis Assoc Disord* 1998;**2**:(12 Suppl): S1–S6.

McLean PJ, Kawamata H, Shariff S, Hewett J, Sharma N, Ueda K, Breakefield XO, Hyman BT. TorsinA and heat shock proteins act as molecular chaperones: suppression of alpha-synuclein aggregation. *J Neurochem* 2002;**83**:846–854.

McNaught KS, Jenner P. Proteasomal function is impaired in substantia nigra in Parkinson's disease. *Neurosci Lett* 2001;**297**:191–194.

Meehan S, Berry Y, Luisi B, Dobson CM, Carver JA, MacPhee CE. Amyloid fibril formation by lens crystallin proteins and its implications for cataract formation. *J Biol Chem* 2004;**279**:3413–3419.

Mosser DD, Morimoto RI. Molecular chaperones and the stress of oncogenesis. *Oncogene* 2004;**23**:2907–2918.

Murata S, Yashiroda H, Tanaka K. Molecular mechanisms of proteasome assembly. *Nat Rev Mol Cell Biol* 2009;**10**:104–115.

Murphy DD, Rueter SM, Trojanowski JQ, Lee VM. Synucleins are developmentally expressed, and alpha-synuclein regulates the size of the presynaptic vesicular pool in primary hippocampal neurons. *J Neurosci* 2000;**20**:3214–3220.

Narberhaus F. Alpha-crystallin-type heat shock proteins: socializing minichaperones in the context of a multichaperone network. *Microbiol Mol Biol Rev* 2002;**66**:64–93. table of contents.

Nemes Z, Devreese B, Steinert PM, Van Beeumen J, Fesus L. Cross-linking of ubiquitin, HSP27, parkin, and alpha-synuclein by gamma-glutamyl-epsilon-lysine bonds in Alzheimer's neurofibrillary tangles. *Faseb J* 2004;**18**:1135–1137.

Ninkina N, Peters O, Millership S, Salem H, van der Putten H, Buchman VL. {gamma}-synucleinopathy—neurodegeneration associated with overexpression of the mouse protein. *Hum Mol Genet* 2009;**18**:1779–1794..

Ohtake H, Limprasert P, Fan Y, Onodera O, Kakita A, Takahashi H, Bonner LT, Tsuang DW, Murray IV, Lee VM, *et al.* Beta-synuclein gene alterations in dementia with Lewy bodies. *Neurology* 2004;**63**:805–811.

Olanow CW, McNaught KS. Ubiquitin-proteasome system and Parkinson's disease. *Mov Disord* 2006;**21**:1806–1823.

Ostrerova N, Petrucelli L, Farrer M, Mehta N, Choi P, Hardy J, Wolozin B. alpha-Synuclein shares physical and functional homology with 14-3-3 proteins. *J Neurosci* 1999;**19**:5782–5791.

Ostrerova-Golts N, Petrucelli L, Hardy J, Lee JM, Farer M, Wolozin B. The A53T alpha-synuclein mutation increases iron-dependent aggregation and toxicity. *J Neurosci* 2000;**20**:6048–6054.

Pan ZZ, Bruening W, Giasson BI, Lee VM, Godwin AK. Gamma-synuclein promotes cancer cell survival and inhibits stress- and chemotherapy drug-induced apoptosis by modulating MAPK pathways. *J Biol Chem* 2002;**277**:35050–35060.

Park JY, Lansbury PT Jr. Beta-synuclein inhibits formation of alpha-synuclein protofibrils: a possible therapeutic strategy against Parkinson's disease. *Biochemistry* 2003;**42**:3696–3700.

Pasinetti GM, Eberstein JA. Metabolic syndrome and the role of dietary lifestyles in Alzheimer's disease. *J Neurochem* 2008;**106**:1503–1514.

Perng MD, Wen SF, van den IP, Prescott AR, Quinlan RA. Desmin aggregate formation by R120G alphaB-crystallin is caused by altered filament interactions and is dependent upon network status in cells. *Mol Biol Cell* 2004;**15**:2335–2346.

Perrin RJ, Woods WS, Clayton DF, George JM. Interaction of human alpha-Synuclein and Parkinson's disease variants with phospholipids. Structural analysis using site-directed mutagenesis. *J Biol Chem* 2000;**275**:34393–34398.

Polymeropoulos MH, Lavedan C, Leroy E, Ide SE, Dehejia A, Dutra A, Pike B, Root H, Rubenstein J, Boyer R, *et al.* Mutation in the alpha-synuclein gene identified in families with Parkinson's disease. *Science (New York, NY)* 1997;**276**:2045–2047.

Puig A, Gilbert HF. Protein disulfide isomerase exhibits chaperone and anti-chaperone activity in the oxidative refolding of lysozyme. *J Biol Chem* 1994a;**269**:7764–7771.

Puig A, Gilbert HF. Anti-chaperone behavior of BiP during the protein disulfide isomerase-catalyzed refolding of reduced denatured lysozyme. *J Biol Chem* 1994b;**269**:25889–25896.

Prusiner SB. The prion diseases. *Brain Pathol* 1998;**8**:499–513.

Rath A, Davidson AR, Deber CM. The structure of "unstructured" regions in peptides and proteins: role of the polyproline II helix in protein folding and recognition. *Biopolymers* 2005;**80**:179–185.

Ravikumar B, Vacher C, Berger Z, Davies JE, Luo S, Oroz LG, Scaravilli F, Easton DF, Duden R, O'Kane CJ, *et al.* Inhibition of mTOR induces autophagy and reduces toxicity of polyglutamine expansions in fly and mouse models of Huntington disease. *Nat Genet* 2004;**36**:585–595.

Rockenstein E, Hansen LA, Mallory M, Trojanowski JQ, Galasko D, Masliah E. Altered expression of the synuclein family mRNA in Lewy body and Alzheimer's disease. *Brain Res* 2001;**914**:48–56.

Saitoh T, Sundsmo M, Roch JM, Kimura N, Cole G, Schubert D, Oltersdorf T, Schenk DB. Secreted form of amyloid beta protein precursor is involved in the growth regulation of fibroblasts. *Cell* 1989;**58**:615–622.

Scheuermann S, Hambsch B, Hesse L, Stumm J, Schmidt C, Beher D, Bayer TA, Beyreuther K, Multhaup G. Homodimerization of amyloid precursor protein and its implication in the amyloidogenic pathway of Alzheimer's disease. *J Biol Chem* 2001;**276**:33923–33929.

Sharon R, Bar-Joseph I, Frosch MP, Walsh DM, Hamilton JA, Selkoe DJ. The formation of highly soluble oligomers of alpha-synuclein is regulated by fatty acids and enhanced in Parkinson's disease. *Neuron* 2003;**37**:583–595.

Shinohara H, Inaguma Y, Goto S, Inagaki T, Kato K. Alpha B crystallin and HSP28 are enhanced in the cerebral cortex of patients with Alzheimer's disease. *J Neurol Sci* 1993;**119**:203–208.

Singleton AB, Farrer M, Johnson J, Singleton A, Hague S, Kachergus J, Hulihan M, Peuralinna T, Dutra A, Nussbaum R, *et al.* alpha-Synuclein locus triplication causes Parkinson's disease. *Science (New York, NY)* 2003;**302**: 841.

Snyder H, Mensah K, Hsu C, Hashimoto M, Surgucheva IG, Festoff B, Surguchov A, Masliah E, Matouschek A, Wolozin B. beta-Synuclein reduces proteasomal inhibition by alpha-synuclein but not gamma-synuclein. *J Biol Chem* 2005;**280**:7562–7569.

Snyder H, Mensah K, Theisler C, Lee J, Matouschek A, Wolozin B. Aggregated and monomeric alpha-synuclein bind to the S 6′ proteasomal protein and inhibit proteasomal function. *J Biol Chem* 2003;**278**:11753–11759.

Soba P, Eggert S, Wagner K, Zentgraf H, Siehl K, Kreger S, Lower A, Langer A, Merdes G, Paro R, *et al.* Homo- and heterodimerization of APP family members promotes intercellular adhesion. *EMBO J* 2005;**24**:3624–3634.

Souza JM, Giasson BI, Lee VM, Ischiropoulos H. Chaperone-like activity of synucleins. *FEBS Lett* 2000;**474**:116–119.

Surgucheva I, Ninkina N, Buchman VL, Grasing K, Surguchov A. Protein aggregation in retinal cells and approaches to cell protection. *Cell Mol Neurobiol* 2005;**25**:1051–1066.

Thal LJ. Therapeutics and mild cognitive impairment: current status and future directions. *Alzheimer Dis Assoc Disord* 2003;**2**:(17 Suppl): S69–S71.

Tian R, Nie CL, He RQ. Chaperone-like manner of human neuronal tau towards lactate dehydrogenase. *Neurochem Res* 2004;**29**:1863–1872.

Trojanowski JQ, Goedert M, Iwatsubo T, Lee VM. Fatal attractions: abnormal protein aggregation and neuron death in Parkinson's disease and Lewy body dementia. *Cell Death Differ* 1998;**5**:832–837.

Tuli A, Sharma M, Capek HL, Naslavsky N, Caplan S, Solheim JC. Mechanism for amyloid precursor-like protein 2 enhancement of major histocompatibility complex class I molecule degradation. *J Biol Chem* 2009;**284**:276–307.

Uversky VN, Li J, Souillac P, Millett IS, Doniach S, Jakes R, Goedert M, Fink AL. Biophysical properties of the synucleins and their propensities to fibrillate: inhibition of alpha-synuclein assembly by beta- and gamma-synucleins. *J Biol Chem* 2002;**277**:11970–11978.

Vicart P, Caron A, Guicheney P, Li Z, Prevost MC, Faure A, Chateau D, Chapon F, Tome F, Dupret JM, *et al.* A missense mutation in the alphaB-crystallin chaperone gene causes a desmin-related myopathy. *Nat Genet* 1998;**20**:92–95.

Vila M, Ramonet D, Perier C. Mitochondrial alterations in Parkinson's disease: new clues. *J Neurochem* 2008;**107**:317–328.

Wakabayashi K, Yoshimoto M, Tsuji S, Takahashi H. Alpha-synuclein immunoreactivity in glial cytoplasmic inclusions in multiple system atrophy. *Neurosci Lett* 1998;**249**:180–182.

Walsh DM, Minogue AM, Sala Frigerio C, Fadeeva JV, Wasco W, Selkoe DJ. The APP family of proteins: similarities and differences. *Biochem Soc Trans* 2007;**35**:416–420.

Wang YL, Takeda A, Osaka H, Hara Y, Furuta A, Setsuie R, Sun YJ, Kwon J, Sato Y, Sakurai M, *et al.* Accumulation of beta- and gamma-synucleins in the ubiquitin carboxyl-terminal hydrolase L1-deficient gad mouse. *Brain Res* 2004;**1019**:1–9.

Webb JL, Ravikumar B, Atkins J, Skepper JN, Rubinsztein DC. Alpha-Synuclein is degraded by both autophagy and the proteasome. *J Biol Chem* 2003;**278**:25009–25013.

Wei J, Fujita M, Nakai M, Waragai M, Sekigawa A, Sugama S, Takenouchi T, Masliah E, Hashimoto M. Protective role of endogenous gangliosides for lysosomal pathology in a cellular model of synucleinopathies. *Am J Pathol* 2009;**174**:1891–1909.

Wei J, Fujita M, Nakai M, Waragai M, Watabe K, Akatsu H, Rockenstein E, Masliah E, Hashimoto M. Enhanced lysosomal pathology caused by beta-synuclein mutants linked to dementia with Lewy bodies. *J Biol Chem* 2007;**282**:28904–28914.

Weinreb PH, Zhen W, Poon AW, Conway KA, Lansbury PT Jr. NACP, a protein implicated in Alzheimer's disease and learning, is natively unfolded. *Biochemistry* 1996;**35**:13709–13715.

Wen FC, Li YH, Tsai HF, Lin CH, Li C, Liu CS, Lii CK, Nukina N, Hsieh M. Down-regulation of heat shock protein 27 in neuronal cells and non-neuronal cells expressing mutant ataxin-3. *FEBS Lett* 2003;**546**:307–314.

Wilhelmus MM, Otte-Holler I, Wesseling P, de Waal RM, Boelens WC, Verbeek MM. Specific association of small heat shock proteins with the pathological hallmarks of Alzheimer's disease brains. *Neuropathol Appl Neurobiol* 2006;**32**:119–130.

Witt SN. Hsp70 molecular chaperones and Parkinson's disease. *Biopolymers* 2010; **93**:218–228.

Wood SJ, Wypych J, Steavenson S, Louis JC, Citron M, Biere AL. alpha-synuclein fibrillogenesis is nucleation-dependent. Implications for the pathogenesis of Parkinson's disease. *J Biol Chem* 1999;**274**:19509–19512.

Wyttenbach A, Sauvageot O, Carmichael J, Diaz-Latoud C, Arrigo AP, Rubinsztein DC. Heat shock protein 27 prevents cellular polyglutamine toxicity and suppresses the increase of reactive oxygen species caused by huntingtin. *Hum Mol Genet* 2002;**11**:1137–1151.

Yamin G, Munishkina LA, Karymov MA, Lyubchenko YL, Uversky VN, Fink AL. Forcing nonamyloidogenic beta-synuclein to fibrillate. *Biochemistry* 2005;**44**:9096–9107.

Yoshimoto M, Iwai A, Kang D, Otero DA, Xia Y, Saitoh T. NACP, the precursor protein of the non-amyloid beta/A4 protein (A beta) component of Alzheimer disease amyloid, binds A beta and stimulates A beta aggregation. *Proc Natl Acad Sci U S A* 1995;**92**:9141–9145.

Youngman LD, Park JY, Ames BN. Protein oxidation associated with aging is reduced by dietary restriction of protein or calories. *Proc Natl Acad Sci U S A* 1992;**89**:9112–9116.

Zarranz JJ, Alegre J, Gomez-Esteban JC, Lezcano E, Ros R, Ampuero I, Vidal L, Hoenicka J, Rodriguez O, Atares B, *et al.* The new mutation, E46K, of alpha-synuclein causes Parkinson and Lewy body dementia. *Ann Neurol* 2004;**55**:164–173.

Zhang W, Wang T, Pei Z, Miller DS, Wu X, Block ML, Wilson B, Zhang W, Zhou Y, Hong JS, *et al.* Aggregated alpha-synuclein activates microglia: a process leading to disease progression in Parkinson's disease. *FASEB J* 2005;**19**:533–542.

5

THE UBIQUITIN–PROTEASOME SYSTEM IN NEURODEGENERATIVE DISEASES: MORE THAN THE USUAL SUSPECTS

Anne Bertolotti
MRC Laboratory of Molecular Biology, Cambridge, UK

5.1 UBIQUITINATED PROTEIN DEPOSITS: THE HALLMARK OF NEURODEGENERATIVE DISEASES

Accumulation of ubiquitinated proteinaceous deposits is one of the major hallmarks of most neurodegenerative diseases such as Alzheimer's disease (AD), Parkinson's disease (PD), prion disorders, amyotrophic lateral sclerosis (ALS), and polyglutamine (polyQ) expansion disorders. With the exception of the dominantly inherited polyglutamine disorders, these diseases are mostly sporadic, and usually strike in mid- or late life. In all these diseases, progressive accumulation of aberrantly folded proteins provokes dysfunction of a subset of neurons, eventually leading to their death. These disorders are clinically distinct, as they are caused by the progressive dysfunction of a specific group of neurons.

Protein Chaperones and Protection from Neurodegenerative Diseases, First Edition.
Edited by Stephan N. Witt.

Table 5.1 Proteinaceous deposits in neurodegenerative diseases

Disease	Protein Deposits	Major Component
Alzheimer's	Extracellular plaques	Aβ42
	Neurofibrillary tangles (cytoplasmic)	Hyperphosphorylated Tau
Parkinson's	Lewy bodies (cytoplasmic)	α-Synuclein
Tauopathies	Neurofibrillary tangles	Hyperphosphorylated Tau
SBMA	Nuclear inclusions	Androgen receptor
Huntington's	Inclusions	Mutant huntingtin
SCA-2	Increased cytoplasmic concentration	Ataxin-2
SCA-3	Nuclear inclusions	Ataxin-3
SCA-6	Nuclear inclusions	CACNA1A
SCA-7	Nuclear inclusions	Ataxin-7
SCA-17	Nuclear inclusions	TBP
DRPLA	Nuclear inclusions	Atrophin-1
Prion	Extracellular plaques	PrPSc
fALS	Lewy body-like inclusions (cytoplasmic)	SOD1
Sporadic ALS FTLD	Lewy body-like inclusions (cytoplasmic)	TDP-43
Sporadic ALS	Cytoplasmic inclusions	TLS/FUS

SMBA, spinal bulbar muscular atrophy; SCA, spinocerebellar ataxia; DRPLA, dentatorubropallidoluysian atrophy; fALS, familial amyotrophic lateral sclerosis; FTLD, frontotemporal lobar degeneration. Polyglutamine diseases are highlighted in gray.

The protein deposits characteristic of these diseases originate from a few normally soluble proteins (Table 5.1), with a variety of primary sequences and native structures. Yet, the different disease-causing proteins have one common property: they all have the propensity to assemble in the common amyloid filaments characterized by a cross-β structure. These amyloid-like deposits are found associated with ubiquitin in neurons of diseased brains. Because the pathological conformers of diverse proteins associated with neurodegenerative disease share common molecular characteristics, some common principles might underlie the molecular origin of neurodegeneration.

5.1.1 Protein Deposits in Neurodegenerative Diseases

AD is the most frequent neurodegenerative disease affecting 35 million people worldwide. This disease is characterized by a progressive loss of memory. The histopathological hallmarks of AD are plaques and neurofibrillary tangles in the brain (for a review, see Goedert and Spillantini, 2006). The major component of extracellular deposits is the amyloid-β 42 peptide (Aβ42), an aberrant proteolytic product of the amyloid precursor protein (APP), a single transmembrane protein (for a review, see Haass and Selkoe, 2007). Oligomerization of Aβ42 elicits a series of cellular dysfunctions, known as the *amyloid cascade*, at the origin of AD (for a review, see Hardy and Selkoe, 2002). Mutations associated with the rare cases of inherited AD have been found both in APP and in its proteolytic enzymes and increase the production of the pathogenic peptide (for a review, see

Goedert and Spillantini, 2006). Neurofibrillary tangles, which are intracellular deposits composed of hyperphosphorylated Tau assembled in paired helical filaments, represent the second neuropathological hallmark of AD (for a review, see Goedert and Spillantini, 2006). Tau inclusions have also been observed in several other neurodegenerative diseases collectively known as *tauopathies*, in the absence of Aβ42 plaques (for a review, see Ballatore *et al.*, 2007). Antibodies have been generated against paired helical filaments isolated from the brain and were found to cross-react with ubiquitin (Mori *et al.*, 1987). Antibodies directed against ubiquitin also strongly label the intracellular tangles (Perry *et al.*, 1987). Since ubiquitin marks proteins for degradation, the observation that the intracellular deposits in AD were ubiquitinated has revealed a link between these pathological hallmarks of AD and the ubiquitin–proteasome system, almost 20 years ago. The neuritic plaques were also revealed with anti-ubiquitin antibodies (Cole and Timiras, 1987).

PD is the second most frequent neurodegenerative disorder and affects 1% of individuals over 65 years of age. The hallmark of the disease is the Lewy bodies, mostly composed of α-synuclein (Spillantini *et al.*, 1997). Ubiquitin was found in Lewy bodies (Kuzuhara *et al.*, 1988) long before the discovery of the nature of their major component, α-synuclein. Ubiquitin reactivity in Lewy bodies had been observed using antibodies generated against paired helical filaments, the intracellular hallmark of AD, (Kuzuhara *et al.*, 1988; Mori *et al.*, 1987). This observation highlighted one common trait in the two distinct intracellular deposits characterizing the most frequent neurodegenerative disorders.

Polyglutamine expansion (PolyQ) disorders, such as Huntington's disease (HD), are a group of nine inherited neurodegenerative diseases. These diseases are caused by an expansion of a CAG repeat in nine different genes. The CAG expansion is translated in a polyglutamine stretch in the otherwise unrelated proteins and causes their aggregation (Table 5.1). With the exception of spinal bulbar muscular atrophy (SBMA), which is recessive, HD, the six spinocerebellar ataxia (SCA), and dentatorubropallidoluysian atrophy (DRPLA) are dominantly inherited. These disorders have distinct clinical features. However, they share similar characteristics. The polyQ disorders usually strike late in life and the age of onset inversely correlates with the length of the polyQ repeat (for a review, see Zoghbi and Orr, 2000). PolyQ diseases are progressive and are caused by the dysfunction of a subset of neurons. However, the mutated proteins are usually expressed in most cells during the entire life of individuals with the mutation. Max Perutz found that polyQ peptides form β-sheets and proposed that polyQ expansion causes aggregation of proteins (Perutz *et al.*, 1994). Ubiquitin-positive intraneuronal inclusions were subsequently found in mice transgenic for the first exon of huntingtin (Davies *et al.*, 1997). Similar inclusions containing an amino-terminal fragment of Huntingtin encompassing the polyQ expansion were found in postmortem brain tissue from HD patients (DiFiglia *et al.*, 1997). Likewise, affected regions from SCA-3 brain contain inclusions of protein ataxin-3, and these inclusions are ubiquitin-positive (Paulson *et al.*, 1997). Furthermore, in transgenic mice expressing an expanded ataxin-1, the number of Purkinje cells containing inclusions increases gradually over time (Skinner *et al.*, 1997). Thus, a common pathological feature of the polyQ expansion disorders is the deposition of the mutated proteins in neurons (Table 5.1; for a review, see Bauer and Nukina, 2009). Very often, the polyQ inclusions are ubiquitinated.

Spongiform encephalopathies, such as Creutzfeldt–Jakob disease, are a group of infectious neurodegenerative diseases. In human spongiform encephalopathies, the plaques also contain ubiquitin (Ironside *et al.*, 1993). The infectious agent is the protein PrP in a pathogenic conformation known as PrPSc. The infectious PrPSc propagates by converting the cellular and benign conformer PrP^{c} into a misfolded, toxic protein (for a review, see Prusiner, 1996). While the infectious nature of PrPSc is a unique property, recent studies have revealed that the misfolding pathology characteristic of Alzheimer's, Parkinson's and Huntington's disease and ALS can be transmitted experimentally in animal or cellular models (Meyer-Luehmann *et al.*, 2006; Desplats *et al.*, 2009; Ren *et al.*, 2009; Clavaguera *et al.*, 2009; Munch *et al.*, 2011). In addition to seeding aggregation of the intracellular, normally soluble mutant SOD1 protein, SOD1 aggregates indefinitely propagate their altered conformation and are transmissible in neuronal cells, just like the mamalian prion (Munch *et al.*, 2011).

Like other neurodegenerative diseases, ALS is a progressive disease. It is caused by the degeneration of motor neurons leading to a progressive motor weakness. Most cases are sporadic, but rare forms of the disorder can be inherited. About 20% of the familial cases (familial forms of amyotrophic lateral sclerosis, fALS) are caused by mutations in the gene encoding copper–zinc superoxide dismutase-1 (SOD1) (for a review, see Valentine *et al.*, 2005). SOD1 is highly structured and, *a priori*, one of the proteins least likely to be involved in a misfolding disease. However, more than 140, mostly missense, mutations in the *SOD1* gene cause aggregation in fALS (for reviews, see Bruijn *et al.*, 2004; Valentine *et al.*, 2005). The remarkable diversity of the effects of these mutations on SOD1 properties has suggested that they may promote aggregation through a variety of mechanisms. Some fALS-causing SOD1 mutations cause structural alterations of the protein but, intriguingly, many mutations have little or no measurable effect on its biophysical properties (Shaw and Valentine, 2007). Recently, we found that the diverse ALS-causing SOD1 mutants provoke aggregation by increasing the propensity of the affected proteins to expose hydrophobic surfaces, thereby reconciling the seemingly diverse effects of the SOD1 mutations in one unifying mechanism (Münch and Bertolotti, *et al.*, 2010). SOD1 inclusions in human fALS are ubiquitin positive (Kato *et al.*, 1996). However, it is not clear what the target of ubiquitination is in ALS inclusions since dissociation of SOD1 containing inclusions from transgenic mice revealed that their major component is the unmodified, non-ubiquitinated SOD1 protein (Shaw *et al.*, 2008). The hallmark of sporadic ALS as well as in a subset of frontotemporal dementias is the presence of Tau, SOD1, and α-synuclein negative but ubiquitin-positive inclusions (Hodges *et al.*, 2004; Leigh *et al.*, 1991; Lipton *et al.*, 2004). The transactive response DNA-binding protein-43 (TDP-43) was identified as the major pathological component of these inclusions (Sreedharan *et al.*, 2008; Van Deerlin *et al.*, 2008). In addition, mutations have been found in the gene encoding TDP-43 in familial and sporadic ALS (Sreedharan *et al.*, 2008).

The latest addition to the small group of proteins that form abnormal inclusions in neurodegenerative diseases is the protein TLS/FUS (translocated in liposarcomas/fusion). Mutations in TLS/FUS encoding gene cause aggregation of the protein and fALS (Kwiatkowski *et al.*, 2009).

Thus, in the past 15 years, research in many different laboratories has revealed that the accumulation of protein deposits in neurons is a feature common to distinct pathologies.

5.1.2 Misfolded Proteins Cause Neurodegenerative Diseases by a Gain of Toxic Function

While the mechanism by which misfolded proteins are toxic to neurons remains to be elucidated, it is now clear that misfolded proteins kill neurons by a gain of toxic properties. This comes from several lines of evidence. The rare familial forms of neurodegenerative diseases are mostly dominantly inherited, and often, the mutated gene encodes inclusion components (Table 5.1). Furthermore, the mutated proteins have the tendency to misfold and aggregate. For most diseases, transgenic animal models have been generated by overexpression of a mutant allele of the gene causing familial forms of neurodegenerative diseases. These models recapitulated essential features of the diseases. Overexpression of Aβ42, mutant APP, mutant presenilin, or a combination of them provokes of amyloid pathology in mice (for a review, see Radde *et al.*, 2008). Expression of mutant Tau (for a review, see Gotz *et al.*, 2007), overexpression of wild-type α-synuclein (Masliah *et al.*, 2000), SCA-1 (Burright *et al.*, 1995), a fragment of mutant huntingtin (Mangiarini *et al.*, 1996), different SOD1 mutants (for a review, see Valentine *et al.*, 2005), or TDP-34 (Wegorzewska *et al.*, 2009) provokes symptoms in mice that are reminiscent of the human diseases.

Because protein inclusions were identified as neuropathological features in AD, PD, prion diseases, ALS, and polyQ disorders, it was first thought that proteinaceous deposits had a causal role in neurodegeneration. Animal and cellular models of neurodegenerative diseases have challenged this belief by revealing a poor correlation between the load of amyloid plaques and the severity of the symptoms (for a review, see Ross and Poirier, 2004). This has initiated the intense debate regarding the role of protein inclusions in neurodegeneration. The exact nature of the toxic entity causing neurodegeneration remains unclear. The current belief is that misfolded proteins are more toxic than the mature amyloid. However, since multiple species may coexist and are likely to be in equilibrium with one another, it may be very difficult to identify the disease-causing species. Regardless of the oligomeric nature of the species triggering neurodegeneration, it is now clear that accumulation of misfolded proteins causes neurodegeneration. Yet, how they kill neurons remains to be established.

Cells have proteolytic systems that normally dispose of unwanted proteins. The fact that inclusions in most neurodegenerative diseases are very often revealed with antibodies specific to ubiquitin has suggested long ago that the proteasome should be blamed for not removing the potentially harmful proteins. Such deposits may further impair proteolysis and provoke a plethora of dysfunctions leading to neurodegeneration.

5.2 THE PROTEASOME

Thousands of proteins are required to ensure the function of cells and organisms and each protein carries out specific functions. The abundance of the various proteins required for proper cell function can vary tremendously, and even a given protein can be present at different levels in different cell types and under different conditions. Protein levels are tightly regulated at every level during protein synthesis and also through their controlled turnover. Protein degradation is a tightly regulated process that plays major roles in virtually all aspects of cell function. Not surprisingly, perturbation of protein degradation has pleiotropic and detrimental consequences on cell function.

5.2.1 Targeting Proteins for Proteasomal Degradation

The proteasome is a large protein complex that degrades a large number of cellular proteins in a tightly controlled manner. One important level of control in protein breakdown is achieved by specific marking of the proteins to be degraded, by the covalent attachment of the 76-amino acid protein ubiquitin (for a review, see Hershko, 2005). The carboxy-terminal glycine residue of ubiquitin is attached to the ε-amino group of lysine residues in proteins destined to be degraded. Ubiquitination requires the tightly controlled actions of a cascade of enzymes. The ubiquitin-activating enzyme E1 transfers ubiquitin to one of the several ubiquitin-conjugating enzymes, also called E2. E2s act in concert with the E3 ubiquitin ligases to transfer ubiquitin onto protein substrates (for a review, see Hershko and Ciechanover, 1998). The specificity of the system is conferred by the E3 enzymes, which recognize the substrate (Figure 5.1). The action of E1, E2,

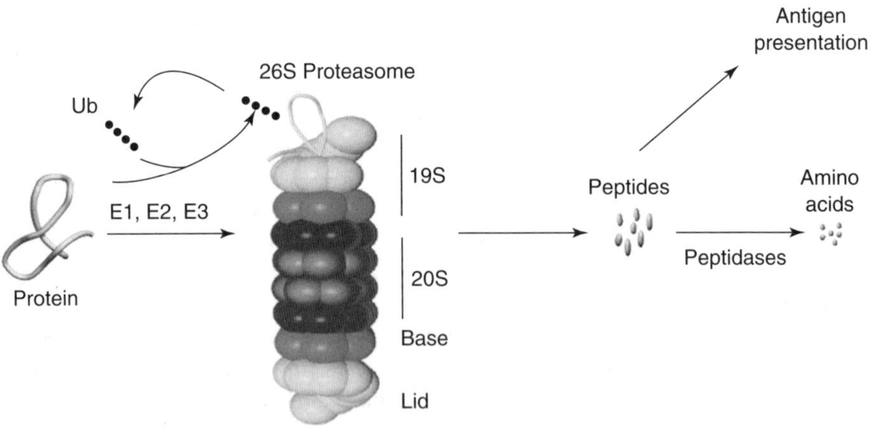

Figure 5.1 The ubiquitin-proteasome pathway. Proteins are targeted to the proteasome for degradation predominantly following cycles of covalent addition of ubiquitin (Ub) mediated by E1, E2, and E3 enzymes. The proteasome, composed of a catalytic core (20S) flanked by regulatory particles (19S), degrades proteins into peptides, further processed into amino acids by peptidases.

and E3 is reiterated many times to produce polyubiquitinated chains except when preassembled polyubiquitinated chains are transferred *en bloc* to the substrate (Li *et al.*, 2007). Polyubiquitination is usually the recognition marker for proteasomal degradation. However, there are some exceptions to this rule (for a review, see Jariel-Encontre *et al.*, 2008). The short-lived ornithine decarboxylase (ODC) is the best characterized protein degraded in a ubiquitin-independent manner (Hoyt and Coffino, 2004). Other proteins, such as the cyclin-dependent kinase inhibitor p21, do not require ubiquitin for their degradation by the proteasome (Asher *et al.*, 2006). How proteins can be targeted for degradation and recognized by the proteasome independently of ubiquitin is unclear.

5.2.2 The 26S Proteasome

The proteasome is a proteolytic machine built on the basis of a design and operating on the principles established by bacterial proteases (for a review, see Striebel *et al.*, 2009). It consists of a proteolytic cylinder, known as the *core particle* (CP) or *20S proteasome*, composed of four seven-subunit rings. The α-rings are outside and the inner β-rings contain the proteolytic subunits, with the active sites facing the interior of the chamber (Lowe *et al.*, 1995). Three of the inner ring β-type subunits, β1, β2, and β5, contain proteolytic active sites. Proteolysis is mediated by nucleophilic attack of the substrate by the amino-terminal threonine of each catalytic subunit (Seemuller *et al.*, 1995). The proteasome can cleave a broad range of peptide sequences, but the study of the degradation of short peptides has revealed that each proteolytic subunit exhibits some preference: β1 prefers to cleave after acidic residues, β2 after basic residues, and β5 after hydrophobic residues (Borissenko and Groll, 2007; Kim and Arvan, 1998). These specificities are referred to as caspase-like for β1, trypsin-like for β2, and chymotrypsin-like for β5. Interferon γ induces the expression of alternative proteolytic subunits β1i, β2i, and β5i (for a review, see Goldberg *et al.*, 2002), resulting in the immunoproteasome with altered cleavage specificities.

The interior channel of the proteasome is too narrow to accommodate folded proteins. Therefore, access to proteolytic sites of the proteasome is restricted to unfolded proteins. This feature is similar to the bacterial proteases. Thus, the structural design of the barrel-shaped proteases ensures that proteolysis is compartmentalized. To be degraded by the bacterial protease ClpP or the eukaryotic proteasome, protein substrates need to be unfolded to access the narrow pore of the proteolytic particles. Unfolding is controlled by AAA+ ATPases bound to one or both ends of the proteolytic barrel. In bacteria, the homohexameric AAA+ ATPase ClpA controls access and translocation of substrates to the proteolytic channel (for a review, see Striebel *et al.*, 2009). The base of the regulatory particle of the proteasome, also called 19S particle, contains six ATPases (Rpt1–6 for regulatory particle triple-A protein) that are thought to unfold proteins through adenosine triphosphate (ATP) hydrolysis, thereby controlling their translocation to the proteolytic channel of the proteasome. Furthermore, the AAA+ ATPases open the narrow proteolytic channel to allow substrate entry (for a review, see

Finley, 2009). In addition to the features conserved in the bacterial proteases, the proteasome contains additional components that serve important regulatory functions: substrate recognition is performed by the ubiquitin receptors Rpn10 and Rpn13 and deubiquitination of the substrate, prior to its engagement to proteolysis, is performed by Rpn11 (for a review, see Finley, 2009). Ubiquitin is not degraded with the substrate but removed prior to the degradation of its targets and recycled (Figure 5.1). The 26S proteasome is an assembly of the 20S catalytic particle flanked at one or both ends by a regulatory particle. Proteasome degradation produces short peptides that are usually efficiently processed into amino acids by cellular peptidases, except when they are presented by the major histocompatibility complex (MHC) class I complex (for a review, see Rock *et al.*, 2004).

5.2.3 Alternative Regulatory Particles

In addition to the 19S regulatory particle, mammalian cells have three alternative regulatory particles: PA28α, β(REGα, β), PA28γ (REGγ), and PA200. The common function of the distinct proteasome activators is to bind to the 20S proteasome and to activate degradation of substrates by opening the gate of the 20S narrow proteolytic chamber (for a review, see Rechsteiner and Hill, 2005). The 19S proteasome is the only regulatory particle that contains ubiquitin receptors, a deubiquitinating enzyme, and ATPases. PA28γ (REGγ) is highly expressed in the brain and localized in the nucleus and REGγ proteasomes have been found to degrade small and loosely folded proteins such as p21 in a ubiquitin- and ATP-independent manner (for a review, see Mao *et al.*, 2008). Although PA28α and $-\beta$ (REGα, $-\beta$) are expressed in many cells, they are induced by interferon and have immune function (for a review, see Rechsteiner and Hill, 2005). Because the 20S catalytic particle can bind regulatory particles at one or both ends, and since multiple regulatory particles coexist in cells, different populations of proteasomes can be generated. It is likely that the abundance of specific complexes may vary under different conditions (Shibatani *et al.*, 2006).

Selective degradation by the proteasome is achieved by tight control of each step in the process: targeting and recognition of the substrate, deubiquitination, unfolding, translocation, and proteolysis (Figure 5.1). Consequently, failure to degrade a specific protein can be due to a defect at any of these critical steps.

5.3 ASSESSING PROTEASOME DEGRADATION

5.3.1 Ubiquitination

Since ubiquitin marks proteins for degradation, analysis of the ubiquitination status of a protein has been used to determine whether a protein is degraded by the proteasome. A protein of interest can be immunoprecipitated and the immunoprecipitate can be probed with ubiquitin antibodies. Care must be taken during cell lysis to avoid deubiquitination of the substrates. To improve detection, various epitope-tagged ubiquitins have been engineered. Immunoprecipitation or immunoblots can then be carried out with the tag-specific antibody (for a review, see Rechsteiner and Hill, 2005). Recently, mass spectrometry has been used for quantitative or qualitative analysis of protein ubiquitination (Kirkpatrick *et al.*, 2005).

While degradation of proteasome substrates usually requires their ubiquitination, ubiquitination does not necessarily lead to degradation. Ubiquitin has seven lysine residues, each of which can serve for the linkage of another ubiquitin. Thus, the complexity of polyubiquitin chains is potentially enormous. A proteomic analysis of ubiquitin linkages in yeast has revealed that K48 and K11 linkages are the most abundant (28%) and K63 linkages represented 17% of the ubiquitin chain; other linkages were also found albeit with lower abundance (Xu *et al.*, 2009). K48-linked polyubiquitin is considered as the canonical signal for degradation (Chau *et al.*, 1989), while K63-linked chains serve nonproteolytic function such as DNA repair, endocytosis, and signal transduction (for a review, see Chen and Sun, 2009). The broadly used polyubiquitin monoclonal antibody FK2 does not discriminate between K48 and K63 chains (Fujimuro and Yokosawa, 2005). Thus, immunodetection of polyubiquitination with the FK2 antibody does not necessarily indicate that a protein is destined for proteasomal degradation.

Proteasome inhibition and many forms of cellular stresses provoke the accumulation of polyubiquitinated proteins. Thus, accumulation of polyubiquitinated proteins is regarded as a symptom of impaired proteasomal degradation. However, the accuracy of this diagnostic tool has been questioned by recent observations. Inactivation of ATG5, a key component of autophagic degradation in the nervous system, leads to accumulation of polyubiquitinated proteins and to neurodegeneration (Hara *et al.*, 2006). Autophagy is a degradation system that has evolved to allow cells to adapt to nutrient starvation (for a review, see Nakatogawa *et al.*, 2009). In this process, a portion of the cytoplasm is sequestered into a double-membrane-bound structure, the autophagosome, which then fuses with the lysosome/vacuole, in order to degrade its contents. Autophagy is induced upon demand but may also have some constitutive functions (Hara *et al.*, 2006): it is thought to be involved in the degradation of long-lived proteins as well as cellular organelles, whereas the proteasome degrades short-lived proteins. Since protein aggregates are most likely resistant to proteasome degradation because they cannot be easily unfolded, it has been proposed that they might be cleared by autophagy (for a review, see Winslow and Rubinsztein, 2008).

Surprisingly, young autophagy-deficient neurons accumulate detergent-soluble polyubiquinitated proteins, whereas older neurons compromised for autophagic degradation accumulate detergent-insoluble, polyubiquitinated inclusions (Hara *et al.*, 2006). The reason for this is not clear. One possible interpretation to these puzzling findings may be that when one degradation system is compromised, the other is overwhelmed. Indeed, induction of autophagy rescues the toxic effects of proteasome inhibition (Pandey *et al.*, 2007). Recent findings have revealed molecular links between autophagy and proteasome degradation and therefore suggest an alternative explanation to the observation that polyubiquitinated proteins accumulate when autophagy is compromised. The proteins p62 and NBR1 have the bivalent properties of binding simultaneously both polyubiquitinated proteins and autophagosome components. This led to the hypothesis that ubiquitin can also serve as a signal to target proteins for autophagic degradation (for a review, see Kirkin *et al.*, 2009). How one signal can target proteins to two different machineries remains to be elucidated. Nevertheless, it appears that polyubiquitination serves multiple functions. While proteasome inhibition systematically increases the abundance of polyubiquitinated proteins, monitoring the levels of polyubiquitinated proteins is not sufficient to diagnose proteasome dysfunction.

5.3.2 Proteasome Inhibitors

Several classes of proteasome inhibitors have been developed and are broadly used. These inhibitors inhibit the degradation of the vast majority of cellular proteins (Rock *et al.*, 1994). Since the proteasome controls numerous essential cellular processes, inhibition of the proteasome is detrimental to proper cell function. Thus, the proteasome has been made a drug target and proteasome inhibitors have been developed and are used in the clinic to selectively kill cancer cells (for a review, see Navon and Ciechanover, 2009). The inhibitor PS-341 (also known as *Bortezomib* or *Velcade*) is used to treat patients with multiple myeloma, a malignancy characterized by the uncontrolled proliferation of plasma cells (for a review, see Navon and Ciechanover, 2009). It is possible that proteasome inhibitors will be useful to treat other forms of cancer. However, while inhibiting the proteasome reduces the proliferation of cancer cells, it has adverse effects on neurons: patients treated with Bortezomib suffer from major neuropathic pain (Schiff *et al.*, 2009).

Proteasome inhibition has been used as an experimental paradigm to determine whether a protein is degraded by the proteasome. Proteasome inhibition leads to the stabilization of substrates, often in their polyubiquitinated form. The half-life of a substrate can also be monitored in cells lacking key components of the ubiquitin–proteasome pathway. Cells thermosensitive to E1 have been used to determine whether a protein is degraded by a ubiquitin-dependent or ubiquitin-independent pathway (for a review, see Jariel-Encontre *et al.*, 2008). Yeast cells lacking E2 or E3 are available and can be utilized to screen for the series of enzymes involved in the degradation of a given substrate. Similar experiments can be carried out in mammalian cells where RNA interference can be used to specifically reduce the abundance of selected E2 or E3. Ultimately, *in vitro* reconstitution of the degradation of a protein substrate using purified components can be used to validate other observations. Thus, unambiguous indication that a protein is degraded by the proteasome comes from multiple and converging lines of evidence.

5.3.3 Assessment of Proteasome Activity Using Small Fluorogenic Substrates

Small fluorogenic peptide substrates have been developed and provide a very convenient, sensitive, and rapid method to measure proteasome activity (for a review, see Kisselev and Goldberg, 2005). These substrates are three to four oligopeptides attached to a fluorogenic group in their carboxy-termini. The most frequently used fluorophore fused to a proteasome substrate is 7-amino-4-methylcoumarin (amc). Upon proteolytic cleavage by the proteasome, the fluorophore is released, and this produces fluorescence. Several substrates are available to measure the activities of the three active sites of the proteasomes. Suc-LLVY-amc is often used to assay chymotrypsin-like activity; Ac-RLR-amc is a substrate of the trypsin-like subunits and Ac-nLPnLD-amc is a substrate of the caspase-like sites (for a review, see Kisselev and Goldberg, 2005). The small fluorogenic peptide substrates are extremely reliable to measure the activity of purified proteasome. They are also broadly utilized to assess proteasome activity in crude extracts. However, in a cell extract, other proteases or peptidases can cleave the fluorogenic peptides. One way to reduce the contribution of lysosomal proteases in the proteolysis assay is to remove intracellular organelles during cell lysis. In addition, degradation of fluorogenic proteasome substrates needs to be performed at neutral pH to reduce the activity of lysosomal proteases since lysosomal proteases function at an acidic pH (for a review, see Kisselev and Goldberg, 2005). Control reactions are routinely performed in the presence of proteasome inhibitors. In this case, Epoxomycin must be used because so far, no target other than the proteasome has been found for this inhibitor, while other proteasome inhibitors can also inhibit other proteases (for a review, see Kisselev and Goldberg, 2005).

5.4 MONITORING PROTEASOME DYSFUNCTION

5.4.1 Accumulation of Polyubiquitinated Proteins and Degradation of Fluorogenic Peptide Substrates

The tools developed to monitor proteasome function can also be used to assess dysfunction of the proteasome. An increase in the level of polyubiquitinated proteins monitored by immunoblots or by mass spectrometry is often used to diagnose proteasome malfunction. As indicated in a previous section, ubiquitination also serves functions other than marking a protein for degradation. Thus, other readouts should be considered to validate the observations. Proteasome activity is also often evaluated in various cells and tissue lysates by quantification of the degradation of exogenous fluorogenic peptide substrates.

5.4.2 Proteasome Reporters

Different proteasome reporters have been developed in the past 10 years and are broadly used to assess proteasome dysfunction in cells or tissues. These reporters are destabilized versions of the green fluorescent protein (GFP) that are constitutively degraded unless proteasome degradation is compromised. Fusion of a noncleavable ubiquitin, where the carboxy-terminal glycine has been replaced by a valine, to the amino-terminus of GFP (Ubi-G76V-GFP) converts GFP into a short-lived protein (Dantuma *et al.*, 2000). The uncleavable ubiquitin moiety serves as an acceptor for polyubiquitin chains that target the reporter protein for degradation. The blockade of proteasomal degradation by a variety of inhibitors leads to stabilization of the reporter and augmentation of fluorescence. The fluorescence increase can be visualized by fluorescence microscopy and quantified by flow cytometry or using fluorescence plate readers. The destabilized protein is not detected on immunoblots, unless proteasomal degradation is compromised. Thus, it is believed that the reporter behaves like endogenous proteasome substrates.

Other reporters have been designed by fusing GFP with different degrons. A screen performed in yeast, designed to search for signals that target proteins for proteasomal degradation has led to the identification of short sequences that efficiently destabilize heterologous proteins (Gilon *et al.*, 1998). One such sequence, the CL1 degradation signal, efficiently destabilizes GFP in mammalian cells, when fused to its carboxy-terminus (Bence *et al.*, 2001). Proteasome inhibition stabilizes GFP-CL1 (Bence *et al.*, 2001). Both Ubi-G76V-GFP and GFP-CL1 are degraded by the ubiquitin–proteasome system. The ODC is degraded by a proteasome-dependent, ubiquitin-independent route (Hoyt and Coffino, 2004). The 37 carboxy-terminal amino acids of ODC serve as a degradation signal, and this sequence functions as an autonomous, ubiquitin-independent degron (Hoyt *et al.*, 2005). The ZsProSensor-1 (Clontech) is a fusion of ZsGreen with the mouse ODC degradation domain and is a useful proteasome reporter. Ubi-G76V-GFP and GFP-CL1 are stabilized by proteasome dysfunction and most likely also by deficiencies in their ubiquitination pathways. In contrast, ZsGreen-ODC is degraded by a ubiquitin-independent pathway and is therefore unlikely to be stabilized by alteration in ubiquitination pathways. Thus, ZsGreen-ODC may be a more direct reporter of proteasome dysfunction than the ubiquitin-dependent reporters. However, it is unclear whether degradation of ZsGreen-ODC requires additional components than the proteasome. Both the Ubi-G76V-GFP and the GFP-CL1 have been used to generate transgenic mice (Bove *et al.*, 2006; Lindsten *et al.*, 2003).

5.4.3 Monitoring the Degradation of Endogenous Proteins

The assays described above are potent and convenient. However, they all have their caveats: polyubiquitination does not necessarily lead to degradation, the fluorogenic substrate peptides can report on proteasomal activity in cell lysates but not in intact cells, and to be used, the proteasome reporters need to be overexpressed in cells or animals. Therefore, it is also useful to look at the levels of short-lived endogenous proteins to report on proteasome activity or dysfunction. We find that monitoring the abundance of the proteins p21 and E2-F1 provides robust and sensitive readouts for proteasomal activity (Rousseau *et al.*, 2009).

5.5 THE UBIQUITIN–PROTEASOME SYSTEM IN NEURODEGENERATIVE DISEASES: FACTS AND DISCREPANCIES

5.5.1 Proteasome Inhibitors Increase Inclusion Formation

Neurodegenerative diseases typically strike late in life. In the inherited forms of the disease, the mutated proteins are expressed throughout the life of the individual harboring a mutant allele, and often in many different cells. This implies that the mutated, aggregation-prone protein is expressed in a benign, presumably soluble state for many years and acquires toxic properties overtime. Even in the most severe mouse models of HD, inclusions of the mutant huntingtin fragment with as many as 150Q are only detected after a few weeks of age (Davies *et al.*, 1997). This suggests that murine neuronal cells can cope with a fragment of huntingtin causing juvenile forms of the disease in humans, at least for some time. Similarly, mutant SOD1 is soluble in the motor neurons of mice in the early weeks of age but gradually misfolds with time (Johnston *et al.*, 2000). Thus, misfolding and aggregation of the mutated protein causing neurodegeneration can be avoided, at least in young cells.

Proteasome inhibitors enhance aggregation of the amino-terminal fragment of huntingtin that builds inclusions in HD patients (Wyttenbach *et al.*, 2000). SOD1 mutants expressed in 293T cells are soluble, just like the wild-type protein (Johnston *et al.*, 2000). This is surprising because 293T cells can express very high levels of proteins. Thus, even under such experimental conditions, the aggregation-prone SOD1 mutants are soluble. Treatment of 293T cells expressing soluble SOD1 mutants with proteasome inhibitors triggers their aggregation. This treatment had no noticeable effects on the wild-type protein, despite the fact that it was expressed at much higher levels than the mutants (Johnston *et al.*, 2000).

5.5.2 Proteasome Inhibition Can Cause Neurodegeneration

The fact that inclusions in neurodegenerative diseases are often ubiquitinated together with the observation that proteasome inhibition enhances their formation has suggested that dysfunction of the proteasome could be at the origin of neurodegeneration.

Further support for this model came from various lines of investigation. Similar to the experiments done in cell culture, infusion of a selective proteasome inhibitor into the substantia nigra of rats triggered the formation of inclusion bodies containing α-synuclein, the hallmark of PD. These animals developed neurodegeneration (McNaught *et al.*, 2002), suggesting that targeting proteasome inhibition to the specific cells affected in PD can recapitulate some aspects of the disease. However, the reproducibility of these experiments has been questioned by another group (Bove *et al.*, 2006).

Clear evidence that proteasome inhibition can provoke neurodegeneration came from a study in flies. The expression of a dominant negative and temperature-sensitive mutant of the β2 subunit of the proteasome in the *Drosophila* eye provokes neurodegeneration of the retina when the flies are exposed to the nonpermissive temperature (Pandey *et al.*, 2007). Similarly, in mice, conditional inactivation of a proteasome subunit in selected neurons provokes proteasome dysfunction, accumulation of polyubiquitin conjugates, Lewy body-like inclusions, and neurodegeneration (Bedford *et al.*, 2008). Together, these findings clearly show that perturbation of proteasome function can be sufficient to provoke neurodegeneration. However, these studies do not address whether dysfunction of the proteasome is at the origin of neurodegenerative diseases. The genetic manipulation of both flies and mice lead to the inactivation of essential proteasome subunits, and it is unlikely that such a drastic ablation of proteasome function occurs during the course of neurodegenerative diseases. As will be discussed later, strong evidence supporting the hypothesis that proteasome dysfunction is at the origin of neurodegeneration is still missing, although this attractive model has been examined in many labs.

5.5.3 Can the Proteasome Digest PolyQ Peptides?

Proteasome inhibitors are classically used as an experimental paradigm to determine whether a protein is degraded by the proteasome. The fact that proteasome inhibitors stabilize many aggregation-prone proteins such as polyQ-containing proteins and SOD1 mutants has suggested that these disease-causing proteins are normally degraded by the proteasome. The proteasome digests proteins and releases short peptides that are rapidly hydrolyzed in amino acids by peptidases, unless presented on MHC class I molecules. On the basis of the digestibility of short peptides, the three classes of catalytic subunits of the proteasome were found to exhibit different specificities. The β1 caspase-like subunits prefer to cleave after acidic residues, the β2 chymotrypsin-like subunits after basic residues, and the β5 chymotrypsin-like subunits after hydrophobic residues. Thus, based on the different specificities of the proteolytic subunits of the proteasome, it was unclear whether the proteasome can cleave within the polyQ stretch. Fred Goldberg and colleagues analyzed the degradation of polyQ-containing peptides or proteins *in vitro*, using purified proteasomes and characterized the proteolytic products by mass spectrometry (Venkatraman *et al.*, 2004). Synthetic polyQ peptides are made soluble by the addition of two flanking lysines at each terminus of the peptides (Perutz *et al.*, 1994). Purified proteasomes can cleave non-polyubiquitinated polypeptides, and this can be used *in vitro* to assess the degradation of peptides or small proteins. The as-purified 20S is in a closed conformation but can be activated by the addition of small amounts of sodium dodecyl sulfate (SDS). Activated 20S eukaryotic proteasomes were found capable of a single cut after the first glutamine of 10-glutamine long peptides (Venkatraman *et al.*, 2004). Eukaryotic 26S proteasome performed similarly to the 20S proteasome on 20-glutamine long peptide. The proteolytic subunits of the archeal proteasome had relaxed specificity compared to the eukaryotic one and were found to efficiently process polyQ peptides: they cleaved within the polyQ stretch (Venkatraman *et al.*, 2004). The experiments were repeated using myoglobin fused to a stretch of 35Q and led essentially to the same conclusion: the eukaryotic proteasome could digest myoglobin but spares the polyQ stretch, whereas the archeal proteasome could cleave within the polyQ stretch (Venkatraman *et al.*, 2004). This work has been very influential because it immediately suggested a mechanism by which polyQ-containing protein could gradually become overwhelming for nerve cells: proteasome activity could decline over time and its intrinsically poor ability to digest polyQ-containing peptides may lead to the release of pure polyQ peptides, lacking the flanking sequences removed by proteasomal proteolysis. In fact, the sequences flanking the polyQ stretch in huntingtin have evolved to antagonize aggregation of the polyQ stretch (Dehay and Bertolotti, 2006; Duennwald *et al.*, 2006). Pure polyQ peptides, devoid of additional sequences, have an intrinsically high aggregation propensity (Perutz *et al.*, 1994), but the proline-rich sequences immediately adjacent to the polyQ stretch antagonize aggregation and toxicity of the protein profoundly (Dehay and Bertolotti, 2006; Duennwald *et al.*, 2006). Removal of such protective sequences by the proteasome, while leaving the polyQ stretch intact, could release aggregation-prone, pure polyQ peptides. The model proposed by Goldberg predicts that inclusions in HD contain polyQ peptides devoid of flanking regions. However, inclusions in HD patients contain a fragment of huntingtin with sequences additional to the polyQ stretch: the 17 amino-terminal amino acids as well as the proline-rich region are present in inclusions from HD patients (Lunkes *et al.*, 2002) (and our unpublished results). While various proteolytic events are involved in the production of the amino-terminal fragment of huntingtin that builds up inclusions in patients, it is not clear whether inclusions contain polyQ fragments lacking flanking sequences. However, it remains possible that such fragments are produced and elicit inclusion formation but they have not been found yet. A recent study, published by Rechsteiner and colleagues, contradicted the findings of the Goldberg group. Using a similar approach, Rechsteiner and colleagues found that a eukaryotic proteasome (activated by the proteasome activator REGγ) cleaved at multiple sites in polyQ-containing peptides (Pratt and Rechsteiner, 2008). Could it be that the different proteasome activators influence the specificity of the proteolytic subunits?

Similar to the *in vitro* data, conflicting reports questioned whether polyQ-containing proteins can be efficiently degraded in cells. PolyQ derivatives carrying a GFP moiety in their amino-termini were efficiently degraded in cells, whether they contained 103 or 25 glutamines (Michalik and Van Broeckhoven, 2004). In contrast, expanded polyQ fused to the yellow fluorescent protein (YFP) in its carboxy-terminus was a poor proteasome substrate (Holmberg *et al.*, 2004).

In the two *in vitro* studies, many of the steps that are required for efficient degradation of proteins in their cellular context are missing ubiquitination, targeting followed by deubiquitination and unfolding. In the two studies in cells cited above, the degradation of pure polyQ stretch fused to GFP or YFP has been studied. These artificial proteins most certainly lack the regions of the natural polyQ-containing proteins that normally target them to the proteasome. Thus, it would be worthwhile repeating these studies with the natural, untagged polyQ-containing proteins. We found that the amino-terminal fragments of huntingtin exhibit a similar half-life in 293T cells, whether they contain a normal or a pathogenic expansion (Rousseau *et al.*, 2004), suggesting that in these cells the degradation machinery that degrades them handles them with similar processivity.

5.5.4 Engineering Efficient Degradation of PolyQ-Containing Peptides in Cells

The conflicting studies described above raised the following question: Is the proteasome a limiting factor in the degradation of polyQ-containing proteins?

As cited earlier, several lines of evidence have suggested that huntingtin is targeted to the proteasome by a ubiquitin-dependent route—inclusions are ubiquitinated and blockage of proteasome degradation by proteasome inhibitors increase accumulation and aggregation of mutant huntingtin. Thus, if there is a factor limiting huntingtin degradation, it could be the proteasome itself or one of the many components of the ubiquitin–proteasome pathway.

To gain insight into the rate-limiting factor in huntingtin degradation, we targeted huntingtin to the proteasome using a ubiquitin-independent degron. Huntingtin with 73 glutamines causes juvenile forms of HD in human. The huntingtin amino-terminal fragment with 73 glutamines was fused to the ubiquitin-independent degron of ODC (Figure 5.2). This degron converted the aggregation-prone protein with 73 glutamines into an evanescent protein, except when proteasomal degradation was compromised by treatment with proteasome inhibitors (Rousseau *et al.*, 2009). Introduction of a point mutation

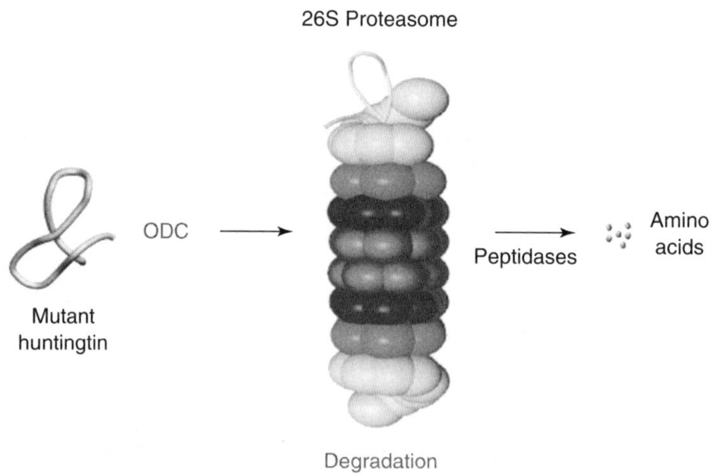

Figure 5.2 Mutant huntingtin is efficiently degraded by the proteasome, when targeted via a ubiquitin-independent route. Addition of the ubiquitin-independent degron of ornithine decarboxylase (ODC) to mutant huntingtin converts the aggregation-prone protein into an evanescent protein. (*A full color version of this figure appears in the color plate section.*)

in the ODC degron abolished its destabilizing effect on huntingtin. Since this degron was added carboxy-terminally to huntingtin, degradation ought to be initiated by the carboxy-terminus of huntingtin ODC. If the proteasome would cleave once in the polyQ stretch, then a huntingtin fragment with an intact amino-terminus followed by a 72Q-long stretch would be released. We searched for polyQ-containing fragments, both in the soluble and in the aggregated fraction after solubilization with formic acid and failed to detect them (Rousseau *et al.*, 2009). These results reveal that the cells can efficiently process long polyQ-containing peptides. Thus, neither the proteasome nor the downstream peptidases are rate-limiting in the cells we tested. Rather, the pathway by which an aggregation-prone protein is targeted to the proteasome controls its fate. The finding that the ODC degron can efficiently promote the clearance of mutant huntingtin fragment indicates that improving the delivery of aggregation-prone proteins to the proteasome could facilitate their clearance.

5.5.5 Improving Targeting Increases Degradation

Even if the proteasome is not the rate-limiting factor in polyQ degradation in cells, other components of the ubiquitin–proteasome system may be. The E2 or E3 enzymes that target aggregation-prone proteins to the proteasome have not been clearly identified. Such enzymes may be rate-limiting. There is indeed quite an abundant literature on the topic. E2-25K was identified as an interactor of huntingtin in an yeast two-hybrid screen (Kalchman *et al.*, 1996). Reducing the levels of E2-25K was later found to reduce rather than increase the levels of huntingtin aggregates (de Pril *et al.*, 2007). Increasing the level of the E3 ligase E6-AP increases degradation of mutant huntingtin, whereas reducing the abundance of this ligase has the reverse effect (Mishra *et al.*, 2008). Similar manipulations were performed on the E3 ligase Hrd1 and led to similar effects: increasing the levels of Hrd1 facilitated degradation of mutant huntingtin, whereas reducing it stabilized the protein (Yang *et al.*, 2007). It has been reported that dorfin exhibits such function on mutant SOD1: dorfin ubiquitinated SOD1 mutants and improved their degradation (Niwa *et al.*, 2002). Overexpression of the E3 ligase gp78 was found to promote the ubiquitination and degradation of SOD1 and ataxin-3, whereas knockdown of gp78 had the reverse effects (Ying *et al.*, 2009). Parkin is an E3 ligase associated with PD. Mutations in the parkin gene cause autosomal recessive inherited juvenile parkinsonism. Several potential substrates have been reported (for a review, see Adachi *et al.*, 2007). Overexpression of Parkin in cells was found to reduce aggregation of polyQ-containing proteins (Tsai *et al.*, 2003).

Different groups have reported that the E3 ligase carboxyl terminus of Hsc70-interacting protein (CHIP) facilitates the degradation of polyQ-containing proteins: overexpression of CHIP promoted degradation of polyQ-containing proteins (Miller *et al.*, 2005). HD mice with only one functional allele of CHIP display an accelerated phenotype suggesting that CHIP may also modulate neurodegeneration (Miller *et al.*, 2005). CHIP may also participate in the clearance of

Tau (Dickey *et al.*, 2006), mutant androgen receptor (Adachi *et al.*, 2007), and ataxin-1 (Al-Ramahi *et al.*, 2006). CHIP was therefore considered as one of the primary ligases for the clearance of aggregation-prone proteins. CHIP is certainly an important ligase because cells lacking CHIP have a compromised ability to survive heat stress (Dai *et al.*, 2003), revealing that under severe stress condition the function of CHIP is essential. Surprisingly, cells lacking CHIP degraded polyQ-containing proteins at the same rate (Morishima *et al.*, 2008)(and our unpublished observations). Similarly, the glucocorticoid receptor is another classical CHIP "substrate" degraded at the same rate in cells containing or lacking CHIP (Morishima *et al.*, 2008). This reveals that CHIP is not essential for the degradation of these "substrates" and suggests functional redundancy in the E3 ligases. However, CHIP function in the clearance of heat-damaged proteins cannot be compensated by other ligases (Morishima *et al.*, 2008) (and our unpublished observations). Another possible interpretation of the inconsistencies observed between the CHIP knockout cells and the cells overexpressing CHIP could be that the overexpressed ligases have a relaxed specificity compared to the endogenous ones. Even then, an artificial increase of the ligases could facilitate the clearance of aggregation-prone proteins and may have a beneficial therapeutic outcome. Since several groups have reported that different ligases can promote the degradation of polyQ-containing proteins, it appears likely that these proteins are normally targeted to the proteasome for degradation.

These studies show the following common feature: increasing the abundance of selective E3 ligases promotes the degradation of proteasome substrates, suggesting that the abundance of E3 ligases may be limiting in cells.

5.6 PROTEASOME IMPAIRMENT BY EXPANDED POLYQ-CONTAINING PROTEINS

Impairment of the proteolytic function of the proteasome has been reported to increase inclusion formation. Conversely, accumulation of aggregation-prone proteins was also reported to perturb proteasomal degradation. The first evidence that polyQ-containing proteins can impair proteasomal degradation came from the observation that expression of exon 1 of huntingtin with 103Q caused the accumulation of the GFP-CL1 reporter in cells containing aggregates (Bence *et al.*, 2001). Later, the same group reported that stabilization of the proteasome reporter GFP-CL1 precedes inclusion formation and occurs even in compartments that do not contain the aggregation-prone polyQ-containing protein (Bennett *et al.*, 2005). This indicated that the impairment of the proteasomal degradation provoked by the polyQ-containing proteins was not due to the sequestration of proteasome components in inclusions. Similarly, stabilization of p53 was observed in cells expressing huntingtin exon 1 with 150Q fused to GFP, in contrast to cells expressing the nonexpanded form of the protein (Jana *et al.*, 2001). The chymotrypsin-like activity of the proteasome was reduced upon expression of the Q150 derivative, in contrast to the nonexpanded allele (Jana *et al.*, 2001).

Similar observations were made with mutant SCA-1 in cells (Park *et al.*, 2005) and the mutant androgen receptor.

On the basis of the degradation of fluorogenic peptide substrates, it has been reported that extracts of brain tissue from deceased HD patients exhibited a reduced proteasomal activity (Seo *et al.*, 2004). Another study reported the opposite effect: increased proteasome activities in HD brains (Valera *et al.*, 2005). These results are contradicting, but it is worth stressing that it is difficult to preserve the integrity of human postmortem brain tissues. The study of the mouse models then became essential to clarify the involvement of proteasomal dysfunction in neurodegeneration. Intriguingly, the Ubi-G76V-GFP proteasome reporter was found to accumulate in the course of the degeneration of the retina of SCA-7 mice, but this effect was attributed to an increase in the levels of the transcript that consequently caused an increase in protein levels (Bowman *et al.*, 2005). The activity of the proteasome was evaluated on retina lysates by monitoring the degradation of fluorogenic peptide substrates. Since the chymotrypsin-like activity of the proteasome was unaltered in the retinal extracts of SCA-7 mice, it was concluded that neurodegeneration occurs without impairment of proteasomal degradation in SCA-7 mice (Bowman *et al.*, 2005). Similarly, evidence of impaired proteasome degradation was searched for in models of HD by monitoring the proteolytic activity of the tissue extracts on fluorogenic proteasome peptide substrates and the levels of the proteasome reporter GFP-CL1 (Bett *et al.*, 2006, 2009). These analyses revealed no impairment of the proteasome in a mouse model of HD. Ubiquitin metabolism was however altered in these mice (Bett *et al.*, 2008). After these studies, it appeared that while proteasome dysfunction is detected in cells upon (massive) expression of polyQ-containing proteins, this feature is not detectable in animal models of neurodegeneration caused by expanded polyQ proteins. More recently, it was found that polyubiquitin-marked proteins accumulate in an HD mouse model as well as in HD patients (Bennett *et al.*, 2007), but the nature of the polyubiquitinated proteins was not determined. Accumulation of ubiquitin conjugates was reported in HD mice in a separate study, and there too, this occurred independently of proteasome impairment monitored by the stability of proteasome reporters (Maynard *et al.*, 2009).

The proteasome dysfunction hypothesis has also been investigated in the context of other neurodegenerative diseases. Proteasome dysfunction has been reported in cells overexpressing SOD1 (Maat-Schieman *et al.*, 2007) and in a few cells of a mouse model of ALS (Cheroni *et al.*, 2009). PrP oligomers were reported to inhibit the 26S proteasome *in vitro*, and prion infection led to the stabilization of the Ubi-G76V-GFP reporter in a few cells (Kristiansen *et al.*, 2007).

The different studies discussed above have produced discrepancies that await clarification. Monitoring the levels of several endogenous proteasome substrates might be a useful approach to clarify the links between impairment of proteasomal degradation and accumulation of aggregation-prone proteins.

5.7 PROTEASOME ATPases AND INCLUSION FORMATION: MORE THAN THE USUAL SUSPECTS

5.7.1 Nucleation of Mutant Huntingtin Aggregation

Aggregation of polyQ has been extensively studied *in vitro* using small synthetic peptides and is thought to occur by nucleated growth polymerization (Chen *et al.*, 2002). In this model, the rate-limiting step in amyloidogenesis is the conversion of the unstructured and soluble polyQ monomer into the β-sheet conformer. Whether the polyQ peptide adopts a β-sheet conformation as a monomer or in the context of a larger assembly has been questioned in a theoretical study (Vitalis *et al.*, 2009). In contrast to the pure polyQ peptide scenario, in inclusions from HD patients, huntingtin amino-terminal fragments contain sequences in addition to the polyQ stretch, including the proline-rich region (Lunkes *et al.*, 2002), which antagonizes aggregation and toxicity (Dehay and Bertolotti, 2006) and an amino-terminal region that also modulates aggregation (Thakur *et al.*, 2009).

The finding that the proline-rich region in huntingtin antagonizes aggregation led us to propose that the proline-rich region might impose some structural constraints to the polyQ stretch and lock it in a soluble conformation, at least for some time (Dehay and Bertolotti, 2006). A study of polyQ peptides flanked by proline residues supported our earlier interpretation: polyQ peptides are unstructured but flanking proline residues prevent the formation of the β-sheet structure by imposing a polyproline type II helical structure (Darnell *et al.*, 2007). This has been further validated by a structural analysis of the amino-terminal fragment of huntingtin, which revealed that both the amino- and the carboxy-terminal flanking regions in huntingtin adopt a helical structure that propagates to the polyQ segment (Kim *et al.*, 2009). Thus, the amino-terminal fragment of huntingtin is structured. This finding may explain well why cells can express large amounts of mutant huntingtin in a soluble state (Rousseau *et al.*, 2009). This most likely reflects the disease situation. While it is clear that the polyQ expansion causes aggregation of mutant huntingtin, the mutant protein needs many years to perturb neurons in the disease context. Indeed, HD is mostly a late-onset disease, like the other neurodegenerative diseases, while the mutated protein is expressed, but harmless, for many years. We have been intrigued by the observation that only a relatively small number of cells form inclusions in different models, even upon expression of the amino-terminal fragment of huntingtin with 73Q, an expansion causing juvenile HD in humans (Stine *et al.*, 1993). In mice, expressing an amino-terminal fragment of huntingtin with 150Q, an expansion that causes the most severe forms of the disease in human, inclusion formation takes several weeks (Mangiarini *et al.*, 1996). Biochemical analyses confirmed that cells can indeed accumulate a large amount of soluble mutant huntingtin with 73Q (Rousseau *et al.*, 2009), suggesting that aggregation requires not only a high protein concentration but also additional events. Thus, while polyQ peptides inevitably aggregate *in vitro* without assistance, the situation is different in the context of the whole amino-terminal fragment of huntingtin and in the cellular environment.

5.7.2 Misfolding and Aggregation of Mutant Huntingtin is Assisted by Proteasomal ATPases

Yeast cells expressing mutant huntingtin exhibit some features that are similar to mammalian cells: mutant huntingtin derivative with 103Q is soluble in most yeast cells but forms a large inclusion in a small subset of cells (Krobitsch and Lindquist, 2000). Intriguingly, the formation of polyQ inclusions in yeast cells requires the presence of the chaperone Hsp104 (Krobitsch and Lindquist, 2000), which is also required for the propagation of yeast prions (Chernoff *et al.*, 1995). While chaperones are widely conserved throughout evolution, there is no mammalian ortholog of Hsp104. (See Chapter 7 for additional information on Hsp104.) The typical feature of Hsp104 and the related bacterial ATPase ClpB proteins is the presence of two AAA+ domains and a middle region or M-domain, which adopts a coiled-coil structure, essential for the function of Hsp104/ClpB (Martin *et al.*, 2004; Schirmer *et al.*, 2004). We have examined the AAA+ ATPases in available databases, aiming to identify human proteins containing both a coiled-coil and an AAA+ domain. We found that only the ATPase subunits of the 19S proteasome, Rpt6, Rpt4, Rpt3, Rpt2, and Rpt1, exhibit these specific features. This prompted us to investigate whether Rpts influence inclusion formation.

Strikingly, we found that the chaperone subunits of the proteasome colocalize with HD inclusions, both in primary neurons and in postmortem brain tissues from HD patients, in absence of the 20S subunits (Rousseau *et al.*, 2009). This suggested that the ATPases might exist independently of the proteolytic particle, a hypothesis we confirmed by biochemical analyses (Rousseau *et al.*, 2009). It has been previously recognized that some of the 19S ATPases have a nonproteolytic function in transcription, DNA repair, and chromatin remodeling (Ferdous *et al.*, 2001, 2007; Gonzalez *et al.*, 2002; Lassot *et al.*, 2007; Lee *et al.*, 2005; Swaffield *et al.*, 1992, 1995). Furthermore, the proteasomal ATPases exist as a complex, independent of the 20S proteolytic particle. This complex is known as APIS (ATPase proteins independent of 20S) (Gonzalez *et al.*, 2002) or free 19S-like complex (Lassot *et al.*, 2007).

We found that when 19S ATPases function independently of the 20S proteolytic complex, they facilitate the conversion of soluble mutant huntingtin into the misfolded, aggregation-prone protein. An artificial increase of the levels of the free-19S complex increased misfolding and aggregation of mutant huntingtin and SCA-3, while reducing the levels of Rpt4 or Rpt6 reduced inclusions (Lassot *et al.*, 2007). *In vitro* reconstitution experiments revealed that purified 19S particles promote the formation of huntingtin inclusions in a catalytic, ATP-dependent manner (Rousseau *et al.*, 2009). One likely interpretation of these observations is that the 19S ATPases unfold mutant huntingtin and mutant SCA-3, and if unfolding is not tightly coupled to proteolysis, unfolded intermediates are released and these species have an extremely high aggregation propensity (Figure 5.3).

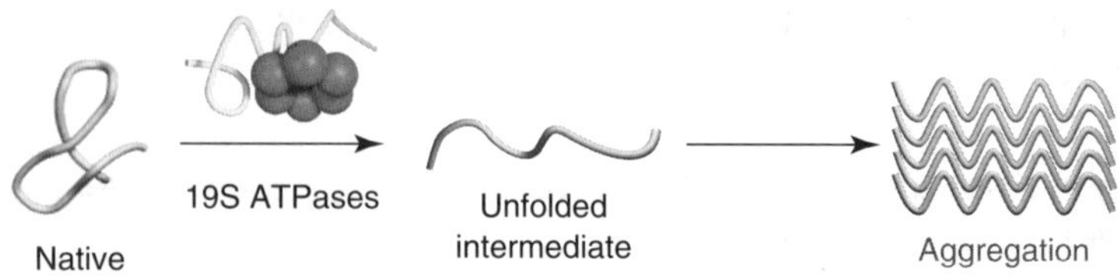

Figure 5.3 Misfolding and aggregation of polyQ-containing proteins is facilitated by 19S ATPases. When the 19S ATPases function independently of proteolysis, unfolding is uncoupled to degradation, releasing unfolded intermediates with high aggregation propensity.

Thus, aggregation of proteins with polyQ expansion requires a transition from a soluble to an aggregation-prone conformation, a process, which in cells, is facilitated by proteasomal unfoldases. This conversion is reminiscent of the pathogenic switch of PrP. The amyloid structure can therefore be considered as a pathogenic conformer. Like any other folding reaction, the formation of the amyloid fold requires the assistance of a peculiar set of chaperones. In the future, it will be interesting to see whether the 19S ATPases are involved in the formation of protein deposits formed by normally soluble, globular proteins.

5.7.3 Is Cytosolic Exposure Required for Inclusion Formation?

Protein deposits in neurodegenerative diseases originate from diverse proteins, produced in different subcellular compartments with specific quality control machineries, and accumulate either intra- or extracellularly. This raised the question whether the subcellular environment could somehow modulate aggregation propensity of aggregation-prone proteins. To address this question, we targeted the expression of mutant huntingtin to the endoplasmic reticulum (ER) or to the mitochondria (Rousseau *et al.*, 2004). While mutant huntingtin with a pathological polyQ expansion aggregated in the cytosol and in the nucleus, its aggregation was abolished when targeted to the ER or to the mitochondria. In the ER, mutant huntingtin did not elicit the unfolded protein response, indicating that it was not recognized as a misfolded protein by the ER quality control system. Once retrogradely transported outside the ER, the aggregation-prone polyglutamine-containing protein was ubiquitinated and recovered its ability to aggregate. Both in the ER and in the mitochondrial matrix, mutant huntingtin derivatives were expressed at very high levels and exhibited an extended half-life, compared to the nucleo-cytosolic derivatives. Preventing aggregation of a polyQ-containing protein of pathological length, while achieving high expression levels, is a striking finding. It indicates that polyglutamine aggregation is a property restricted to the nucleo-cytosolic compartment and suggests the existence of compartment-specific cofactors promoting or preventing aggregation of pathological proteins (Rousseau *et al.*, 2004).

The age of onset of HD is largely determined by the length of the polyQ expansion: the longer the expansion, the earlier the age of onset. However, patients with identical repeat length develop the disease at various ages. This suggests that there are some potent modifiers of disease onset. Our efforts to understand the formation of intracellular inclusions characteristic of HD have led to the identification of modifiers of this process. We found that both the sequences surrounding the aggregation-prone polyQ region and the cellular context profoundly influence mutant huntingtin aggregation.

This work has revealed that aggregation is not an inevitable consequence of the polyglutamine expansion but is strongly influenced by factors other than the polyQ stretch.

Some of the principles underlying the formation of polyQ deposits may apply to other proteins. The mammalian Prion protein PrP^c is a cell surface-anchored glycoprotein that is synthesized in the ER. What controls the switch from the benign to the infectious conformation is unclear. Intriguingly, the idea of a factor assisting PrP^c conversion was proposed long ago (Telling *et al.*, 1995). The pathogenic conversion of PrP may be initiated in the cytosol (Ma and Lindquist, 2002). Likewise, huntingtin aggregation is restricted to the nucleo-cystosolic compartment (Rousseau *et al.*, 2004).

It is intriguing that aggregation takes place in the cellular compartments that contain the proteasome. In light of our recent observations on the role of the proteasomal unfoldases in assisting misfolding and aggregation of polyQ-containing proteins, it is tempting to speculate that the conversion of PrP^c into PrPSc may be facilitated by the 19S ATPases.

5.8 CONCLUDING NOTE

Despite intense scrutiny and lot of efforts from many different laboratories, it remains unclear whether dysfunction of the proteolytic function of the proteasome underlies neurodegeneration.

Other components of the ubiquitin–proteasome system than the proteasome itself have been found to play important roles in the metabolism of aggregation-prone proteins associated with neurodegenerative diseases. The proteasomal ATPases have been long known to perform nonproteolytic function and were recently found to catalyze the formation of polyQ-containing protein aggregates. In addition, different studies have converged in revealing that manipulating the targeting of an aggregation-prone protein to the proteasome can dramatically influence its fate. Thus, a broader examination of more components of the ubiquitin–proteasome system than the proteasome itself may provide important clues to understand how aggregation-prone proteins trigger neurodegeneration.

ACKNOWLEDGMENT

I am grateful to Amila Suraweera and Christian Munch for discussions and comments on the manuscript and Graham Lingley for illustrations.

REFERENCES

Adachi H, Waza M, Tokui K, Katsuno M, Minamiyama M, Tanaka F, Doyu M, Sobue G. CHIP overexpression reduces mutant androgen receptor protein and ameliorates phenotypes of the spinal and bulbar muscular atrophy transgenic mouse model. *J Neurosci* 2007;**27**:5115–5126.

Al-Ramahi I, Lam YC, Chen HK, de Gouyon B, Zhang M, Perez AM, Branco J, de Haro M, Patterson C, Zoghbi HY, Botas J. CHIP protects from the neurotoxicity of expanded and wild-type ataxin-1 and promotes their ubiquitination and degradation. *J Biol Chem* 2006;**281**:26714–26724.

Asher G, Reuven N, Shaul Y. 20S proteasomes and protein degradation "by default". *Bioessays* 2006;**28**:844–849.

Ballatore C, Lee VM, Trojanowski JQ. Tau-mediated neurodegeneration in Alzheimer's disease and related disorders. *Nat Rev Neurosci* 2007;**8**:663–672.

Bauer PO, Nukina N. The pathogenic mechanisms of polyglutamine diseases and current therapeutic strategies. *J Neurochem* 2009;**110**:1737–1765.

Bedford L, Hay D, Devoy A, Paine S, Powe DG, Seth R, Gray T, Topham I, Fone K, Rezvani N, Mee M, Soane T, Layfield R, Sheppard PW, Ebendal T, Usoskin D, Lowe J, Mayer RJ. Depletion of 26S proteasomes in mouse brain neurons causes neurodegeneration and Lewy-like inclusions resembling human pale bodies. *J Neurosci* 2008;**28**:8189–8198.

Bence NF, Sampat RM, Kopito RR. Impairment of the ubiquitin-proteasome system by protein aggregation. *Science* 2001;**292**:1552–1555.

Bennett EJ, Bence NF, Jayakumar R, Kopito RR. Global impairment of the ubiquitin-proteasome system by nuclear or cytoplasmic protein aggregates precedes inclusion body formation. *Mol Cell* 2005;**17**:351–365.

Bennett EJ, Shaler TA, Woodman B, Ryu KY, Zaitseva TS, Becker CH, Bates GP, Schulman H, Kopito RR. Global changes to the ubiquitin system in Huntington's disease. *Nature* 2007;**448**:704–708.

Bett JS, Benn CL, Ryu KY, Kopito RR, Bates GP. The polyubiquitin Ubc gene modulates histone H2A monoubiquitylation in the R6/2 mouse model of Huntington's disease. *J Cell Mol Med* 2008;**13**:2645–2657.

Bett JS, Cook C, Petrucelli L, Bates GP. The ubiquitin-proteasome reporter GFPu does not accumulate in neurons of the R6/2 transgenic mouse model of Huntington's disease. *PLoS One* 2009;**4**: e5128.

Bett JS, Goellner GM, Woodman B, Pratt G, Rechsteiner M, Bates GP. Proteasome impairment does not contribute to pathogenesis in R6/2 Huntington's disease mice: exclusion of proteasome activator REGgamma as a therapeutic target. *Hum Mol Genet* 2006;**15**:33–44.

Borissenko L, Groll M. 20S proteasome and its inhibitors: crystallographic knowledge for drug development. *Chem Rev* 2007;**107**:687–717.

Bove J, Zhou C, Jackson-Lewis V, Taylor J, Chu Y, Rideout HJ, Wu DC, Kordower JH, Petrucelli L, Przedborski S. Proteasome inhibition and Parkinson's disease modeling. *Ann Neurol* 2006;**60**:260–264.

Bowman AB, Yoo SY, Dantuma NP, Zoghbi HY. Neuronal dysfunction in a polyglutamine disease model occurs in the absence of ubiquitin-proteasome system impairment and inversely correlates with the degree of nuclear inclusion formation. *Hum Mol Genet* 2005;**14**:679–691.

Bruijn LI, Miller TM, Cleveland DW. Unraveling the mechanisms involved in motor neuron degeneration in ALS. *Annu Rev Neurosci* 2004;**27**:723–749.

Burright EN, Clark HB, Servadio A, Matilla T, Feddersen RM, Yunis WS, Duvick LA, Zoghbi HY, Orr HT. SCA1 transgenic mice: a model for neurodegeneration caused by an expanded CAG trinucleotide repeat. *Cell* 1995;**82**:937–948.

Chau V, Tobias JW, Bachmair A, Marriott D, Ecker DJ, Gonda DK, Varshavsky A. A multiubiquitin chain is confined to specific lysine in a targeted short-lived protein. *Science* 1989;**243**:1576–1583.

Chen S, Ferrone FA, Wetzel R. Huntington's disease age-of-onset linked to polyglutamine aggregation nucleation. *Proc Natl Acad Sci U S A* 2002;**99**:11884–11889.

Chen ZJ, Sun LJ. Nonproteolytic functions of ubiquitin in cell signaling. *Mol Cell* 2009;**33**:275–286.

Chernoff YO, Lindquist SL, Ono B, Inge-Vechtomov SG, Liebman SW. Role of the chaperone protein Hsp104 in propagation of the yeast prion-like factor [psi+]. *Science* 1995;**268**:880–884.

Cheroni C, Marino M, Tortarolo M, Veglianese P, De Biasi S, Fontana E, Zuccarello LV, Maynard CJ, Dantuma NP, Bendotti C. Functional alterations of the ubiquitin-proteasome system in motor neurons of a mouse model of familial amyotrophic lateral sclerosis. *Hum Mol Genet* 2009;**18**:82–96.

Clavaguera F, Bolmont T, Crowther RA, Abramowski D, Frank S, Probst A, Fraser G, Stalder AK, Beibel M, Staufenbiel M, Jucker M, Goedert M, Tolnay M. Transmission and spreading of tauopathy in transgenic mouse brain. *Nat Cell Biol* 2009;**11**:909–913.

Cole GM, Timiras PS. Ubiquitin-protein conjugates in Alzheimer's lesions. *Neurosci Lett* 1987;**79**:207–212.

Dai Q, Zhang C, Wu Y, McDonough H, Whaley RA, Godfrey V, Li HH, Madamanchi N, Xu W, Neckers L, Cyr D, Patterson C. CHIP activates HSF1 and confers protection against apoptosis and cellular stress. *EMBO J* 2003;**22**:5446–5458.

Dantuma NP, Lindsten K, Glas R, Jellne M, Masucci MG. Short-lived green fluorescent proteins for quantifying ubiquitin/proteasome-dependent proteolysis in living cells. *Nat Biotechnol* 2000;**18**:538–543.

Darnell G, Orgel JP, Pahl R, Meredith SC. Flanking polyproline sequences inhibit beta-sheet structure in polyglutamine segments by inducing PPII-like helix structure. *J Mol Biol* 2007;**374**:688–704.

Davies SW, Turmaine M, Cozens BA, DiFiglia M, Sharp AH, Ross CA, Scherzinger E, Wanker EE, Mangiarini L, Bates GP. Formation of neuronal intranuclear inclusions underlies the neurological dysfunction in mice transgenic for the HD mutation. *Cell* 1997;**90**:537–548.

Dehay B, Bertolotti A. Critical role of the proline-rich region in Huntingtin for aggregation and cytotoxicity in yeast. *J Biol Chem* 2006;**281**:35608–35615.

de Pril R, Fischer DF, Roos RA, van Leeuwen FW. Ubiquitin-conjugating enzyme E2-25K increases aggregate formation and cell death in polyglutamine diseases. *Mol Cell Neurosci* 2007;**34**:10–19.

Desplats P, Lee HJ, Bae EJ, Patrick C, Rockenstein E, Crews L, Spencer B, Masliaha E, Leeb SJ. Inclusion formation and neuronal cell death through neuron-to-neuron transmission of α-synuclein. *Proc Natl Acad Sci U S A* 2009;**106**:13010–13015.

Dickey CA, Yue M, Lin WL, Dickson DW, Dunmore JH, Lee WC, Zehr C, West G, Cao S, Clark AM, Caldwell GA, Caldwell KA, Eckman C, Patterson C, Hutton M, Petrucelli L. Deletion of the ubiquitin ligase CHIP leads to the accumulation, but not the aggregation, of both endogenous phospho- and caspase-3-cleaved tau species. *J Neurosci* 2006;**26**:6985–6996.

DiFiglia M, Sapp E, Chase KO, Davies SW, Bates GP, Vonsattel JP, Aronin N. Aggregation of huntingtin in neuronal intranuclear inclusions and dystrophic neurites in brain. *Science* 1997;**277**:1990–1993.

Duennwald ML, Jagadish S, Muchowski PJ, Lindquist S. Flanking sequences profoundly alter polyglutamine toxicity in yeast. *Proc Natl Acad Sci U S A* 2006;**103**: 11045–11050.

Ferdous A, Gonzalez F, Sun L, Kodadek T, Johnston SA. The 19S regulatory particle of the proteasome is required for efficient transcription elongation by RNA polymerase II. *Mol Cell* 2001;**7**:981–991.

Ferdous A, Sikder D, Gillette T, Nalley K, Kodadek T, Johnston SA. The role of the proteasomal ATPases and activator monoubiquitylation in regulating Gal4 binding to promoters. *Genes Dev* 2007;**21**:112–123.

Finley D. Recognition and processing of ubiquitin-protein conjugates by the proteasome. *Annu Rev Biochem* 2009;**78**:477–513.

Fujimuro M, Yokosawa H. Production of antipolyubiquitin monoclonal antibodies and their use for characterization and isolation of polyubiquitinated proteins. *Methods Enzymol* 2005;**399**:75–86.

Gilon T, Chomsky O, Kulka RG. Degradation signals for ubiquitin system proteolysis in *Saccharomyces cerevisiae*. *EMBO J* 1998;**17**:2759–2766.

Goedert M, Spillantini MG. A century of Alzheimer's disease. *Science* 2006;**314**: 777–781.

Goldberg AL, Cascio P, Saric T, Rock KL. The importance of the proteasome and subsequent proteolytic steps in the generation of antigenic peptides. *Mol Immunol* 2002;**39**:147–164.

Gonzalez F, Delahodde A, Kodadek T, Johnston SA. Recruitment of a 19S proteasome subcomplex to an activated promoter. *Science* 2002;**296**:548–550.

Gotz J, Deters N, Doldissen A, Bokhari L, Ke Y, Wiesner A, Schonrock N, Ittner LM. A decade of tau transgenic animal models and beyond. *Brain Pathol* 2007;**17**:91–103.

Haass C, Selkoe DJ. Soluble protein oligomers in neurodegeneration: lessons from the Alzheimer's amyloid beta-peptide. *Nat Rev Mol Cell Biol* 2007;**8**:101–112.

Hara T, Nakamura K, Matsui M, Yamamoto A, Nakahara Y, Suzuki-Migishima R, Yokoyama M, Mishima K, Saito I, Okano H, Mizushima N. Suppression of basal autophagy in neural cells causes neurodegenerative disease in mice. *Nature* 2006; **441**:885–889.

Hardy J, Selkoe DJ. The amyloid hypothesis of Alzheimer's disease: progress and problems on the road to therapeutics. *Science* 2002;**297**:353–356.

Hershko A. The ubiquitin system for protein degradation and some of its roles in the control of the cell division cycle. *Cell Death Differ* 2005;**12**:1191–1197.

Hershko A, Ciechanover A. The ubiquitin system. *Annu Rev Biochem* 1998;**67**:425–479.

Hodges JR, Davies RR, Xuereb JH, Casey B, Broe M, Bak TH, Kril JJ, Halliday GM. Clinicopathological correlates in frontotemporal dementia. *Ann Neurol* 2004;**56**:399–406.

Holmberg CI, Staniszewski KE, Mensah KN, Matouschek A, Morimoto RI. Inefficient degradation of truncated polyglutamine proteins by the proteasome. *EMBO J* 2004;**23**:4307–4318.

Hoyt MA, Coffino P. Ubiquitin-free routes into the proteasome. *Cell Mol Life Sci* 2004;**61**:1596–1600.

Hoyt MA, Zhang M, Coffino P. Probing the ubiquitin/proteasome system with ornithine decarboxylase, a ubiquitin-independent substrate. *Methods Enzymol* 2005;**398**: 399–413.

Ironside JW, McCardle L, Hayward PA, Bell JE. Ubiquitin immunocytochemistry in human spongiform encephalopathies. *Neuropathol Appl Neurobiol* 1993;**19**:134–140.

Jana NR, Zemskov EA, Wang G, Nukina N. Altered proteasomal function due to the expression of polyglutamine-expanded truncated N-terminal huntingtin induces apoptosis by caspase activation through mitochondrial cytochrome c release. *Hum Mol Genet* 2001;**10**:1049–1059.

Jariel-Encontre I, Bossis G, Piechaczyk M. Ubiquitin-independent degradation of proteins by the proteasome. *Biochim Biophys Acta* 2008;**1786**:153–177.

Johnston JA, Dalton MJ, Gurney ME, Kopito RR. Formation of high molecular weight complexes of mutant Cu, Zn-superoxide dismutase in a mouse model for familial amyotrophic lateral sclerosis. *Proc Natl Acad Sci U S A* 2000;**97**:12571–12576.

Kalchman MA, Graham RK, Xia G, Koide HB, Hodgson JG, Graham KC, Goldberg YP, Gietz RD, Pickart CM, Hayden MR. Huntingtin is ubiquitinated and interacts with a specific ubiquitin-conjugating enzyme. *J Biol Chem* 1996;**271**:19385–19394.

Kato S, Shimoda M, Watanabe Y, Nakashima K, Takahashi K, Ohama E. Familial amyotrophic lateral sclerosis with a two base pair deletion in superoxide dismutase 1: gene multisystem degeneration with intracytoplasmic hyaline inclusions in astrocytes. *J Neuropathol Exp Neurol* 1996;**55**:1089–1101.

Kim PS, Arvan P. Endocrinopathies in the family of endoplasmic reticulum (ER) storage diseases: disorders of protein trafficking and the role of ER molecular chaperones. *Endocr Rev* 1998;**19**:173–202.

Kim MW, Chelliah Y, Kim SW, Otwinowski Z, Bezprozvanny I. Secondary structure of Huntingtin amino-terminal region. *Structure* 2009;**17**:1205–1212.

Kirkin V, McEwan DG, Novak I, Dikic I. A role for ubiquitin in selective autophagy. *Mol Cell* 2009;**34**:259–269.

Kirkpatrick DS, Denison C, Gygi SP. Weighing in on ubiquitin: the expanding role of mass-spectrometry-based proteomics. *Nat Cell Biol* 2005;**7**:750–757.

Kisselev AF, Goldberg AL. Monitoring activity and inhibition of 26S proteasomes with fluorogenic peptide substrates. *Methods Enzymol* 2005;**398**:364–378.

Kristiansen M, Deriziotis P, Dimcheff DE, Jackson GS, Ovaa H, Naumann H, Clarke AR, van Leeuwen FW, Menendez-Benito V, Dantuma NP, Portis JL, Collinge J, Tabrizi SJ. Disease-associated prion protein oligomers inhibit the 26S proteasome. *Mol Cell* 2007;**26**:175–188.

Krobitsch S, Lindquist S. Aggregation of huntingtin in yeast varies with the length of the polyglutamine expansion and the expression of chaperone proteins. *Proc Natl Acad Sci U S A* 2000;**97**:1589–1594.

Kuzuhara S, Mori H, Izumiyama N, Yoshimura M, Ihara Y. Lewy bodies are ubiquitinated. A light and electron microscopic immunocytochemical study. *Acta Neuropathol* 1988;**75**:345–353.

Kwiatkowski TJJ, Bosco DA, Leclerc AL, Tamrazian E, Vanderburg CR, Russ C, Davis A, Gilchrist J, Kasarskis EJ, Munsat T, Valdmanis P, Rouleau GA, Hosler BA, Cortelli P, de Jong PJ, Yoshinaga Y, Haines JL, Pericak-Vance MA, Yan J, Ticozzi N, Siddique T, McKenna-Yasek D, Sapp PC, Horvitz HR, Landers JE, Brown RHJ. Mutations in the FUS/TLS gene on chromosome 16 cause familial amyotrophic lateral sclerosis. *Science* 2009;**323**:1205–1208.

Lassot I, Latreille D, Rousset E, Sourisseau M, Linares LK, Chable-Bessia C, Coux O, Benkirane M, Kiernan RE. The proteasome regulates HIV-1 transcription by both proteolytic and nonproteolytic mechanisms. *Mol Cell* 2007;**25**:369–383.

Lee D, Ezhkova E, Li B, Pattenden SG, Tansey WP, Workman JL. The proteasome regulatory particle alters the SAGA coactivator to enhance its interactions with transcriptional activators. *Cell* 2005;**123**:423–436.

Leigh PN, Whitwell H, Garofalo O, Buller J, Swash M, Martin JE, Gallo JM, Weller RO, Anderton BH. Ubiquitin-immunoreactive intraneuronal inclusions in amyotrophic lateral sclerosis. Morphology, distribution, and specificity. *Brain* 1991;**114**:775–788.

Li W, Tu D, Brunger AT, Ye Y. A ubiquitin ligase transfers preformed polyubiquitin chains from a conjugating enzyme to a substrate. *Nature* 2007;**446**:333–337.

Lindsten K, Menendez-Benito V, Masucci MG, Dantuma NP. A transgenic mouse model of the ubiquitin/proteasome system. *Nat Biotechnol* 2003;**21**:897–902.

Lipton AM, White CLr, Bigio EH. Frontotemporal lobar degeneration with motor neuron disease-type inclusions predominates in 76 cases of frontotemporal degeneration. *Acta Neuropathol* 2004;**108**:379–385.

Lowe J, Stock D, Jap B, Zwickl P, Baumeister W, Huber R. Crystal structure of the 20S proteasome from the archaeon *T. acidophilum* at 3.4 A resolution. *Science* 1995;**268**:533–539.

Lunkes A, Lindenberg KS, Ben-Haiem L, Weber C, Devys D, Landwehrmeyer GB, Mandel JL, Trottier Y. Proteases acting on mutant huntingtin generate cleaved products that differentially build up cytoplasmic and nuclear inclusions. *Mol Cell* 2002;**10**:259–269.

Ma J, Lindquist S. Conversion of PrP to a self-perpetuating PrPSc-like conformation in the cytosol. *Science* 2002;**298**:1785–1788.

Maat-Schieman M, Roos R, Losekoot M, Dorsman J, Welling-Graafland C, Hegeman-Kleinn I, Broeyer F, Breuning M, van Duinen S. Neuronal intranuclear and neuropil inclusions for pathological assessment of Huntington's disease. *Brain Pathol* 2007;**17**:31–37.

Mangiarini L, Sathasivam K, Seller M, Cozens B, Harper A, Hetherington C, Lawton M, Trottier Y, Lehrach H, Davies SW, Bates GP. Exon 1 of the HD gene with an expanded CAG repeat is sufficient to cause a progressive neurological phenotype in transgenic mice. *Cell* 1996;**87**:493–506.

Mao I, Liu J, Li X, Luo H. REGgamma, a proteasome activator and beyond? *Cell Mol Life Sci* 2008;**65**:3971–3980.

Martin J, Gruber M, Lupas AN. Coiled coils meet the chaperone world. *Trends Biochem Sci* 2004;**29**:455–458.

Masliah E, Rockenstein E, Veinbergs I, Mallory M, Hashimoto M, Takeda A, Sagara Y, Sisk A, Mucke L. Dopaminergic loss and inclusion body formation in alpha-synuclein mice: implications for neurodegenerative disorders. *Science* 2000;**287**:1265–1269.

Maynard CJ, Bottcher C, Ortega Z, Smith R, Florea BI, Diaz-Hernandez M, Brundin P, Overkleeft HS, Li JY, Lucas JJ, Dantuma NP. Accumulation of ubiquitin conjugates in a polyglutamine disease model occurs without global ubiquitin/proteasome system impairment. *Proc Natl Acad Sci U S A* 2009;**106**:13986–13991.

McNaught KS, Bjorklund LM, Belizaire R, Isacson O, Jenner P, Olanow CW. Proteasome inhibition causes nigral degeneration with inclusion bodies in rats. *Neuroreport* 2002;**13**:1437–1441.

Meyer-Luehmann M, Coomaraswamy J, Bolmont T, Kaeser S, Schaefer C, Kilger E, Neuenschwander A, Abramowski D, Frey P, Jaton AL, Vigouret JM, Paganetti P, Walsh DM, Mathews PM, Ghiso J, Staufenbiel M, Walker LC, Jucker M. Exogenous induction of cerebral beta-amyloidogenesis is governed by agent and host. *Science* 2006;**313**:1781–1784.

Michalik A, Van Broeckhoven C. Proteasome degrades soluble expanded polyglutamine completely and efficiently. *Neurobiol Dis* 2004;**16**:202–211.

Miller VM, Nelson RF, Gouvion CM, Williams A, Rodriguez-Lebron E, Harper SQ, Davidson BL, Rebagliati MR, Paulson HL. CHIP suppresses polyglutamine aggregation and toxicity *in vitro* and *in vivo*. *J Neurosci* 2005;**25**:9152–9161.

Mishra A, Dikshit P, Purkayastha S, Sharma J, Nukina N, Jana NR. E6-AP promotes misfolded polyglutamine proteins for proteasomal degradation and suppresses polyglutamine protein aggregation and toxicity. *J Biol Chem* 2008;**283**:7648–7656.

Mori H, Kondo J, Ihara Y. Ubiquitin is a component of paired helical filaments in Alzheimer's disease. *Science* 1987;**235**:1641–1644.

Morishima Y, Wang AM, Yu Z, Pratt WB, Osawa Y, Lieberman AP. CHIP deletion reveals functional redundancy of E3 ligases in promoting degradation of both signaling proteins and expanded glutamine proteins. *Hum Mol Genet* 2008;**17**:3942–3952.

Munch C, Bertolotti A. Exposure of hydrophobic surfaces initiates aggregation of diverse ALS-causing superoxide dismutase-1 mutants. *J Mol Biol* 2010;**399**:512–525.

Munch C, O'Brien J, Bertolotti A. Prion-like propagation of mutant superoxide dismutase-1 misfolding in neuronal cells. *Proc Natl Acad Sci U S A* 2011.

Nakatogawa H, Suzuki K, Kamada Y, Ohsumi Y. Dynamics and diversity in autophagy mechanisms: lessons from yeast. *Nat Rev Mol Cell Biol* 2009;**10**:458–467.

Navon A, Ciechanover A. The 26S proteasome: from basic mechanisms to drug targeting. *J Biol Chem* 2009;**284**:33713–33718.

Niwa J, Ishigaki S, Hishikawa N, Yamamoto M, Doyu M, Murata S, Tanaka K, Taniguchi N, Sobue G. Dorfin ubiquitylates mutant SOD1 and prevents mutant SOD1-mediated neurotoxicity. *J Biol Chem* 2002;**277**:36793–36798.

Pandey UB, Nie Z, Batlevi Y, McCray BA, Ritson GP, Nedelsky NB, Schwartz SL, DiProspero NA, Knight MA, Schuldiner O, Padmanabhan R, Hild M, Berry DL, Garza D, Hubbert CC, Yao TP, Baehrecke EH, Taylor JP. HDAC6 rescues neurodegeneration and provides an essential link between autophagy and the UPS. *Nature* 2007;**447**:859–863.

Park Y, Hong S, Kim SJ, Kang S. Proteasome function is inhibited by polyglutamine-expanded ataxin-1, the SCA1 gene product. *Mol Cells* 2005;**19**:23–30.

Paulson HL, Perez MK, Trottier Y, Trojanowski JQ, Subramony SH, Das SS, Vig P, Mandel JL, Fischbeck KH, Pittman RN. Intranuclear inclusions of expanded polyglutamine protein in spinocerebellar ataxia type 3. *Neuron* 1997;**19**:333–344.

Perry G, Friedman R, Shaw G, Chau V. Ubiquitin is detected in neurofibrillary tangles and senile plaque neurites of Alzheimer disease brains. *Proc Natl Acad Sci U S A* 1987;**84**:3033–3036.

Perutz MF, Johnson T, Suzuki M, Finch JT. Glutamine repeats as polar zippers: their possible role in inherited neurodegenerative diseases. *Proc Natl Acad Sci U S A* 1994; **91**:5355–5358.

Pratt G, Rechsteiner M. Proteasomes cleave at multiple sites within polyglutamine tracts: activation by PA28 {gamma}(K188E). *J Biol Chem* 2008;**283**:12919–12925.

Prusiner SB. Human prion diseases and neurodegeneration. *Curr Top Microbiol Immunol* 1996;**207**:1–17.

Radde R, Duma C, Goedert M, Jucker M. The value of incomplete mouse models of Alzheimer's disease. *Eur J Nucl Med Mol Imaging* 2008;**35** Suppl 1: S70–S74.

Rechsteiner M, Hill CP. Mobilizing the proteolytic machine: cell biological roles of proteasome activators and inhibitors. *Trends Cell Biol* 2005;**15**:27–33.

Ren PH, Lauckner JE, Kachirskaia I, Heuser JE, Melki R, Kopito RR. Cytoplasmic penetration and persistent infection of mammalian cells by polyglutamine aggregates. *Nat Cell Biol* 2009;**11**:219–225.

Rock KL, Gramm C, Rothstein L, Clark K, Stein R, Dick L, Hwang D, Goldberg AL. Inhibitors of the proteasome block the degradation of most cell proteins and the generation of peptides presented on MHC class I molecules. *Cell* 1994;**78**:761–771.

Rock KL, York IA, Goldberg AL. Post-proteasomal antigen processing for major histocompatibility complex class I presentation. *Nat Immunol* 2004;**5**:670–677.

Ross CA, Poirier MA. Protein aggregation and neurodegenerative disease. *Nat Med* 2004;**10** Suppl:S10–S17.

Rousseau E, Dehay B, Ben-Haiem L, Trottier Y, Morange M, Bertolotti A. Targeting expression of expanded polyglutamine proteins to the endoplasmic reticulum or mitochondria prevents their aggregation. *Proc Natl Acad Sci U S A* 2004;**101**:9648–9653.

Rousseau E, Kojima R, Hoffner G, Djian P, Bertolotti A. Misfolding of proteins with a polyglutamine expansion is facilitated by proteasomal chaperones. *J Biol Chem* 2009;**284**:1917–1929.

Schiff D, Wen PY, van den Bent MJ. Neurological adverse effects caused by cytotoxic and targeted therapies. *Nat Rev Clin Oncol* 2009;**6**:596–603.

Schirmer EC, Homann OR, Kowal AS, Lindquist S. Dominant gain-of-function mutations in Hsp104p reveal crucial roles for the middle region. *Mol Biol Cell* 2004;**15**: 2061–2072.

Seemuller E, Lupas A, Stock D, Lowe J, Huber R, Baumeister W. Proteasome from *Thermoplasma acidophilum*: a threonine protease. *Science* 1995;**268**:579–582.

Seo H, Sonntag KC, Isacson O. Generalized brain and skin proteasome inhibition in Huntington's disease. *Ann Neurol* 2004;**56**:319–328.

Shaw BF, Lelie HL, Durazo A, Nersissian AM, Xu G, Chan PK, Gralla EB, Tiwari A, Hayward LJ, Borchelt DR, Valentine JS, Whitelegge JP. Detergent-insoluble aggregates associated with amyotrophic lateral sclerosis in transgenic mice contain primarily full-length, unmodified superoxide dismutase-1. *J Biol Chem* 2008;**283**:8340–8350.

Shaw BF, Valentine JS. How do ALS-associated mutations in superoxide dismutase 1 promote aggregation of the protein? *Trends Biochem Sci* 2007;**32**:78–85.

Shibatani T, Carlson EJ, Larabee F, McCormack AL, Fruh K, Skach WR. Global organization and function of mammalian cytosolic proteasome pools: implications for PA28 and 19S regulatory complexes. *Mol Biol Cell* 2006;**17**:4962–4971.

Skinner PJ, Koshy BT, Cummings CJ, Klement IA, Helin K, Servadio A, Zoghbi HY, Orr HT. Ataxin-1 with an expanded glutamine tract alters nuclear matrix-associated structures. *Nature* 1997;**389**:971–974.

Spillantini MG, Schmidt ML, Lee VM, Trojanowski JQ, Jakes R, Goedert M. Alpha-synuclein in Lewy bodies. *Nature* 1997;**388**(6645): 839–840.

Sreedharan J, Blair IP, Tripathi VB, Hu X, Vance C, Rogelj B, Ackerley S, Durnall JC, Williams KL, Buratti E, Baralle F, de Belleroche J, Mitchell JD, Leigh PN, Al-Chalabi A, Miller CC, Nicholson G, Shaw CE. TDP-43 mutations in familial and sporadic amyotrophic lateral sclerosis. *Science* 2008;**319**:1668–1672.

Stine OC, Pleasant N, Franz ML, Abbott MH, Folstein SE, Ross CA. Correlation between the onset age of Huntington's disease and length of the trinucleotide repeat in IT-15. *Hum Mol Genet* 1993;**2**:1547–1549.

Striebel F, Kress W, Weber-Ban E. Controlled destruction: AAA+ ATPases in protein degradation from bacteria to eukaryotes. *Curr Opin Struct Biol* 2009;**19**:209–217.

Swaffield JC, Bromberg JF, Johnston SA. Alterations in a yeast protein resembling HIV Tat-binding protein relieve requirement for an acidic activation domain in GAL4. *Nature* 1992;**357**:698–700.

Swaffield JC, Melcher K, Johnston SA. A highly conserved ATPase protein as a mediator between acidic activation domains and the TATA-binding protein. *Nature* 1995;**374**:88–91.

Telling GC, Scott M, Mastrianni J, Gabizon R, Torchia M, Cohen FE, DeArmond SJ, Prusiner SB. Prion propagation in mice expressing human and chimeric PrP transgenes implicates the interaction of cellular PrP with another protein. *Cell* 1995;**83**:79–90.

Thakur AK, Jayaraman M, Mishra R, Thakur M, Chellgren VM, Byeon IJ, Anjum DH, Kodali R, Creamer TP, Conway JF, Gronenborn AM, Wetzel R. Polyglutamine disruption of the huntingtin exon 1N terminus triggers a complex aggregation mechanism. *Nat Struct Mol Biol* 2009;**16**:380–389.

Tsai YC, Fishman PS, Thakor NV, Oyler GA. Parkin facilitates the elimination of expanded polyglutamine proteins and leads to preservation of proteasome function. *J Biol Chem* 2003;**278**:22044–22055.

Valentine JS, Doucette PA, Zittin Potter S. Copper-zinc superoxide dismutase and amyotrophic lateral sclerosis. *Annu Rev Biochem* 2005;**74**:563–593.

Valera AG, Diaz-Hernandez M, Hernandez F, Ortega Z, Lucas JJ. The ubiquitin-proteasome system in Huntington's disease. *Neuroscientist* 2005;**11**:583–594.

Van Deerlin VM, Leverenz JB, Bekris LM, Bird TD, Yuan W, Elman LB, Clay D, Wood EM, Chen-Plotkin AS, Martinez-Lage M, Steinbart E, McCluskey L, Grossman M, Neumann M, Wu IL, Yang WS, Kalb R, Galasko DR, Montine TJ, Trojanowski JQ, Lee VM, Schellenberg GD, Yu CE. TARDBP mutations in amyotrophic lateral sclerosis with TDP-43 neuropathology: a genetic and histopathological analysis. *Lancet Neurol* 2008;**7**:409–416.

Venkatraman P, Wetzel R, Tanaka M, Nukina N, Goldberg AL. Eukaryotic proteasomes cannot digest polyglutamine sequences and release them during degradation of polyglutamine-containing proteins. *Mol Cell* 2004;**14**:95–104.

Vitalis A, Lyle N, Pappu RV. Thermodynamics of beta-sheet formation in polyglutamine. *Biophys J* 2009;**97**:303–311.

Wegorzewska I, Bell S, Cairns NJ, Miller TM, Baloh RH. TDP-43 mutant transgenic mice develop features of ALS and frontotemporal lobar degeneration. *Proc Natl Acad Sci U S A* 2009;**106**:18809–18814.

Winslow AR, Rubinsztein DC. Autophagy in neurodegeneration and development. *Biochim Biophys Acta* 2008;**1782**:723–729.

Wyttenbach A, Carmichael J, Swartz J, Furlong RA, Narain Y, Rankin J, Rubinsztein DC. Effects of heat shock, heat shock protein 40 (HDJ-2), and proteasome inhibition on protein aggregation in cellular models of Huntington's disease. *Proc Natl Acad Sci U S A* 2000;**97**:2898–2903.

Xu P, Duong DM, Seyfried NT, Cheng D, Xie Y, Robert J, Rush J, Hochstrasser M, Finley D, Peng J. Quantitative proteomics reveals the function of unconventional ubiquitin chains in proteasomal degradation. *Cell* 2009;**137**:133–145.

Yang H, Zhong X, Ballar P, Luo S, Shen Y, Rubinsztein DC, Monteiro MJ, Fang S. Ubiquitin ligase Hrd1 enhances the degradation and suppresses the toxicity of polyglutamine-expanded huntingtin. *Exp Cell Res* 2007;**313**:538–550.

Ying Z, Wang H, Fan H, Zhu X, Zhou J, Fei E, Wang G. Gp78, an ER associated E3, promotes SOD1 and ataxin-3 degradation. *Hum Mol Genet* 2009;**18**:4268–4281.

Zoghbi HY, Orr HT. Glutamine repeats and neurodegeneration. *Annu Rev Neurosci* 2000;**23**:217–247.

6

REGULATION OF THE POLYGLUTAMINE ANDROGEN RECEPTOR BY THE Hsp90/Hsp70-BASED CHAPERONE MACHINERY

Andrew P. Lieberman

Department of Pathology, University of Michigan Medical School, Ann Arbor, MI, USA

William B. Pratt

Department of Pharmacology, University of Michigan Medical School, Ann Arbor, MI, USA

6.1 INTRODUCTION

The androgen receptor is a member of the superfamily of ligand-activated nuclear receptor transcription factors and plays important roles in human development and disease (Roy *et al.*, 2001; Shen and Coetzee, 2005; Zitzmann and Nieschlag, 2003). The receptor's functional domains include a highly conserved central DNA-binding domain through which the activated receptor binds androgen-responsive DNA elements. Carboxy-terminal ligand binding and amino-terminal transactivating domains flank this region. Ligand binding to the receptor's carboxy-terminal domain promotes its translocation from the

Protein Chaperones and Protection from Neurodegenerative Diseases, First Edition.
Edited by Stephan N. Witt.

cytoplasm to the nucleus and initiates hormone response. The receptor's central DNA-binding domain targets it to promoter regions of androgen-responsive genes. The androgen receptor's amino-terminal transactivation domain, which is unusually large and shares little similarity with functionally homologous regions of other nuclear hormone receptors, recruits a diversity of factors that coordinate transcriptional activity. Interactions between the amino- and carboxy-terminal domains occur for androgen receptor as for other receptors and are particularly important for ligand binding and transcriptional regulation.

Because the *androgen receptor* gene is on the X chromosome, many naturally occurring mutations, such as those causing androgen insensitivity syndrome, are expressed phenotypically in males (Gottlieb *et al.*, 2004). These have provided extensive structure/function information and have contributed to our understanding of androgen receptor action. Among sequence variants in the *androgen receptor* gene, the one that has attracted considerable attention is the length of a CAG repeat in the region encoding the receptor's amino-terminal transactivating domain (Zitzmann and Nieschlag, 2003). The glutamine tract encoded by this repeat is polymorphic in length in the normal population, containing between 9 and 37 residues. Pathologic expansions of this tract to 40 or more glutamines cause Kennedy's disease, or spinal and bulbar muscular atrophy—a slowly progressive degenerative disorder that affects only men (La Spada *et al.*, 1991). In this chapter, we review what we have come to know about the mechanisms by which the expanded glutamine androgen receptor causes disease. As studies in cellular and animal models have established that pathways regulating the degradation of the mutant protein are potent modifiers of the disease phenotype, we will focus particularly on the data defining the role of the Hsp90/Hsp70-based chaperone machinery in this process. For a broader discussion of mechanisms implicated in the pathogenesis of CAG repeat disorders, several excellent reviews are available (Orr and Zoghbi, 2007; Zoghbi and Orr, 2000).

6.2 KENNEDY'S DISEASE

The adult onset neurodegenerative disorders include a diverse collection of chronic, progressive diseases that show selective vulnerability of distinct neuronal populations and accumulate abnormally processed or mutant proteins that misfold and aggregate. Among these disorders are the ones caused by expansions of CAG/glutamine tracts (Orr and Zoghbi, 2007; Zoghbi and Orr, 2000). Kennedy's disease, a member of this group, is a progressive neuromuscular disorder with clinical onset in adolescence to adulthood that is characterized initially by muscle cramps and elevated serum creatine kinase (Katsuno *et al.*, 2006a; Sperfeld *et al.*, 2002). These myopathic features commonly precede muscle weakness, which inevitably develops as the disease progresses and is most severe in the proximal limb and bulbar muscles (Katsuno *et al.*, 2006a; Sperfeld *et al.*, 2002). The clinical features of Kennedy's disease correlate with a loss of motor neurons in the brainstem and spinal cord and with marked myopathic and neurogenic changes in skeletal muscle (Kennedy *et al.*, 1968; Sobue *et al.*, 1989). Patients may also develop a subclinical sensory neuropathy. The age of onset, bulbar involvement, and X-linked inheritance distinguish Kennedy's disease from other forms of motor neuron disease, such as autosomal forms of spinal muscular atrophy. Also a distinguishing feature of Kennedy's disease is that the affected males exhibit signs of partial androgen insensitivity, including gynecomastia, testicular atrophy, and decreased fertility (Arbizu *et al.*, 1983; Nagashima *et al.*, 1988). The cause of this disease is an expansion of a CAG repeat in the first exon of the *androgen receptor* gene (La Spada *et al.*, 1991). The expanded glutamine tract promotes hormone-dependent androgen receptor unfolding and aggregation, and leads to both a toxic gain of function and a partial loss of normal androgen receptor function (Chamberlain *et al.*, 1994; Fischbeck *et al.*, 1999; Kazemi-Esfarjani *et al.*, 1995; Lieberman *et al.*, 2002; Merry *et al.*, 1998; Mhatre *et al.*, 1993).

Kennedy's disease is one of the nine degenerative disorders, including Huntington's disease and several autosomal dominant spinocerebellar ataxias, caused by CAG/glutamine tract expansions. As a group, these diseases share several important features that suggest the occurrence of common mechanisms underlying neuronal dysfunction and degeneration. Each is inherited in an autosomal dominant pattern, except for Kennedy's disease, which occurs only in males (Zoghbi and Orr, 2000) because of androgen-dependent toxicity (Katsuno *et al.*, 2002; Schmidt *et al.*, 2002; Sobue *et al.*, 1993). These diseases share a similar age of onset and rate of progression, and their inheritance is characterized by genetic anticipation, wherein longer repeats are associated with an earlier onset and more severe phenotype (La Spada *et al.*, 1992). The presence of an expanded glutamine tract in a diverse set of widely expressed proteins also leads to a similar pathology—aggregates of the misfolded, disease-causing proteins in neuronal nuclei, cytoplasm, or both (Abdullah *et al.*, 1998; Becher *et al.*, 1998; David *et al.*, 1998; Davies *et al.*, 1997; Paulson *et al.*, 1997; Mauger *et al.*, 1999; Skinner *et al.*, 1997). Pathology reminiscent of that occurring in the polyglutamine diseases has been described in other disorders as well, such as neuronal intranuclear inclusion disease, for which the genetic basis is not currently known (Lieberman *et al.*, 1998, 1999). These protein aggregates, like those occurring in Alzheimer's and Parkinson's disease, are pathologic hallmarks, yet their involvement in disease pathogenesis remains controversial. Data from experimental models of Kennedy's disease (Li *et al.*, 2007) and other CAG repeat and protein aggregation disorders suggest that soluble oligomeric or monomeric species of the misfolded, mutant proteins may be most toxic.

6.3 CAG REPEAT LENGTH AND ANDROGEN RECEPTOR FUNCTION

The consequence of the glutamine tract expansion on androgen receptor function has been carefully evaluated to determine the extent to which loss-of-function contributes to the Kennedy's disease phenotype. The androgen receptor is expressed by lower motor neurons, and its activation mediates trophic effects on these cells (Kurz *et al.*, 1986; Kujawa *et al.*, 1991; Nordeen *et al.*, 1985; Yu, 1989), leading to the notion that loss of neurotrophic support might contribute to selective neuronal vulnerability. Studies in cell culture have shown that there is an inverse correlation between the length of the glutamine tract and the ability of the androgen receptor to activate exogenous reporters or endogenous androgen-responsive genes (Chamberlain *et al.*, 1994; Irvine *et al.*, 2000; Kazemi-Esfarjani *et al.*, 1995; Lieberman *et al.*, 2002; Mhatre *et al.*, 1993). This loss-of-function is associated with decreased expression of the expanded glutamine androgen receptor protein, mediated in part by its enhanced degradation by the proteasome (Lieberman *et al.*, 2002). The length of the glutamine tract also impacts the androgen receptor's transcriptional efficacy. This functional effect likely reflects a combination of factors, including influences on receptor dimerization and interactions with transcriptional coactivators such as the Ras-related protein RAN/ARA24 (Buchanan *et al.*, 2004; Hsiao *et al.*, 1999; Wang *et al.*, 2004). Although partial loss of androgen receptor function may occur in patients with Kennedy's disease, it is unlikely to play a primary role in the degenerative phenotype. This conclusion is based on the fact that loss-of-function mutations in the human *androgen receptor* gene lead to androgen insensitivity syndrome and not neuromuscular disease. In additional, transgenic overexpression of the expanded glutamine androgen receptor in mice reproduces many aspects of the Kennedy's disease phenotype in the face of continued expression of the normal, endogenous allele.

6.4 MODELS OF KENNEDY'S DISEASE

On the basis of the experimental work in cellular and animal models, several general principles have emerged that guide our understanding of disease pathogenesis. Amino-terminal fragments of the androgen receptor exhibit glutamine length-dependent toxicity (Abel *et al.*, 2001; Merry *et al.*, 1998) and may model amino-terminal fragments of the androgen receptor that are thought to accumulate in disease. Toxicity is accompanied by numerous downstream sequelae, including activation of the unfolded protein response (Thomas *et al.*, 2005), disruption of intracellular transport (Katsuno *et al.*, 2006b; Morfini *et al.*, 2006; Piccioni *et al.*, 2002; Szebenyi *et al.*, 2003), mitochondrial dysfunction, transcriptional dysregulation (McCampbell *et al.*, 2000, 2001), and alternations in RNA splicing (Yu *et al.*, 2009). Model systems expressing the full-length androgen receptor protein establish that the receptor's toxicity is dependent upon the presence of androgen (Chevalier-Larsen *et al.*, 2004; Katsuno *et al.*, 2002; McManamny *et al.*, 2002; Sopher *et al.*, 2004) and is regulated, in part, by hormone-dependent retrograde transport that is dependent upon Hsp90 function (Thomas *et al.*, 2004, 2006). This ligand dependence of the disease phenotype provides a mechanistic explanation for the minimal effects of the mutant *androgen receptor* allele in females (Sobue *et al.*, 1993; Schmidt *et al.*, 2002) and offers an important therapeutic approach for treating men with this disorder (Chevalier-Larsen *et al.*, 2004; Katsuno *et al.*, 2003).

Mouse models have been particularly informative experimental systems for studying the pathogenesis of Kennedy's disease. Among the initial models developed were transgenic mice in which the prion protein promoter or neurofilament promoter drove expression of an amino-terminal fragment of the mutant receptor (Abel *et al.*, 2001) or the androgen receptor promoter drove expression of a long glutamine tract (Adachi *et al.*, 2001). Both approaches produced neuronal dysfunction but failed to model hormone-dependent selective vulnerability of motor neurons. These mice confirmed the toxicity of long glutamine tracts to neurons and demonstrated that important aspects of the Kennedy's disease phenotype required expression of the full-length mutant protein. Similarly, more recent models in which transgenic expression of the full-length mutant androgen receptor was driven by the chicken beta-actin (Katsuno *et al.*, 2002), cytomegalovirus (CMV) (McManamny *et al.*, 2002), or prion protein promoters (Chevalier-Larsen *et al.*, 2004), or the endogenous *androgen receptor* promoter as part of a YAC transgene (Sopher *et al.*, 2004) exhibited androgen-dependent motor neuron degeneration. Partial rescue of aged, symptomatic males by surgical castration revealed that neuronal dysfunction contributes to the phenotype (Chevalier-Larsen *et al.*, 2004). These concepts led to recent clinical trials of anti-androgen therapy in Kennedy's disease patients (Banno *et al.*, 2009).

Additional efforts to model Kennedy's disease in mice used gene targeting to generate mice with humanized *androgen receptor* alleles encoding glutamine tracts of varying lengths (Albertelli *et al.*, 2006; Yu *et al.*, 2006a). The strategy to create these mice was based on the significant sequence similarity between the mouse and human androgen receptor. These proteins are virtually identical in their carboxy-terminal ligand-binding domains and central DNA-binding domains but differ by ~15% in amino acid sequence in the amino-terminal transactivating domains. Since the amino-terminal domain contains the CAG repeat and is encoded by exon one, much of the first exon of the *androgen receptor* gene was swapped between mice and humans by homologous recombination in mouse embryonic stem cells. This manipulation yielded a humanized *androgen receptor* gene under the control of endogenous mouse regulatory elements, ensuring that the androgen receptor protein was expressed at normal levels and in appropriate cell types.

Mice with targeted *androgen receptor* genes containing 12, 21, or 48 CAG repeats were similar to wild-type, indicating that humanizing the androgen receptor did not disrupt its function (Albertelli *et al.*, 2006). Males with these targeted *androgen receptor* alleles were fertile and exhibited no behavioral abnormalities, although analyses did reveal more subtle changes in the expression of androgen-responsive genes, supporting the contention that an elongated glutamine tract decreased androgen receptor function. In contrast, male mice with a humanized *androgen receptor* gene containing 113 CAG repeats (AR113Q) modeled both the systemic and hormone-dependent neuromuscular manifestations of Kennedy's disease (Yu *et al.*, 2006a, b). Systemic pathology in AR113Q males included testicular atrophy and decreased fertility that became more severe with age (Yu *et al.*, 2006a). AR113Q males also developed androgen-dependent forelimb weakness and exhibited morphologic and gene expression alterations in limb muscle that suggested both neurogenic and myopathic effects, similar to the mixed features described in the muscles of patients with Kennedy's disease (Yu *et al.*, 2006b). Skeletal muscle pathology in AR113Q mice included groups of angulated small fibers, rounded large fibers with internally displaced nuclei, and androgen receptor immunoreactive intranuclear inclusions. These changes occurred between 3–5 months and were accompanied by electrophysiological and gene expression changes indicative of both neurogenic and myopathic changes. This muscle pathology preceded changes in spinal cord by ~15 months, indicating that the effects in skeletal muscle occurred early and that lower motor neuron loss and spinal cord gliosis were late manifestations of the disease process. Similarly, distal axonal degeneration occurs as an early pathogenic event in mouse models of amyotrophic lateral sclerosis (Fischer and Glass, 2007) suggesting that this may be a common, initial lesion in motor neuron disease.

AR113Q mice revealed a myopathic contribution to the Kennedy's disease phenotype that suggests a role for skeletal muscle in disease pathogenesis (Jordan and Lieberman, 2008; Lieberman and Robins, 2008). This notion is supported by recent data from transgenic mice that overexpress the wild-type androgen receptor only in muscle and develop hormone-dependent denervation (Johansen *et al.*, 2009; Monks *et al.*, 2007). This effect of skeletal muscle pathology on motor neurons may be mediated by a loss of trophic support through decreased production of muscle-derived factors (Monks *et al.*, 2007; Yu *et al.*, 2006b). Support for this concept is provided by experiments in which transgenic overexpression of insulin growth factor-1 (IGF-1) in skeletal muscle ameliorates the phenotype of Kennedy's disease transgenic mice (Palazzolo *et al.*, 2009). These data suggest that noncell autonomous factors can regulate motor neuron function and survival in these model systems, highlighting a novel approach to therapeutic intervention.

6.5 REGULATION OF ANDROGEN RECEPTOR DEGRADATION BY Hsp90 AND Hsp70

Whether acting in motor neurons or skeletal muscle cells, the complexity of downstream mechanisms triggered by the polyglutamine androgen receptor that contributes to toxicity in Kennedy's disease poses a therapeutic challenge. Numerous potential targets have been identified, including the transcriptional and splicing machineries, intracellular trafficking pathways, and mitochondria, with strong experimental evidence supporting a role for divergent pathways as mediators of toxicity. This plurality of mechanisms suggests that one strategy to efficiently obtain therapeutic effects is to directly target the expanded glutamine androgen receptor protein itself.

Manipulating endogenous cellular systems that regulate androgen receptor protein degradation has proven effective in alleviating toxicity in several models of

Kennedy's disease. For the androgen receptor, the major chaperones involved in protein quality control decisions are Hsp90 and Hsp70, which act together in a multichaperone machinery to regulate the function, trafficking, and turnover of a wide variety of signaling proteins (Pratt and Toft, 2003). Over the past decade, both advances in our understanding of how Hsp90 interacts with proteins and the discovery of the role of chaperone-dependent E3 ligases in protein ubiquitination have contributed to a general model of how Hsp90 and Hsp70 work together to select for degradation of both aberrant proteins, such as the expanded glutamine androgen receptor and nonmutant proteins that have undergone toxic damage. Although the Hsp90 chaperone machinery also affects the function and trafficking of the androgen receptor (Pratt and Toft, 2003), we focus here on the way the machinery functions in protein quality control.

We propose that Hsp90 and Hsp70 have essentially opposing roles in the triage of damaged proteins, in that Hsp70 promotes substrate ubiquitination whereas Hsp90 inhibits ubiquitination. Such opposing effects of the two chaperones have been demonstrated in experiments utilizing purified ubiquitinating enzymes and an Hsp90-regulated signaling protein (Peng *et al.*, 2009). In this model, we envision that as proteins undergo toxic damage, their ligand-binding clefts open to expose hydrophobic residues as the initial step in unfolding. An expanded glutamine tract in the amino terminus of the androgen receptor destabilizes the ligand-bound receptor and promotes unfolding. The Hsp90 chaperone machinery regulates signaling proteins by modulating ligand-binding clefts (reviewed in Pratt *et al.*, 2008, 2010). When cleft opening is such that Hsp90 can no longer interact with the protein to inhibit ubiquitination, the E3 ligases interacting with substrate-bound Hsp70 target ubiquitin-charged E2 enzymes to the nascently unfolding substrate. In this way, the Hsp90/Hsp70-based chaperone machinery may function as a comprehensive protein management system for quality control of damaged proteins.

6.6 THE CHAPERONE MACHINERY

Hsp70 and Hsp90 are conserved, abundant, and essential proteins of eukaryotic cells where they are present in the cytoplasm and nucleus, with paralogs being present in mitochondria and the endoplasmic reticulum. Both chaperones have adenosine triphosphate (ATP) binding sites and possess intrinsic ATPase activity that regulates their conformation. In each case, the ATP-bound conformation has a low affinity for binding hydrophobic peptides, and ATP hydrolysis is accompanied by a conformational change to a state with high affinity for binding hydrophobic peptides (reviewed in Hartl and Hayer-Hartl, 2002; Picard, 2002). The conformational changes that occur during the ATPase cycle of Hsp90 have been reviewed by Wandinger *et al.* (2008). Also, both Hsp70 and Hsp90 possess EEVD motifs at the carboxy termini, which are binding sites for tetratricopeptide repeat (TPR) domains. The TPR co-chaperones that bind to the chaperones play a number of roles in the activation and trafficking of signaling proteins (Pratt and Toft, 2003; Pratt *et al.*, 2004; Pearl and Prodromou, 2006; Wandinger *et al.*, 2008). However, we focus our discussion here on one TPR co-chaperone, CHIP (carboxy-terminus of Hsp70 interacting protein), which is a chaperone-binding E3 ubiquitin ligase (Cyr *et al.*, 2002) that is important for quality control of damaged proteins.

In contrast to the classic model of chaperones interacting with unfolded proteins to facilitate their refolding, the Hsp90 chaperone machinery acts on prefolded proteins in their native conformations to assist the opening and stabilization of ligand-binding clefts (Pratt *et al.*, 2008). The Hsp90 chaperone machine acts on proteins in a manner that is not dependent upon protein sequence, size, or structure. Rather, Hsp90 interacts with regions where protein folding clefts merge with the charged, hydrophilic surface of the protein. Folding clefts are a general topological feature of proteins in native conformation, and many of these hydrophobic clefts must open to permit access of ligands, such as steroids, ATP, and so on, to binding sites in the protein's interior. Indeed, access of the ligand to the steroid-binding clefts of nuclear receptors, including the glucocorticoid receptor and androgen receptor, is dependent upon the Hsp90 chaperone machinery, and the inhibition of Hsp90 prevents high affinity ligand binding (Pratt and Toft, 2003; Pratt *et al.*, 2008).

In the absence of the chaperone machinery, these binding clefts are dynamic, in that they shift to varying extents between closed and open states. When clefts open during this molecular breathing process, hydrophobic residues of the protein interior are exposed to solvent, and continued opening may progress to nascent stages of unfolding. The extent to which the binding cleft is open determines ligand access and protein function, but these clefts are inherent sites of conformational instability. The chaperone machinery assists cleft opening, and Hsp90 binding to the protein stabilizes the open cleft, impeding further unfolding and Hsp70-dependent ubiquitination (Pratt *et al.*, 2010).

The androgen receptor is a classical Hsp90 "client" protein, a group that includes many transcription factors and protein kinases, that turn over rapidly in the absence of Hsp90 stabilization. These client proteins have metastable clefts that, in the absence of Hsp90 binding, have a high tendency to further unfold, leading to protein degradation. The classical client proteins are assembled into complexes with Hsp90 that are stable enough to be isolated and analyzed biochemically. Although we call these "stable" Hsp90 heterocomplexes, they are constantly undergoing cycles of assembly and disassembly in the cytoplasm and nucleoplasm (Pratt and Toft, 2003). These metastable cleft proteins are quite profoundly stabilized when they are complexed with Hsp90; therefore, the client proteins are under stringent Hsp90 regulation.

The assembly of stable steroid receptor–Hsp90 heterocomplexes proceeds through an ordered series of events in which Hsp70 first binds to the client protein and primes the substrate for interaction with Hsp90 (assembly is presented in detail in Pratt and Toft, 2003). In stable assembly, both Hsp70 and Hsp90 must complete at least one ATPase cycle, and Hsp90 in the final complex is in the ATP-bound state. A third protein required for stable assembly is p23—an Hsp90 co-chaperone that binds to the ATP-dependent conformation of Hsp90 and stabilizes the client protein–Hsp90 complex. Other proteins participate in the assembly machinery but only Hsp70, Hsp90, and p23 are required for stable cycling of steroid receptors with Hsp90 (Pratt and Toft, 2003).

6.7 THE CHAPERONE MACHINERY AND PROTEASOMAL DEGRADATION

It has been widely accepted that the ubiquitin–proteasome system is involved in the degradation of damaged and aberrant proteins, but it has not been clear how these damaged proteins are recognized and targeted for ubiquitination (Hershko and Ciechanover, 1998). In the mid-1990s, it was shown that Hsp70 and its co-chaperone Hsp40 are required for ubiquitin-dependent degradation of short-lived and abnormal proteins (Bercovich *et al.*, 1997; Lee *et al.*, 1996). Overexpression of Hsp70 or Hsp40 decreases the level of abnormal proteins and improves viability in cellular models of certain neurodegenerative diseases characterized by the accumulation of aberrant proteins, such as Kennedy's, Huntington's, and Parkinson's diseases (Bailey *et al.*, 2002; Jana *et al.*, 2000; Klucken *et al.*, 2004). Overexpression of these chaperones also ameliorates the disease phenotype in *Drosophila* and mouse models of several of these diseases, including a mouse model of Kennedy's disease (Adachi *et al.*, 2003; Auluck *et al.*, 2002; Chan *et al.*, 2000; Warrick *et al.*, 1999; reviewed in Muchowski and Wacker, 2005).

Hsp90 was connected to ubiquitin-dependent degradation in studies using benzoquinone ansamycins that are quite specific Hsp90 inhibitors. The first of these ansamycins, herbimycin A, was found to reverse v-src transformation of cells to a normal phenotype (Uehara *et al.*, 1985, 1986), and it was then used as a protein-tyrosine kinase inhibitor (Uehara and Fukazawa, 1991). In a paper that led to an explosion of work on Hsp90, Whitesell *et al.* (1994) showed that the target of the ansamycins, herbimycin A, and geldanamycin is Hsp90, not v-src. The treatment with geldanamycin led to a disruption of the v-src–Hsp90 heterocomplex and loss of v-src protein from cells (Whitesell *et al.*, 1994). Subsequently, Sepp-Lorenzino *et al.* (1995) showed that herbimycin-induced degradation of several receptor tyrosine kinases occurred via the ubiquitin–proteasome pathway.

In the ensuing years, Hsp90 inhibition was shown to promote the proteasomal degradation of several hundred proteins (Neckers, 2007). The ansamycin antibiotics, such as geldanamycin, bind in the nucleotide binding pocket near the amino-terminus of Hsp90 (Prodromou *et al.*, 1997; Stebbins *et al.*, 1997). This ATP-binding site is structurally unique to the very small GHKL family whose ATP-binding domains contain four common motifs that define a "Bergerat fold" for binding ATP (Bergerat *et al.*, 1997; Dutta and Inouye, 2000). As most of these proteins are bacterial, geldanamycin effects are quite specific for inhibition of Hsp90 family proteins in eukaryotes. A variety of Hsp90 inhibitors that are pharmacologically and toxicologically more appropriate for human use have been synthesized, and one of them, 17-allylamino-17-demethoxygeldanamycin (17-AAG), is in phase II clinical trial for treatment of cancer. 17-AAG also promotes polyglutamine androgen receptor protein degradation and ameliorates the phenotype in a transgenic mouse model of Kennedy's disease (Waza *et al.*, 2005).

The mechanism by which geldanamycin promotes client protein degradation has attracted considerable attention. Because Hsp90 binding to the transcription factor heat shock factor 1 (HSF1) maintains it in an inactive state (Zou *et al.*, 1998) and treatment of cells with geldanamycin induces an HSF1-dependent stress response (Auluck and Bonini, 2002; Bagatell *et al.*, 2000; Hay *et al.*, 2004; Sittler *et al.*, 2001; Zou *et al.*, 1998), it is often proposed that geldanamycin alleviates the phenotype and accumulation of misfolded proteins in neurodegenerative disease models by inducing a stress response (Auluck and Bonini, 2002; Hay *et al.*, 2004; Muchowski and Wacker, 2005; Sittler *et al.*, 2001). However, as geldanamycin promotes proteasomal degradation of the polyglutamine androgen receptor protein in *Hsf1* $^{-/-}$ cells that cannot mount a stress response, this explanation is not valid (Thomas *et al.*, 2006).

The observation that treatment with Hsp90 inhibitors promotes the degradation of Hsp90 client proteins via the ubiquitin–proteasome pathway (Sepp-Lorenzino *et al.*, 1995; Whitesell *et al.*, 1994) is explained most simply by the notion that Hsp90 binding to a client protein inhibits its degradation, and that inhibitors such as geldanamycin relieve this inhibition by preventing cycling with Hsp90. Support for this model is provided by the observation that stabilizing the Hsp90 chaperone complex with the polyglutamine androgen receptor by overexpressing the co-chaperone p23 decreases ligand-dependent unfolding and aggregation (Thomas *et al.*, 2006). Chaperone effects on client protein turnover are consistent with the two essential components of the chaperone machinery having opposing effects, with Hsp70 promoting ubiquitination and Hsp90 stabilizing the protein against degradation.

6.8 CHAPERONE-DEPENDENT UBIQUITIN–PROTEIN LIGASES

Ubiquitination occurs via three sequential steps catalyzed by activating (E1), conjugating (E2), and ligase (E3) enzymes, and protein ubiquitination is used to trigger a wide variety of physiological processes (Glickman and Ciechanover, 2002). A number of E3 ligases play a key role in ubiquitin-mediated protein degradation by serving as the specific recognition factors in the cascade. In some cases, recognition is aided by various chaperones, but it is not known how many E3s function in a chaperone-dependent manner although several clearly do (Hatakeyama *et al.*, 2004). Two E3s are known to interact with the chaperones of interest here. Parkin is an E3 ligase (Zhang *et al.*, 2000) that is targeted to substrate by Hsp70 (Tsai *et al.*, 2003), but it was the discovery of CHIP by Ballinger *et al.* (1999) that germinated a tremendous expansion in our understanding of chaperone-dependent ubiquitination.

Most of the mechanistic information about CHIP comes from the laboratory of Cam Patterson where CHIP was discovered, and Patterson and his colleagues have written several reviews on CHIP function (Cyr *et al.*, 2002; Hohfeld *et al.*, 2001; McDonough and Patterson, 2003). CHIP is a 35-kDa U-box E3 ligase, and it binds via an amino-terminal TPR domain to both Hsc/Hsp70 and Hsp90 (Ballinger *et al.*, 1999; Connell *et al.*, 2001). CHIP possesses a carboxy-terminal U-box that interacts with the UBCH5 family of E2 ubiquitin conjugating enzymes (Jiang *et al.*, 2001). The fact that CHIP binds to both Hsp90 and Hsp70 led to the proposal that both chaperones could target CHIP to the substrate (Connell *et al.*, 2001; Cyr *et al.*, 2002). However, using neuronal nitric oxide synthase (nNOS) as a substrate, it has been shown that CHIP-dependent ubiquitination is promoted by Hsp70 and inhibited by Hsp90 (Peng *et al.*, 2009).

The suggestion that Hsp90 is involved in targeting the substrate for ubiquitination cannot account for geldanamycin-induced client protein degradation. Geldanamycin binding to Hsp90 uniformly results in client protein destabilization (Isaacs *et al.*, 2003), but geldanamycin prevents client protein cycling with Hsp90 (Pratt and Toft, 2003). The stable client protein heterocomplexes are assembled in a stepwise fashion through an initial ATP-dependent priming interaction with Hsp70 followed by a second ATP-dependent interaction with Hsp90 (Pratt and Toft, 2003). The interaction with Hsp90 is blocked by geldanamycin, leaving the Hsp70-bound client protein to be ubiquitinated in a CHIP-dependent manner. This explanation is very different from a study of luciferase refolding where it was concluded that geldanamycin binding shifts Hsp90 from refolding to degradation mode (Schneider *et al.*, 1996).

Our conclusion that Hsp90-bound CHIP does not target the client protein for CHIP-dependent ubiquitination does not mean that CHIP interaction with the TPR acceptor site has no effect on Hsp90 function. CHIP binding to the TPR acceptor site on Hsp70 inhibits the chaperone's ATPase activity, thereby increasing the reactivation of luciferase after thermal inactivation in cells (Kampinga *et al.*, 2003). This action requires the TPR domain but not the U-box domain of CHIP. It is possible that CHIP binding to the TPR acceptor site on Hsp90 has a similar effect on Hsp90 function that is independent of its E3 ligase action. Indeed, two other TPR domain proteins that bind to Hsp90, HOP, and Cdc37 have been shown to inhibit its ATPase cycle (Pearl and Prodromou, 2006). Despite the observations with geldanamycin, several reports (Dickey *et al.*, 2007) have concluded that an Hsp90–CHIP complex selectively degrades various Hsp90 client proteins.

Overexpression of CHIP has been shown to increase ubiquitination and degradation of many established Hsp90 client proteins, including the glucocorticoid receptor (Connell *et al.*, 2001), p53 (Esser *et al.*, 2005), and ErbB-2 (Zhou *et al.*, 2003). Additionally, CHIP is found in aggregates of aberrant proteins involved in neurodegenerative diseases, such as α-synuclein and polyglutamine proteins, including the androgen receptor, huntingtin, ataxin-1, and ataxin-3 (Al-Ramahi *et al.*, 2006; Jana *et al.*, 2005; Miller *et al.*, 2005; Shin *et al.*, 2005; Thomas *et al.*, 2004). Overexpression of CHIP suppresses aggregation and protein levels in cellular disease models (Al-Ramahi *et al.*, 2006; Jana *et al.*, 2005; Miller *et al.*, 2005; Shin *et al.*, 2005). The importance of CHIP in the neuronal response to aberrant, misfolded proteins is emphasized by observations in animal models of disease in which CHIP levels have been manipulated. Notably, overexpression of CHIP in *Drosophila* and mouse polyglutamine disease models, including Kennedy's disease transgenic mice, suppresses toxicity (Adachi *et al.*, 2007; Jana *et al.*, 2005), whereas haploinsufficiency for CHIP in Huntington's disease transgenic mice accelerates the disease phenotype (Al-Ramahi *et al.*, 2006).

Because CHIP has received so much attention, it is often regarded as the most important E3 ligase involved in chaperone-dependent ubiquitination and degradation. CHIP is clearly important, but it should be noted that there is redundancy between CHIP and some other E3 ligases. Although overexpression of CHIP has been shown to promote proteasomal degradation of a wide variety of normal and aberrant proteins, overexpression of one E3 ligase could favor a normally minor pathway of ubiquitination. Support for the conclusion that there is functional redundancy among E3 ligases comes from the observation that both the glucocorticoid receptor and polyglutamine androgen receptor are degraded at the same rate in CHIP$^{-/-}$ and CHIP$^{+/+}$ mouse embryonic fibroblasts treated with geldanamycin (Morishima *et al.*, 2008). CHIP$^{-/-}$ cytosol also has the same ability as CHIP$^{+/+}$ cytosol to ubiquitinate a CHIP substrate. Although in this case, the E3 ligases that are acting redundantly to CHIP have not been identified, overexpression of either CHIP or parkin promotes the degradation of nNOS (Morishima *et al.*, 2008) and polyglutamine-expanded ataxin-3 (Jana *et al.*, 2005; Morishima *et al.*, 2008; Tsai *et al.*, 2003), consistent with the notion that multiple E3 ligases function in a redundant manner. These examples of functional redundancy certainly show that CHIP-dependent degradation of damaged and aberrant proteins is not exclusive, but it does not diminish the widespread enthusiasm for CHIP as a central player in protein triage.

6.9 SMALL MOLECULE INHIBITORS OF Hsp70 TO PROBE TRIAGE DECISIONS

Most of what is known about Hsp70′s role in the degradation of polyglutamine-expanded proteins comes from Hsp70 overexpression experiments. To enhance mechanistic understanding of Hsp70-dependent processes in general, it would be useful to have a small molecule inhibitor of Hsp70, much as geldanamycin has been so useful in probing Hsp90-dependent effects. However, few inhibitors targeted to Hsp70 have been identified. To this end, the Gestwicki laboratory recently employed a high-throughput chemical screen to identify compounds that inhibit Hsp70 ATPase activity. An inhibitor identified in the compound library was methylene blue, which was shown to interact with purified Hsp70 by nuclear magnetic resonance (NMR) spectroscopy (Jinwal *et al.*, 2009).

Methylene blue was used to probe Hsp70-dependent effects in well-established systems of increasing complexity, from the purified Hsp90/Hsp70-based chaperone machinery to the physiological ubiquitinating system of reticulocyte lysate.

In this work, we found that methylene blue abrogates the generation of steroid binding activity of the glucocorticoid receptor by specifically inhibiting the Hsp70 component of the Hsp90/Hsp70-based multiprotein chaperone machinery (Wang *et al.*, 2010). Methylene blue was next used as a tool to probe the pathway regulating ubiquitination of nNOS. Using the classic system that was originally used to resolve the components of the ubiquitin–protein ligase pathway (Hershko *et al.*, 1983), nNOS ubiquitination by the DE52-retained fraction of rabbit reticulocyte lysate was found to be inhibited by methylene blue in an Hsp70-dependent manner (Wang *et al.*, 2010).

Using methylene blue as a tool, the role of Hsp70 in controlling the turnover of the expanded glutamine androgen receptor was explored in a cellular system. The Hsp90/Hsp70-based chaperone machinery binds to the carboxy-terminal domain of the receptor, and similar to its action on the glucocorticoid receptor, regulates opening of the steroid-binding cleft to permit ligand binding (Pratt and Toft, 2003; Pratt *et al.*, 2008). In this system, Hsp90 functions to prevent androgen receptor unfolding, whereas treatment of these cells with geldanamycin or radicicol promotes androgen receptor degradation. Here, we have presented a model in which unfolding of the androgen receptor leads to Hsp70-dependent degradation through the ubiquitin–proteasome pathway. Consistent with this model, methylene blue prevented degradation of the expanded glutamine androgen receptor and promoted the accumulation of aggregated species (Wang *et al.*, 2010).

Methylene blue had contrasting effects on the degradation of an amino-terminal fragment of the polyglutamine androgen receptor. Methylene blue promoted degradation of truncated forms of the receptor, thereby ameliorating glutamine length-dependent toxicity (Wang *et al.*, 2010). These androgen receptor fragments include the glutamine tract flanked by ~50 amino acids and therefore lack the ligand-binding domain (Merry *et al.*, 1998). In the absence of this domain, these proteins are not Hsp90 clients whose stability is regulated by the Hsp90/Hsp70-based chaperone machinery. Truncated fragments of the huntingtin protein are primarily degraded by macroautophagy (Ravikumar *et al.*, 2004, 2005), and it is likely that these androgen receptor fragments are handled similarly. Notably, methylene blue promoted induction and flux through macroautophagy—a lysosomal protein quality control pathway. This homeostatic response likely reflects impairment of Hsp70-dependent degradation through the ubiquitin–proteasome pathway. A similar induction of macroautophagy has been observed in *Drosophila* mutants with impaired proteasome function (Pandey *et al.*, 2007) suggesting that degradation of truncated androgen receptor fragments was facilitated by the compensatory activation of the pathway through which it is normally degraded.

6.10 PROTEIN REFOLDING VERSUS CLEFT STABILIZATION

Although it was originally thought that Hsp90 and Hsp70 acted to promote refolding of proteins that aggregated in stressed cells, it has subsequently become clear that Hsp70 promotes the proteasomal degradation of unfolded proteins. In contrast, Hsp90 acts to stabilize client proteins, and the inhibition of Hsp90 promotes their degradation. The model of chaperone function discussed here is distinct from the traditional view of chaperone action that is based on the assumption of protein refolding. In assays *in vitro* where catalytic activity of an enzyme, such as luciferase, is inactivated by mild heating and recovery of activity is promoted by chaperones, it has been assumed that Hsp90 and Hsp70 act by promoting the refolding of an unfolded protein. We suggest a reinterpretation of these findings in light of the alternative model presented here.

Our model is based on the study of steroid receptors where the focus has been on ligand-binding activity. When the Hsp90-free progesterone receptor, for example, is submitted to mild heating, it loses its ligand-binding activity, and the Hsp90/Hsp70-based chaperone machinery and ATP are required to maintain ligand-binding activity (Smith, 1993). Similarly, the glucocorticoid receptor immediately loses its ligand-binding activity when Hsp90 is dissociated at 0°C. The loss of ligand-binding activity reflects the collapse of the hydrophobic ligand-binding cleft, and the Hsp90 chaperone machinery is required to open and stabilize the cleft in a ligand-binding state (Pratt *et al.*, 2008). The analogous event with luciferase would be that mild heating promotes collapse of the ATP-binding cleft, and that dynamic cycling with the chaperone favors the open state of the cleft, promoting restoration of luciferase activity. Therefore, this traditional assay of refolding may not be assaying refolding at all, but rather it may be assaying the chaperone's ability to dynamically stabilize a partially unfolded, or an open cleft, state of the protein. This would explain why so many proteins that can bind both hydrophobic and charged regions possess some luciferase reactivating (chaperone) activity.

By considering that Hsp90 and Hsp70 act together as a machinery to modulate ligand-binding clefts in properly folded proteins, we have taken a very different approach to explaining how these two chaperones may interact with aberrant proteins or proteins undergoing toxic stress. Using the refolding analysis that is a fundamental assumption of the chaperone field, one is forced into a model in which the chaperones interact with an unfolded state of the protein substrate to promote its proper refolding. In contrast, the ligand-binding cleft model discussed here predicts that Hsp90 interacts with a protein to stabilize it by preventing inordinate cleft opening that would yield unfolding. The presence of an expanded glutamine tract promotes ligand-induced unfolding of the mutant androgen receptor of Kennedy's disease. When Hsp90 fails to maintain native folding, Hsp70-dependent ubiquitination is favored. This model predicts that the Hsp90/Hsp70-based chaperone machinery plays a central role in androgen receptor protein triage decisions. Strategies to manipulate the activity of this machinery with small molecules offer great promise for treating Kennedy's disease and other protein aggregation neurodegenerative disorders by facilitating the degradation of these toxic proteins.

ACKNOWLEDGMENT

Preparation of this chapter was supported by National Institutes of Health (grant NS055746) and the McKnight Foundation.

REFERENCES

Abdullah A, Trifiro MA, Panet-Raymond V, Alvarado C, de Tourreil S, Frankel D, Schipper HM, Pinsky L. Spinobulbar muscular atrophy: polyglutamine-expanded androgen receptor is proteolytically resistant *in vitro* and processed abnormally in transfected cells. *Hum Mol Genet* 1998;**7**:379–384.

Abel A, Walcott J, Woods J, Duda J, Merry DE. Expression of expanded repeat androgen receptor produces neurologic disease in transgenic mice. *Hum Mol Genet* 2001;**10**:107–116.

Adachi H, Katsuno M, Minamiyama M, Sang C, Pagoulatos G, Angelidis C, Kusakabe M, Yoshiki A, Kobayashi Y, Doyu M, Sobue G. Heat shock protein 70 chaperone overexpression ameliorates phenotypes of the spinal and bulbar muscular atrophy transgenic mouse model by reducing nuclear-localized mutant androgen receptor protein. *J Neurosci* 2003;**23**:2203–2211.

Adachi H, Kume A, Li M, Nakagomi Y, Niwa H, Do J, Sang C, Kobayashi Y, Doyu M, Sobue G. Transgenic mice with an expanded CAG repeat controlled by the human AR promoter show polyglutamine nuclear inclusions and neuronal dysfunction without neuronal cell death. *Hum Mol Genet* 2001;**10**:1039–1048.

Adachi H, Waza M, Tokui K, Katsuno M, Minamiyama M, Tanaka F, Doyu M, Sobue G. CHIP overexpression reduces mutant androgen receptor protein and ameliorates phenotypes of the spinal and bulbar muscular atrophy transgenic mouse model. *J Neurosci* 2007;**27**:5115–5126.

Al-Ramahi I, Lam YC, Chen HK, de Gouyon B, Zhang M, Perez AM, Branco J, de Haro M, Patterson C, Zoghbi HY, Botas J. CHIP protects from the neurotoxicity of expanded and wild-type ataxin-1 and promotes their ubiquitination and degradation. *J Biol Chem* 2006;**281**:26714–26724.

Albertelli MA, Scheller A, Brogley M, Robins DM. Replacing the mouse androgen receptor with human alleles demonstrates glutamine tract length-dependent effects on physiology and tumorigenesis in mice. *Mol Endocrinol* 2006;**20**: 1248–1260.

Arbizu T, Santamaria J, Gomez JM, Quilez A, Serra JP. A family with adult spinal and bulbar muscular atrophy, X-linked inheritance and associated testicular failure. *J Neurol Sci* 1983;**59**:371–382.

Auluck PK, Bonini NM. Pharmacological prevention of Parkinson disease in Drosophila. *Nat Med* 2002;**8**:1185–1186.

Auluck PK, Chan HY, Trojanowski JQ, Lee VM, Bonini NM. Chaperone suppression of alpha-synuclein toxicity in a Drosophila model for Parkinson's disease. *Science* 2002;**295**:865–868.

Bagatell R, Paine-Murrieta GD, Taylor CW, Pulcini EJ, Akinaga S, Benjamin IJ, Whitesell L. Induction of a heat shock factor 1-dependent stress response alters the cytotoxic activity of hsp90-binding agents. *Clin Cancer Res* 2000;**6**:3312–3318.

Bailey CK, Andriola IF, Kampinga HH, Merry DE. Molecular chaperones enhance the degradation of expanded polyglutamine repeat androgen receptor in a cellular model of spinal and bulbar muscular atrophy. *Hum Mol Genet* 2002;**11**:515–523.

Ballinger CA, Connell P, Wu Y, Hu Z, Thompson LJ, Yin LY, Patterson C. Identification of CHIP, a novel tetratricopeptide repeat-containing protein that interacts with heat shock proteins and negatively regulates chaperone functions. *Mol Cell Biol* 1999;**19**:4535–4545.

Banno H, Katsuno M, Suzuki K, Takeuchi Y, Kawashima M, Suga N, Takamori M, Ito M, Nakamura T, Matsuo K, Yamada S, Oki Y, Adachi H, Minamiyama M, Waza M, Atsuta N, Watanabe H, Fujimoto Y, Nakashima T, Tanaka F, Doyu M, Sobue G. Phase 2 trial of leuprorelin in patients with spinal and bulbar muscular atrophy. *Ann Neurol* 2009;**65**:140–150.

Becher MW, Kotzuk JA, Sharp AH, Davies SW, Bates GP, Price DL, Ross CA. Intranuclear neuronal inclusions in Huntington's disease and dentatorubral and pallidoluysian atrophy: correlation between the density of inclusions and IT15 CAG triplet repeat length. *Neurobiol Dis* 1998;**4**:387–397.

Bercovich B, Stancovski I, Mayer A, Blumenfeld N, Laszlo A, Schwartz AL, Ciechanover A. Ubiquitin-dependent degradation of certain protein substrates *in vitro* requires the molecular chaperone Hsc70. *J Biol Chem* 1997;**272**:9002–9010.

Bergerat A, de Massy B, Gadelle D, Varoutas PC, Nicolas A, Forterre P. An atypical topoisomerase II from Archaea with implications for meiotic recombination. *Nature* 1997;**386**:414–417.

Buchanan G, Yang M, Cheong A, Harris JM, Irvine RA, Lambert PF, Moore NL, Raynor M, Neufing PJ, Coetzee GA, Tilley WD. Structural and functional consequences of glutamine tract variation in the androgen receptor. *Hum Mol Genet* 2004;**13**:1677–1692.

Chamberlain NL, Driver ED, Miesfeld RL. The length and location of CAG trinucleotide repeats in the androgen receptor N-terminal domain affect transactivation function. *Nucleic Acids Res* 1994;**22**:3181–3186.

Chan HY, Warrick JM, Gray-Board GL, Paulson HL, Bonini NM. Mechanisms of chaperone suppression of polyglutamine disease: selectivity, synergy and modulation of protein solubility in Drosophila. *Hum Mol Genet* 2000;**9**:2811–2820.

Chevalier-Larsen ES, O'Brien CJ, Wang H, Jenkins SC, Holder L, Lieberman AP, Merry DE. Castration restores function and neurofilament alterations of aged symptomatic males in a transgenic mouse model of spinal and bulbar muscular atrophy. *J Neurosci* 2004;**24**:4778–4786.

Connell P, Ballinger CA, Jiang J, Wu Y, Thompson LJ, Hohfeld J, Patterson C. The co-chaperone CHIP regulates protein triage decisions mediated by heat-shock proteins. *Nat Cell Biol* 2001;**3**:93–96.

Cyr DM, Hohfeld J, Patterson C. Protein quality control: U-box-containing E3 ubiquitin ligases join the fold. *Trends Biochem Sci* 2002;**27**:368–375.

David G, Durr A, Stevanin G, Cancel G, Abbas N, Benomar A, Belal S, Lebre AS, Abada-Bendib M, Grid D, Holmberg M, Yahyaoui M, Hentati F, Chkili T, Agid Y, Brice A. Molecular and clinical correlations in autosomal dominant cerebellar ataxia with progressive macular dystrophy (SCA7). *Hum Mol Genet* 1998;**7**:165–170.

Davies SW, Turmaine M, Cozens BA, DiFiglia M, Sharp AH, Ross CA, Scherzinger E, Wanker EE, Mangiarini L, Bates GP. Formation of neuronal intranuclear inclusions underlies the neurological dysfunction in mice transgenic for the HD mutation. *Cell* 1997;**90**:537–548.

Dickey CA, Kamal A, Lundgren K, Klosak N, Bailey RM, Dunmore J, Ash P, Shoraka S, Zlatkovic J, Eckman CB, Patterson C, Dickson DW, Nahman NS, Hutton M, Burrows F, Petrucelli L Jr. The high-affinity HSP90-CHIP complex recognizes and selectively degrades phosphorylated tau client proteins. *J Clin Invest* 2007;**117**:648–658.

Dutta R, Inouye M. GHKL, an emergent ATPase/kinase superfamily. *Trends Biochem Sci* 2000;**25**:24–28.

Esser C, Scheffner M, Hohfeld J. The chaperone-associated ubiquitin ligase CHIP is able to target p53 for proteasomal degradation. *J Biol Chem* 2005;**280**:27443–27448.

Fischbeck KH, Lieberman A, Bailey CK, Abel A, Merry DE. Androgen receptor mutation in Kennedy's disease. *Philos Trans R Soc Lond B Biol Sci* 1999;**354**:1075–1078.

Fischer LR, Glass JD. Axonal degeneration in motor neuron disease. *Neurodegener Dis* 2007;**4**:431–442.

Glickman MH, Ciechanover A. The ubiquitin-proteasome proteolytic pathway: destruction for the sake of construction. *Physiol Rev* 2002;**82**:373–428.

Gottlieb B, Beitel LK, Wu JH, Trifiro M. The androgen receptor gene mutations database (ARDB): 2004 update. *Hum Mutat* 2004;**23**:527–533.

Hartl FU, Hayer-Hartl M. Molecular chaperones in the cytosol: from nascent chain to folded protein. *Science* 2002;**295**:1852–1858.

Hatakeyama S, Matsumoto M, Yada M, Nakayama KI. Interaction of U-box-type ubiquitin-protein ligases (E3s) with molecular chaperones. *Genes Cells* 2004;**9**:533–548.

Hay DG, Sathasivam K, Tobaben S, Stahl B, Marber M, Mestril R, Mahal A, Smith DL, Woodman B, Bates GP. Progressive decrease in chaperone protein levels in a mouse model of Huntington's disease and induction of stress proteins as a therapeutic approach. *Hum Mol Genet* 2004;**13**:1389–1405.

Hershko A, Ciechanover A. The ubiquitin system. *Annu Rev Biochem* 1998;**67**:425–479.

Hershko A, Heller H, Elias S, Ciechanover A. Components of ubiquitin-protein ligase system. Resolution, affinity purification, and role in protein breakdown. *J Biol Chem* 1983;**258**:8206–8214.

Hohfeld J, Cyr DM, Patterson C. From the cradle to the grave: molecular chaperones that may choose between folding and degradation. *EMBO Rep* 2001;**2**:885–890.

Hsiao PW, Lin DL, Nakao R, Chang C. The linkage of Kennedy's neuron disease to ARA24, the first identified androgen receptor polyglutamine region-associated coactivator. *J Biol Chem* 1999;**274**:20229–20234.

Irvine RA, Ma H, Yu MC, Ross RK, Stallcup MR, Coetzee GA. Inhibition of p160-mediated coactivation with increasing androgen receptor polyglutamine length. *Hum Mol Genet* 2000;**9**:267–274.

Isaacs JS, Xu W, Neckers L. Heat shock protein 90 as a molecular target for cancer therapeutics. *Cancer Cell* 2003;**3**:213–217.

Jana NR, Dikshit P, Goswami A, Kotliarova S, Murata S, Tanaka K, Nukina N. Co-chaperone CHIP associates with expanded polyglutamine protein and promotes their degradation by proteasomes. *J Biol Chem* 2005;**280**:11635–11640.

Jana NR, Tanaka M, Wang G, Nukina N. Polyglutamine length-dependent interaction of Hsp40 and Hsp70 family chaperones with truncated N-terminal huntingtin: their role in suppression of aggregation and cellular toxicity. *Hum Mol Genet* 2000;**9**:2009–2018.

Jiang J, Ballinger CA, Wu Y, Dai Q, Cyr DM, Hohfeld J, Patterson C. CHIP is a U-box-dependent E3 ubiquitin ligase: identification of Hsc70 as a target for ubiquitylation. *J Biol Chem* 2001;**276**:42938–42944.

Jinwal UK, Miyata Y, Koren J, Jones JR, Trotter JH, Chang L, O'Leary J, Morgan D, Lee DC, Shults CL, Rousaki A, Weeber EJ, Zuiderweg ER, Gestwicki JE, Dickey CA III. Chemical manipulation of hsp70 ATPase activity regulates tau stability. *J Neurosci* 2009;**29**:12079–12088.

Johansen JA, Yu Z, Mo K, Monks DA, Lieberman AP, Breedlove SM, Jordan CL. Recovery of function in a myogenic mouse model of spinal bulbar muscular atrophy. *Neurobiol Dis* 2009;**34**:113–120.

Jordan CL, Lieberman AP. Spinal and bulbar muscular atrophy: a motoneuron or muscle disease? *Curr Opin Pharmacol* 2008;**8**:752–758.

Kampinga HH, Kanon B, Salomons FA, Kabakov AE, Patterson C. Overexpression of the cochaperone CHIP enhances Hsp70-dependent folding activity in mammalian cells. *Mol Cell Biol* 2003;**23**:4948–4958.

Katsuno M, Adachi H, Doyu M, Minamiyama M, Sang C, Kobayashi Y, Inukai A, Sobue G. Leuprorelin rescues polyglutamine-dependent phenotypes in a transgenic mouse model of spinal and bulbar muscular atrophy. *Nat Med* 2003;**9**:768–773.

Katsuno M, Adachi H, Kume A, Li M, Nakagomi Y, Niwa H, Sang C, Kobayashi Y, Doyu M, Sobue G. Testosterone reduction prevents phenotypic expression in a transgenic mouse model of spinal and bulbar muscular atrophy. *Neuron* 2002;**35**:843–854.

Katsuno M, Adachi H, Waza M, Banno H, Suzuki K, Tanaka F, Doyu M, Sobue G. Pathogenesis, animal models and therapeutics in spinal and bulbar muscular atrophy (SBMA). *Exp Neurol* 2006a;**200**:8–18.

Katsuno M, Adachi H, Minamiyama M, Waza M, Tokui K, Banno H, Suzuki K, Onoda Y, Tanaka F, Doyu M, Sobue G. Reversible disruption of dynactin 1-mediated retrograde axonal transport in polyglutamine-induced motor neuron degeneration. *J Neurosci* 2006b;**26**:12106–12117.

Kazemi-Esfarjani P, Trifiro MA, Pinsky L. Evidence for a repressive function of the long polyglutamine tract in the human androgen receptor: possible pathogenetic relevance for the (CAG)n-expanded neuronopathies. *Hum Mol Genet* 1995;**4**:523–527.

Kennedy WR, Alter M, Sung JH. Progressive proximal spinal and bulbar muscular atrophy of late onset. A sex-linked recessive trait. *Neurology* 1968;**18**:671–680.

Klucken J, Shin Y, Masliah E, Hyman BT, McLean PJ. Hsp70 reduces alpha-synuclein aggregation and toxicity. *J Biol Chem* 2004;**279**:25497–25502.

Kujawa KA, Emeric E, Jones KJ. Testosterone differentially regulates the regenerative properties of injured hamster facial motoneurons. *J Neurosci* 1991;**11**:3898–3906.

Kurz EM, Sengelaub DR, Arnold AP. Androgens regulate the dendritic length of mammalian motoneurons in adulthood. *Science* 1986;**232**:395–398.

La Spada AR, Roling DB, Harding AE, Warner CL, Spiegel R, Hausmanowa-Petrusewicz I, Yee WC, Fischbeck KH. Meiotic stability and genotype-phenotype correlation of the trinucleotide repeat in X-linked spinal and bulbar muscular atrophy. *Nat Genet* 1992;**2**:301–304.

La Spada AR, Wilson EM, Lubahn DB, Harding AE, Fischbeck KH. Androgen receptor gene mutations in X-linked spinal and bulbar muscular atrophy. *Nature* 1991;**352**:77–79.

Lee DH, Sherman MY, Goldberg AL. Involvement of the molecular chaperone Ydj1 in the ubiquitin-dependent degradation of short-lived and abnormal proteins in *Saccharomyces cerevisiae*. *Mol Cell Biol* 1996;**16**:4773–4781.

Li M, Chevalier-Larsen ES, Merry DE, Diamond MI. Soluble androgen receptor oligomers underlie pathology in a mouse model of spinobulbar muscular atrophy. *J Biol Chem* 2007;**282**:3157–3164.

Lieberman AP, Harmison G, Strand AD, Olson JM, Fischbeck KH. Altered transcriptional regulation in cells expressing the expanded polyglutamine androgen receptor. *Hum Mol Genet* 2002;**11**:1967–1976.

Lieberman AP, Robins DM. The androgen receptor's CAG/glutamine tract in mouse models of neurological disease and cancer. *J Alzheimers Dis* 2008;**14**:247–255.

Lieberman AP, Robitaille Y, Trojanowski JQ, Dickson DW, Fischbeck KH. Polyglutamine-containing aggregates in neuronal intranuclear inclusion disease. *Lancet* 1998;**351**:884.

Lieberman AP, Trojanowski JQ, Leonard DG, Chen KL, Barnett JL, Leverenz JB, Bird TD, Robitaille Y, Malandrini A, Fischbeck KH. Ataxin 1 and ataxin 3 in neuronal intranuclear inclusion disease. *Ann Neurol* 1999;**46**:271–273.

Mauger C, Del-Favero J, Ceuterick C, Lubke U, van Broeckhoven C, Martin J. Identification and localization of ataxin-7 in brain and retina of a patient with cerebellar ataxia type II using anti-peptide antibody. *Brain Res Mol Brain Res* 1999;**74**:35–43.

McCampbell A, Taye AA, Whitty L, Penney E, Steffan JS, Fischbeck KH. Histone deacetylase inhibitors reduce polyglutamine toxicity. *Proc Natl Acad Sci U S A* 2001; **98**:15179–15184.

McCampbell A, Taylor JP, Taye AA, Robitschek J, Li M, Walcott J, Merry D, Chai Y, Paulson H, Sobue G, Fischbeck KH. CREB-binding protein sequestration by expanded polyglutamine. *Hum Mol Genet* 2000;**9**:2197–2202.

McDonough H, Patterson C. CHIP: a link between the chaperone and proteasome systems. *Cell Stress Chaperones* 2003;**8**:303–308.

McManamny P, Chy HS, Finkelstein DI, Craythorn RG, Crack PJ, Kola I, Cheema SS, Horne MK, Wreford NG, O'Bryan MK, De Kretser DM, Morrison JR. A mouse model of spinal and bulbar muscular atrophy. *Hum Mol Genet* 2002;**11**:2103–2111.

Merry DE, Kobayashi Y, Bailey CK, Taye AA, Fischbeck KH. Cleavage, aggregation and toxicity of the expanded androgen receptor in spinal and bulbar muscular atrophy. *Hum Mol Genet* 1998;**7**:693–701.

Mhatre AN, Trifiro MA, Kaufman M, Kazemi-Esfarjani P, Figlewicz D, Rouleau G, Pinsky L. Reduced transcriptional regulatory competence of the androgen receptor in X-linked spinal and bulbar muscular atrophy. *Nat Genet* 1993;**5**:184–188.

Miller VM, Nelson RF, Gouvion CM, Williams A, Rodriguez-Lebron E, Harper SQ, Davidson BL, Rebagliati MR, Paulson HL. CHIP suppresses polyglutamine aggregation and toxicity *in vitro* and *in vivo*. *J Neurosci* 2005;**25**:9152–9161.

Monks DA, Johansen JA, Mo K, Rao P, Eagleson B, Yu Z, Lieberman AP, Breedlove SM, Jordan CL. Overexpression of wild-type androgen receptor in muscle recapitulates polyglutamine disease. *Proc Natl Acad Sci U S A* 2007;**104**:18259–18264.

Morfini G, Pigino G, Szebenyi G, You Y, Pollema S, Brady ST. JNK mediates pathogenic effects of polyglutamine-expanded androgen receptor on fast axonal transport. *Nat Neurosci* 2006;**9**:907–916.

Morishima Y, Wang AM, Yu Z, Pratt WB, Osawa Y, Lieberman AP. CHIP deletion reveals functional redundancy of E3 ligases in promoting degradation of both signaling proteins and expanded glutamine proteins. *Hum Mol Genet* 2008;**17**:3942–3952.

Muchowski PJ, Wacker JL. Modulation of neurodegeneration by molecular chaperones. *Nat Rev Neurosci* 2005;**6**:11–22.

Nagashima T, Seko K, Hirose K, Mannen T, Yoshimura S, Arima R, Nagashima K, Morimatsu Y. Familial bulbo-spinal muscular atrophy associated with testicular atrophy and sensory neuropathy (Kennedy-Alter-Sung syndrome). Autopsy case report of two brothers. *J Neurol Sci* 1988;**87**:141–152.

Neckers L. Heat shock protein 90: the cancer chaperone. *J Biosci* 2007;**32**:517–530.

Nordeen EJ, Nordeen KW, Sengelaub DR, Arnold AP. Androgens prevent normally occurring cell death in a sexually dimorphic spinal nucleus. *Science* 1985;**229**:671–673.

Orr HT, Zoghbi HY. Trinucleotide repeat disorders. *Annu Rev Neurosci* 2007;**30**:575–621.

Palazzolo I, Stack C, Kong L, Musaro A, Adachi H, Katsuno M, Sobue G, Taylor JP, Sumner CJ, Fischbeck KH, Pennuto M. Overexpression of IGF-1 in muscle attenuates disease in a mouse model of spinal and bulbar muscular atrophy. *Neuron* 2009;**63**:316–328.

Pandey UB, Nie Z, Batlevi Y, McCray BA, Ritson GP, Nedelsky NB, Schwartz SL, DiProspero NA, Knight MA, Schuldiner O, Padmanabhan R, Hild M, Berry DL, Garza D, Hubbert CC, Yao TP, Baehrecke EH, Taylor JP. HDAC6 rescues neurodegeneration and provides an essential link between autophagy and the UPS. *Nature* 2007;**447**:859–863.

Paulson HL, Perez MK, Trottier Y, Trojanowski JQ, Subramony SH, Das SS, Vig P, Mandel JL, Fischbeck KH, Pittman RN. Intranuclear inclusions of expanded polyglutamine protein in spinocerebellar ataxia type 3. *Neuron* 1997;**19**:333–344.

Pearl LH, Prodromou C. Structure and mechanism of the Hsp90 molecular chaperone machinery. *Annu Rev Biochem* 2006;**75**:271–294.

Peng HM, Morishima Y, Clapp KM, Lau M, Pratt WB, Osawa Y. Dynamic cycling with Hsp90 stabilizes neuronal nitric oxide synthase through calmodulin-dependent inhibition of ubiquitination. *Biochemistry* 2009;**48**:8483–8490.

Picard D. Heat-shock protein 90, a chaperone for folding and regulation. *Cell Mol Life Sci* 2002;**59**:1640–1648.

Piccioni F, Pinton P, Simeoni S, Pozzi P, Fascio U, Vismara G, Martini L, Rizzuto R, Poletti A. Androgen receptor with elongated polyglutamine tract forms aggregates that alter axonal trafficking and mitochondrial distribution in motor neuronal processes. *FASEB J* 2002;**16**:1418–1420.

Pratt WB, Galigniana MD, Harrell JM, DeFranco DB. Role of hsp90 and the hsp90-binding immunophilins in signalling protein movement. *Cell Signal* 2004;**16**:857–872.

Pratt WB, Morishima Y, Osawa Y. The Hsp90 chaperone machinery regulates signaling by modulating ligand binding clefts. *J Biol Chem* 2008;**283**:22885–22889.

Pratt WB, Morishima Y, Peng HM, Osawa Y. Proposal for a role of the Hsp90/Hsp70-based chaperone machinery in making triage decisions when proteins undergo oxidative and toxic damage. *Exp Bio Med (Maywood)* 2010;**235**:278–289

Pratt WB, Toft DO. Regulation of signaling protein function and trafficking by the hsp90/hsp70-based chaperone machinery. *Exp Biol Med (Maywood)* 2003;**228**: 111–133.

Prodromou C, Roe SM, O'Brien R, Ladbury JE, Piper PW, Pearl LH. Identification and structural characterization of the ATP/ADP-binding site in the Hsp90 molecular chaperone. *Cell* 1997;**90**:65–75.

Ravikumar B, Acevedo-Arozena A, Imarisio S, Berger Z, Vacher C, O'Kane CJ, Brown SD, Rubinsztein DC. Dynein mutations impair autophagic clearance of aggregate-prone proteins. *Nat Genet* 2005;**37**:771–776.

Ravikumar B, Vacher C, Berger Z, Davies JE, Luo S, Oroz LG, Scaravilli F, Easton DF, Duden R, O'Kane CJ, Rubinsztein DC. Inhibition of mTOR induces autophagy and reduces toxicity of polyglutamine expansions in fly and mouse models of Huntington disease. *Nat Genet* 2004;**36**:585–595.

Roy AK, Tyagi RK, Song CS, Lavrovsky Y, Ahn SC, Oh TS, Chatterjee B. Androgen receptor: structural domains and functional dynamics after ligand-receptor interaction. *Ann N Y Acad Sci* 2001;**949**:44–57.

Schmidt BJ, Greenberg CR, Allingham-Hawkins DJ, Spriggs EL. Expression of X-linked bulbospinal muscular atrophy (Kennedy disease) in two homozygous women. *Neurology* 2002;**59**:770–772.

Schneider C, Sepp-Lorenzino L, Nimmesgern E, Ouerfelli O, Danishefsky S, Rosen N, Hartl FU. Pharmacologic shifting of a balance between protein refolding and degradation mediated by Hsp90. *Proc Natl Acad Sci U S A* 1996;**93**:14536–14541.

Sepp-Lorenzino L, Ma Z, Lebwohl DE, Vinitsky A, Rosen N. Herbimycin A induces the 20S proteasome- and ubiquitin-dependent degradation of receptor tyrosine kinases. *J Biol Chem* 1995;**270**:16580–16587.

Shen HC, Coetzee GA. The androgen receptor: unlocking the secrets of its unique transactivation domain. *Vitam Horm* 2005;**71**:301–319.

Shin Y, Klucken J, Patterson C, Hyman BT, McLean PJ. The co-chaperone carboxyl terminus of Hsp70-interacting protein (CHIP) mediates alpha-synuclein degradation decisions between proteasomal and lysosomal pathways. *J Biol Chem* 2005;**280**:23727–23734.

Sittler A, Lurz R, Lueder G, Priller J, Lehrach H, Hayer-Hartl MK, Hartl FU, Wanker EE. Geldanamycin activates a heat shock response and inhibits huntingtin aggregation in a cell culture model of Huntington's disease. *Hum Mol Genet* 2001;**10**:1307–1315.

Skinner PJ, Koshy BT, Cummings CJ, Klement IA, Helin K, Servadio A, Zoghbi HY, Orr HT. Ataxin-1 with an expanded glutamine tract alters nuclear matrix-associated structures. *Nature* 1997;**389**:971–974.

Smith DF. Dynamics of heat shock protein 90-progesterone receptor binding and the disactivation loop model for steroid receptor complexes. *Mol Endocrinol* 1993;**7**:1418–1429.

Sobue G, Doyu M, Kachi T, Yasuda T, Mukai E, Kumagai T, Mitsuma T. Subclinical phenotypic expressions in heterozygous females of X-linked recessive bulbospinal neuronopathy. *J Neurol Sci* 1993;**117**:74–78.

Sobue G, Hashizume Y, Mukai E, Hirayama M, Mitsuma T, Takahashi A. X-linked recessive bulbospinal neuronopathy. A clinicopathological study. *Brain* 1989;**112**:(Pt 1): 209–232.

Sopher BL, Thomas PS, LaFevre-Bernt MA, Holm IE, Wilke SA, Ware CB, Jin LW, Libby RT, Ellerby LM, La Spada AR Jr. Androgen receptor YAC transgenic mice recapitulate SBMA motor neuronopathy and implicate VEGF164 in the motor neuron degeneration. *Neuron* 2004;**41**:687–699.

Sperfeld AD, Karitzky J, Brummer D, Schreiber H, Haussler J, Ludolph AC, Hanemann CO. X-linked bulbospinal neuronopathy: Kennedy disease. *Arch Neurol* 2002;**59**:1921–1926.

Stebbins CE, Russo AA, Schneider C, Rosen N, Hartl FU, Pavletich NP. Crystal structure of an Hsp90-geldanamycin complex: targeting of a protein chaperone by an antitumor agent. *Cell* 1997;**89**:239–250.

Szebenyi G, Morfini GA, Babcock A, Gould M, Selkoe K, Stenoien DL, Young M, Faber PW, MacDonald ME, McPhaul MJ, Brady ST. Neuropathogenic forms of huntingtin and androgen receptor inhibit fast axonal transport. *Neuron* 2003;**40**:41–52.

Thomas M, Dadgar N, Aphale A, Harrell JM, Kunkel R, Pratt WB, Lieberman AP. Androgen receptor acetylation site mutations cause trafficking defects, misfolding, and aggregation similar to expanded glutamine tracts. *J Biol Chem* 2004;**279**:8389–8395.

Thomas M, Harrell JM, Morishima Y, Peng HM, Pratt WB, Lieberman AP. Pharmacologic and genetic inhibition of hsp90-dependent trafficking reduces aggregation and promotes degradation of the expanded glutamine androgen receptor without stress protein induction. *Hum Mol Genet* 2006;**15**:1876–1883.

Thomas M, Yu Z, Dadgar N, Varambally S, Yu J, Chinnaiyan AM, Lieberman AP. The unfolded protein response modulates toxicity of the expanded glutamine androgen receptor. *J Biol Chem* 2005;**280**:21264–21271.

Tsai YC, Fishman PS, Thakor NV, Oyler GA. Parkin facilitates the elimination of expanded polyglutamine proteins and leads to preservation of proteasome function. *J Biol Chem* 2003;**278**:22044–22055.

Uehara Y, Fukazawa H. Use and selectivity of herbimycin A as inhibitor of protein-tyrosine kinases. *Methods Enzymol* 1991;**201**:370–379.

Uehara Y, Hori M, Takeuchi T, Umezawa H. Screening of agents which convert 'transformed morphology' of Rous sarcoma virus-infected rat kidney cells to 'normal morphology': identification of an active agent as herbimycin and its inhibition of intracellular src kinase. *Jpn J Cancer Res* 1985;**76**:672–675.

Uehara Y, Hori M, Takeuchi T, Umezawa H. Phenotypic change from transformed to normal induced by benzoquinonoid ansamycins accompanies inactivation of p60src in rat kidney cells infected with Rous sarcoma virus. *Mol Cell Biol* 1986;**6**:2198–2206.

Wandinger SK, Richter K, Buchner J. The Hsp90 chaperone machinery. *J Biol Chem* 2008;**283**:18473–18477.

Wang Q, Udayakumar TS, Vasaitis TS, Brodie AM, Fondell JD. Mechanistic relationship between androgen receptor polyglutamine tract truncation and androgen-dependent transcriptional hyperactivity in prostate cancer cells. *J Biol Chem* 2004;**279**:17319–17328.

Wang AM, Morishima Y, Clapp KM, Peng HM, Pratt WB, Gestwicki JE, Liebermant AP, Inhibition of Hsp70 by methylene blue affects signaling protein function and ubiquitination and modulates polyglutamine protein degradation. *J Biol Chem* 2010;**285**:15714–15723.

Warrick JM, Chan HY, Gray-Board GL, Chai Y, Paulson HL, Bonini NM. Suppression of polyglutamine-mediated neurodegeneration in Drosophila by the molecular chaperone HSP70. *Nat Genet* 1999;**23**:425–428.

Waza M, Adachi H, Katsuno M, Minamiyama M, Sang C, Tanaka F, Inukai A, Doyu M, Sobue G. 17-AAG, an Hsp90 inhibitor, ameliorates polyglutamine-mediated motor neuron degeneration. *Nat Med* 2005;**11**:1088–1095.

Whitesell L, Mimnaugh EG, De Costa B, Myers CE, Neckers LM. Inhibition of heat shock protein HSP90-pp60v-src heteroprotein complex formation by benzoquinone ansamycins: essential role for stress proteins in oncogenic transformation. *Proc Natl Acad Sci U S A* 1994;**91**:8324–8328.

Yu WH. Administration of testosterone attenuates neuronal loss following axotomy in the brain-stem motor nuclei of female rats. *J Neurosci* 1989;**9**:3908–3914.

Yu Z, Dadgar N, Albertelli M, Scheller A, Albin RL, Robins DM, Lieberman AP. Abnormalities of germ cell maturation and sertoli cell cytoskeleton in androgen receptor 113 CAG knock-in mice reveal toxic effects of the mutant protein. *Am J Pathol* 2006a;**168**:195–204.

Yu Z, Dadgar N, Albertelli M, Gruis K, Jordan C, Robins DM, Lieberman AP. Androgen-dependent pathology demonstrates myopathic contribution to the Kennedy disease phenotype in a mouse knock-in model. *J Clin Invest* 2006b;**116**:2663–2672.

Yu Z, Wang AM, Robins DM, Lieberman AP. Altered RNA splicing contributes to skeletal muscle pathology in Kennedy disease knock-in mice. *Dis Model Mech* 2009;**2**:500–507.

Zhang Y, Gao J, Chung KK, Huang H, Dawson VL, Dawson TM. Parkin functions as an E2-dependent ubiquitin- protein ligase and promotes the degradation of the synaptic vesicle-associated protein, CDCrel-1. *Proc Natl Acad Sci U S A* 2000;**97**:13354–13359.

Zhou P, Fernandes N, Dodge IL, Reddi AL, Rao N, Safran H, DiPetrillo TA, Wazer DE, Band V, Band H. ErbB2 degradation mediated by the co-chaperone protein CHIP. *J Biol Chem* 2003;**278**:13829–13837.

Zitzmann M, Nieschlag E. The CAG repeat polymorphism within the androgen receptor gene and maleness. *Int J Androl* 2003;**26**:76–83.

Zoghbi HY, Orr HT. Glutamine repeats and neurodegeneration. *Annu Rev Neurosci* 2000;**23**:217–247.

Zou J, Guo Y, Guettouche T, Smith DF, Voellmy R. Repression of heat shock transcription factor HSF1 activation by HSP90 (HSP90 complex) that forms a stress-sensitive complex with HSF1. *Cell* 1998;**94**:471–480.

7

AMYLOID REMODELING BY Hsp104

James Shorter

Department of Biochemistry and Biophysics, University of Pennsylvania School of Medicine, Philadelphia, PA, USA

7.1 PROTEIN MISFOLDING IS A UNIVERSAL PROBLEM

Life demands that proteins fold into very precise functional structures. Functional native structure is enciphered by primary sequence (Anfinsen, 1973; Englander *et al.*, 2007). However, native structures are dynamic systems composed of sophisticated networks of weak, mutually supportive contacts that are difficult to establish simultaneously during folding (Bartlett and Radford, 2009; Englander *et al.*, 2007). Thus, folding energy landscapes are often rugged and create challenges for successful folding (Bartlett and Radford, 2009). Polypeptides can become trapped in non-native intermediate states or become diverted into off-pathway states. Even after the completion of folding, cooperative units of native structure, termed *foldons*, repeatedly unfold and refold (Englander *et al.*, 2007). Moreover, mutation or errors in transcription or translation can yield polypeptides that are less able to form functional structures (Dobson, 2003; Lee *et al.*, 2006). Environmental stress can also disrupt protein folding (Parsell and Lindquist, 1993). Consequently, proteins can fail to fold or fail to remain correctly folded. These failures increase the risk of aggregation. The highly crowded macromolecular environment that cells are forced to maintain to function optimally further accentuates this risk (Dobson, 2003; Ellis and Minton,

Protein Chaperones and Protection from Neurodegenerative Diseases, First Edition.
Edited by Stephan N. Witt.

2006). Therefore, sophisticated protein homeostasis (proteostasis) systems have evolved, which ensure that polypeptides can effectively acquire, maintain, and reacquire their functional native structure or be eliminated should folding become too improbable (Balch *et al.*, 2008; Powers *et al.*, 2009).

To ensure a folding-competent state, molecular chaperones contact the nascent polypeptide even before translation is complete (Kramer *et al.*, 2009). Following translation, molecular chaperones prevent aggregation and assist polypeptides in acquiring their native form (Young *et al.*, 2004). The ubiquitin–proteasome system degrades any terminally misfolded forms (Varshavsky, 2005; Vembar and Brodsky, 2008). However, aggregated proteins resist proteasomal degradation (Bence *et al.*, 2001) but can be catabolized by autophagy (Cuervo, 2008). Finally, sophisticated disaggregases reverse protein aggregation. Disaggregation can be coupled to degradation (Bieschke *et al.*, 2009; Cohen *et al.*, 2006; Murray *et al.*, 2010) or renaturation (Doyle and Wickner, 2009; Glover and Lum, 2009; Shorter, 2008; Weibezahn *et al.*, 2005).

Once individuals reach postreproductive age, these proteostatic safeguards decline inexorably, and errors in protein folding can arise with devastating sequelae (Cohen *et al.*, 2006; Cuervo, 2008; Morimoto, 2006; Skovronsky *et al.*, 2006). A pernicious and recurring problem is that the functional native structure is not always the lowest free energy form (Englander *et al.*, 2007). Rather, many proteins, irrespective of primary sequence, can spontaneously form generic, cross-β polymers of even lower free energy, termed *amyloid* (Dobson, 2003; Englander *et al.*, 2007).

7.2 AMYLOID CONFORMERS CAN BE PATHOGENIC, PROTECTIVE, OR BENEFICIAL

Amyloidogenesis of various specific proteins is linked with a legion of devastating disorders, including Alzheimer's disease (AD), Parkinson's disease (PD), Huntington's disease (HD), type II diabetes, prion diseases, and various cardiovascular and systemic amyloidoses (Caughey and Lansbury, 2003; Cushman *et al.*, 2010; Kholova and Niessen, 2005; Skovronsky *et al.*, 2006; Taylor *et al.*, 2002). There are no effective treatments for any of these conditions. Furthermore, a severe risk factor for these diseases is aging (Morimoto, 2006). Indeed, because natural selection acts less powerfully on genetic variation expressed at postreproductive age, many genes may harbor "late-expressing" harmful mutations (Medawar, 1952). Some of these mutations may predispose proteins to forming amyloids or prions (infectious amyloids) in the environment of an aging individual where the proteostasis network is in decline. Several examples are found in the mammalian prion protein (PrP) (Kong *et al.*, 2004). As life spans are extended through improvements in medicine and public health, these disorders will inevitably increase in prevalence. Indeed, they threaten to become one among the most intractable barriers to living longer, more fulfilling lives.

Amyloids possess a "cross-β" form in which the strands of the β-sheets align perpendicular to the fiber axis (Nelson and Eisenberg, 2006; Sunde and

Blake, 1997; Sunde *et al.*, 1997). The amyloid fold is extremely stable and resists disruption by various denaturing conditions, including protease digestion, detergents, chaotropes, and high temperatures (Eisenberg *et al.*, 2006; Knowles *et al.*, 2007; Smith *et al.*, 2006). This extreme stability makes amyloid difficult to eliminate. Indeed, in systemic amyloidoses, amyloid accumulation can be so severe that tissue architecture becomes mechanically disrupted (Merlini and Westermark, 2004). The ends of amyloid fibers capture other copies of the same protein and convert them to the cross-β structure. Once initiated, this self-templating or "seeding" process can convert all the copies of a given protein to the amyloid fold (Lansbury and Caughey, 1995; Nelson and Eisenberg, 2006). Steric effects usually cause proteins to lose functionality in the amyloid state (Baxa *et al.*, 2002). This "loss-of-function" contributes to pathogenesis in some disorders (Forman *et al.*, 2004). Furthermore, by depleting other cellular components that coprecipitate, amyloids can also cause other proteins to lose functionality (Chen *et al.*, 2005).

In various diseases, however, the quantity of amyloid deposits can be minimal and their presence can correlate with cell survival (Arrasate *et al.*, 2004; Cohen *et al.*, 2006; de Calignon *et al.*, 2010; Dobson, 2003). These findings have generated proposals that amyloid forms may be relatively benign and reflect a cellular defense mechanism that sequesters toxic soluble species (Bucciantini *et al.*, 2002; Kayed *et al.*, 2003). This benefit might outweigh the cost of these space-occupying lesions. Indeed, the soluble oligomeric species that assemble during the distinctive lag phase of amyloid formation can be highly toxic and share a generic conformation, which is distinct from fibers and independent of primary sequence (Bucciantini *et al.*, 2002; Haass and Selkoe, 2007; Kayed *et al.*, 2003; Lashuel *et al.*, 2002; Lesne *et al.*, 2006). These shared features of amyloidogenesis indicate that effective therapeutics might have broad applicability (Skovronsky *et al.*, 2006). Despite these similarities, however, a major unresolved issue concerns how the amyloidogenesis of different proteins can confer the selective neuronal cell death that distinguishes various neurodegenerative disorders (Cushman *et al.*, 2010; Skovronsky *et al.*, 2006).

Recent studies, however, suggest that the amyloid state is unlikely to be invariably benign. In a mouse AD model, amyloid-β (Aβ) plaques can form rapidly and mediate pathology (Meyer-Luehmann *et al.*, 2008). Amyloid might also slowly release toxic misfolded species. For example, natural lipids can destabilize amyloid fibers and liberate toxic oligomers (Martins *et al.*, 2008). Another issue concerns the ability of amyloidogenic proteins to fold into multiple structurally distinct amyloid forms or "strains," which confer distinct phenotypes (Legname *et al.*, 2006; Safar *et al.*, 1998; Tanaka *et al.*, 2006). Beyond sharing the cross-β amyloid form, little is known about the underlying atomic structures of these distinct strains or how structural polymorphism enciphers distinct phenotypes or disease states (Wiltzius *et al.*, 2009). Distinct ensembles of strains form depending on the environment (e.g., pH, temperature). Strains are distinguished by distinct intermolecular contacts between fiber protomers and different lengths of primary sequence sequestered in cross-β structure (Krishnan and Lindquist, 2005; Roberts *et al.*, 2009; Tessier and Lindquist, 2007; Toyama *et al.*, 2007). Different strains of Aβ40 fibers, which are connected with AD, and polyglutamine (polyQ), which are connected to HD, confer different levels of toxicity (Nekooki-Machida *et al.*, 2009; Petkova *et al.*, 2005). Some strains are relatively benign, whereas others are highly toxic (Nekooki-Machida *et al.*, 2009; Petkova *et al.*, 2005). Intriguingly, toxic strains are more abundant in brain regions with the most neurodegeneration in mouse models of HD (Nekooki-Machida *et al.*, 2009). These data suggest that local proteostatic buffers or expression levels might create strain biases *in situ*. It will be critical to determine how strain variation correlates with affected brain regions in various neurodegenerative amyloidoses. Identifying which strains are toxic and which are benign will help inform potential targeted therapies.

Thus, amyloid can be either detrimental or benign, depending on the precise strain. It is therefore not surprising that benign amyloids have been captured during evolution for functional, adaptive purposes (Fowler *et al.*, 2007; Shorter and Lindquist, 2005b). For example, Pmel17 amyloids function in melanosome formation (Berson *et al.*, 2003; Fowler *et al.*, 2006; Watt *et al.*, 2009). Amyloid forms of cytoplasmic polyadenylation element-binding protein (CPEB) might function in synapse stabilization, which promotes long-term memory formation (Si *et al.*, 2003, 2010). In yeast, many proteins can form infectious amyloids, termed *prions*, which provide a vast reservoir of heritable phenotypic variation that can be advantageous under diverse environmental conditions (Alberti *et al.*, 2009; Griswold and Masel, 2009; King and Masel, 2007; Shorter and Lindquist, 2005b; True and Lindquist, 2000; Tyedmers *et al.*, 2008). In these cases, the proteostasis network ensures that benign amyloid conformers assemble instead of toxic intermediates or strains (Douglas *et al.*, 2008; Shorter and Lindquist, 2004; Treusch *et al.*, 2009). An accurate understanding of how amyloids have been exploited for beneficial purposes will likely yield important insights into how to safely eliminate toxic amyloid fibers and preamyloid oligomers.

There are no cures or effective treatments for any of the neurodegenerative amyloidoses confronting humankind. Therapies remain palliative in nature and do not antagonize the underlying causative continuum of amyloid forms or cytotoxic oligomers. A seminal therapeutic advance will come with the ability to enhance proteostasis to eliminate entire spectra of toxic amyloid strains and preamyloid oligomers, while leaving beneficial amyloid structures unperturbed. Here, we discuss the notion of enhancing mammalian proteostasis with a protein disaggregase from yeast, heat shock protein (Hsp104), which can rapidly resolve amyloid conformers and preamyloid oligomers (Shorter, 2008). First, however, I will introduce Hsp104 and consider the mechanistic basis of its activity.

7.3 Hsp104 IS AN AAA+ ATPase WITH PROTEIN DISAGGREGASE ACTIVITY

Hsp104 is an *A*TPase *A*ssociated with diverse *A*ctivities (AAA+) protein (Erzberger and Berger, 2006; Neuwald *et al.*, 1999), which enhances yeast survival several 1000-fold after a variety of environmental stresses that induce protein aggregation, including heat and chemical shock (Sanchez and Lindquist, 1990; Sanchez *et al.*, 1992). Orthologs in bacteria (ClpB) and plants (Hsp101) provide selective advantages of a similar magnitude (Queitsch *et al.*, 2000; Squires *et al.*, 1991). This adaptive benefit of Hsp104 lies in its ability to rescue proteins from denatured aggregates and restore them to native structure and function. This extraordinary activity is coordinated by the Hsp70 chaperone system (Cashikar *et al.*, 2005; Glover and Lindquist, 1998; Goloubinoff *et al.*, 1999; Mogk *et al.*, 1999; Parsell *et al.*, 1993, 1994b; Weibezahn *et al.*, 2004). The Hsp70 chaperone system helps deliver aggregated substrates to Hsp104 (Glover and Lindquist, 1998; Tessarz *et al.*, 2008). Once an unfolded substrate is released from the aggregate, the Hsp70 chaperone system also promotes refolding (Doyle *et al.*, 2007b; Glover and Lindquist, 1998). This salvage and rapid renaturation of proteins obviates the severe energetic costs of protein degradation and *de novo* biosynthesis that would otherwise be required to eliminate and replace the aggregated protein. Thus, cells can recover rapidly from environmental stresses that induce protein aggregation.

Hsp104 can be divided into five domains: an N-terminal domain, a first AAA+ nucleotide-binding domain (NBD1), a coiled-coil middle domain, a second AAA+ nucleotide-binding domain (NBD2), and a short C-terminal domain (Doyle and Wickner, 2009). Like many AAA+ proteins, Hsp104 is only active as a hexamer, which forms upon adenosine diphosphate (ADP) or adenosine triphosphate (ATP) binding to NBD2 (Parsell *et al.*, 1994a; Schirmer *et al.*, 1998, 2001). Unfortunately, there is no atomic resolution structure of the Hsp104 hexamer. However, cryo-electron microscopy and single particle reconstruction have revealed that the Hsp104 hexamer is a three-tiered ring structure that envelops a large central cavity or channel (Wendler and Saibil, 2010; Wendler *et al.*, 2007, 2009).

The monomeric structure of the *T. thermophilus* ortholog, tClpB, has been solved (Lee *et al.*, 2003). Using this structure, the Hsp104 monomer has been homology modeled and fitted as rigid bodies into electron density envelopes (Wendler and Saibil, 2010; Wendler *et al.*, 2007, 2009). These studies have revealed that a small ring of N-terminal domains forms the top tier of the hexamer, whereas expanded rings of NBD1 and NBD2 form the middle and lower tiers, respectively (Wendler *et al.*, 2007, 2009). The distinctive middle domain, which is composed of two antiparallel coiled-coil motifs reminiscent of a two-bladed propeller (Lee *et al.*, 2003), intercalates between NBD1 and NBD2 in the wall of the hexamer (Wendler *et al.*, 2007, 2009). This hexameric model of Hsp104 differs markedly from a hexameric model advanced for tClpB, where the coiled-coil domains protrude laterally from the surface of the hexameric ring (Lee *et al.*, 2003, 2007). There are several potential explanations for these differences, which are discussed in detail elsewhere (Wendler and Saibil, 2010).

Hsp104 and orthologs use energy from ATP binding and hydrolysis to translocate polypeptides from the aggregate surface across the central channel to solution (Lum *et al.*, 2004, 2008; Schlieker *et al.*, 2004; Shorter and Lindquist,

2005a; Weibezahn *et al.*, 2004). However, we are only beginning to understand the mechanistic and structural basis of this activity. Simultaneous elimination of ATPase activity at *both* NBDs abolishes disaggregase activity (Doyle *et al.*, 2007b; Parsell *et al.*, 1991; Shorter and Lindquist, 2004). Hsp104 initially engages misfolded substrates when NBD1 is in an ATP-bound conformation (Bosl *et al.*, 2005; Schaupp *et al.*, 2007). Both NBDs catalyze ATP hydrolysis cooperatively, and allosteric communication occurs within and between NBD1 and NBD2 (Cashikar *et al.*, 2002; Doyle *et al.*, 2007b; Hattendorf and Lindquist, 2002; Schaupp *et al.*, 2007; Schirmer *et al.*, 2001). In the absence of substrate, NBD1 makes the major contribution to ATPase activity ($k_{cat} \sim 76$ min^{-1}, $K_M \sim 170$ μM, $n_h = 2.3$) but has a lower affinity for nucleotide compared to NBD2 ($k_{cat} \sim 0.27$ min^{-1}, $K_M \sim 4.7$ μM, $n_h = 1.6$) (Hattendorf and Lindquist, 2002). Despite these advances, little is known about precisely how allosteric regulation of ATP hydrolysis within and between NBD1 and NBD2 is coupled to the substrate binding, unfolding, and translocation that are required for disaggregation. Indeed, how individual subunits within the hexamer collaborate to coordinate protein disaggregation remains obscure.

Several insights have been afforded by artificially inducing disaggregation activity in ClpB and Hsp104 in the absence of Hsp70 and Hsp40. For example, dissolution of denatured aggregates by ClpB and Hsp104 can be triggered with specific mixtures of ATP and ATPγS, a slowly hydrolyzable ATP analog (Doyle *et al.*, 2007a, 2007b; Hoskins *et al.*, 2009). Alternatively, mutation of conserved AAA+ motifs (Walker A, Walker B, or sensor-1) at one nucleotide-binding domain (NBD) to slow ATP hydrolysis at that site can also elicit disaggregase or substrate unfolding activity in the absence of Hsp70 and Hsp40 (Doyle *et al.*, 2007a, 2007b; Hoskins *et al.*, 2009; Schaupp *et al.*, 2007). ClpB hexamers exchange protomers rapidly, which might enable recycling of monomers should disaggregation stall or fail (Haslberger *et al.*, 2008; Werbeck *et al.*, 2008). This rapid exchange of monomers facilitates titration experiments with mutant subunits to assess how easily a hexamer can be inactivated, for example, by one or two mutant subunits (Crampton *et al.*, 2006). A particularly useful mutant bears Walker B mutations in both NBDs (Weibezahn *et al.*, 2003). This mutant can engage substrate, bind nucleotide, and form hexamers, and is solely defective in ATP hydrolysis (Weibezahn *et al.*, 2003). In situations where ClpB is activated in the absence of Hsp70 and Hsp40, ClpB hexamers are relatively insensitive to this mutant, suggesting that ClpB subunits can act via a probabilistic mechanism to promote some disaggregation events (Hoskins *et al.*, 2009).

However, the disaggregation of denatured aggregates is most effective when coordinated by Hsp70 and Hsp40. Hsp70 and Hsp40 are required to present denatured aggregated substrates to Hsp104 (Glover and Lindquist, 1998; Tessarz *et al.*, 2008; Weibezahn *et al.*, 2004) and likely play some role in coordinating Hsp104 ATPase cycling (Doyle *et al.*, 2007a, 2007b; Hoskins *et al.*, 2009). In some cases, the substrate itself (e.g., amyloid) can impose the requisite changes (Doyle *et al.*, 2007b; Shorter and Lindquist, 2004). A small fraction of double Walker B mutant subunits, perhaps as little as one per hexamer, poisons ClpB-mediated disaggregation of denatured aggregates, which is coordinated by the Hsp70 chaperone system (Haslberger *et al.*, 2008; Hoskins *et al.*, 2009; Werbeck *et al.*, 2008).

Collectively, these data suggest that a cooperative division of labor among the 12 AAA+ domains drives protein disaggregation. One subset slowly hydrolyzes ATP to facilitate substrate binding, whereas another subset rapidly hydrolyzes ATP to promote substrate unfolding and translocation. This division of labor is adaptable, and precisely how it is established can vary depending on the substrate and the presence of the Hsp70 system. For example, under the conditions where disaggregation is artificially elicited by mutation or by mixture of ATP and ATPγS, the subunits that hydrolyze ATP are determined on some probabilistic basis (Hoskins *et al.*, 2009). By contrast, in the presence of Hsp70 and Hsp40, the division of labor is coordinated such that individual subunits must hydrolyze ATP in a concerted or sequential manner to drive disaggregation (Haslberger *et al.*, 2008; Hoskins *et al.*, 2009; Werbeck *et al.*, 2008). Such concerted or sequential intersubunit collaboration is considerably more effective in driving disaggregation than the probabilistic mode (Doyle *et al.*, 2007a, b; Hoskins *et al.*, 2009).

How does this cooperative division of labor promote substrate translocation across the central channel? The N- and C-terminal domains may help bind substrates and cofactors (Barnett *et al.*, 2005; Cashikar *et al.*, 2002; Mackay *et al.*, 2008). However, critical substrate interactions are mediated by an α-helical insertion in NBD1 and a β-hairpin insertion in NBD2, located before helix α 2 in the αβ subdomain in both NBDs (Lum *et al.*, 2004, 2008; Schlieker *et al.*, 2004; Tessarz *et al.*, 2008; Weibezahn *et al.*, 2004). Short, highly conserved loops, KYKG in NBD1 and GYVG in NBD2, project into the channel (Wendler *et al.*, 2009). Of particular importance is the tyrosine residue in these loops, as mutation of this residue to alanine disrupts substrate interactions and disaggregation activity *in vitro* and *in vivo* (Lum *et al.*, 2004, 2008; Tessarz *et al.*, 2008). Mutation of the NBD2 loop tyrosine confers the most drastic effects *in vivo* and phenocopies deletion of Hsp104 (Lum *et al.*, 2004). More conservative substitutions of the NBD1 or NBD2 loop tyrosines, such as phenylalanine and tryptophan, maintain partial functionality (Cashikar *et al.*, 2002; Hung and Masison, 2006). Dynamic rearrangements of channel loop tyrosines, which are proposed to "grip" the substrate, synchronized with ATPase cycling likely provide a series of motions that translocate substrates across the channel.

Cryo-electron microscopy reconstructions of Hsp104 hexamers in the presence of ATPγS, ATP, and ADP have provided structural insight into the conformational changes that facilitate substrate translocation (Wendler *et al.*, 2009). This study employed the NBD2 sensor-1 mutant, Hsp104^{N728A}, which is able to disaggregate denatured aggregates without Hsp70 or Hsp40 (Doyle *et al.*, 2007b), but is defective in prion disaggregation and provides only limited thermotolerance *in vivo* (Hattendorf and Lindquist, 2002; Shorter and Lindquist, 2004). Reconstructions with imposed sixfold symmetry reveal that ATP binding and hydrolysis induce large domain movements in NBD1 that impart a peristaltic mechanism for substrate translocation. The extremely large size of the Hsp104 channel compared to other AAA+ proteins might enable the translocation of exposed loops or more than one polypeptide, rather than having to search for exposed N- or C-termini of individual polypeptides (Haslberger *et al.*, 2008; Wendler *et al.*, 2007). Upon ATP binding, the NBD1 substrate-binding KYKG motifs move up toward the N-terminal end of the channel and are poised to receive the substrate. Upon ATP hydrolysis, NBD1 generates a large motion that displaces the KYKG motif from the N-terminal end to the center of the channel. Simultaneously, the NBD2 substrate-binding GYVG motifs rotate into the center of the channel to receive the substrate translocated by NBD1. Subsequent ATP binding to NBD1 then moves the NBD1 KYKG motifs back up toward the N-terminal entrance, while simultaneously moving the NBD2 GYVG motifs down toward the C-terminal end of the channel. Thus, the NBD2 GYVG motif is able to exert a pulling force without ATP hydrolysis by NBD2. The ADP state of the hexamer suggests that ATP hydrolysis at NBD2 might induce a dramatic rotation of this domain that would eject substrate. These interdependent motions of NBD1 and NBD2 ensure continuous substrate handling during disaggregation (Wendler *et al.*, 2009). Throughout the Hsp104 ATPase cycle, the coiled-coil middle domain, which distinguishes Hsp104 and orthologs from all other AAA+ proteins, appears to play a critical structural role that facilitates the dramatic rotations of NBD1 and NBD2 that forcibly drive substrate translocation (Wendler and Saibil, 2010; Wendler *et al.*, 2007, 2009).

7.4 Hsp104 HAS A POWERFUL AMYLOID-REMODELING ACTIVITY

Hsp104 possesses an unusually powerful amyloid-remodeling activity and couples ATP hydrolysis to the rapid deconstruction of amyloid forms of Sup35 and Ure2, two yeast PrPs (Narayanan *et al.*, 2006; Savistchenko *et al.*, 2008; Shorter and Lindquist, 2004, 2006, 2008). Curiously, bacterial homologs appear to lack the ability to remodel amyloid (Shorter and Lindquist, 2004; Tipton *et al.*, 2008). Importantly, even a brief overexpression of Hsp104 is sufficient to eliminate Sup35 prions (Chernoff *et al.*, 1995). At lower concentrations, Hsp104 fragments yeast prions, which ensures their inheritance through successive generations (Chernoff *et al.*, 1995; Kryndushkin *et al.*, 2003; Patino *et al.*, 1996; Paushkin *et al.*, 1996; Shorter and Lindquist, 2004, 2006). This ability to tightly regulate amyloid conformers endows yeast with another massive selective advantage: the ability to employ prions as metastable switches in protein function (Alberti *et al.*, 2009; Halfmann *et al.*, 2010; Shorter and Lindquist, 2005b). Indeed, yeast exploits prions as a vast reservoir of heritable phenotypic variation, which can be advantageous in diverse environments (Alberti *et al.*, 2009; Griswold and Masel, 2009; King and Masel, 2007; Shorter and Lindquist, 2005b; True and Lindquist, 2000; Tyedmers *et al.*, 2008).

Dissolution of amyloid structure by Hsp104 does not require Hsp70 and Hsp40 (Narayanan *et al.*, 2006; Savistchenko *et al.*, 2008; Shorter and Lindquist, 2004, 2006). However, the presence of the Hsp70 chaperone system can ameliorate

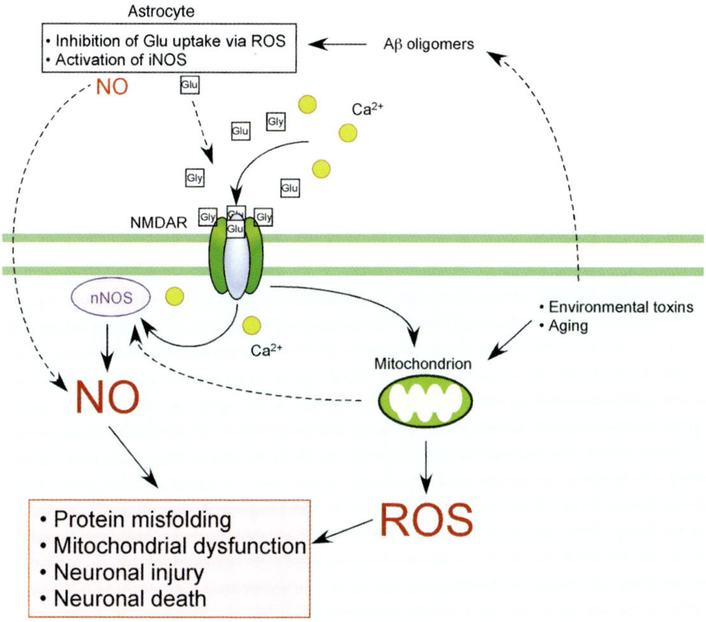

Figure 2.1 For caption see page 69.

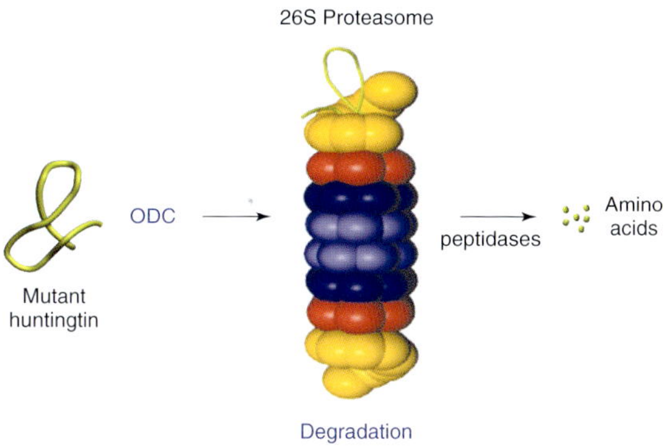

Figure 5.2 For caption see page 194.

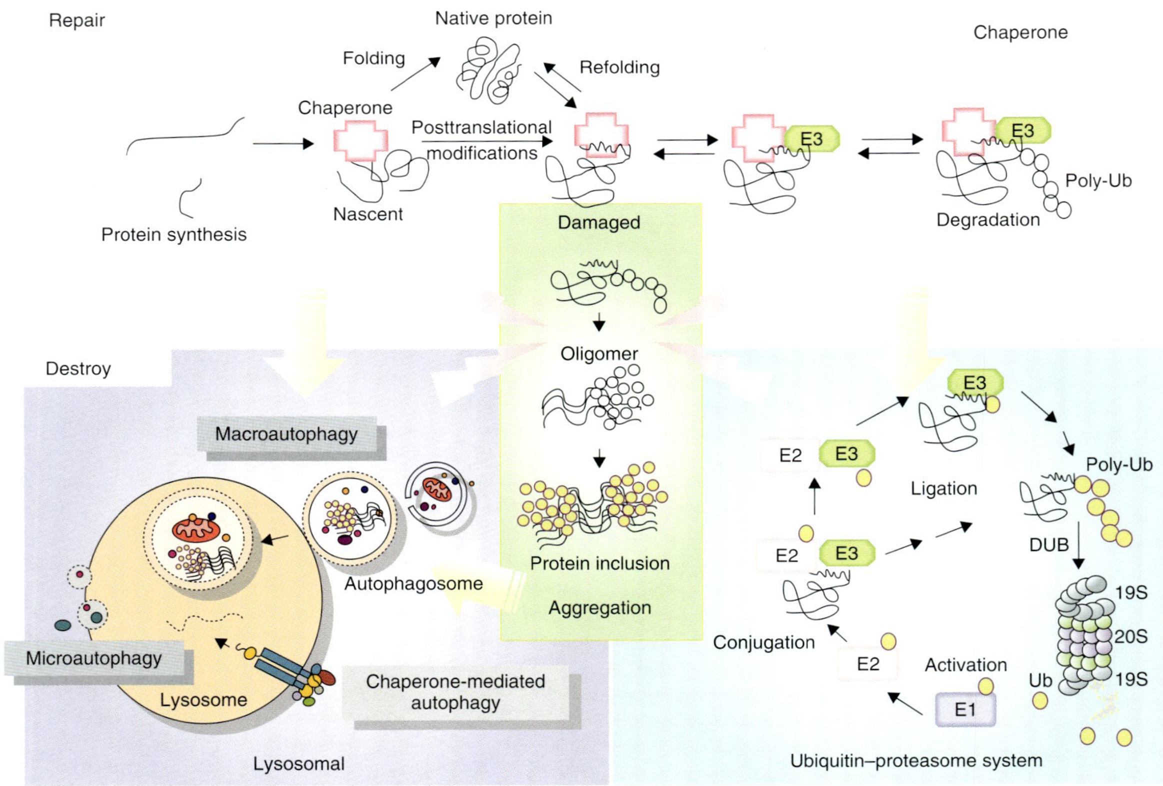

Figure 3.1 For caption see page 103.

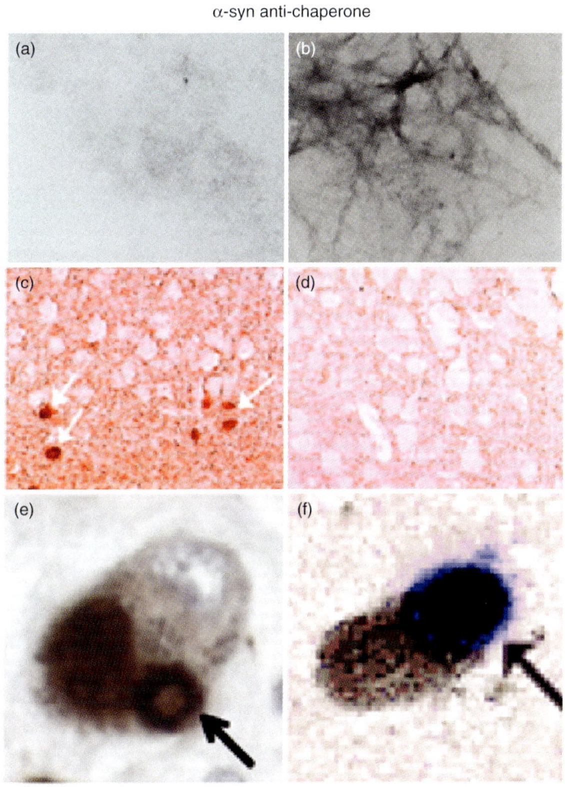

Figure 4.3 For caption see page 144.

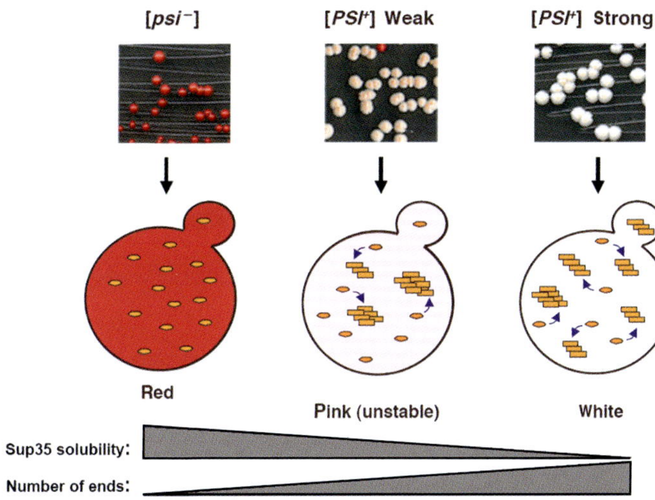

Figure 9.2 For caption see page 281.

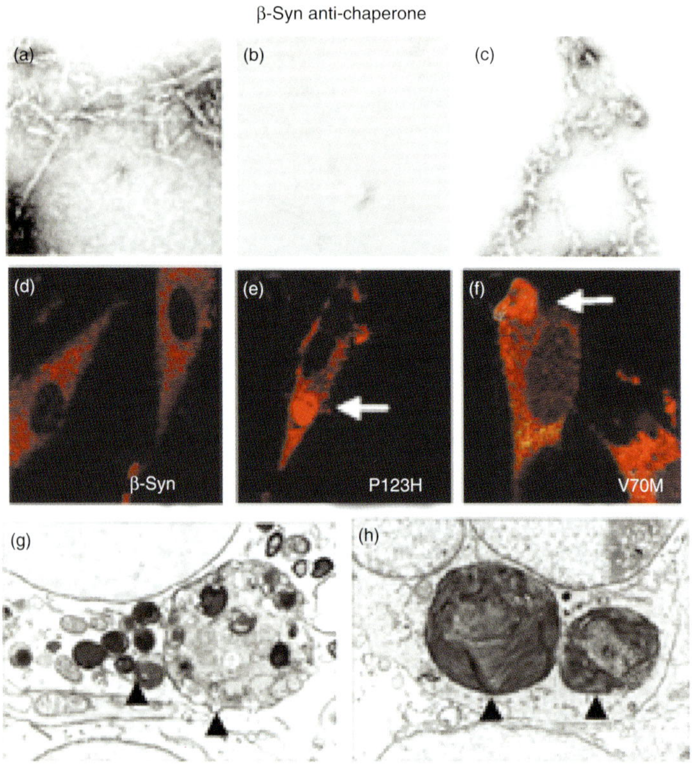

Figure 4.5 For caption see page 152.

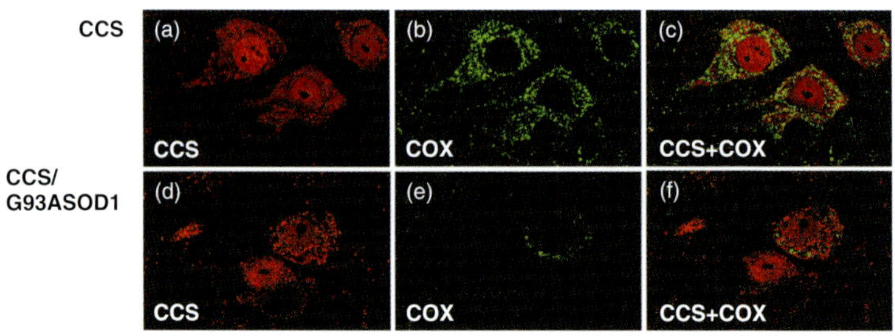

Figure 10.2 For caption see page 346.

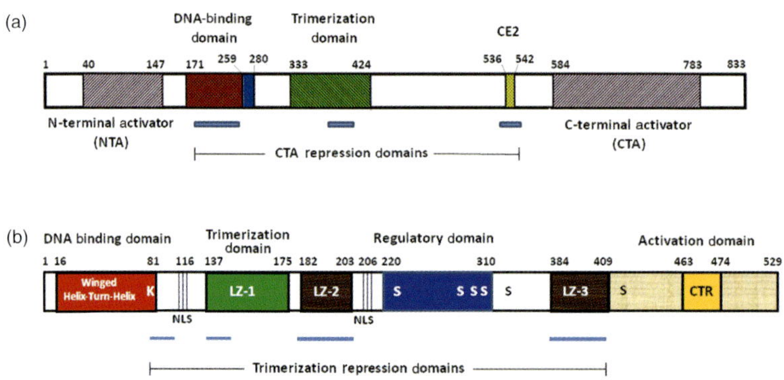

Figure 12.2 For caption see page 389.

Hsp104 activity against various amyloids *in vitro* (Lo Bianco *et al.*, 2008; Shorter and Lindquist, 2008; Sweeny and Shorter, 2008) and *in vivo* (Chernoff *et al.*, 1999; Higurashi *et al.*, 2008; Tipton *et al.*, 2008). Hsp104 also resolves preamyloid oligomers of Sup35 (Shorter and Lindquist, 2004, 2006), which adopt a generic conformation shared by many disease-associated amyloidogenic proteins (Kayed *et al.*, 2003).

Perplexingly, no clear metazoan homolog or analog of Hsp104 has been identified. Moreover, no activity that couples protein disaggregation to renaturation has been identified in metazoa. Initial attempts to isolate an analogous disaggregase by biochemical fractionation of mammalian cytosol have been unsuccessful (Mosser *et al.*, 2004). Crude homogenates from *C. elegans* and mouse are able to slowly disaggregate Aβ40 and Aβ42 fibers (Bieschke *et al.*, 2009; Cohen *et al.*, 2006; Murray *et al.*, 2010). However, disaggregation is invariably coupled to degradation unless protease inhibitors are added (Bieschke *et al.*, 2009; Cohen *et al.*, 2006; Murray *et al.*, 2010). Compared to Hsp104, this disaggregation activity is relatively slow. Moreover, it displays an unusual resistance to inactivation by high temperature and pH (Bieschke *et al.*, 2009; Murray *et al.*, 2010). Identification of the metazoan factor(s) that promote disaggregation will be extremely illuminating and might enable therapeutic manipulations (Bieschke *et al.*, 2009; Cohen *et al.*, 2006; Murray *et al.*, 2010).

Regardless of the identity of these putative metazoan disaggregases, the ability of Hsp104 to rapidly disassemble the generic cross-β forms of various yeast prions as well as the shared generic structure of preamyloid oligomers raises the possibility of unleashing Hsp104 on metazoan systems to prevent or reverse various amyloidoses (Shorter, 2008). An agent that reverses the formation of amyloid fibers and preamyloid oligomers would antagonize multiple recalcitrant pathological events that likely synergize to various degrees in the etiology of diverse amyloid disorders: (i) the toxic gain-of-function of amyloid or preamyloid oligomers; (ii) the loss-of-function of the protein sequestered in misfolded forms; and (iii) the depletion of various essential proteins that might coaggregate with the disease-associated polypeptide.

Initial efforts to introduce Hsp104 into metazoan systems have been extremely encouraging. Despite being a yeast protein, Hsp104 is well tolerated in metazoan systems and confers no noticeable toxicity. Indeed, expression of Hsp104 in several mammalian cell lines increases their resistance to stresses that promote protein aggregation (Dandoy-Dron *et al.*, 2006; Mosser *et al.*, 2004). Furthermore, Hsp104 synergizes with the mammalian Hsp70 system to resolve denatured aggregates (Glover and Lindquist, 1998; Mosser *et al.*, 2004; Schaupp *et al.*, 2007). Remarkably, Hsp104 protects mammalian cells from several diverse protein-misfolding events, including polyQ aggregation associated with HD (Carmichael *et al.*, 2000; Perrin *et al.*, 2007) and poly (A)-binding protein 2 misfolding associated with oculopharyngeal muscular dystrophy (Bao *et al.*, 2002). Expression of Hsp104 in *C. elegans* or rodents counters polyQ toxicity (Dandoy-Dron *et al.*, 2006; Satyal *et al.*, 2000; Vacher *et al.*, 2005). Transgenic mice that express Hsp104 are grossly normal (Dandoy-Dron *et al.*, 2006;

Vacher *et al.*, 2005). Moreover, Hsp104 expression reduced polyQ aggregation and prolonged the life span of an HD mouse model by ~20% (Vacher *et al.*, 2005). These studies suggest that Hsp104 can enhance metazoan proteostasis and counter protein aggregation and amyloidogenesis. In the remainder of this chapter, I consider two recent applications of Hsp104 to the amyloidogenic events that distinguish AD and PD.

7.5 Hsp104 AND AD

Beyond minor symptomatic relief, there are no effective treatments (Roberson and Mucke, 2006) for AD, the most common fatal neurodegenerative disorder, which afflicts ~35 million people (Prince *et al.*, 2009). AD is characterized by gross diffuse atrophy of the brain and neurodegeneration in the cerebral cortex and certain subcortical regions (Wenk, 2003). The defining pathological lesions are intracellular neurofibrillary tangles composed of amyloid forms of the microtubule-binding protein tau (Skovronsky *et al.*, 2006) and extracellular neuritic plaques composed primarily of amyloid forms of the Aβ peptides: Aβ42 and Aβ40 (Glenner and Wong, 1984; Iwatsubo *et al.*, 1994; Masters *et al.*, 1985). Several potential treatments are in clinical trials (Roberson and Mucke, 2006) and several small molecules have been isolated that inhibit (Gestwicki *et al.*, 2004) or even reverse Aβ42 fibrillization (Wang *et al.*, 2008). The effects of Hsp104 on Aβ42 amyloidogenesis have recently been tested (Arimon *et al.*, 2008).

Aβ40 and Aβ42 interact with Hsp104 directly and modulate its ATPase activity (Arimon *et al.*, 2008; Cashikar *et al.*, 2002; Schirmer and Lindquist, 1997). The interaction between Hsp104 and Aβ42 very potently inhibited *de novo* Aβ42 fibrillization, even when Hsp104 was at concentrations 1000-fold lower than Aβ42 (Arimon *et al.*, 2008). This substoichiometric inhibition indicates that Hsp104 might selectively antagonize an obligate intermediate that nucleates Aβ42 fibrillization. In support of this concept, Hsp104 antagonized the conversion of Aβ42 oligomers into fibers and interacted directly with Aβ42 oligomers (Arimon *et al.*, 2008). These inhibitory activities were observed in the presence of ATPγS, a slowly hydrolyzable ATP analog and in the presence of the double Walker B mutant Hsp104$^{E285Q:E687Q}$, which is able to bind, but not hydrolyze, ATP. Therefore, it would appear that this potent inhibition does not require ATP hydrolysis. ATP-restricted Hsp104 is likely to bind to Aβ42 conformers and passively inhibit amyloidogenesis. Indeed, similar observations have been made with Sup35. Inhibition of *de novo* Sup35 fibrillization by high concentrations of Hsp104 can occur without ATP hydrolysis (Shorter and Lindquist, 2004, 2006). In the presence of AMP-PNP (adenosine 5'-(β, γ-imido)triphosphate) or AMP-PCP (β, γ-Methyleneadenosine 5'-triphosphate) two nonhydrolyzable ATP analogs, Hsp104 inhibits the maturation of molten Sup35 oligomers and thereby prevents fiber nucleation (Shorter and Lindquist, 2004, 2006). Importantly, Hsp104 also potently inhibited fibrillization that was seeded by preformed Aβ42 fibers (Arimon *et al.*, 2008). Consistent with these data, Hsp104 inhibited Aβ42 fibrillization when added during lag phase or assembly phase (Arimon *et al.*, 2008). The ability to inhibit seeded assembly is a key finding, because from a therapeutic standpoint, any treatment is only likely to be applied after substantial accumulation of seeding-competent amyloid forms.

Intriguingly, despite the apparent interaction between Hsp104 and Aβ42, Hsp104 did not disassemble Aβ42 fibers or oligomers (Arimon *et al.*, 2008). This might reflect a requirement for the Hsp70 system, which can improve Hsp104-mediated amyloid disassembly (Higurashi *et al.*, 2008; Lo Bianco *et al.*, 2008; Shorter and Lindquist, 2008; Sweeny and Shorter, 2008; Tipton *et al.*, 2008). Alternatively, the Aβ42 fiber strain that formed under the conditions employed might be resistant to Hsp104. Indeed, it is conceivable that all strains of Aβ42 fibers are refractory to Hsp104, since they are substrates that Hsp104 never ordinarily encounters. Nevertheless, the ability of Hsp104 to bind Aβ42 monomers and inhibit seeded assembly, coupled to the fact that amyloids exchange monomers very slowly via a soluble pool (Carulla *et al.*, 2005), might enable Hsp104 to slowly shift the equilibrium away from the assembled fibrous state. Thus, Hsp104 might slowly resolve Aβ42 fibers over a time frame longer than those thus far explored (Arimon *et al.*, 2008).

These *in vitro* findings are promising (Arimon *et al.*, 2008). However, extension to cell culture and animal models is needed for validation. Neuroblastoma cell lines have been widely used to assess the toxicity of Aβ fibers and oligomers (Kayed *et al.*, 2003; Petkova *et al.*, 2005). This system might be readily adapted to test whether the Hsp104–Aβ42 interactions reduce toxicity to cultured neurons. Another issue is that the majority of Aβ42 fibers are extracellular in AD, which may make them challenging targets for Hsp104. However, intraneuronal Aβ42 is also found in AD and may contribute to disease progression (Grundke-Iqbal *et al.*, 1989; LaFerla *et al.*, 2007; Wertkin *et al.*, 1993). It is possible that Hsp104 might be efficacious against intraneuronal pools of misfolded Aβ42.

7.6 Hsp104 AND PD

There are no efficacious treatments for PD, the most common neurodegenerative movement disorder, which afflicts several million people worldwide (Dorsey *et al.*, 2007). PD is due to a severe and selective devastation of dopaminergic neurons from the substantia nigra pars compacta although neuropathology extends into other regions of the brain (Braak *et al.*, 2003). Intracellular inclusions termed *Lewy bodies* and *Lewy neurites*, which are composed of amyloid forms of the small presynaptic protein, α-synuclein (α-syn), are the signature lesion of PD (Spillantini *et al.*, 1997). Although PD is most frequently a sporadic disorder, mutations in α-syn (e.g., A30P, A53T, E46K) and duplication or triplication of the wild-type gene are linked with early-onset PD in rare familial forms of the disease (Moore *et al.*, 2005). α-Syn function is uncertain but may play a key regulatory role in dopamine release from synaptic vesicle pools (Abeliovich *et al.*, 2000; Gitler and Shorter, 2007; Larsen *et al.*, 2006). Pure α-syn readily accesses amyloid forms *in vitro*, which bear remarkable similarities to α-syn fibers isolated from synucleinopathy patients (Crowther *et al.*, 2000; Spillantini *et al.*, 1998).

In vitro, Hsp104 potently inhibited the fibrillization of α-syn and the early-onset PD-linked variants: A30P, A53T, and E46K (Lo Bianco *et al.*, 2008). Inhibition was most effective in the presence of ATP. In the presence of AMP-PNP, Hsp104 failed to inhibit α-syn fibrillization (Lo Bianco *et al.*, 2008). Furthermore, a double Walker A mutant, $Hsp104^{K218T:K620T}$, which cannot bind or hydrolyze ATP, also failed to inhibit assembly (Lo Bianco *et al.*, 2008). Hsp104 coupled ATP hydrolysis to the disassembly of toxic oligomers composed of α-syn A30P (Lo Bianco *et al.*, 2008). Hsp104 also coupled ATPase activity to the disassembly of α-syn fibers (Lo Bianco *et al.*, 2008). Disassembly was enhanced by the mammalian Hsp70 system, and in particular, by the specific combination of Hsc70 and Hdj2 (Lo Bianco *et al.*, 2008). All α-syn variant fibers were effectively disassembled, except for the E46K PD-linked mutant, which was highly resistant to Hsp104 (Lo Bianco *et al.*, 2008). This might indicate that α-syn E46K forms a different strain of amyloid. Indeed, α-syn E46K fibers tend to form compact bundles and meshwork arrays not observed with wild-type α-syn (Choi *et al.*, 2004; Greenbaum *et al.*, 2005). Nonetheless, this battery of remodeling activities suggested that Hsp104 might effectively buffer α-syn misfolding and toxicity *in vivo*.

Unfortunately, the development of PD therapies has been hindered by a paucity of animal models that successfully recreate the progressive and selective degeneration of dopaminergic neurons and formation of phosphorylated α-syn inclusions. However, a rat PD model based on the lentiviral-mediated expression of human α-syn A30P in the substantia nigra has successfully recapitulated these key phenotypes (Lo Bianco *et al.*, 2002, 2004). Thus, Hsp104 and α-syn A30P were expressed simultaneously in the rat substantia nigra using the lentiviral delivery system. Remarkably, Hsp104 reduced the formation of phosphorylated α-syn A30P inclusions and prevented nigrostriatal dopaminergic neurodegeneration (Lo Bianco *et al.*, 2008). Thus, Hsp104 is able to buffer α-syn A30P misfolding and toxicity in the physiological arena of the mammalian substantia nigra.

While these results are promising, several questions remain that must be addressed in subsequent studies. First, it has not yet been possible to express Hsp104 after α-syn has already aggregated, which is a situation that might mimic more closely any potential treatment. Thus, whether Hsp104 can reverse α-syn aggregation in the setting of the rat substantia nigra remains unclear. Another issue concerns the release of a large pulse of soluble α-syn from Lewy bodies in surviving neurons. Such a pulse might be detrimental since high levels of soluble α-syn can inhibit synaptic vesicle release and perturb other membrane trafficking events (Gitler and Shorter, 2007; Gitler *et al.*, 2008; Larsen *et al.*, 2006). However, this situation is likely to be preferable to the persistence of toxic α-syn conformers. Finally and most importantly, further study is needed to assess any dangers of long-term Hsp104 expression in the mammalian brain.

7.7 DEVELOPMENT OF SUBSTRATE-OPTIMIZED Hsp104 VARIANTS AS POTENTIAL THERAPIES

The foregoing sections increase optimism that Hsp104 may have therapeutic potential for antagonizing or even reversing the specific amyloidogenic events connected with AD and PD. However, the amyloid-remodeling activity of Hsp104 might also have applications that extend beyond various neurodegenerative amyloidoses. For example, the peptide hormone, amylin, forms amyloid inclusions in the endocrine pancreas in 90% of patients with type II diabetes (Cooper *et al.*, 1987; Kahn *et al.*, 1999; Maloy *et al.*, 1981) and likely exacerbates β-cell failure (Janson *et al.*, 1999; Lorenzo *et al.*, 1994). Another potential amyloid target is provided by fragments of prostatic acidic phosphatase, an abundant component of semen. Amyloid forms of these peptides can drastically potentiate human immunodeficiency virus (HIV) infection by $\sim 10^5$-fold, whereas soluble forms of these peptides have no effect (Munch *et al.*, 2007). Thus, these amyloid species represent a novel target for preventing sexual transmission of HIV.

The ability of Hsp104 to prevent or reverse the formation of nonamyloid, disease-associated aggregates should also be considered. For example, two devastating neurodegenerative disorders: amyotrophic lateral sclerosis (ALS) and frontotemporal lobar degeneration with ubiquitin-positive inclusions are connected with the formation of nonamyloid, aggregated species of a conserved heterogeneous nuclear ribonucleoprotein (hnRNP), transactive response DNA-binding protein-43 (TDP-43) (Johnson *et al.*, 2009; Kerman *et al.*, 2010; Kwong *et al.*, 2008; Neumann *et al.*, 2006). Other forms of ALS are associated with the nonamyloid aggregation of another hnRNP, Fused in sarcoma (FUS) (Kwiatkowski *et al.*, 2009; Vance *et al.*, 2009). Both FUS and TDP-43 may represent promising targets for Hsp104 because they both contain a domain that resembles a yeast prion domain (Cushman *et al.*, 2010).

Nevertheless, a daunting array of issues must be addressed if Hsp104 is to be developed as a therapeutic agent. Not least is the issue that gene therapy would seem to be required to introduce Hsp104 as a therapeutic agent. (See Chapter 12 for more information on gene therapy.) Gene therapy has produced encouraging preclinical outcomes for several disorders including congenital blindness (Bainbridge *et al.*, 2008; Hacein-Bey-Abina *et al.*, 2002; Maguire *et al.*, 2008). However, technical and safety issues continue to restrict translation to the clinic. Indeed, gene therapy approaches to treat neurodegenerative amyloidoses remain in very early developmental stages, and considerable caution is needed at this time. However, initial studies suggest that gene therapy in the adult brain might be safe for various neurodegenerative disorders, including PD (Feigin *et al.*, 2007; Kaplitt *et al.*, 2007; Stoessl, 2007). Thus, even though we await several key advances in gene therapy before any Hsp104 gene therapy (or any other gene therapy) becomes feasible, it remains important to develop solutions to amyloid problems and to test these solutions both *in vitro* and in animal models.

Another issue concerns the fact that existing Hsp104 specificity or activity is unlikely to be optimal against substrates that it never ordinarily encounters, such as α-syn or Aβ42. Indeed, disassembly of α-syn fibers requires considerably larger amounts of Hsp104 than disassembly of Sup35 or Ure2, two natural prion substrates (Lo Bianco *et al.*, 2008; Shorter and Lindquist, 2006). Even for the natural substrates Sup35 and Ure2, high concentrations of Hsp104 are required to reverse amyloid formation (Shorter and Lindquist, 2006). Acting at lower concentrations, Hsp104 fragments Sup35 and Ure2 prions, which generates more fiber ends that can convert soluble copies of the protein to the prion form (Shorter and Lindquist, 2006). Thus, an important therapeutic consideration is to express Hsp104 above a certain threshold that reduces and does not exacerbate the amyloid burden. Indeed, fragmenting fibers might initially be detrimental because short amyloid fibers can be more toxic than long fibers, at least in cell culture (Xue *et al.*, 2009, 2010).

Hsp104 is likely to be a generalist since it must disaggregate large portions of the yeast proteome after environmental stress. Regarding amyloid conformers, it seems likely that Hsp104 might be adapted to remodel cross-β structures composed of the uncharged polar residues that distinguish the prion domains of many proteins in yeast (Alberti *et al.*, 2009). However, Hsp104 is able to propagate HET-s prions in yeast, which harbor a prion domain that is very distinct to those of other yeast prions (Taneja *et al.*, 2007). Promiscuous disaggregation activity might also be undesirable in a therapeutic setting. Ideally, a therapeutic disaggregase would selectively eliminate toxic strains and misfolded species, and not eradicate benign strains or even beneficial amyloids such as CPEB prions, which might encode long-term memory (Shorter and Lindquist, 2005b; Si *et al.*, 2003, 2010). Thus, an important goal is to engineer or evolve Hsp104 variants with enhanced and selective ability to eradicate specific amyloid or aggregated conformers. Ultimately, designer disaggregases might be developed to annihilate purely toxic conformers unique to each particular disease. This might require a very different tailored Hsp104 variant for each disease. Nonetheless, the development of Hsp104-based disaggregases dedicated to the resolution of select proteins or protein conformations remains an important future goal that will simultaneously facilitate a deeper understanding of how this intriguing protein disaggregase operates.

ACKNOWLEDGMENT

Work in the Shorter lab is supported by: an NIH Director's New Innovator Award (1DP2OD002177-01), an Ellison Medical Foundation New Scholar in Aging Award, an NINDS grant (1R21NS067354-0110), and a University of Pennsylvania Diabetes and Endocrinology Research Center Pilot and Feasibility grant (J.S.).

REFERENCES

Abeliovich A, Schmitz Y, Farinas I, Choi-Lundberg D, Ho WH, Castillo PE, Shinsky N, Verdugo JM, Armanini M, Ryan A, *et al.* Mice lacking alpha-synuclein display functional deficits in the nigrostriatal dopamine system. *Neuron* 2000;**25**:239–252.

Alberti S, Halfmann R, King O, Kapila A, Lindquist S. A systematic survey identifies prions and illuminates sequence features of prionogenic proteins. *Cell* 2009;**137**:146–158.

Anfinsen CB. Principles that govern the folding of protein chains. *Science* 1973; **181**:223–230.

Arimon M, Grimminger V, Sanz F, Lashuel HA. Hsp104 targets multiple intermediates on the amyloid pathway and suppresses the seeding capacity of Abeta fibrils and protofibrils. *J Mol Biol* 2008;**384**:1157–1173.

Arrasate M, Mitra S, Schweitzer ES, Segal MR, Finkbeiner S. Inclusion body formation reduces levels of mutant huntingtin and the risk of neuronal death. *Nature* 2004;**431**:805–810.

Bainbridge JW, Smith AJ, Barker SS, Robbie S, Henderson R, Balaggan K, Viswanathan A, Holder GE, Stockman A, Tyler N, *et al.* Effect of gene therapy on visual function in Leber's congenital amaurosis. *N Engl J Med* 2008;**358**:2231–2239.

Balch WE, Morimoto RI, Dillin A, Kelly JW. Adapting proteostasis for disease intervention. *Science* 2008;**319**:916–919.

Bao YP, Cook LJ, O'Donovan D, Uyama E, Rubinsztein DC. Mammalian, yeast, bacterial, and chemical chaperones reduce aggregate formation and death in a cell model of oculopharyngeal muscular dystrophy. *J Biol Chem* 2002;**277**:12263–12269.

Barnett ME, Nagy M, Kedzierska S, Zolkiewski M. The amino-terminal domain of ClpB supports binding to strongly aggregated proteins. *J Biol Chem* 2005; **280**:34940–34945.

Bartlett AI, Radford SE. An expanding arsenal of experimental methods yields an explosion of insights into protein folding mechanisms. *Nat Struct Mol Biol* 2009;**16**:582–588.

Baxa U, Speransky V, Steven AC, Wickner RB. Mechanism of inactivation on prion conversion of the *Saccharomyces cerevisiae* Ure2 protein. *Proc Natl Acad Sci USA* 2002;**99**:5253–5260.

Bence NF, Sampat RM, Kopito RR. Impairment of the ubiquitin-proteasome system by protein aggregation. *Science* 2001;**292**:1552–1555.

Berson JF, Theos AC, Harper DC, Tenza D, Raposo G, Marks MS. Proprotein convertase cleavage liberates a fibrillogenic fragment of a resident glycoprotein to initiate melanosome biogenesis. *J Cell Biol* 2003;**161**:521–533.

Bieschke J, Cohen E, Murray A, Dillin A, Kelly JW. A kinetic assessment of the *C. elegans* amyloid disaggregation activity enables uncoupling of disassembly and proteolysis. *Protein Sci* 2009;**18**:2231–2241.

Bosl B, Grimminger V, Walter S. Substrate binding to the molecular chaperone Hsp104 and its regulation by nucleotides. *J Biol Chem* 2005;**280**:38170–38176.

Braak H, Del Tredici K, Rub U, de Vos RA, Jansen Steur EN, Braak E. Staging of brain pathology related to sporadic Parkinson's disease. *Neurobiol Aging* 2003;**24**:197–211.

Bucciantini M, Giannoni E, Chiti F, Baroni F, Formigli L, Zurdo J, Taddei N, Ramponi G, Dobson CM, Stefani M. Inherent toxicity of aggregates implies a common mechanism for protein misfolding diseases. *Nature* 2002;**416**:507–511.

Carmichael J, Chatellier J, Woolfson A, Milstein C, Fersht AR, Rubinsztein DC. Bacterial and yeast chaperones reduce both aggregate formation and cell death in mammalian cell models of Huntington's disease. *Proc Natl Acad Sci USA* 2000;**97**:9701–9705.

Carulla N, Caddy GL, Hall DR, Zurdo J, Gairi M, Feliz M, Giralt E, Robinson CV, Dobson CM. Molecular recycling within amyloid fibrils. *Nature* 2005;**436**:554–558.

Cashikar AG, Duennwald M, Lindquist SL. A chaperone pathway in protein disaggregation: Hsp26 alters the nature of protein aggregates to facilitate reactivation by Hsp104. *J Biol Chem* 2005;**280**:23869–23875.

Cashikar AG, Schirmer EC, Hattendorf DA, Glover JR, Ramakrishnan MS, Ware DM, Lindquist SL. Defining a pathway of communication from the C-terminal peptide binding domain to the N-terminal ATPase domain in a AAA protein. *Mol Cell* 2002;**9**:751–760.

Caughey B, Lansbury PT. Protofibrils, pores, fibrils, and neurodegeneration: separating the responsible protein aggregates from the innocent bystanders. *Annu Rev Neurosci* 2003;**26**:267–298.

Chen Q, Liu JB, Horak KM, Zheng H, Kumarapeli AR, Li J, Li F, Gerdes AM, Wawrousek EF, Wang X. Intrasarcoplasmic amyloidosis impairs proteolytic function of proteasomes in cardiomyocytes by compromising substrate uptake. *Circ Res* 2005;**97**:1018–1026.

Chernoff YO, Lindquist SL, Ono B, Inge-Vechtomov SG, Liebman SW. Role of the chaperone protein Hsp104 in propagation of the yeast prion-like factor [PSI^+]. *Science* 1995;**268**:880–884.

Chernoff YO, Newnam GP, Kumar J, Allen K, Zink AD. Evidence for a protein mutator in yeast: role of the Hsp70-related chaperone ssb in formation, stability, and toxicity of the [PSI^+] prion. *Mol Cell Biol* 1999;**19**:8103–8112.

Choi W, Zibaee S, Jakes R, Serpell LC, Davletov B, Crowther RA, Goedert M. Mutation E46K increases phospholipid binding and assembly into filaments of human alpha-synuclein. *FEBS Lett* 2004;**576**:363–368.

Cohen E, Bieschke J, Perciavalle RM, Kelly JW, Dillin A. Opposing activities protect against age-onset proteotoxicity. *Science* 2006;**313**:1604–1610.

Cooper GJ, Willis AC, Clark A, Turner RC, Sim RB, Reid KB. Purification and characterization of a peptide from amyloid-rich pancreases of type 2 diabetic patients. *Proc Natl Acad Sci USA* 1987;**84**:8628–8632.

Crampton DJ, Mukherjee S, Richardson CC. DNA-induced switch from independent to sequential dTTP hydrolysis in the bacteriophage T7 DNA helicase. *Mol Cell* 2006;**21**:165–174.

Crowther RA, Daniel SE, Goedert M. Characterisation of isolated alpha-synuclein filaments from substantia nigra of Parkinson's disease brain. *Neurosci Lett* 2000; **292**:128–130.

Cuervo AM. Autophagy and aging: keeping that old broom working. *Trends Genet* 2008;**24**:604–612.

Cushman M, Johnson BS, King OD, Gitler AD, Shorter J. Prion-like disorders: blurring the divide between transmissibility and infectivity. *J Cell Sci* 2010;**123**:1191–1201.

Dandoy-Dron F, Bogdanova A, Beringue V, Bailly Y, Tovey MG, Laude H, Dron M. Infection by ME7 prion is not modified in transgenic mice expressing the yeast chaperone Hsp104 in neurons. *Neurosci Lett* 2006;**405**:181–185.

de Calignon A, Fox LM, Pitstick R, Carlson GA, Bacskai BJ, Spires-Jones TL, Hyman BT. Caspase activation precedes and leads to tangles. *Nature* 2010;**464**:1201–1204.

Dobson CM. Protein folding and misfolding. *Nature* 2003;**426**:884–890.

Dorsey ER, Constantinescu R, Thompson JP, Biglan KM, Holloway RG, Kieburtz K, Marshall FJ, Ravina BM, Schifitto G, Siderowf A, *et al.* Projected number of people with Parkinson disease in the most populous nations, 2005 through 2030. *Neurology* 2007;**68**:384–386.

Douglas PM, Treusch S, Ren HY, Halfmann R, Duennwald ML, Lindquist S, Cyr DM. Chaperone-dependent amyloid assembly protects cells from prion toxicity. *Proc Natl Acad Sci USA* 2008;**105**:7206–7211.

Doyle SM, Hoskins JR, Wickner S. Inaugural article: collaboration between the ClpB AAA+ remodeling protein and the DnaK chaperone system. *Proc Natl Acad Sci USA* 2007a;**104**:11138–11144.

Doyle SM, Shorter J, Zolkiewski M, Hoskins JR, Lindquist S, Wickner S. Asymmetric deceleration of ClpB or Hsp104 ATPase activity unleashes protein-remodeling activity. *Nat Struct Mol Biol* 2007b;**14**:114–122.

Doyle SM, Wickner S. Hsp104 and ClpB: protein disaggregating machines. *Trends Biochem Sci* 2009;**34**:40–48.

Eisenberg D, Nelson R, Sawaya MR, Balbirnie M, Sambashivan S, Ivanova MI, Madsen AO, Riekel C. The structural biology of protein aggregation diseases: Fundamental questions and some answers. *Acc Chem Res* 2006;**39**:568–575.

Ellis RJ, Minton AP. Protein aggregation in crowded environments. *Biol Chem* 2006;**387**:485–497.

Englander SW, Mayne L, Krishna MM. Protein folding and misfolding: mechanism and principles. *Q Rev Biophys* 2007;**40**:287–326.

Erzberger JP, Berger JM. Evolutionary relationships and structural mechanisms of AAA+ proteins. *Annu Rev Biophys Biomol Struct* 2006;**35**:93–114.

Feigin A, Kaplitt MG, Tang C, Lin T, Mattis P, Dhawan V, During MJ, Eidelberg D. Modulation of metabolic brain networks after subthalamic gene therapy for Parkinson's disease. *Proc Natl Acad Sci USA* 2007;**104**:19559–19564.

Forman MS, Trojanowski JQ, Lee VM. Neurodegenerative diseases: a decade of discoveries paves the way for therapeutic breakthroughs. *Nat Med* 2004;**10**: 1055–1063.

Fowler DM, Koulov AV, Alory-Jost C, Marks MS, Balch WE, Kelly JW. Functional amyloid formation within mammalian tissue. *PLoS Biol* 2006;**4**:e6.

Fowler DM, Koulov AV, Balch WE, Kelly JW. Functional amyloid - from bacteria to humans. *Trends Biochem Sci* 2007;**32**:217–224.

Gestwicki JE, Crabtree GR, Graef IA. Harnessing chaperones to generate small-molecule inhibitors of amyloid beta aggregation. *Science* 2004;**306**:865–869.

Gitler AD, Bevis BJ, Shorter J, Strathearn KE, Hamamichi S, Su LJ, Caldwell KA, Caldwell GA, Rochet JC, McCaffery JM, *et al.* The Parkinson's disease protein alpha-synuclein disrupts cellular Rab homeostasis. *Proc Natl Acad Sci USA* 2008;**105**:145–150.

Gitler AD, Shorter J. Prime time for alpha-synuclein. *J Neurosci* 2007;**27**:2433–2434.

Glenner GG, Wong CW. Alzheimer's disease: initial report of the purification and characterization of a novel cerebrovascular amyloid protein. *Biochem Biophys Res Commun* 1984;**120**:885–890.

Glover JR, Lindquist S. Hsp104, Hsp70, and Hsp40: a novel chaperone system that rescues previously aggregated proteins. *Cell* 1998;**94**:73–82.

Glover JR, Lum R. Remodeling of protein aggregates by Hsp104. *Protein Pept Lett* 2009;**16**:587–597.

Goloubinoff P, Mogk A, Zvi AP, Tomoyasu T, Bukau B. Sequential mechanism of solubilization and refolding of stable protein aggregates by a bichaperone network. *Proc Natl Acad Sci USA* 1999;**96**:13732–13737.

Greenbaum EA, Graves CL, Mishizen-Eberz AJ, Lupoli MA, Lynch DR, Englander SW, Axelsen PH, Giasson BI. The E46K mutation in alpha-synuclein increases amyloid fibril formation. *J Biol Chem* 2005;**280**:7800–7807.

Griswold CK, Masel J. Complex adaptations can drive the evolution of the capacitor [*PSI*$^+$], even with realistic rates of yeast sex. *PLoS Genet* 2009;**5**:e1000517.

Grundke-Iqbal I, Iqbal K, George L, Tung YC, Kim KS, Wisniewski HM. Amyloid protein and neurofibrillary tangles coexist in the same neuron in Alzheimer disease. *Proc Natl Acad Sci USA* 1989;**86**:2853–2857.

Haass C, Selkoe DJ. Soluble protein oligomers in neurodegeneration: lessons from the Alzheimer's amyloid beta-peptide. *Nat Rev Mol Cell Biol* 2007;**8**:101–112.

Hacein-Bey-Abina S, Le Deist F, Carlier F, Bouneaud C, Hue C, De Villartay JP, Thrasher AJ, Wulffraat N, Sorensen R, Dupuis-Girod S, *et al.* Sustained correction of X-linked severe combined immunodeficiency by *ex vivo* gene therapy. *N Engl J Med* 2002;**346**:1185–1193.

Halfmann R, Alberti S, Lindquist S. Prions, protein homeostasis, and phenotypic diversity. *Trends Cell Biol* 2010;**20**:125–133.

Haslberger T, Zdanowicz A, Brand I, Kirstein J, Turgay K, Mogk A, Bukau B. Protein disaggregation by the AAA+ chaperone ClpB involves partial threading of looped polypeptide segments. *Nat Struct Mol Biol* 2008;**15**:641–650.

Hattendorf DA, Lindquist SL. Cooperative kinetics of both Hsp104 ATPase domains and interdomain communication revealed by AAA sensor-1 mutants. *EMBO J* 2002;**21**:12–21.

Higurashi T, Hines JK, Sahi C, Aron R, Craig EA. Specificity of the J-protein Sis1 in the propagation of 3 yeast prions. *Proc Natl Acad Sci USA* 2008;**105**:16596–16601.

Hoskins JR, Doyle SM, Wickner S. Coupling ATP utilization to protein remodeling by ClpB, a hexameric AAA+ protein. *Proc Natl Acad Sci USA* 2009;**106**: 22233–22238.

Hung GC, Masison DC. N-terminal domain of yeast Hsp104 chaperone is dispensable for thermotolerance and prion propagation but necessary for curing prions by Hsp104 overexpression. *Genetics* 2006;**173**:611–620.

Iwatsubo T, Odaka A, Suzuki N, Mizusawa H, Nukina N, Ihara Y. Visualization of A beta 42(43) and A beta 40 in senile plaques with end-specific A beta monoclonals: evidence that an initially deposited species is A beta 42(43). *Neuron* 1994; **13**:45–53.

Janson J, Ashley RH, Harrison D, McIntyre S, Butler PC. The mechanism of islet amyloid polypeptide toxicity is membrane disruption by intermediate-sized toxic amyloid particles. *Diabetes* 1999;**48**:491–498.

Johnson BS, Snead D, Lee JJ, McCaffery JM, Shorter J, Gitler AD. TDP-43 is intrinsically aggregation-prone and ALS-linked mutations accelerate aggregation and increase toxicity. *J Biol Chem* 2009;**284**:20329–20339.

Kahn SE, Andrikopoulos S, Verchere CB. Islet amyloid: a long-recognized but underappreciated pathological feature of type 2 diabetes. *Diabetes* 1999;**48**:241–253.

Kaplitt MG, Feigin A, Tang C, Fitzsimons HL, Mattis P, Lawlor PA, Bland RJ, Young D, Strybing K, Eidelberg D, *et al.* Safety and tolerability of gene therapy with an adeno-associated virus (AAV) borne GAD gene for Parkinson's disease: an open label, phase I trial. *Lancet* 2007;**369**:2097–2105.

Kayed R, Head E, Thompson JL, McIntire TM, Milton SC, Cotman CW, Glabe CG. Common structure of soluble amyloid oligomers implies common mechanism of pathogenesis. *Science* 2003;**300**:486–489.

Kerman A, Liu HN, Croul S, Bilbao J, Rogaeva E, Zinman L, Robertson J, Chakrabartty A. Amyotrophic lateral sclerosis is a non-amyloid disease in which extensive misfolding of SOD1 is unique to the familial form. *Acta Neuropathol* 2010;**119**:335–344.

Kholova I, Niessen HW. Amyloid in the cardiovascular system: a review. *J Clin Pathol* 2005;**58**:125–133.

King OD, Masel J. The evolution of bet-hedging adaptations to rare scenarios. *Theor Popul Biol* 2007;**72**:560–575.

Knowles TP, Fitzpatrick AW, Meehan S, Mott HR, Vendruscolo M, Dobson CM, Welland ME. Role of intermolecular forces in defining material properties of protein nanofibrils. *Science* 2007;**318**:1900–1903.

Kong Q, Surewicz WK, Peterson RB, Zou W, Chen SG, Gambetti P. Inherited prion diseases. In: Prusiner SB, editor. *Prion biology and diseases*. New York: Cold Spring Harbor Laboratory Press; 2004. p 673–776.

Kramer G, Boehringer D, Ban N, Bukau B. The ribosome as a platform for co-translational processing, folding and targeting of newly synthesized proteins. *Nat Struct Mol Biol* 2009;**16**:589–597.

Krishnan R, Lindquist SL. Structural insights into a yeast prion illuminate nucleation and strain diversity. *Nature* 2005;**435**:765–772.

Kryndushkin DS, Alexandrov IM, Ter-Avanesyan MD, Kushnirov VV. Yeast [PSI^+] prion aggregates are formed by small Sup35 polymers fragmented by Hsp104. *J Biol Chem* 2003;**278**:49636–49643.

Kwiatkowski TJ Jr, Bosco DA, Leclerc AL, Tamrazian E, Vanderburg CR, Russ C, Davis A, Gilchrist J, Kasarskis EJ, Munsat T, *et al.* Mutations in the FUS/TLS gene on chromosome 16 cause familial amyotrophic lateral sclerosis. *Science* 2009;**323**:1205–1208.

Kwong LK, Uryu K, Trojanowski JQ, Lee VM. TDP-43 proteinopathies: neurodegenerative protein misfolding diseases without amyloidosis. *Neurosignals* 2008;**16**:41–51.

LaFerla FM, Green KN, Oddo S. Intracellular amyloid-beta in Alzheimer's disease. *Nat Rev Neurosci* 2007;**8**:499–509.

Lansbury PT Jr, Caughey B. The chemistry of scrapie infection: implications of the 'ice 9' metaphor. *Chem Biol* 1995;**2**:1–5.

Larsen KE, Schmitz Y, Troyer MD, Mosharov E, Dietrich P, Quazi AZ, Savalle M, Nemani V, Chaudhry FA, Edwards RH, *et al.* Alpha-synuclein overexpression in PC12 and chromaffin cells impairs catecholamine release by interfering with a late step in exocytosis. *J Neurosci* 2006;**26**:11915–11922.

Lashuel HA, Hartley D, Petre BM, Walz T, Lansbury PT Jr. Neurodegenerative disease: amyloid pores from pathogenic mutations. *Nature* 2002;**418**:291.

Lee JW, Beebe K, Nangle LA, Jang J, Longo-Guess CM, Cook SA, Davisson MT, Sundberg JP, Schimmel P, Ackerman SL. Editing-defective tRNA synthetase causes protein misfolding and neurodegeneration. *Nature* 2006;**443**:50–55.

Lee S, Choi JM, Tsai FT. Visualizing the ATPase cycle in a protein disaggregating machine: structural basis for substrate binding by ClpB. *Mol Cell* 2007;**25**:261–271.

Lee S, Sowa ME, Watanabe YH, Sigler PB, Chiu W, Yoshida M, Tsai FT. The structure of ClpB: a molecular chaperone that rescues proteins from an aggregated state. *Cell* 2003;**115**:229–240.

Legname G, Nguyen HO, Peretz D, Cohen FE, DeArmond SJ, Prusiner SB. Continuum of prion protein structures enciphers a multitude of prion isolate-specified phenotypes. *Proc Natl Acad Sci USA* 2006;**103**:19105–19110.

Lesne S, Koh MT, Kotilinek L, Kayed R, Glabe CG, Yang A, Gallagher M, Ashe KH. A specific amyloid-beta protein assembly in the brain impairs memory. *Nature* 2006;**440**:352–357.

Lo Bianco C, Ridet JL, Schneider BL, Deglon N, Aebischer P. alpha-Synucleinopathy and selective dopaminergic neuron loss in a rat lentiviral-based model of Parkinson's disease. *Proc Natl Acad Sci USA* 2002;**99**:10813–10818.

Lo Bianco C, Schneider BL, Bauer M, Sajadi A, Brice A, Iwatsubo T, Aebischer P. Lentiviral vector delivery of parkin prevents dopaminergic degeneration in an alpha-synuclein rat model of Parkinson's disease. *Proc Natl Acad Sci USA* 2004;**101**:17510–17515.

Lo Bianco C, Shorter J, Regulier E, Lashuel H, Iwatsubo T, Lindquist S, Aebischer P. Hsp104 antagonizes alpha-synuclein aggregation and reduces dopaminergic degeneration in a rat model of Parkinson disease. *J Clin Invest* 2008;**118**:3087–3097.

Lorenzo A, Razzaboni B, Weir GC, Yankner BA. Pancreatic islet cell toxicity of amylin associated with type-2 diabetes mellitus. *Nature* 1994;**368**:756–760.

Lum R, Niggemann M, Glover JR. Peptide and protein binding in the axial channel of Hsp104. Insights into the mechanism of protein unfolding. *J Biol Chem* 2008;**283**:30139–30150.

Lum R, Tkach JM, Vierling E, Glover JR. Evidence for an unfolding/threading mechanism for protein disaggregation by *Saccharomyces cerevisiae* Hsp104. *J Biol Chem* 2004;**279**:29139–29146.

Mackay RG, Helsen CW, Tkach JM, Glover JR. The C-terminal extension of *Saccharomyces cerevisiae* Hsp104 plays a role in oligomer assembly. *Biochemistry* 2008;**47**:1918–1927.

Maguire AM, Simonelli F, Pierce EA, Pugh EN Jr, Mingozzi F, Bennicelli J, Banfi S, Marshall KA, Testa F, Surace EM, *et al.* Safety and efficacy of gene transfer for Leber's congenital amaurosis. *N Engl J Med* 2008;**358**:2240–2248.

Maloy AL, Longnecker DS, Greenberg ER. The relation of islet amyloid to the clinical type of diabetes. *Hum Pathol* 1981;**12**:917–922.

Martins IC, Kuperstein I, Wilkinson H, Maes E, Vanbrabant M, Jonckheere W, Van Gelder P, Hartmann D, D'Hooge R, De Strooper B, *et al.* Lipids revert inert Abeta amyloid fibrils to neurotoxic protofibrils that affect learning in mice. *EMBO J* 2008;**27**:224–233.

Masters CL, Simms G, Weinman NA, Multhaup G, McDonald BL, Beyreuther K. Amyloid plaque core protein in Alzheimer disease and Down syndrome. *Proc Natl Acad Sci USA* 1985;**82**:4245–4249.

Medawar PB. *An unsolved problem of biology*. London: H.K. Lewis; 1952.

Merlini G, Westermark P. The systemic amyloidoses: clearer understanding of the molecular mechanisms offers hope for more effective therapies. *J Int Med* 2004;**255**:159–178.

Meyer-Luehmann M, Spires-Jones TL, Prada C, Garcia-Alloza M, de Calignon A, Rozkalne A, Koenigsknecht-Talboo J, Holtzman DM, Bacskai BJ, Hyman BT. Rapid appearance and local toxicity of amyloid-beta plaques in a mouse model of Alzheimer's disease. *Nature* 2008;**451**:720–724.

Mogk A, Tomoyasu T, Goloubinoff P, Rudiger S, Roder D, Langen H, Bukau B. Identification of thermolabile *Escherichia coli* proteins: prevention and reversion of aggregation by DnaK and ClpB. *EMBO J* 1999;**18**:6934–6949.

Moore DJ, West AB, Dawson VL, Dawson TM. Molecular pathophysiology of Parkinson's disease. *Annu Rev Neurosci* 2005;**28**:57–87.

Morimoto RI. Stress, aging, and neurodegenerative disease. *N Engl J Med* 2006; **355**:2254–2255.

Mosser DD, Ho S, Glover JR. *Saccharomyces cerevisiae* Hsp104 enhances the chaperone capacity of human cells and inhibits heat stress-induced proapoptotic signaling. *Biochemistry* 2004;**43**:8107–8115.

Munch J, Rucker E, Standker L, Adermann K, Goffinet C, Schindler M, Wildum S, Chinnadurai R, Rajan D, Specht A, *et al.* Semen-derived amyloid fibrils drastically enhance HIV infection. *Cell* 2007;**131**:1059–1071.

Murray AN, Solomon JP, Wang YJ, Balch WE, Kelly JW. Discovery and characterization of a mammalian amyloid disaggregation activity. *Protein Sci* 2010;**19**:836–846.

Narayanan S, Walter S, Reif B. Yeast prion-protein, sup35, fibril formation proceeds by addition and substraction of oligomers. *Chembiochem* 2006;**7**:757–765.

Nekooki-Machida Y, Kurosawa M, Nukina N, Ito K, Oda T, Tanaka M. Distinct conformations of *in vitro* and *in vivo* amyloids of huntingtin-exon1 show different cytotoxicity. *Proc Natl Acad Sci USA* 2009;**106**:9679–9684.

Nelson R, Eisenberg D. Structural models of amyloid-like fibrils. *Adv Protein Chem* 2006;**73**:235–282.

Neumann M, Sampathu DM, Kwong LK, Truax AC, Micsenyi MC, Chou TT, Bruce J, Schuck T, Grossman M, Clark CM, *et al.* Ubiquitinated TDP-43 in frontotemporal lobar degeneration and amyotrophic lateral sclerosis. *Science* 2006;**314**: 130–133.

Neuwald AF, Aravind L, Spouge JL, Koonin EV. AAA+: a class of chaperone-like ATPases associated with the assembly, operation, and disassembly of protein complexes. *Genome Res* 1999;**9**:27–43.

Parsell DA, Kowal AS, Lindquist S. *Saccharomyces cerevisiae* Hsp104 protein. Purification and characterization of ATP-induced structural changes. *J Biol Chem* 1994a;**269**:4480–4487.

Parsell DA, Kowal AS, Singer MA, Lindquist S. Protein disaggregation mediated by heat-shock protein Hsp104. *Nature* 1994b;**372**:475–478.

Parsell DA, Lindquist S. The function of heat-shock proteins in stress tolerance: degradation and reactivation of damaged proteins. *Annu Rev Genet* 1993;**27**:437–496.

Parsell DA, Sanchez Y, Stitzel JD, Lindquist S. Hsp104 is a highly conserved protein with two essential nucleotide-binding sites. *Nature* 1991;**353**:270–273.

Parsell DA, Taulien J, Lindquist S. The role of heat-shock proteins in thermotolerance. *Philos Trans R Soc Lond B Biol Sci* 1993;**339**:279–285.

Patino MM, Liu JJ, Glover JR, Lindquist S. Support for the prion hypothesis for inheritance of a phenotypic trait in yeast. *Science* 1996;**273**:622–626.

Paushkin SV, Kushnirov VV, Smirnov VN, Ter-Avanesyan MD. Propagation of the yeast prion-like [PSI^+] determinant is mediated by oligomerization of the SUP35-encoded polypeptide chain release factor. *EMBO J* 1996;**15**:3127–3134.

Perrin V, Regulier E, Abbas-Terki T, Hassig R, Brouillet E, Aebischer P, Luthi-Carter R, Deglon N. Neuroprotection by Hsp104 and Hsp27 in lentiviral-based rat models of Huntington's disease. *Mol Ther* 2007;**15**:903–911.

Petkova AT, Leapman RD, Guo Z, Yau WM, Mattson MP, Tycko R. Self-propagating, molecular-level polymorphism in Alzheimer's beta-amyloid fibrils. *Science* 2005;**307**:262–265.

Powers ET, Morimoto RI, Dillin A, Kelly JW, Balch WE. Biological and chemical approaches to diseases of proteostasis deficiency. *Annu Rev Biochem* 2009;**78**:959–991.

Prince M, Jackson J, Ferri CP, Sousa R, Albanese E, Ribeiro WS, Honyashiki M. Alzheimer's disease international world alzheimer report. In: International AsD (ed); 2009. p 1–96.

Queitsch C, Hong SW, Vierling E, Lindquist S. Heat shock protein 101 plays a crucial role in thermotolerance in *Arabidopsis*. *Plant Cell* 2000;**12**:479–492.

Roberson ED, Mucke L. 100 years and counting: prospects for defeating Alzheimer's disease. *Science* 2006;**314**:781–784.

Roberts BE, Duennwald ML, Wang H, Chung C, Lopreiato NP, Sweeny EA, Knight MN, Shorter J. A synergistic small-molecule combination directly eradicates diverse prion strain structures. *Nat Chem Biol* 2009;**5**:936–946.

Safar J, Wille H, Itri V, Groth D, Serban H, Torchia M, Cohen FE, Prusiner SB. Eight prion strains have PrP(Sc) molecules with different conformations. *Nat Med* 1998;**4**:1157–1165.

Sanchez Y, Lindquist SL. HSP104 required for induced thermotolerance. *Science* 1990;**248**:1112–1115.

Sanchez Y, Taulien J, Borkovich KA, Lindquist S. Hsp104 is required for tolerance to many forms of stress. *EMBO J* 1992;**11**:2357–2364.

Satyal SH, Schmidt E, Kitagawa K, Sondheimer N, Lindquist S, Kramer JM, Morimoto RI. Polyglutamine aggregates alter protein folding homeostasis in *Caenorhabditis elegans*. *Proc Natl Acad Sci USA* 2000;**97**:5750–5755.

Savistchenko J, Krzewska J, Fay N, Melki R. Molecular chaperones and the assembly of the prion Ure2p *in vitro*. *J Biol Chem* 2008;**283**:15732–15739.

Schaupp A, Marcinowski M, Grimminger V, Bosl B, Walter S. Processing of proteins by the molecular chaperone hsp104. *J Mol Biol* 2007;**370**:674–686.

Schirmer EC, Lindquist S. Interactions of the chaperone Hsp104 with yeast Sup35 and mammalian PrP. *Proc Natl Acad Sci USA* 1997;**94**:13932–13937.

Schirmer EC, Queitsch C, Kowal AS, Parsell DA, Lindquist S. The ATPase activity of Hsp104, effects of environmental conditions and mutations. *J Biol Chem* 1998;**273**:15546–15552.

Schirmer EC, Ware DM, Queitsch C, Kowal AS, Lindquist SL. Subunit interactions influence the biochemical and biological properties of Hsp104. *Proc Natl Acad Sci USA* 2001;**98**:914–919.

Schlieker C, Weibezahn J, Patzelt H, Tessarz P, Strub C, Zeth K, Erbse A, Schneider-Mergener J, Chin JW, Schultz PG, *et al.* Substrate recognition by the AAA+ chaperone ClpB. *Nat Struct Mol Biol* 2004;**11**:607–615.

Shorter J. Hsp104: a weapon to combat diverse neurodegenerative disorders. *Neurosignals* 2008;**16**:63–74.

Shorter J, Lindquist S. Hsp104 catalyzes formation and elimination of self-replicating Sup35 prion conformers. *Science* 2004;**304**:1793–1797.

Shorter J, Lindquist S. Navigating the ClpB channel to solution. *Nat Struct Mol Biol* 2005a;**12**:4–6.

Shorter J, Lindquist S. Prions as adaptive conduits of memory and inheritance. *Nat Rev Genet* 2005b;**6**:435–450.

Shorter J, Lindquist S. Destruction or potentiation of different prions catalyzed by similar Hsp104 remodeling activities. *Mol Cell* 2006;**23**:425–438.

Shorter J, Lindquist S. Hsp104, Hsp70 and Hsp40 interplay regulates formation, growth and elimination of Sup35 prions. *EMBO J* 2008;**27**:2712–2724.

Si K, Choi YB, White-Grindley E, Majumdar A, Kandel ER. *Aplysia* CPEB can form prion-like multimers in sensory neurons that contribute to long-term facilitation. *Cell* 2010;**140**:421–435.

Si K, Lindquist S, Kandel ER. A neuronal isoform of the *Aplysia* CPEB has prion-like properties. *Cell* 2003;**115**:879–891.

Skovronsky DM, Lee VM-Y, Trojanowski JQ. Neurodegenerative diseases: new concepts of pathogenesis and their therapeutic implications. *Annu Rev Pathol Mech Dis* 2006;**1**:151–170.

Smith JF, Knowles TP, Dobson CM, Macphee CE, Welland ME. Characterization of the nanoscale properties of individual amyloid fibrils. *Proc Natl Acad Sci USA* 2006;**103**:15806–15811.

Spillantini MG, Crowther RA, Jakes R, Cairns NJ, Lantos PL, Goedert M. Filamentous alpha-synuclein inclusions link multiple system atrophy with Parkinson's disease and dementia with Lewy bodies. *Neurosci Lett* 1998;**251**:205–208.

Spillantini MG, Schmidt ML, Lee VM, Trojanowski JQ, Jakes R, Goedert M. Alpha-synuclein in Lewy bodies. *Nature* 1997;**388**:839–840.

Squires CL, Pedersen S, Ross BM, Squires C. ClpB is the *Escherichia coli* heat shock protein F84.1. *J Bacteriol* 1991;**173**:4254–4262.

Stoessl AJ. Gene therapy for Parkinson's disease: early data. *Lancet* 2007;**369**: 2056–2058.

Sunde M, Blake C. The structure of amyloid fibrils by electron microscopy and X-ray diffraction. *Adv Protein Chem* 1997;**50**:123–159.

Sunde M, Serpell LC, Bartlam M, Fraser PE, Pepys MB, Blake CC. Common core structure of amyloid fibrils by synchrotron X-ray diffraction. *J Mol Biol* 1997;**273**: 729–739.

Sweeny EA, Shorter J. Prion proteostasis: Hsp104 meets its supporting cast. *Prion* 2008;**2**:135–140.

Tanaka M, Collins SR, Toyama BH, Weissman JS. The physical basis of how prion conformations determine strain phenotypes. *Nature* 2006;**442**:585–589.

Taneja V, Maddelein ML, Talarek N, Saupe SJ, Liebman SW. A non-Q/N-rich prion domain of a foreign prion, [Het-s], can propagate as a prion in yeast. *Mol Cell* 2007;**27**:67–77.

Taylor JP, Hardy J, Fischbeck KH. Toxic proteins in neurodegenerative disease. *Science* 2002;**296**:1991–1995.

Tessarz P, Mogk A, Bukau B. Substrate threading through the central pore of the Hsp104 chaperone as a common mechanism for protein disaggregation and prion propagation. *Mol Microbiol* 2008;**68**:87–97.

Tessier PM, Lindquist S. Prion recognition elements govern nucleation, strain specificity and species barriers. *Nature* 2007;**447**:556–561.

Tipton KA, Verges KJ, Weissman JS. *In Vivo* monitoring of the prion replication cycle reveals a critical role for Sis1 in delivering substrates to Hsp104. *Mol Cell* 2008;**32**:584–591.

Toyama BH, Kelly MJ, Gross JD, Weissman JS. The structural basis of yeast prion strain variants. *Nature* 2007;**449**:233–237.

Treusch S, Cyr DM, Lindquist S. Amyloid deposits: protection against toxic protein species? *Cell Cycle* 2009;**8**:1668–1674.

True HL, Lindquist SL. A yeast prion provides a mechanism for genetic variation and phenotypic diversity. *Nature* 2000;**407**:477–483.

Tyedmers J, Madariaga ML, Lindquist S. Prion switching in response to environmental stress. *PLoS Biol* 2008;**6**:e294.

Vacher C, Garcia-Oroz L, Rubinsztein DC. Overexpression of yeast hsp104 reduces polyglutamine aggregation and prolongs survival of a transgenic mouse model of Huntington's disease. *Hum Mol Genet* 2005;**14**:3425–3433.

Vance C, Rogelj B, Hortobagyi T, De Vos KJ, Nishimura AL, Sreedharan J, Hu X, Smith B, Ruddy D, Wright P, *et al.* Mutations in FUS, an RNA processing protein, cause familial amyotrophic lateral sclerosis type 6. *Science* 2009;**323**:1208–1211.

Varshavsky A. Regulated protein degradation. *Trends Biochem Sci* 2005;**30**:283–286.

Vembar SS, Brodsky JL. One step at a time: endoplasmic reticulum-associated degradation. *Nat Rev Mol Cell Biol* 2008;**9**:944–957.

Wang H, Duennwald ML, Roberts BE, Rozeboom LM, Zhang YL, Steele AD, Krishnan R, Su LJ, Griffin D, Mukhopadhyay S, *et al.* Direct and selective elimination of specific prions and amyloids by 4,5-dianilinophthalimide and analogs. *Proc Natl Acad Sci USA* 2008;**105**:7159–7164.

Watt B, van Niel G, Fowler DM, Hurbain I, Luk KC, Stayrook SE, Lemmon MA, Raposo G, Shorter J, Kelly JW, *et al.* N-terminal domains elicit formation of functional Pmel17 amyloid fibrils. *J Biol Chem* 2009;**284**:35543–35555.

Weibezahn J, Schlieker C, Bukau B, Mogk A. Characterization of a trap mutant of the AAA+ chaperone ClpB. *J Biol Chem* 2003;**278**:32608–32617.

Weibezahn J, Schlieker C, Tessarz P, Mogk A, Bukau B. Novel insights into the mechanism of chaperone-assisted protein disaggregation. *Biol Chem* 2005;**386**:739–744.

Weibezahn J, Tessarz P, Schlieker C, Zahn R, Maglica Z, Lee S, Zentgraf H, Weber-Ban EU, Dougan DA, Tsai FT, *et al.* Thermotolerance requires refolding of aggregated proteins by substrate translocation through the central pore of ClpB. *Cell* 2004;**119**:653–665.

Wendler P, Saibil HR. Cryo-electron microscopy structures of Hsp100 proteins-crowbars in or out. *Biochem Cell Biol* 2010;**88**:89–96.

Wendler P, Shorter J, Plisson C, Cashikar AG, Lindquist S, Saibil HR. Atypical AAA+ subunit packing creates an expanded cavity for disaggregation by the protein-remodeling factor Hsp104. *Cell* 2007;**131**:1366–1377.

Wendler P, Shorter J, Snead D, Plisson C, Clare DK, Lindquist S, Saibil HR. Motor mechanism for protein threading through Hsp104. *Mol Cell* 2009;**34**:81–92.

Wenk GL. Neuropathologic changes in Alzheimer's disease. *J Clin Psychiatry* 2003;**64** Suppl 9:7–10.

Werbeck ND, Schlee S, Reinstein J. Coupling and dynamics of subunits in the hexameric AAA+ chaperone ClpB. *J Mol Biol* 2008;**378**:178–190.

Wertkin AM, Turner RS, Pleasure SJ, Golde TE, Younkin SG, Trojanowski JQ, Lee VM. Human neurons derived from a teratocarcinoma cell line express solely the 695-amino acid amyloid precursor protein and produce intracellular beta-amyloid or A4 peptides. *Proc Natl Acad Sci USA* 1993;**90**:9513–9517.

Wiltzius JJ, Landau M, Nelson R, Sawaya MR, Apostol MI, Goldschmidt L, Soriaga AB, Cascio D, Rajashankar K, Eisenberg D. Molecular mechanisms for protein-encoded inheritance. *Nat Struct Mol Biol* 2009;**16**:973–978.

Xue WF, Hellewell AL, Gosal WS, Homans SW, Hewitt EW, Radford SE. Fibril fragmentation enhances amyloid cytotoxicity. *J Biol Chem* 2009;**284**:34272–34282.

Xue WF, Hellewell AL, Hewitt EW, Radford SE. Fibril fragmentation in amyloid assembly and cytotoxicity: when size matters. *Prion* 2010;**4**:20–25.

Young JC, Agashe VR, Siegers K, Hartl FU. Pathways of chaperone-mediated protein folding in the cytosol. *Nat Rev Mol Cell Biol* 2004;**5**:781–791.

8

CHAPERONE-DEPENDENT AMYLOID ASSEMBLY AND PRION TOXICITY

Daniel W. Summers, Katie J. Wolfe, and Douglas M. Cyr
Department of Cell and Developmental Biology, School of Medicine, University of North Carolina, Chapel Hill, NC, USA

ABBREVIATIONS

APP, amyloid precursor protein
SDS, sodium dodecyl sulfate
HSF, heat shock factor
UPS, ubiqutin–proteasome system
Gln/Asn, glutamine/asparagines
CPEB, cytoplasmic polyadenylation element binding protein

8.1 INTRODUCTION

Protein misfolding and aggregation are the hallmarks of a wide range of neurodegenerative disorders (Carrell and Lomas, 1997). For decades, the accumulation of protein aggregates in the form of inclusion bodies or β-rich amyloid-like assemblies was considered the major neurotoxic insult responsible for neuronal dysfunction and cell death (Fiala, 2007). However, this point is still a matter of

Protein Chaperones and Protection from Neurodegenerative Diseases, First Edition.
Edited by Stephan N. Witt.

controversy and the toxic culprit in many of these diseases is unknown. Amyloid assembly has been studied extensively in the context of the cleaved amyloid precursor protein (APP) fragment Aβ(1-42) that accumulates in extracellular amyloid plaques in Alzheimer's disease (Haass and Selkoe, 2007). In addition to Aβ(1-42), numerous proteins display some propensity to form amyloid-like fibrils (Chiti and Dobson, 2006; Sipe and Cohen, 2000). The accumulation of amyloid-like fibrils is reported in a wide range of neurodegenerative diseases including Alzheimer's, Huntington's, Parkinson's, and prion diseases (Carrell and Lomas, 1997). Assembly of amyloid fibrils occurs through nucleated polymerization where fibrillization is initially very slow, but the formation of an amyloid "seed" stimulates rapid fibril polymerization (Jarrett and Lansbury, 1993). Structures of amyloid-like fibrils are enriched in β-sheets that are stacked perpendicular to a fibril axis, an arrangement called a *cross-beta spine conformation* (Nelson *et al.*, 2005; Sawaya *et al.*, 2007). Amyloid-like fibrils are characterized *in vitro* and *in vivo* by a series of criteria including recognition by the indicator dyes Congo Red or Thioflavin T, resistance to protease digestion, and insolubility in ionic detergents such as sodium dodecyl sulfate (SDS) (Chiti and Dobson, 2006).

Molecular chaperones are well-recognized modifiers of numerous neurodegenerative disorders including diseases characterized by amyloid formation (Muchowski and Wacker, 2005). Not only are molecular chaperones commonly localized in inclusion bodies but overexpression of specific molecular chaperones also suppresses neurodegeneration in a variety of disease model systems (Adachi *et al.*, 2003; Chan *et al.*, 2000; Cummings *et al.*, 2001). Molecular chaperones cooperate to refold nonnative proteins or target chronically misfolded proteins for degradation. Deficiencies in protein quality control machinery due to aging or disease render cells more susceptible to the accumulation of protein aggregates (Balch *et al.*, 2008).

Molecular chaperones conventionally protect cells from proteotoxic insult by binding exposed, hydrophobic peptides in a misfolded protein conformer and preventing aberrant protein aggregation (Hartl and Hayer-Hartl, 2002). However, chaperone-mediated suppression of neurodegeneration does not always correlate with a decrease in aggregation (Chan *et al.*, 2000; Warrick *et al.*, 1999). Indeed, several recent studies suggest that enhancing the assembly of amyloid-like aggregates is protective in model systems for Alzheimer's disease and Huntington's disease (Behrends *et al.*, 2006; Bodner *et al.*, 2006; Cohen *et al.*, 2006, 2009; Douglas *et al.*, 2008, 2009b). Interestingly, as discussed in this chapter, some of these studies demonstrate a fundamental role for molecular chaperones in promoting assembly of amyloid-like fibrils. Thus, molecular chaperones utilize multiple pathways to protect cells from proteotoxic stress (Figure 8.1). First, molecular chaperones can hold misfolded proteins in a soluble state for refolding or degradation. Alternatively, or in conjunction with the first pathway, molecular chaperones can promote aggregation and the formation of amyloid-like assemblies. The goal of either pathway is to prevent the accumulation of cytotoxic, soluble conformers that appear to induce cell death in many neurodegenerative disorders (Hartley *et al.*, 1999; Kayed *et al.*, 2003; Lambert *et al.*, 1998; Roher

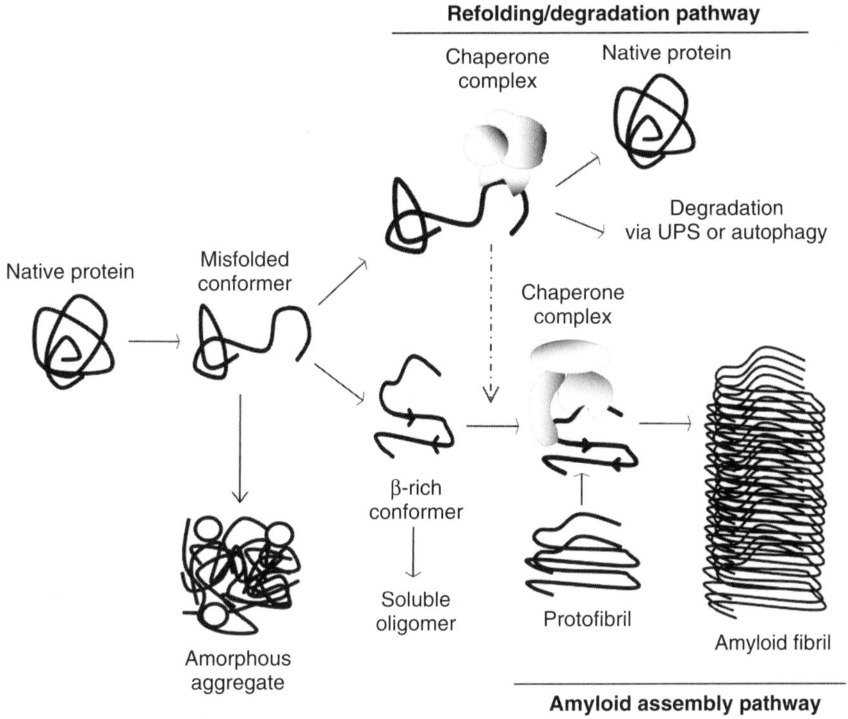

Figure 8.1 Molecular chaperones partition misfolded proteins between distinct quality control pathways. If a native protein misfolds, protein quality control machinery can partition this nonnative protein between several distinct pathways. Molecular chaperones bind exposed, hydrophobic motifs in the nonnative conformer and promote refolding or target this protein for degradation via the ubiquitin–proteasome system (UPS) or autophagy. How molecular chaperones partition nonnative substrates between these two endpoints is unclear, though chaperones often cooperate in discrete complexes that, depending on the nature of the substrate interaction and the presence or the absence of particular cofactors, might discriminate between refolding or degradation endpoints. If molecular chaperones do not hold the nonnative protein in a soluble state, then it might associate with other nonnative proteins through hydrophobic interactions and accumulate as a large amorphous aggregate. Some nonnative proteins display a strong propensity to assemble into β-sheet-rich amyloid-like conformers. Specific molecular chaperones recognize this β-sheet-rich conformer and promote the assembly of large, insoluble amyloid-like fibrils. If the amyloid assembly is inefficient, then an off-pathway, oligomeric or intermediate protofibril protein species might accumulate. Furthermore, if the refolding/degradation pathway fails, then the amyloid assembly pathway might serve as a backup quality control mechanism.

et al., 1996). How specific molecular chaperone networks act to direct nonnative proteins toward different fates remains unclear. This chapter will examine several examples in which amyloid formation is benign or cytoprotective in disease model systems, describe a specific example of how an Hsp40 molecular chaperone promotes formation of amyloid-like aggregates as a protective mechanism in prion toxicity, and finally highlight cellular pathways that promote amyloid assembly for a functional role in cell biology.

8.2 AMYLOID FORMATION IN DISEASE

While the appearance of amyloid-like plaques in neurodegenerative disorders has garnered much attention, recent studies highlight the significance of a soluble oligomer or intermediate as a toxic protein species in many of these diseases (Haass and Selkoe, 2007). Although the exact nature of the toxic, soluble oligomer is still unclear (Glabe, 2008), the accumulation of a soluble Aβ(1-42) oligomeric protein species has been correlated with the onset of dementia in Alzheimer's disease (Lue *et al.*, 1999; McLean *et al.*, 1999). Furthermore, an antibody raised against Aβ(1-42) recognizes a soluble oligomeric protein species that forms from several different amyloid-like proteins, suggesting a common structural conformation independent of amino acid sequence (Kayed *et al.*, 2003). As a result, amyloid-like aggregates might be the inert end point of a misfolding event that generates an elusive toxic culprit in many neurodegenerative diseases.

If neurodegeneration in amyloid-linked diseases is primarily a result of soluble, oligomeric protein species, then promoting amyloid formation should protect cells from death by sequestering this toxic conformer in a stable, insoluble aggregate. Indeed, numerous studies support this hypothesis. Transgenic mice expressing human APP with the "Artic" mutation showed greater plaque formation while exhibiting reduced defects in some behavioral categories compared to transgenic mice expressing wild-type human APP (Cheng *et al.*, 2007). Studies in a nematode model of Alzheimer's disease also showed no correlation between aggregate formation and Aβ(1-42)-induced toxicity (Cohen *et al.*, 2006). Instead, increasing activity of the FOXO transcription factor DAF-16 protected worms from Aβ(1-42)-induced toxicity, yet enhanced formation of high molecular mass aggregates. Conversely, reducing heat shock factor (HSF) signaling sensitized worms to Aβ(1-42)-induced toxicity and likewise increased aggregation of Aβ(1-42). Thus, while Aβ(1-42) aggregation does not strictly correlate with toxicity, multiple pathways appear to manage Aβ(1-42) aggregation. One cellular pathway inhibits aggregation (HSF-dependent), while another cellular pathway promotes aggregation (DAF-16 dependent). Similar observations were made in a mouse model of Alzheimer's disease in which reduced insulin-like growth factor signaling (a pathway that inhibits DAF-16 activity in the nematode) increased survival and correlated with the appearance of more dense amyloid plaques and a reduction in free Aβ(1-42) (Cohen *et al.*, 2009). Although the molecular mechanisms underlying these distinct activities are still unclear, these studies suggest that molecular pathways can promote aggregation to protect cells from the accumulation of soluble, cytotoxic protein species.

Another example of protective aggregation comes from a yeast model of Huntington's disease. Huntington's disease is caused by a mutation in the gene encoding the protein huntingtin in which a polyglutamine repeat is expanded beyond 40 amino acids (Zoghbi and Orr, 2000). Polyglutamine expansion beyond this threshold causes a misfolding event and amyloid fibril assembly (Chen and Wetzel, 2001; Scherzinger *et al.*, 1997). In neurons, this process manifests in the appearance of intranuclear inclusions and cell death (DiFiglia *et al.*, 1997). Similar to Alzheimer's disease, the correlation between the accumulation of protein aggregates and cell death is controversial. Studies examining the role of molecular chaperones on polyglutamine assembly utilizing *in vitro* and *in vivo* models have provided interesting insight into this particular question. Hsp70 and Hsp40 molecular chaperones have been identified as suppressors of neurotoxicity in numerous models of polyglutamine expansion disorders (Chan *et al.*, 2000; Cummings *et al.*, 2001; Jana *et al.*, 2000; Kazemi-Esfarjani and Benzer, 2000; Warrick *et al.*, 1999). Overexpression of Hsp70 or Hsp40 molecular chaperones resulted in a decrease in inclusion body formation in many of these studies; however, there are several reports in which aggregate levels were unchanged or even increased (Chan *et al.*, 2000; Warrick *et al.*, 1999; Wyttenbach *et al.*, 2000). *In vitro* studies of huntingtin assembly with atomic force microscopy showed that human Hsp70/Hsp40 molecular chaperones altered the conformation of monomeric, polyglutamine-expanded huntingtin protein species and prevented the formation of spherical or annular oligomers (Wacker *et al.*, 2004). Subsequently, the Hsp40/Hsp70 molecular chaperones were shown to cooperate with the TRiC chaperonin complex to drive huntingtin assembly into higher molecular mass aggregate species and prevent the accumulation of a 200-kDa huntingtin pool that correlated with cytotoxicity (Behrends *et al.*, 2006). Thus, the Hsp40/Hsp70/TRiC chaperonin complex serves as a prime example of a molecular chaperone pathway that promotes protein aggregation as a cytoprotective mechanism.

Altogether, these observations demonstrate that molecular chaperones alter the conformational properties of misfolded polypeptides to either prevent aberrant protein aggregation or drive assembly of nonnative proteins into higher molecular mass species. These studies do not diminish the potential cytotoxic effects of amyloid plaque accumulation. They do suggest that soluble intermediates in the amyloid assembly pathway may be the primary neurotoxic protein species in many neurodegenerative disorders and amyloid plaque formation may induce secondary effects later in disease progression (Fiala, 2007).

8.3 PRION PROPAGATION AND TOXICITY

Perhaps one of the best described examples of chaperone-facilitated amyloid assembly comes from the budding yeast *Saccharomyces cerevisiae. S. cerevisiae* possesses several protein-based heritable elements known collectively as *yeast prions* (Tuite and Lindquist, 1996; Wickner *et al.*, 2004). Yeast prions transmit heritable information through a protein fold that induces conversion of the native protein into the prion conformation with a strong propensity to assemble into amyloid-like fibrils (Shorter and Lindquist, 2005). Propagation of prion elements in yeast is regulated by cellular machinery including molecular chaperones (Tuite and Cox, 2003). How molecular chaperones participate in this process has been extensively studied over the past decade, and these studies have provided fascinating insight into the basic mechanisms in which molecular chaperones recognize and regulate amyloid-like conformers. In this chapter, the discussion is focused on the role of Hsp40 molecular chaperones in prion propagation and suppression of prion toxicity via assembly of amyloid-like aggregates (see also Chapter 9).

Hsp40s are defined by the presence of a highly conserved J-domain that interacts with the Hsp70 nucleotide-binding domain to stimulate adenosine triphosphate (ATP) hydrolysis (Cyr *et al.*, 1994; Qiu *et al.*, 2006; Walsh *et al.*, 2004). ATP hydrolysis induces a conformational change in the Hsp70 substrate-binding domain that increases affinity for the substrate (Langer *et al.*, 1992; Szabo *et al.*, 1994). While the J-domain is sufficient to stimulate ATP hydrolysis by the Hsp70 nucleotide-binding domain, many Hsp40s independently interact with nonnative polypeptides and suppress protein aggregation (Chai *et al.*, 1999; Chan *et al.*, 2000; Cummings *et al.*, 1998; Cyr, 1995). Hsp40s are subdivided on the basis of the presence or absence of domains outside of the J-domain. Type I Hsp40s possess a glycine/phenylalanine-rich domain and a zinc-finger-like region. Type II Hsp40s possess only the glycine/phenylalanine-rich domain, while type III Hsp40s only contain a J-domain (Walsh *et al.*, 2004). Furthermore, Hsp40s demonstrate selective binding preferences (Fan *et al.*, 2004; Lu and Cyr, 1998) and unique cellular functions (Cyr *et al.*, 1994; Walsh *et al.*, 2004), suggesting a significant regulatory step in Hsp70-mediated protein quality control, which lies at the level of Hsp40 co-chaperone binding to a misfolded polypeptide and delivering this substrate to a partner Hsp70 (Summers *et al.*, 2009a).

Hsp40s also differentially regulate the propagation of prions in yeast. For example, overexpression of the type I Hsp40 Ydj1 cures yeast of the [*URE3*] prion (Moriyama *et al.*, 2000) and some variants of [*PSI+*] prions (Kushnirov *et al.*, 2000). These observations are consistent with *in vitro* studies demonstrating that Ydj1 inhibits assembly of Ure2 into amyloid-like fibrils (Lian *et al.*, 2007). Ydj1 likely binds Ure2 and holds the protein in a soluble state, preventing its aggregation (Lian *et al.*, 2007; Savistchenko *et al.*, 2008). Additionally, Ydj1 interacts with the glutamine/asparagine (Gln/Asn)-rich prion domain of Rnq1 specifically in its [*RNQ+*] prion conformation (Summers *et al.*, 2009b). Ydj1 utilizes its zinc-finger-like region and a farnesyl moiety to interact with this prion domain and inhibits the assembly of this fragment into SDS-insoluble aggregates. These features are specific to this type I Hsp40 Ydj1 and likely convey binding selectivity for specific nonnative conformations.

In contrast, the type II Hsp40 Sis1 is required for propagation of [*RNQ+*], [*PSI+*], and [*URE3*] prions (Higurashi *et al.*, 2008; Sondheimer *et al.*, 2001). This dependency was specific for Sis1 and not for other cytosolic Hsp40s, suggesting that Sis1 plays a unique and crucial function in yeast prion propagation (Higurashi *et al.*, 2008). How does Sis1 promote yeast prion propagation? Studies with the prion [*RNQ+*] have revealed some interesting mechanistic details on this process. Sis1 specifically interacts with the Rnq1 protein in the [*RNQ+*] prion conformation in a near 1:1 stoichiometric complex (Lopez *et al.*, 2003; Sondheimer *et al.*, 2001). Sis1 binds a short hydrophobic sequence in the N-terminal, non-prion domain of Rnq1, and mutations in this motif disrupt [*RNQ+*] propagation (Bardill and True, 2009; Douglas *et al.*, 2008), demonstrating that direct chaperone binding is a critical step in prion biogenesis. Depletion of Sis1 from yeast causes a transient increase in the size of Rnq1 oligomers and eventually cures cells of this prion (Aron *et al.*, 2007). Inhibiting Hsp104 activity with guanidinium hydrochloride results in a very similar pattern (Kryndushkin *et al.*, 2003), suggesting that Sis1 and Hsp104 might act via a shared mechanism to propagate yeast prions. Using a substrate-trapping variant of Hsp104, Sis1 was shown to be required for Hsp104 binding to Rnq1 protein and Sup35 (the protein responsible for the [*PSI+*] prion) (Tipton *et al.*, 2008). Thus, Sis1 might bind prion amyloids and recruit Hsp70 and Hsp104 to generate a chaperone complex that shears prion fibrils to generate new prion seeds (Aron *et al.*, 2007; Tipton *et al.*, 2008). Furthermore, as described in more detail below, overexpression of Sis1 increases the level of SDS-insoluble, amyloid-like Rnq1 aggregates (Douglas *et al.*, 2008). Sis1 functions as a homodimer and, therefore, could simultaneously bring one Rnq1 protein in complex with a [*RNQ+*] prion assembly to promote [*RNQ+*] prion elongation. Thus, Sis1 activity in [*RNQ+*] prion shearing and elongation could synergize to promote efficient [*RNQ+*] propagation (Figure 8.2), although these mechanisms require further investigation.

Studies on Sis1-dependent [*RNQ+*] assembly have further implications on amyloid assembly in neurodegenerative disorders. Moderate overexpression of Rnq1 is toxic to yeast, specifically in the presence of a preexisting [*RNQ+*] prion (Douglas *et al.*, 2008). Thus, [*RNQ+*] prions seed conversion of Rnq1 into a toxic protein conformer that accumulates when Rnq1 is overexpressed in yeast. Is the formation of SDS-insoluble, amyloid-like [*RNQ+*] aggregates toxic or, as suggested by studies on Aβ(1-42) and huntingtin discussed above, is there a soluble Rnq1 species that also forms upon Rnq1 overexpression? In the [*RNQ+*] background utilized for these studies, endogenous Rnq1 principally exists in a high molecular mass, SDS-insoluble pool (Douglas *et al.*, 2008). Overexpression of Rnq1 to toxic levels resulted in the formation of a low molecular mass, SDS-soluble pool suggesting that [*RNQ+*] prion assembly is not efficient enough to partition the entire pool of converted Rnq1 into assembled [*RNQ+*] prion if Rnq1 levels are in excess. Supporting this hypothesis, mutating the Sis1-binding site in the Rnq1 non-prion domain (L94A) decreases the efficiency of [*RNQ+*] prion assembly and exacerbates toxicity. Elevating the levels of Sis1 suppresses Rnq1-induced toxicity and correlates with an increase in the accumulation of SDS-insoluble, high molecular mass aggregates. Expression of Rnq1(L94A) was also toxic to yeast in an [*rnq-*] background; however, overexpression of Sis1 could only suppress toxicity of Rnq1(L94A) in the presence of [*RNQ+*] prions, demonstrating that Sis1 acts specifically on the amyloid-like, [*RNQ+*] prion conformer of Rnq1 to promote [*RNQ+*] assembly. Thus, chaperone-dependent formation of amyloid-like prion aggregates is cytoprotective, perhaps by

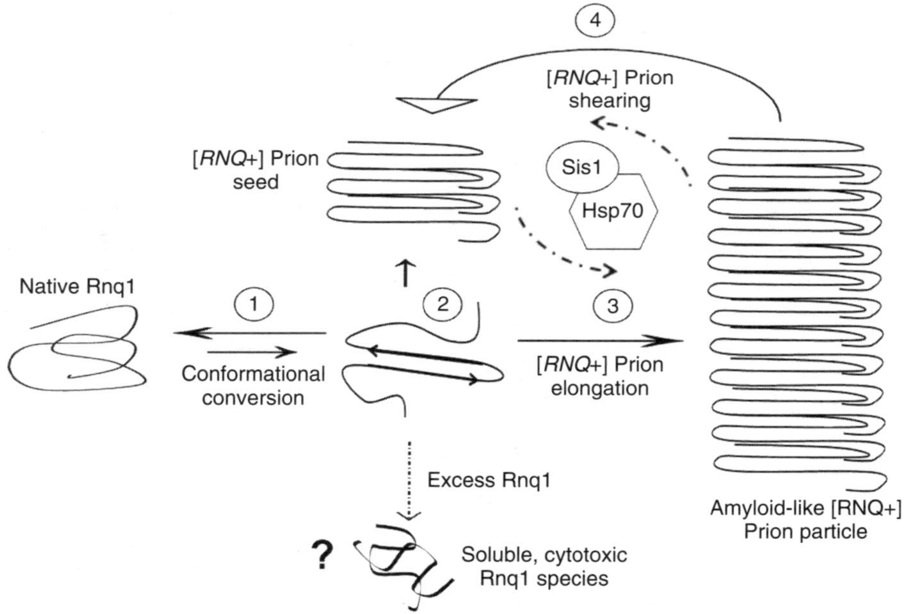

Figure 8.2 Pathway for [*RNQ*+] prion propagation. (1) Native Rnq1 is converted to a β-rich [*RNQ*+] prion conformer. This step is initially unfavorable unless there are preexisting [*RNQ*+] prion seeds to promote conformational conversion. The [*RNQ*+] prion seed is portrayed in this figure as an [*RNQ*+] prion oligomer, though the structure of this protein species is currently unknown. (2) The β-rich [*RNQ*+] prion conformer is added to an [*RNQ*+] prion oligomer, a step that might be facilitated by the Hsp40 Sis1and its cognate Hsp70. (3) Subsequent steps of Rnq1 addition eventually form a large, amyloid-like assembly. (4) The molecular chaperones Sis1 and Hsp70 cooperate with Hsp104 to shear large [*RNQ*+] prion assemblies to form new [*RNQ*+] prion seeds, thereby propagating the [*RNQ*+] prion state.

partitioning soluble, cytotoxic conformers into benign, SDS-insoluble assemblies (Douglas *et al.*, 2009a).

Analysis of Sis1 action on [*RNQ*+] assembly has also revealed how other cellular factors influence amyloid assembly pathways. Surprisingly, elevated Sis1 levels not only suppressed Rnq1 toxicity but also relocalized Rnq1 aggregates to the nucleus (Douglas *et al.*, 2009b). How Sis1 promotes relocalization of Rnq1 to the nucleus and how the nuclear environment influences [*RNQ*+] assembly are still unknown. Conversely, relocalization of polyglutamine-expanded huntingtin to the nucleus exacerbated toxicity in yeast. This effect correlated with a decrease in the level of SDS-insoluble aggregates and an increase in the level of an SDS-soluble, lower molecular mass protein species (Douglas *et al.*, 2009b). How the nuclear environment influences aggregation pathways to such opposing outcomes is unclear. However, these studies demonstrate that chaperone-dependent promotion of protein aggregation constitutes a major cellular pathway to protect cells from proteotoxic insult.

8.4 FUNCTIONAL AMYLOID IN NATURE

If inefficient amyloid assembly results in the formation of cytotoxic protein species, why do cells express proteins with a high propensity to form amyloid-like fibrils? It turns out that amyloid assembly is utilized in a variety of basic cellular processes. Yeast prions represent a compelling example of how driving the conversion into an amyloid assembly pathway impacts cell physiology. The translation termination factor Sup35 is the protein determinant for the yeast prion [*PSI*+] (Tuite and Lindquist, 1996). Inducing conversion of Sup35 from the native [*psi*-] state to the [*PSI*+] prion increases the frequency of translational read-through and results in the expression of novel proteins (Shorter and Lindquist, 2005). Interestingly, the presence of the [*PSI*+] prion can enhance or reduce cell viability under a wide variety of environmental conditions (True and Lindquist, 2000; True *et al.*, 2004) Thus, the [PSI+] prion is an example of a stable epigenetic factor that may promote an adaptive advantage (Shorter and Lindquist, 2005). In addition, three newly identified yeast prions, [SWI+], [OCT+], and [MOT3+], influence transcriptional activity and similarly might regulate adaptive responses via distinct mechanisms from [PSI+] (Alberti *et al.*, 2009; Du *et al.*, 2008; Patel *et al.*, 2009). Inducing conversion of the [PSI+] prion requires the presence of the [RNQ+] prion discussed above (Derkatch *et al.*, 2001; Sondheimer and Lindquist, 2000). While the details underlying this dependency are still unclear, Gln/Asn-rich sequences in the Rnq1 protein might cross-seed conformational conversion and fibrillization of other Gln/Asn-rich proteins including Sup35 (Derkatch *et al.*, 2004; Vitrenko *et al.*, 2007). There is no other biological function identified for Rnq1 except influencing the conversion of other yeast prions. Yet, by controlling the induction of other yeast prions, the [RNQ+] prion might be able to influence adaptation to a dynamic environment (Halfmann *et al.*, 2010). This point will require further investigation.

Examples of functional amyloid assembly are emerging from organisms other than budding yeast. *Escherichia coli* produce a biofilm that is composed of amyloid-like fibrils. Assembly of these fibrils, known as *curli amyloid*, is tightly regulated at several steps by other cellular proteins. Fibrillization of the major curli subunit CsgA requires the minor subunit CsgB, an outer membrane-localized protein that is proposed to serve as a template to nucleate polymerization of CsgA (Hammer *et al.*, 2007; Wang *et al.*, 2008). CsgF, another cellular factor required for curli amyloid biogenesis, is required for the proper localization of CsgB at the cell surface (Nenninger *et al.*, 2009). Thus, multiple cellular factors in *E. coli* promote efficient amyloid assembly and ensure that amyloid fibrillization is spatially sequestered in the extracellular matrix.

In recent years, functional amyloids have been identified in mammals. For example, the protein Pmel17 assembles into amyloid-like fibrils within melanosomes (Fowler *et al.*, 2006). Pmel17 is a transmembrane glycoprotein that is proteolytically cleaved (Kummer *et al.*, 2009) within a post-Golgi membrane compartment to generate a lumenally restricted, aggregation-prone fragment (mα) that rapidly polymerizes into amyloid-like fibrils within this compartment. Similar to curli biogenesis, multiple cellular factors ensure that mα fibril assembly is sequestered from the rest of the cell (Fowler *et al.*, 2006). Interestingly, mα fibrils accelerate melanin synthesis, suggesting that this pathway has evolved a significant role in skin pigmentation (Fowler *et al.*, 2006). Recently, amyloid fibril formation was reported in pancreatic secretory granules within the mouse pituitary (Maji *et al.*, 2009). Specific peptides/prohormones were demonstrated to form amyloid-like fibrils both *in vitro* and within secretory vesicles. Sequestration of hormones within amyloid-like fibrils might help concentrate specific peptides or hormones within individual secretory granules and regulate hormone release because the assembled peptides/hormones could be released as active molecules.

Altogether, these studies suggest that cellular factors actively promote amyloid assembly to serve a functional role. In all cases described thus far, amyloid assembly is tightly regulated and spatially sequestered, which is consistent with the concept that dysregulated amyloid assembly has detrimental effects on cell physiology.

8.5 CONCLUSION AND FUTURE DIRECTIONS

Extensive investigation over the past several decades has revealed the enormous complexity of protein quality control pathways utilized to maintain protein homeostasis (Balch *et al.*, 2008). While protein disaggregation appears to be the preferred mechanism for preventing the accumulation of cytotoxic protein species, pathways also exist that promote the assembly of amyloid-like aggregates that might be a cytoprotective sink for soluble, toxic protein species that escape the refolding/degradation pathway.

Molecular chaperones are an essential component of both the disaggregation pathway and the aggregation-enhancing pathway. Hsp40 molecular chaperones in particular serve on the front-lines of defense to recognize misfolded protein conformers and transfer nonnative clients to other chaperone complexes. How these nonnative protein species are efficiently partitioned between refolding, degradation, or aggregation pathways remains an ongoing question in this field. Furthermore, how cells might cope with the accumulation of amyloid-like fibrils is poorly understood.

Numerous examples have recently emerged where amyloid formation is coupled to a biological function. As a result, molecular chaperones not only protect cells from misfolded protein conformers but may also contribute to diverse cellular pathways via facilitating amyloid assembly. While the majority of data supporting this concept comes from yeast prions, other examples are emerging in evolutionarily diverse organisms from bacteria to mammals. For example, a neuronally specific form of the cytoplasmic polyadenylation element binding protein (CPEB) displays prion-like properties and regulates synaptic activity (Si *et al.*, 2003a, b, 2010). Do specific molecular chaperones assist CPEB assembly and regulate this pathway to prevent the accumulation of aberrant CPEB conformers or perhaps facilitate CPEB assembly into amyloid-like complexes to modulate its activity? How many other cellular proteins form amyloid-like structures and how do molecular chaperones regulate these pathways to prevent the accumulation of cytotoxic protein species? Altogether, cells rely on multiple distinct protein quality control pathways to manage protein misfolding events, and chaperone-dependent amyloid assembly constitutes a surprising yet significant component of this extensive network.

ACKNOWLEDGMENT

D.W.S is supported by a Ruth L. Kirschstein National Research Service Award from the National Institute on Aging (F31AG032790). Work in the lab of D.M.C is supported by the NIH (5R01GM067785-07 and 5R01GM056981-14).

REFERENCES

Adachi H, Katsuno M, Minamiyama M, Sang C, Pagoulatos G, Angelidis C, Kusakabe M, Yoshiki A, Kobayashi Y, Doyu M, Sobue G. Heat shock protein 70 chaperone overexpression ameliorates phenotypes of the spinal and bulbar muscular atrophy transgenic mouse model by reducing nuclear-localized mutant androgen receptor protein. *J Neurosci* 2003;**23**:2203–2211.

Alberti S, Halfmann R, King O, Kapila A, Lindquist S. A systematic survey identifies prions and illuminates sequence features of prionogenic proteins. *Cell* 2009;**137**: 146–158.

Aron R, Higurashi T, Sahi C, Craig EA. J-protein co-chaperone Sis1 required for generation of [RNQ+] seeds necessary for prion propagation. *EMBO J* 2007;**26**:3794–3803.

Balch WE, Morimoto RI, Dillin A, Kelly JW. Adapting proteostasis for disease intervention. *Science* 2008;**319**:916–919.

Bardill JP, True HL. Heterologous prion interactions are altered by mutations in the prion protein Rnq1p. *J Mol Biol* 2009;**388**:583–596.

Behrends C, Langer CA, Boteva R, Bottcher UM, Stemp MJ, Schaffar G, Rao BV, Giese A, Kretzschmar H, Siegers K, Hartl FU. Chaperonin TRiC promotes the assembly of polyQ expansion proteins into nontoxic oligomers. *Mol Cell* 2006;**23**:887–897.

Bodner RA, Outeiro TF, Altmann S, Maxwell MM, Cho SH, Hyman BT, McLean PJ, Young AB, Housman DE, Kazantsev AG. Pharmacological promotion of inclusion formation: a therapeutic approach for Huntington's and Parkinson's diseases. *Proc Natl Acad Sci U S A* 2006;**103**:4246–4251.

Carrell RW, Lomas DA. Conformational disease. *Lancet* 1997;**350**:134–138.

Chai Y, Koppenhafer SL, Bonini NM, Paulson HL. Analysis of the role of heat shock protein (Hsp) molecular chaperones in polyglutamine disease. *J Neurosci* 1999;**19**: 10338–10347.

Chan HY, Warrick JM, Gray-Board GL, Paulson HL, Bonini NM. Mechanisms of chaperone suppression of polyglutamine disease: selectivity, synergy and modulation of protein solubility in Drosophila. *Hum Mol Genet* 2000;**9**:2811–2820.

Chen S, Wetzel R. Solubilization and disaggregation of polyglutamine peptides. *Protein Sci* 2001;**10**:887–891.

Cheng IH, Scearce-Levie K, Legleiter J, Palop JJ, Gerstein H, Bien-Ly N, Puolivali J, Lesne S, Ashe KH, Muchowski PJ, Mucke L. Accelerating amyloid-beta fibrillization reduces oligomer levels and functional deficits in Alzheimer disease mouse models. *J Biol Chem* 2007;**282**:23818–23828.

Chiti F, Dobson CM. Protein misfolding, functional amyloid, and human disease. *Annu Rev Biochem* 2006;**75**:333–366.

Cohen E, Bieschke J, Perciavalle RM, Kelly JW, Dillin A. Opposing activities protect against age-onset proteotoxicity. *Science* 2006;**313**:1604–1610.

Cohen E, Paulsson JF, Blinder P, Burstyn-Cohen T, Du D, Estepa G, Adame A, Pham HM, Holzenberger M, Kelly JW, *et al.* Reduced IGF-1 signaling delays age-associated proteotoxicity in mice. *Cell* 2009;**139**:1157–1169.

Cummings CJ, Mancini MA, Antalffy B, DeFranco DB, Orr HT, Zoghbi HY. Chaperone suppression of aggregation and altered subcellular proteasome localization imply protein misfolding in SCA1. *Nat Genet* 1998;**19**:148–154.

Cummings CJ, Sun Y, Opal P, Antalffy B, Mestril R, Orr HT, Dillmann WH, Zoghbi HY. Over-expression of inducible HSP70 chaperone suppresses neuropathology and improves motor function in SCA1 mice. *Hum Mol Genet* 2001;**10**:1511–1518.

Cyr DM. Cooperation of the molecular chaperone Ydj1 with specific Hsp70 homologs to suppress protein aggregation. *FEBS Lett* 1995;**359**:129–132.

Cyr DM, Langer T, Douglas MG. DnaJ-like proteins: molecular chaperones and specific regulators of Hsp70. *Trends Biochem Sci* 1994;**19**:176–181.

Derkatch IL, Bradley ME, Hong JY, Liebman SW. Prions affect the appearance of other prions: the story of [PIN(+)]. *Cell* 2001;**106**:171–182.

Derkatch IL, Uptain SM, Outeiro TF, Krishnan R, Lindquist SL, Liebman SW. Effects of Q/N-rich, polyQ, and non-polyQ amyloids on the de novo formation of the [PSI+] prion in yeast and aggregation of Sup35 *in vitro*. *Proc Natl Acad Sci U S A* 2004; **101**:12934–12939.

DiFiglia M, Sapp E, Chase KO, Davies SW, Bates GP, Vonsattel JP, Aronin N. Aggregation of huntingtin in neuronal intranuclear inclusions and dystrophic neurites in brain. *Science* 1997;**277**:1990–1993.

Douglas PM, Summers DW, Cyr DM. Molecular chaperones antagonize proteotoxicity by differentially modulating protein aggregation pathways. *Prion* 2009a;**3**:51–58.

Douglas PM, Summers DW, Ren HY, Cyr DM. Reciprocal efficiency of RNQ1 and polyglutamine detoxification in the cytosol and nucleus. *Mol Biol Cell* 2009b;**20**: 4162–4173.

Douglas PM, Treusch S, Ren HY, Halfmann R, Duennwald ML, Lindquist S, Cyr DM. Chaperone-dependent amyloid assembly protects cells from prion toxicity. *Proc Natl Acad Sci U S A* 2008;**105**:7206–7211.

Du Z, Park KW, Yu H, Fan Q, Li L. Newly identified prion linked to the chromatin-remodeling factor Swi1 in Saccharomyces cerevisiae. *Nat Genet* 2008;**40**:460–465.

Fan CY, Lee S, Ren HY, Cyr DM. Exchangeable chaperone modules contribute to specification of type I and type II Hsp40 cellular function. *Mol Biol Cell* 2004;**15**:761–773.

Fiala JC. Mechanisms of amyloid plaque pathogenesis. *Acta Neuropathol* 2007;**114**: 551–571.

Fowler DM, Koulov AV, Alory-Jost C, Marks MS, Balch WE, Kelly JW. Functional amyloid formation within mammalian tissue. *PLoS Biol* 2006;**4**:e6.

Glabe CG. Structural classification of toxic amyloid oligomers. *J Biol Chem* 2008;**283**: 29639–29643.

Haass C, Selkoe DJ. Soluble protein oligomers in neurodegeneration: lessons from the Alzheimer's amyloid beta-peptide. *Nat Rev Mol Cell Biol* 2007;**8**:101–112.

Halfmann R, Alberti S, Lindquist S. Prions, protein homeostasis, and phenotypic diversity. *Trends Cell Biol* 2010;**20**:125–133.

Hammer ND, Schmidt JC, Chapman MR. The curli nucleator protein, CsgB, contains an amyloidogenic domain that directs CsgA polymerization. *Proc Natl Acad Sci U S A* 2007;**104**:12494–12499.

Hartl FU, Hayer-Hartl M. Molecular chaperones in the cytosol: from nascent chain to folded protein. *Science* 2002;**295**:1852–1858.

Hartley DM, Walsh DM, Ye CP, Diehl T, Vasquez S, Vassilev PM, Teplow DB, Selkoe DJ. Protofibrillar intermediates of amyloid beta-protein induce acute electrophysiological changes and progressive neurotoxicity in cortical neurons. *J Neurosci* 1999;**19**:8876–8884.

Higurashi T, Hines JK, Sahi C, Aron R, Craig EA. Specificity of the J-protein Sis1 in the propagation of 3 yeast prions. *Proc Natl Acad Sci U S A* 2008;**105**:16596–16601.

Jana NR, Tanaka M, Wang G, Nukina N. Polyglutamine length-dependent interaction of Hsp40 and Hsp70 family chaperones with truncated N-terminal huntingtin: their role in suppression of aggregation and cellular toxicity. *Hum Mol Genet* 2000;**9**:2009–2018.

Jarrett JT, Lansbury PT Jr. Seeding "one-dimensional crystallization" of amyloid: a pathogenic mechanism in Alzheimer's disease and scrapie? *Cell* 1993;**73**:1055–1058.

Kayed R, Head E, Thompson JL, McIntire TM, Milton SC, Cotman CW, Glabe CG. Common structure of soluble amyloid oligomers implies common mechanism of pathogenesis. *Science* 2003;**300**:486–489.

Kazemi-Esfarjani P, Benzer S. Genetic suppression of polyglutamine toxicity in Drosophila. *Science* 2000;**287**:1837–1840.

Kryndushkin DS, Alexandrov IM, Ter-Avanesyan MD, Kushnirov VV. Yeast [PSI+] prion aggregates are formed by small Sup35 polymers fragmented by Hsp104. *J Biol Chem* 2003;**278**:49636–49643.

Kummer MP, Maruyama H, Huelsmann C, Baches S, Weggen S, Koo EH. Formation of Pmel17 amyloid is regulated by juxtamembrane metalloproteinase cleavage, and the resulting C-terminal fragment is a substrate for gamma-secretase. *J Biol Chem* 2009;**284**:2296–2306.

Kushnirov VV, Kryndushkin DS, Boguta M, Smirnov VN, Ter-Avanesyan MD. Chaperones that cure yeast artificial [PSI+] and their prion-specific effects. *Curr Biol* 2000; **10**:1443–1446.

Lambert MP, Barlow AK, Chromy BA, Edwards C, Freed R, Liosatos M, Morgan TE, Rozovsky I, Trommer B, Viola KL, *et al.* Diffusible, nonfibrillar ligands derived from Abeta1-42 are potent central nervous system neurotoxins. *Proc Natl Acad Sci U S A* 1998;**95**:6448–6453.

Langer T, Lu C, Echols H, Flanagan J, Hayer MK, Hartl FU. Successive action of DnaK, DnaJ and GroEL along the pathway of chaperone-mediated protein folding. *Nature* 1992;**356**:683–689.

Lian HY, Zhang H, Zhang ZR, Loovers HM, Jones GW, Rowling PJ, Itzhaki LS, Zhou JM, Perrett S. Hsp40 interacts directly with the native state of the yeast prion protein Ure2 and inhibits formation of amyloid-like fibrils. *J Biol Chem* 2007;**282**:11931–11940.

Lopez N, Aron R, Craig EA. Specificity of class II Hsp40 Sis1 in maintenance of yeast prion [RNQ+]. *Mol Biol Cell* 2003;**14**:1172–1181.

Lu Z, Cyr DM. Protein folding activity of Hsp70 is modified differentially by the hsp40 co-chaperones Sis1 and Ydj1. *J Biol Chem* 1998;**273**:27824–27830.

Lue LF, Kuo YM, Roher AE, Brachova L, Shen Y, Sue L, Beach T, Kurth JH, Rydel RE, Rogers J. Soluble amyloid beta peptide concentration as a predictor of synaptic change in Alzheimer's disease. *Am J Pathol* 1999;**155**:853–862.

Maji SK, Perrin MH, Sawaya MR, Jessberger S, Vadodaria K, Rissman RA, Singru PS, Nilsson KP, Simon R, Schubert D, *et al.* Functional amyloids as natural storage of peptide hormones in pituitary secretory granules. *Science* 2009;**325**:328–332.

McLean CA, Cherny RA, Fraser FW, Fuller SJ, Smith MJ, Beyreuther K, Bush AI, Masters CL. Soluble pool of Abeta amyloid as a determinant of severity of neurodegeneration in Alzheimer's disease. *Ann Neurol* 1999;**46**:860–866.

Moriyama H, Edskes HK, Wickner RB. [URE3] prion propagation in Saccharomyces cerevisiae: requirement for chaperone Hsp104 and curing by overexpressed chaperone Ydj1p. *Mol Cell Biol* 2000;**20**:8916–8922.

Muchowski PJ, Wacker JL. Modulation of neurodegeneration by molecular chaperones. *Nat Rev Neurosci* 2005;**6**:11–22.

Nelson R, Sawaya MR, Balbirnie M, Madsen AO, Riekel C, Grothe R, Eisenberg D. Structure of the cross-beta spine of amyloid-like fibrils. *Nature* 2005;**435**:773–778.

Nenninger AA, Robinson LS, Hultgren SJ. Localized and efficient curli nucleation requires the chaperone-like amyloid assembly protein CsgF. *Proc Natl Acad Sci U S A* 2009; **106**:900–905.

Patel BK, Gavin-Smyth J, Liebman SW. The yeast global transcriptional co-repressor protein Cyc8 can propagate as a prion. *Nat Cell Biol* 2009;**11**:344–349.

Qiu XB, Shao YM, Miao S, Wang L. The diversity of the DnaJ/Hsp40 family, the crucial partners for Hsp70 chaperones. *Cell Mol Life Sci* 2006;**63**:2560–2570.

Roher AE, Chaney MO, Kuo YM, Webster SD, Stine WB, Haverkamp LJ, Woods AS, Cotter RJ, Tuohy JM, Krafft GA, *et al.* Morphology and toxicity of Abeta-(1-42) dimer derived from neuritic and vascular amyloid deposits of Alzheimer's disease. *J Biol Chem* 1996;**271**:20631–20635.

Savistchenko J, Krzewska J, Fay N, Melki R. Molecular chaperones and the assembly of the prion URE2p *in vitro*. *J Biol Chem* 2008;**283**:15732–15739.

Sawaya MR, Sambashivan S, Nelson R, Ivanova MI, Sievers SA, Apostol MI, Thompson MJ, Balbirnie M, Wiltzius JJ, McFarlane HT, *et al.* Atomic structures of amyloid cross-beta spines reveal varied steric zippers. *Nature* 2007;**447**:453–457.

Scherzinger E, Lurz R, Turmaine M, Mangiarini L, Hollenbach B, Hasenbank R, Bates GP, Davies SW, Lehrach H, Wanker EE. Huntingtin-encoded polyglutamine expansions form amyloid-like protein aggregates *in vitro* and *in vivo*. *Cell* 1997;**90**:549–558.

Shorter J, Lindquist S. Prions as adaptive conduits of memory and inheritance. *Nat Rev Genet* 2005;**6**:435–450.

Si K, Choi YB, White-Grindley E, Majumdar A, Kandel ER. Aplysia CPEB can form prion-like multimers in sensory neurons that contribute to long-term facilitation. *Cell* 2010;**140**:421–435.

Si K, Giustetto M, Etkin A, Hsu R, Janisiewicz AM, Miniaci MC, Kim JH, Zhu H, Kandel ER. A neuronal isoform of CPEB regulates local protein synthesis and stabilizes synapse-specific long-term facilitation in aplysia. *Cell* 2003a;**115**:893–904.

Si K, Lindquist S, Kandel ER. A neuronal isoform of the aplysia CPEB has prion-like properties. *Cell* 2003b;**115**:879–891.

Sipe JD, Cohen AS. Review: history of the amyloid fibril. *J Struct Biol* 2000;**130**:88–98.

Sondheimer N, Lindquist S. Rnq1: an epigenetic modifier of protein function in yeast. *Mol Cell* 2000;**5**:163–172.

Sondheimer N, Lopez N, Craig EA, Lindquist S. The role of Sis1 in the maintenance of the [RNQ+] prion. *EMBO J* 2001;**20**:2435–2442.

Summers DW, Douglas PM, Ramos CH, Cyr DM. Polypeptide transfer from Hsp40 to Hsp70 molecular chaperones. *Trends Biochem Sci* 2009a;**34**:230–233.

Summers DW, Douglas PM, Ren HY, Cyr DM. The type I Hsp40 Ydj1 utilizes a farnesyl moiety and zinc finger-like region to suppress prion toxicity. *J Biol Chem* 2009b;**284**:3628–3639.

Szabo A, Langer T, Schroder H, Flanagan J, Bukau B, Hartl FU. The ATP hydrolysis-dependent reaction cycle of the Escherichia coli Hsp70 system DnaK, DnaJ, and GrpE. *Proc Natl Acad Sci U S A* 1994;**91**:10345–10349.

Tipton KA, Verges KJ, Weissman JS. *In Vivo* monitoring of the prion replication cycle reveals a critical role for Sis1 in delivering substrates to Hsp104. *Mol Cell* 2008;**32**:584–591.

True HL, Berlin I, Lindquist SL. Epigenetic regulation of translation reveals hidden genetic variation to produce complex traits. *Nature* 2004;**431**:184–187.

True HL, Lindquist SL. A yeast prion provides a mechanism for genetic variation and phenotypic diversity. *Nature* 2000;**407**:477–483.

Tuite MF, Cox BS. Propagation of yeast prions. *Nat Rev Mol Cell Biol* 2003;**4**:878–890.

Tuite MF, Lindquist SL. Maintenance and inheritance of yeast prions. *Trends Genet* 1996;**12**:467–471.

Vitrenko YA, Gracheva EO, Richmond JE, Liebman SW. Visualization of aggregation of the Rnq1 prion domain and cross-seeding interactions with Sup35NM. *J Biol Chem* 2007;**282**:1779–1787.

Wacker JL, Zareie MH, Fong H, Sarikaya M, Muchowski PJ. Hsp70 and Hsp40 attenuate formation of spherical and annular polyglutamine oligomers by partitioning monomer. *Nat Struct Mol Biol* 2004;**11**:1215–1222.

Walsh P, Bursac D, Law YC, Cyr D, Lithgow T. The J-protein family: modulating protein assembly, disassembly and translocation. *EMBO Rep* 2004;**5**:567–571.

Wang X, Hammer ND, Chapman MR. The molecular basis of functional bacterial amyloid polymerization and nucleation. *J Biol Chem* 2008;**283**:21530–21539.

Warrick JM, Chan HY, Gray-Board GL, Chai Y, Paulson HL, Bonini NM. Suppression of polyglutamine-mediated neurodegeneration in Drosophila by the molecular chaperone HSP70. *Nat Genet* 1999;**23**:425–428.

Wickner RB, Edskes HK, Roberts BT, Baxa U, Pierce MM, Ross ED, Brachmann A. Prions: proteins as genes and infectious entities. *Genes Dev* 2004;**18**:470–485.

Wyttenbach A, Carmichael J, Swartz J, Furlong RA, Narain Y, Rankin J, Rubinsztein DC. Effects of heat shock, heat shock protein 40 (HDJ-2), and proteasome inhibition on protein aggregation in cellular models of Huntington's disease. *Proc Natl Acad Sci U S A* 2000;**97**:2898–2903.

Zoghbi HY, Orr HT. Glutamine repeats and neurodegeneration. *Annu Rev Neurosci* 2000;**23**:217–247.

9

MODULATION OF AMYLOID PROPAGATION IN YEAST BY Hsp70 AND ITS REGULATORS AND CHAPERONE PARTNERS

Daniel C. Masison

National Institute of Diabetes and Digestive and Kidney Diseases, National Institutes of Health, Bethesda, MD, USA

9.1 BACKGROUND

Protein misfolding is the basis of a large number of diseases, and therefore it presents a considerable clinical problem. Protein conformational disorders can be classified by the association of pathology with (i) accumulation of protein aggregates, as in the amyloidoses, which include Alzheimer's disease, type 2 diabetes, transmissible spongiform encephalopathies (prion diseases), and a number of neurological disorders caused by proteins with expanded polyglutamine tracts; (ii) improper protein trafficking, as in familial hypercholesterolemia and Tay–Sachs disease; or (iii) the inability of proteins to fold into native conformations, as in cystic fibrosis (CF) and amyotrophic lateral sclerosis (ALS) (Thomas *et al.*, 1995). Impaired function or expression of Hsp70, a protein chaperone that helps other proteins to adopt and maintain their native structures, can exacerbate

Protein Chaperones and Protection from Neurodegenerative Diseases, First Edition.
Edited by Stephan N. Witt.

pathology in protein conformational disorders, whereas elevating the expression of Hsp70 reduces the toxic effects of protein misfolding in models of protein folding disease (Chai *et al.*, 1999; Cummings *et al.*, 2001; Gokhale *et al.*, 2005; Klucken *et al.*, 2004; Wacker *et al.*, 2009; Warrick *et al.*, 1999).

A large number of neurological diseases are associated with intra- or extracellular accumulation of amyloid, which is a highly ordered, stable, and insoluble fibrous aggregate of a single type of protein. Amyloid propagates by recruiting the soluble form of the protein and converting it into the amyloid conformation as it joins the fiber ends. Although the many proteins that accumulate as amyloid in these diseases have widely varying amino acid composition and tertiary structures, the overall structure, physical properties, and kinetics of assembly of amyloid fibers formed by them is remarkably similar. This similarity suggests that they share a common protein folding defect and possibly pathogenesis (Prusiner, 2010). As such, factors that moderate pathology caused by one type of protein could have a wide application. The mechanisms underlying pathology in the amyloidoses are uncertain, but a shared view poses that off-pathway folding intermediates or small soluble oligomers produced during amyloid formation contribute to toxicity, whereas larger aggregates and amyloid fibers might be inert or even protective (Douglas *et al.*, 2008; Ferreira *et al.*, 2007). An important discovery providing early support for the notion of a cytoprotective effect of amyloid formation was that Hsp70 reduces pathology without eliminating the accumulation of the protein into large aggregates (Warrick *et al.*, 1999). This finding suggests that Hsp70 might protect cells from amyloidogenic proteins by dismantling smaller toxic aggregates or even promoting their conversion into less toxic higher order aggregates.

The yeast *Saccharomyces cerevisiae* has been a useful cell and genetic model system for studying aggregation and cytotoxicity of human amyloidogenic proteins, such as A-β peptide (Aβ), α-synuclein, and various polyglutamine constructs, and for identifying specific regions of these proteins that contribute to their toxicity (Bagriantsev and Liebman, 2006; Krobitsch and Lindquist, 2000; Li and Harris, 2005; Outeiro and Muchowski, 2004; Willingham *et al.*, 2003; Winderickx *et al.*, 2008). Expression of green fluorescent protein (GFP) fusions of some of these aggregation-prone proteins in yeast has been useful for monitoring aggregation status and other cytological effects (Duennwald *et al.*, 2006; Krobitsch and Lindquist, 2000; Muchowski *et al.*, 2000; Soper *et al.*, 2008). The yeast system has also been proved to be a useful screening tool to identify cellular factors and processes, such as protein chaperones and actin dynamics, that influence aggregation and toxicity of amyloidogenic proteins (Duennwald *et al.*, 2006; Ganusova *et al.*, 2006; Liang *et al.*, 2008; Meriin *et al.*, 2002, 2007; Willingham *et al.*, 2003). Small compounds that modify toxic effects of such proteins were identified using yeast, and compounds identified on the basis of their inhibitory effects on yeast prions also reduce the pathology of mammalian prions (Bach *et al.*, 2006; Tribouillard-Tanvier *et al.*, 2008).

Yeast prions are cellular proteins that have a particularly high propensity to misfold in a way that leads to the formation of amyloid. The most studied yeast prions [PSI^+], [URE3], and [PIN^+]/[RNQ^+] are widely believed to propagate as amyloid forms of the proteins Sup35, Ure2, and Rnq1, respectively (Derkatch *et al.*, 1997; Sondheimer and Lindquist, 2000; Wickner, 1994). These proteins spontaneously form amyloid when purified. In convincing support of amyloid being the infectious material, transfection of yeast cells with amyloid prepared *in vitro* from purified Sup35, Ure2, and Rnq1 produces stably propagating prions in infected cells (Brachmann *et al.*, 2005; King and Diaz-Avalos, 2004; Patel and Liebman, 2007; Tanaka *et al.*, 2004).

In order for prions to propagate stably in a growing yeast population, they must grow (by addition of monomers to the ends of prion polymers) (Collins *et al.*, 2004; Inoue *et al.*, 2001; Satpute-Krishnan and Serio, 2005; Scheibel *et al.*, 2001), replicate (by fragmentation of prion polymers) (Chernoff *et al.*, 1995; Kryndushkin *et al.*, 2003; Paushkin *et al.*, 1996), and be transmitted efficiently to daughter cells (Figure 9.1a). Transmission of prions between dividing and fusing (mating) cells is widely thought to occur through passive diffusion—an inference supported by microscopic evidence (Byrne *et al.*, 2009; Satpute-Krishnan and Serio, 2005). The very stable inheritance of yeast prions in a growing population reflects the normally high number of prions per cell (Byrne *et al.*, 2009). The spontaneous formation of amyloid by purified Sup35, Ure2, and Rnq1 suggests that chaperone activity might not be required for the initial formation and growth of prions *in vivo*.

Different combinations of chaperones, however, can enhance or repress amyloid formation of purified prion proteins (Krzewska and Melki, 2006; Lian *et al.*, 2007; Savistchenko *et al.*, 2008; Shorter and Lindquist, 2004, 2008) suggesting that they have the potential to influence the appearance of prions *in vivo* in both positive and negative ways. Amyloid formation *in vitro* can be accelerated by agitation, which causes fibers to break, producing more templates to recruit the soluble protein. The protein disaggregating molecular chaperone Hsp104, which functions with Hsp40 and Hsp70, can also break Sup35 polymers and accelerate the formation and assembly of Sup35 amyloid *in vitro*, and in certain reaction conditions, the smaller pieces have enhanced infectivity when compared with the untreated polymers (Inoue *et al.*, 2004; Krzewska and Melki, 2006; Shorter and Lindquist, 2004, 2006). Replication of the amyloidogenic yeast prions *in vivo* is thought to occur by fragmentation of prion polymers into more numerous independently propagating pieces. The importance of Hsp104 for this fragmentation (Figure 9.1b) is demonstrated by data showing that prion replication is arrested when Hsp104 function is depleted or inactivated (Chernoff *et al.*, 1995; Eaglestone *et al.*, 2000; Ferreira *et al.*, 2001; Grimminger *et al.*, 2004; Jung and Masison, 2001; Ness *et al.*, 2002; Paushkin *et al.*, 1996).

After the initial discovery that the protein chaperone Hsp104 is a major factor required for prion propagation (Chernoff *et al.*, 1995), it was found that altering the function or abundance of several other chaperones or co-chaperones can influence the propagation of different yeast prions in different ways (Chernoff *et al.*, 1999; Jones *et al.*, 2004; Jung *et al.*, 2000; Kryndushkin and Wickner, 2007; Kryndushkin *et al.*, 2002; Kushnirov *et al.*, 2000; Moriyama *et al.*, 2000;

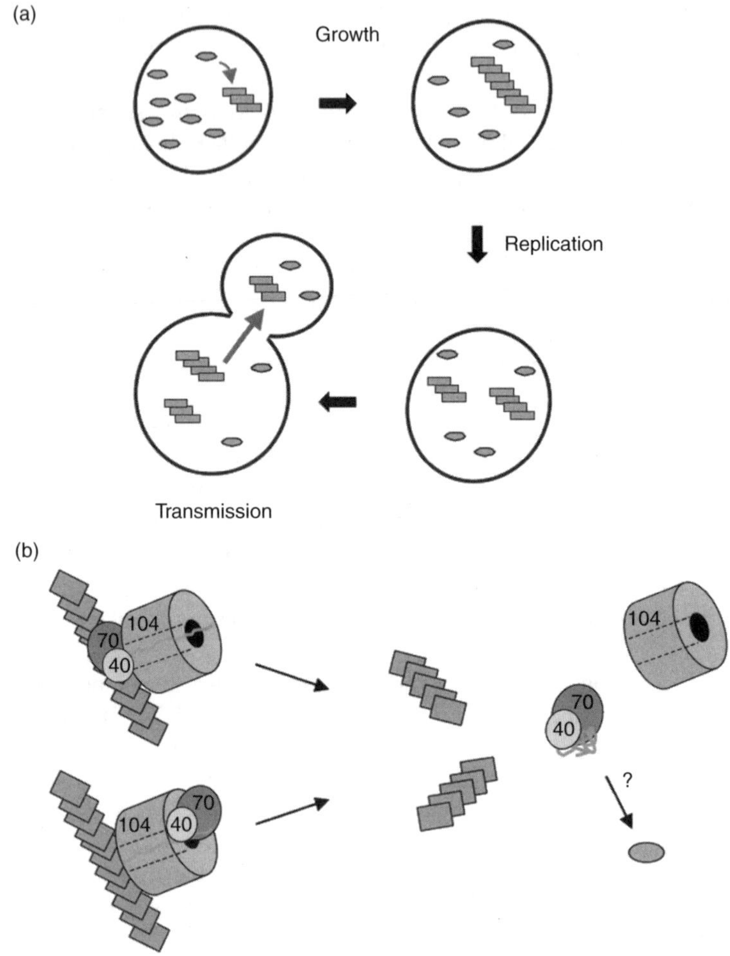

Figure 9.1 Prion propagation *in vivo* requires growth, replication, and transmission of prions. (a) Prions are formed of normal cellular proteins (ovals) that misfold and assemble into amyloid (stacked rectangles), which is a stable highly structured fibrous aggregate held together through intra- and intermolecular hydrogen bonding of β-sheets. Molecular details underlying the initial appearance of prions are uncertain, but once present, prions continue to grow (upper left) by recruiting the soluble form of the protein and converting it into the prion form as it joins the polymer. Replication of yeast prions requires the protein disaggregation activity of the Hsp104 protein chaperone (see panel b) and occurs by fragmentation of polymers into more numerous pieces. The defining characteristic of prions is their infectivity or ability to be transmitted between strains. Transmission of yeast prions occurs vertically from mother to daughter during cell division (lower left), and horizontally through fusion of cells of different strains (mating). (b) One hypothesis for how Hsp104 machinery action fragments prions, which generates new prions from preexisting ones. Hsp40/70 interact with amyloid polymer (upper left) or at exit site of Hsp104 hexamer (lower left), or both (not shown) to facilitate extrusion of monomer from the fiber. This action destabilizes polymers causing them to break into two fragments, each of which can continue to propagate the prion state. It is uncertain if the monomer is completely extruded or if Hsp40/70 assists refolding of extruded protein.

Roberts *et al.*, 2004; Sondheimer *et al.*, 2001). The effects on prion strength and stability phenotypes correlate with changes in the typical numbers of prions per cell and the relative amount of the insoluble prion form, which reflect the influence of chaperones on the growth, replication, aggregation, and transmission of yeast prions (Figure 9.2).

Although the molecular mechanisms by which these chaperones and co-chaperones cause their effects on prions are uncertain, evidence suggests that the effects several co-chaperones have on prion propagation are indirect through their regulation of Hsp70 activity. Hsp70 and its co-chaperones copurify

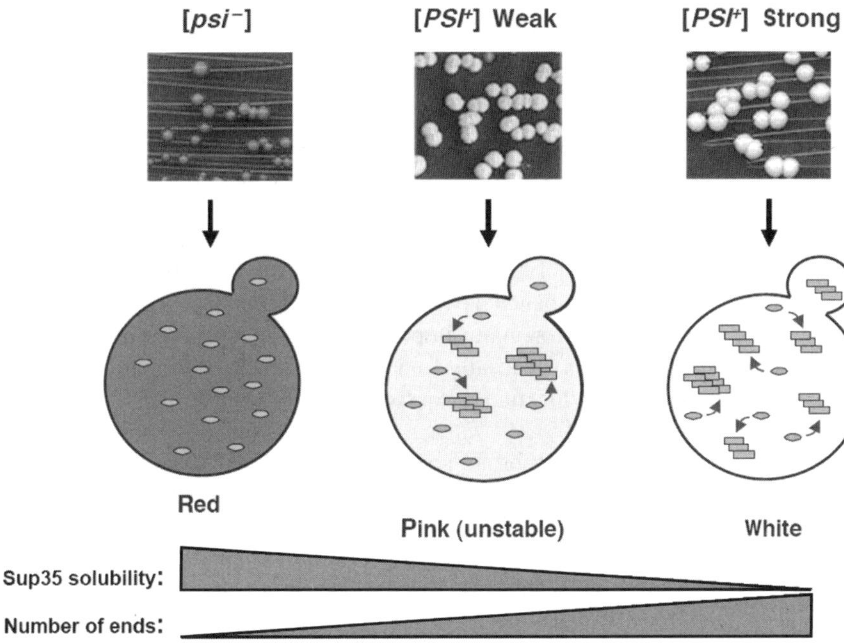

Figure 9.2 Differences in [*PSI*⁺] phenotype can be accounted for by differences in rates of growth, fragmentation, and/or transmission of prion polymers. [*PSI*⁺] is a prion of the Sup35 translation termination factor. When [*PSI*⁺] is present, Sup35 is depleted into prion aggregates. The resulting decrease in translation termination efficiency causes certain mutants of yeast to turn from red (nonsuppressed) to white (suppressed) on an indicator medium (upper panels). The "strength" of prion propagation (indicated at top) is inversely proportional to the degree of red pigmentation, which is directly related to the amount of soluble Sup35. Cells with a normal strong phenotype (right) have many prion polymers that titrate Sup35 by recruiting it to their ends. The center panel shows the weak unstable prion phenotype of the *SSA1-21* mutant (see text). In *SSA1-21* cells, individual prion polymers more readily coalesce into higher order aggregates, which reduces the number of both functional polymer ends and free polymers. Because there are fewer ends to deplete Sup35, more Sup35 remains soluble and cells accumulate some pigment. The reduced number of individual polymers reduces the efficiency with which progeny will inherit prions and gives rise to a detectable frequency of [*psi*⁻] cells (seen as red colonies) in the growing population. (*A full color version of this figure appears in the color plate section.*)

preferentially with aggregated forms of prion protein from yeast cells (Allen *et al.*, 2005; Bagriantsev *et al.*, 2008). Together, these data suggest that Hsp70 acts directly on prions *in vivo*. However, since Hsp70 is a critical component of several chaperone machines, and it has a role in many important cellular processes, it is still unclear if Hsp70 also influences yeast prions indirectly through its effects on chaperone machineries or cellular processes that prions might depend upon for their propagation.

9.2 Hsp70

9.2.1 Hsp70 Has Conserved Structure and Function

Hsp70 is an essential molecular chaperone whose expression is elevated by the exposure of cells to a variety of stressful conditions that cause proteins to become misfolded. Hsp70 acts by binding and releasing short stretches of hydrophobic peptides (Gragerov *et al.*, 1994; Richarme and Kohiyama, 1993). This binding to exposed hydrophobic surfaces on partially unfolded proteins protects cells by preventing nonproductive hydrophobic interactions that can lead to protein aggregation (Schroder *et al.*, 1993), and provides an opportunity for the bound protein to refold properly when it is released. This reversible interaction with polypeptides is also crucial for many important cellular processes in which the proteins being handled are in a partially folded or extended conformation, such as during translation and translocation across membranes (Sousa and Lafer, 2006; Wegrzyn and Deuerling, 2005).

Hsp70 binds substrates with broad specificity and its substrate-binding cycle is regulated by adenosine triphosphate (ATP) hydrolysis and nucleotide exchange (Figure 9.3). Affinity of Hsp70 for a substrate depends on the bound nucleotide. When ATP is bound in its amino-terminal ATPase domain (nucleotide-binding domain, NBD), Hsp70 binds and releases the substrate rapidly. ATP hydrolysis induces a conformational change in the adjacent substrate-binding domain (SBD) and the C-terminal region that stabilizes the interaction by trapping an exposed hydrophobic portion of the unfolded substrate protein in the peptide-binding pocket. Dissociation of adenosine diphosphate (ADP) and subsequent rebinding of ATP restores the low-affinity state and facilitates release of the substrate, thereby completing the cycle. This simple reaction cycle allows Hsp70 to be exceptionally effective at protecting cells from conditions that cause protein destabilization, and the requirement of this activity for many important cellular processes makes Hsp70 essential for cell growth even under optimal conditions.

Because of its important role in many fundamental housekeeping processes, it is not surprising that Hsp70 is highly conserved throughout evolution. Hsp70s from species as diverse as bacteria and mammals share 50% amino acid identity and their three-dimensional structures are essentially superimposable (Flaherty *et al.*, 1990; Harrison *et al.*, 1997; Zhu *et al.*, 1996). Functional conservation of Hsp70 within and across species has also been demonstrated. Any of the four

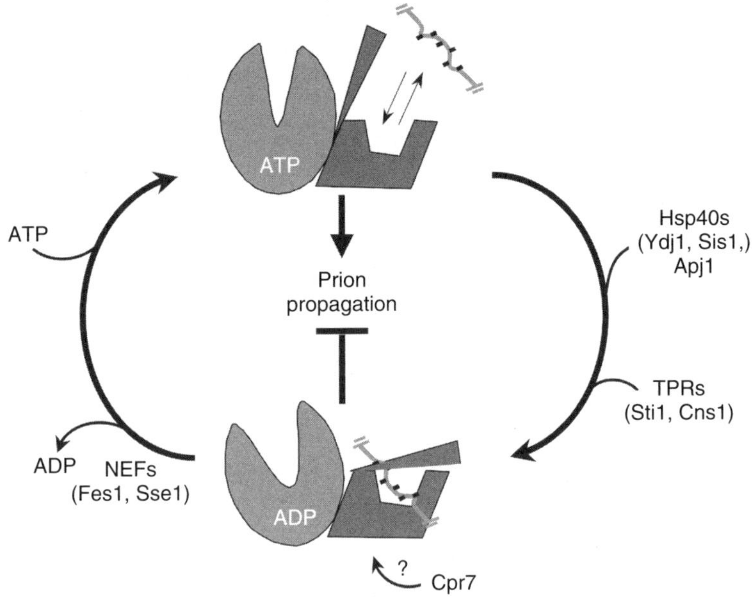

Figure 9.3 Hsp70 reaction cycle. The ATPase domain is depicted in light gray and the substrate-binding domain (SBD) with adjacent "lid" is given in dark gray. The substrate is depicted as an exposed hydrophobic segment (knobbed line) of a larger protein. When ATP is bound, the SBD is open and exchanges substrate rapidly (small arrows). Hydrolysis of ATP, which is stimulated by the indicated co-chaperones on the right, induces a conformational change that closes the lid over the SBD and traps the substrate. Release of ADP, which is stimulated by nucleotide exchange factors (NEFs), allows rebinding of ATP and return to the low-affinity state to facilitate substrate release. Conditions that promote or prolong occupancy of the ADP-bound state, such as enhanced co-chaperone stimulation of ATP hydrolysis or inhibition of NEF activity, inhibit prion propagation (indicated by the bar). Conversely, conditions that promote the ATP-bound state enhance prion propagation (downward arrow). Available evidence suggests an uncertain role? for Cpr7 in stabilizing the ADP-bound state. See text for details.

S. cerevisiae Ssa cytosolic Hsp70s support growth as the sole isoform, *Drosophila* Hsp70 can protect mammalian cells from heat stress, and individual primate, plant, and fungal Hsp70s support growth and prion propagation in *S. cerevisiae* (Pelham *et al.*, 1984; Sharma *et al.*, 2009a; Tutar *et al.*, 2006; Werner-Washburne *et al.*, 1987).

Most organisms encode multiple highly homologous Hsp70 isoforms that are expressed in the cytoplasm and nucleus. In addition to these isoforms, of which some are expressed only under conditions of stress or in specific tissues, Hsp70s are found in the endoplasmic reticulum, mitochondria, and specialized organelles such as chloroplasts. Hsp70s have a common reaction cycle and even widely diverged isoforms share enough functional overlap to complement essential Hsp70 functions (Tutar *et al.*, 2006). In contrast, highly homologous constitutively expressed isoforms can be assigned to subfamilies based on functional distinctions. For example, although the yeast cytosolic Ssa and Ssb subfamilies are about 60% identical and overlap functionally, Ssb proteins cannot provide essential Ssa functions required for cell growth, and Ssa proteins do not complement Ssb function in protein synthesis (Craig *et al.*, 1995; James *et al.*, 1997).

Even nearly identical Hsp70 isoforms within species, such as the 98% identical constitutively expressed Ssa1 and Ssa2, have been found to possess distinct activities (Brown *et al.*, 2000; Roberts *et al.*, 2004; Schwimmer and Masison, 2002; Sharma *et al.*, 2009a; Sharma and Masison, 2008). Similarly, the inducible Hsp70s, Ssa3 and Ssa4, display functional distinctions not only from their constitutively expressed counterparts but also from each other (Sharma *et al.*, 2009a; Sharma and Masison, 2008; Tutar *et al.*, 2006). These observations point to the possibility that inducible isoforms have evolved activities that are particularly helpful under the conditions that induce their expression, rather than simply providing a way to adjust Hsp70 abundance through regulated expression. They also suggest that even highly homologous and functionally redundant Hsp70s have diverged to play specialized roles in the cell under the same environmental conditions. The simultaneous expression of different Hsp70 isoforms therefore allows a broader range and finer tuning of Hsp70 activity for different tasks. Although the mechanistic bases for these differences has yet to be worked out in detail, subtle differences in Hsp70 structure among isoforms could influence intrinsic Hsp70 activities by significantly affecting physical or functional interactions with cofactors that regulate Hsp70 functions in different ways or in particular cellular processes. Different Hsp70 isoforms can influence cell proliferation in a cancer model and whether different Hsp70 isoforms vary in their ability to moderate pathology in models of protein folding disorders remains an interesting and important unanswered question (Daugaard *et al.*, 2007; Rohde *et al.*, 2005).

9.2.2 Hsp70 is Regulated by Co-Chaperones and is a Critical Component of Various Chaperone Machines

The intrinsic ATP hydrolysis and nucleotide exchange activities of Hsp70 are very inefficient, with an ATP turnover rate of about 0.02–0.1 min^{-1} (Ha *et al.*, 1999), and many co-chaperones have evolved to regulate these steps of the cycle and control Hsp70 function (Meimaridou *et al.*, 2009; Szabo *et al.*, 1994; Young *et al.*, 2004) (Figure 9.3). Hsp40s physically interact with Hsp70 to stimulate Hsp70 ATPase activity, and can bind and present substrate to Hsp70. The major cytosolic Hsp40s in yeast are Ydj1, of which a portion is membrane associated (Caplan and Douglas, 1991; Caplan *et al.*, 1992), and the essential Sis1 (Luke *et al.*, 1991). Nucleotide exchange factors (NEFs) accelerate the dissociation of ADP from Hsp70 and thus the transition to the low-affinity ATP-bound state, which facilitates release of the substrate. These cofactors provide additional layers of regulation and specificity to Hsp70 function.

Hsp70 plays a major role in stress protection in microorganisms and plants by cooperating with Hsp100 AAA+ chaperones (Hsp104 in yeast), which act as the major component of a protein disaggregation machine that resolubilizes proteins from aggregates. These chaperones assemble into ring-shaped hexamers that extrude proteins through an axial channel as they extract them from aggregates (Lum *et al.*, 2004; Schlieker *et al.*, 2004; Weibezahn *et al.*, 2004). This disaggregation machinery includes Hsp70 and Hsp40, which appear to recruit the Hsp100 chaperone to the aggregates and facilitate initial extraction of protein monomers (Acebron *et al.*, 2009; Glover and Lindquist, 1998; Goloubinoff *et al.*, 1999; Zietkiewicz *et al.*, 2004, 2006). Hsp70 and Hsp40 could also act at the exit site of the axial channel to aid this process and the refolding of extracted protein (Lum *et al.*, 2004) (Figure 9.1b; also see Chapter 7).

Hsp70 also cooperates with the Hsp90 machinery through interactions with a subset of co-chaperones that contain tetratricopeptide repeat (TPR) domains, which bind conserved residues (EEVD) at the C-terminus of Hsp70 and Hsp90 (Scheufler *et al.*, 2000; Sikorski *et al.*, 1990). Hsp90 is a highly abundant essential protein chaperone that is the central player in a well-characterized folding pathway for a diverse array of "client" proteins such as transcription factors and kinases that directly impact a variety of cellular processes including signaling (Dezwaan and Freeman, 2008; Smith, 2004; Wandinger *et al.*, 2008). Hsp40/70 acts upstream in this pathway by binding substrates and then transferring them to Hsp90. The TPR protein Sti1 facilitates this transfer by simultaneously binding Hsp70 and Hsp90 with distinct TPR domains and physically bridging these chaperones (Chen and Smith, 1998; Song and Masison, 2005). Hsp90 then interacts with other TPR-containing co-chaperones, which replace the Sti1/Hsp40/70 ternary complex after the substrate is transferred. The yeast immunophilin Cpr7, an ortholog of proteins that bind and are inactivated by immunosuppressants like cyclosporin A, and the functionally overlapping Cns1, which was isolated as a multicopy suppressor of growth defects caused by deletion of Cpr7 (Marsh *et al.*, 1998), act at this stage. Sti1, Cpr7, and Cns1 can each regulate Hsp70, independently of Hsp90, in a way that influences [*PSI*$^+$] propagation (Jones *et al.*, 2004; Song and Masison, 2005). TPR proteins on membranes of mitochondria and endoplasmic reticulum act as components of the translocation machinery to both recruit and regulate Hsp70 and Hsp90 for protein transport (Schlegel *et al.*, 2007; Wu and Sha, 2006; Young *et al.*, 2003). Hsp40 and Hsp70 cooperate with translocation complexes on both sides of membranes to facilitate transport of proteins between subcellular compartments.

9.3 Hsp70 AND YEAST PRIONS

9.3.1 Hsp70 Influences Hsp104 Effects on Prions

Before expanding on how Hsp70 affects yeast prions, more background on Hsp104 is warranted because of its primary role in yeast prion propagation and its cooperation with Hsp40 and Hsp70. Although Hsp104 is essential for replication of yeast prions, overexpressing Hsp104 "cures" cells of the [*PSI*$^+$] and [MCA] prions very efficiently (Chernoff *et al.*, 1995; Nemecek *et al.*, 2009). Similar treatment has little or no effect on most other prions and this specificity has remained a puzzle. Because the primary role of Hsp104 is to resolubilize proteins from aggregates, it has been thought that elevating Hsp104 eliminated [*PSI*$^{+}$] prions through a mass action effect of fragmenting prion aggregates until no templates remained to propagate the prion form of the protein. Why [*PSI*$^{+}$] and [MCA] would be particularly sensitive to this activity is difficult to explain. Unexpectedly, overexpressing Hsp104 causes the size of sodium dodecyl sulfate (SDS)-resistant [*PSI*$^{+}$] prion polymers to increase, which could be due to faster assembly or reduced fragmentation (before the cells are cured by the excess Hsp104) (Kryndushkin *et al.*, 2002). Either way, this finding is inconsistent with curing being due simply to the elevated disaggregating activity on Sup35 polymers.

In contrast to synthetically induced Hsp104 expression, natural induction of Hsp104 to high levels by exposing cells to elevated temperature (Sanchez and Lindquist, 1990) does not cure cells of [*PSI*$^{+}$] (Singh *et al.*, 1979; Tuite *et al.*, 1981). However, it does weaken [*PSI*$^{+}$] phenotype by reducing the relative amount of insoluble Sup35 (Jung *et al.*, 2000). It was later found that although [*PSI*$^{+}$] was weaker when cells were grown under stress, the average number of prions per cell was not affected (Eaglestone *et al.*, 1999, 2000). Therefore, the stress does not alter prion seed numbers, and thus Hsp104 function in prion replication. The data imply that the increased amount of soluble Sup35 must originate at the ends of prion polymers, because of a higher rate of disassembly or a slower rate of assembly. Recent data suggest that assembly of Sup35 polymers decreases with increased temperature, which is consistent with the latter explanation (Palhano *et al.*, 2009).

The mechanism of how overexpressing Hsp104 causes cells to lose [*PSI*$^{+}$] remains elusive, but this curing provided a basis for testing the influence of other chaperones. The Hsp70 Ssa1 was first shown to influence prions by its ability to counteract the curing of [*PSI*$^{+}$] by overexpressed Hsp104 (Newnam *et al.*, 1999). Earlier explanations for the lack of curing by stress-induced Hsp104 were that stressful conditions increased overall amounts of aggregated proteins that compete with prions as substrate for Hsp104, or that other Hsps induced by the stress moderate Hsp104 activity and prevent it from completely destroying prion templates (Chernoff *et al.*, 1995; Newnam *et al.*, 1999). The finding that co-induction of Ssa1 diminished the Hsp104 curing effect provided evidence for the latter hypothesis and led to the proposal that Hsp104 disassembles prion polymers, whereas Ssa1 promotes their formation and assembly (Newnam *et al.*, 1999).

This conclusion was bolstered by experiments showing that increasing Ssa1 also increases both the size of [*PSI*$^{+}$] prion polymers and the frequency of *de novo* formation of [*PSI*$^{+}$] when Sup35 is overproduced (Allen *et al.*, 2005). Depleting Hsp104 also causes an increase in the size of Sup35 prion aggregates *in vivo* (Wegrzyn *et al.*, 2001). Additional work identified a prion formed of a mutant Sup35 with a truncated prion-determining region that produced larger than normal aggregates, and a prion "strain" composed of normal Sup35 protein that similarly formed large aggregates (Borchsenius *et al.*, 2001, 2006). Stable propagation of both of these prions depended on the increased expression of

Hsp104, which reduced the size of fluorescent aggregates. In contrast, when Ssa1 expression was elevated both prions were further destabilized and formed even larger aggregates. Although Hsp70 and Hsp104 cooperate in protein disaggregation, these findings lent support to the conclusions that Hsp104 and Hsp70 have defined and opposite effects on prion aggregation.

9.3.2 Overexpressing Different Hsp70s Has Different Effects

Overexpression of Ssa2, Ssa3, and Ssa4 was also assessed for effects on [*PSI*$^+$] and each influenced [*PSI*$^+$] similar to Ssa1 (Allen *et al.*, 2005), indicating that the prion-promoting effects of overexpressing Hsp70 were conserved within this subfamily (Table 9.1). In contrast to the Ssa proteins, however, the effects on [*PSI*$^+$] caused by the overexpressing or deleting the Ssb Hsp70 subfamily points to Ssb being an anti-prion Hsp70. Overexpressing Ssb enhances curing by overexpressed Hsp104, whereas deleting Ssb reduces it (Chernoff *et al.*, 1999). Depleting Ssb also increases the frequency of *de novo* appearance of [*PSI*$^+$] prions (Chernoff *et al.*, 1999), and Ssb was identified in a genetic screen for factors that alone cure [*PSI*$^+$] when overexpressed (Chacinska *et al.*, 2001).

Ssb is primarily associated with the translation of ribosomes to facilitate translation (Nelson *et al.*, 1992) and it was found to have a role in protein degradation, which led to the proposal that it has a "proofreading" function that either restores folding of nascent misfolded proteins or directs them to the proteasome (Ohba, 1997). Interpretation of the effects of Ssb on [*PSI*$^+$] invoked this proofreading function as acting to reduce the formation of misfolded Sup35 that can act as intermediates in amyloid formation. Such intermediates are presumed to be produced either during translation or when Sup35 is extracted from prions by Hsp104, or to promote degradation of misfolded Sup35, or both (Chernoff *et al.*, 1999). Although these hypotheses have not yet been tested experimentally, evidence suggests that Sup35 is neither ubiquitinated nor degraded when Hsp104 is overexpressed (Allen *et al.*, 2007). Nevertheless, ubiquitin has been shown to contribute to prion curing by Hsp104 (Chernova *et al.*, 2003) and so its effects are likely indirect. Even though Ssb has a role in protein degradation, the effects of Ssb and ubiquitin on curing are additive, which suggests that they act in different pathways and speaks of the complexity of the curing mechanism.

In experiments to monitor conservation of prion properties across species, others constructed hybrid prions with the conserved but nonidentical Sup35 prion domain of the unrelated yeast *Pichia methanolica* in place of the same region of *S. cerevisiae* Sup35. Unlike [*PSI*$^+$], prions composed of this fusion protein, designated [*PSI*$^+_{PS}$], are curable by overexpression of either Ssa1 or Ssb1 (Kushnirov *et al.*, 2000). Overexpressing the Hsp40 Ydj1 also cures the hybrid prions, although it has no effect on [*PSI*$^+$], and the combination of elevating Ssa1 and Ydj1 has synergistic inhibitory effects. The same study found that combined overexpression of Ssa1 and Ydj1 also cures cells of a weak variant of [*PSI*$^+$], similar to the prion that is protected by Ssa1 when Hsp104 is overexpressed. These findings are consistent with biochemical data that show Ssa1 combined with Hsp40

Table 9.1 Effects of altered chaperone expression on prion phenotypes

Chaperone/ Co-chaperone	Chaperone Abundance	Effect[a]		References
		[*PSI*$^+$]	[*URE3*]	
Hsp70s				
Ssa1	⇑	⇑	⇓	Newnam *et al.* (1999); Schwimmer and Masison (2002)
Ssa1	⇓	⇓	-	Jones *et al.* (2004)
Ssa2	⇑	⇑	-	Allen *et al.* (2005); Schwimmer and Masison (2002)
Ssa2	⇓	-	⇓	Roberts *et al.* (2004)
Hsp40s				
Ydj1	⇑	-	⇓	Higurashi *et al.* (2008); Lian *et al.* (2007); Moriyama *et al.* (2000); Sharma *et al.* (2009b)
Ydj1	⇓	-	⇓[b]	Higurashi *et al.* (2008); Sharma *et al.* (2009b)
Sis1	⇑	-	-	Higurashi *et al.* (2008); Lian *et al.* (2007)
Sis1	⇓	⇓	⇓	Aron *et al.* (2007); Higurashi *et al.* (2008)
NEFs				
Fes1	⇑	⇑[c]	-	Jones *et al.* (2004); Kryndushkin *et al.* (2008); Lian *et al.* (2007)
Fes1	⇓	⇓	⇓	Jones *et al.* (2004); Kryndushkin *et al.* (2008)
Sse1	⇑	⇑	⇓	Fan *et al.* (2007); Kryndushkin *et al.* (2008)
Sse1	⇓	⇓	⇓	Fan *et al.* (2007); Kryndushkin *et al.* (2008)
TPR proteins				
Sti1	⇑	⇓	-	Fan *et al.* (2007); Jones *et al.* (2004); Lian *et al.* (2007)
Sti1	⇓	⇑[c,d]	?	Fan *et al.* (2007); Jones *et al.* (2004)
Cns1	⇑	⇓	-	Jones *et al.* (2004); Lian *et al.* (2007)
Cpr7	⇑	⇑	-	Jones *et al.* (2004); Lian *et al.* (2007)
Cpr7	⇓	⇑[c]	?	Jones *et al.* (2004)
Hsp104	⇑	⇓	(-)[e]	Chernoff *et al.* (1995)
Hsp104	⇓	⇓	⇓	Chernoff *et al.* (1995)

[a] ⇑, enhances prion strength and/or stability; ⇓, weakens phenotype or eliminates prion; –, alteration has no effect; ?, unknown.
[b] *ydj1* Δ is compatible with [*URE3*] in one study, lethal with Ure2p depletion in another.
[c] improves [*PSI*$^+$] in *SSA1-21* (Hsp70 dominant negative) mutant.
[d] inhibits induction of [*PSI*$^+$] by Sup35NM overexpression.
[e] compared with [*PSI*$^+$], curing of [URE3] is weak and unreliable.

can block the formation and assembly of Sup35 amyloid (Krzewska and Melki, 2006; Shorter and Lindquist, 2008). Whether simultaneous overexpression of Ssa2, Ssa3, or Ssa4 with Ydj1 weakens [*PSI*$^+$] was not tested. Another variant of [*PSI*$^+$] prion formed by Sup35 with GFP fused in frame immediately after the prion-determining region is also curable by Ssa1 (Mathur *et al.*, 2009). Therefore, although Ssa1 overexpression promotes [*PSI*$^+$] propagation in some instances; in other circumstances, it also inhibits propagation of [*PSI*$^+$] and other prions. Such paradoxical findings make it difficult to infer the basis of chaperone involvement in prion propagation with confidence.

These and other findings led to the proposals that rates of the critical processes of growth and replication of prion polymers are different for prions with different structural conformations, that these differences might underlie variations in strength and stability of phenotypes of prion "strains," and that growth and replication of prions can be affected differently by individual and combined chaperones (Kushnirov *et al.*, 2000). If the action of Hsp70 on prions is considered as being direct, then the fact that the same Hsp70 can have opposing effects on the same prion suggests that the way Hsp70 influences prions depends more on how its reaction cycle or substrate specificity is regulated than on its intrinsic Hsp70 activities. The synergistic anti-prion effects of elevating expression of a co-chaperone that regulates Hsp70 is consistent with the notion that elevating co-chaperones influences prions indirectly through altered regulation of Hsp70, an inference that arises frequently.

9.3.3 Yeast Prions Reveal Functional Distinctions among Hsp70 Homologs

Systematic overexpression of Ssa proteins has not been done to study the effects of elevated Hsp70 on prions other than [*PSI*$^+$]. However, in the work to determine if antagonistic interactions between [*PSI*$^+$] and [URE3] in the same cell were due to competition for chaperones, it was found that increasing Ssa1 expression causes cells to lose [URE3], but increasing Ssa2 has no effect on the stability of [URE3] or [*PSI*$^+$] (Schwimmer and Masison, 2002). Therefore, unlike [*PSI*$^+$], [URE3] prions respond differently to overexpression of Ssa1 and Ssa2, which uncovers distinctions between both the prions and the Hsp70 isoforms. Additionally, depleting Ssa1 slightly weakens [*PSI*$^+$] but does not affect [URE3] stability, and depleting Ssa2 does not affect [*PSI*$^+$] but causes cells to lose [URE3] (Jones *et al.*, 2004; Roberts *et al.*, 2004). Moreover, a point mutation that alters but does not inactivate Ssa2 causes [URE3] to become very unstable (Roberts *et al.*, 2004). These data show that Ssa1 and Ssa2 can preferentially promote [*PSI*$^+$] and [URE3], respectively, and uncover a requirement for an Ssa2-specific Hsp70 function for stable [URE3] propagation. Overall, these data for Ssa1 and Ssa2 show that the way altered Hsp70 function influences amyloid propagation *in vivo* can vary significantly depending upon the isoform of Hsp70 being expressed.

Remarkably, Ssa1 and Ssa2 are 98% identical, indicating that very small structural differences can confer significant distinctions in Hsp70 function.

A functional difference between Ssa1 and Ssa2 was first described for a vesicular prevacuolar protein degradation pathway for enzymes involved in carbohydrate metabolism after a switch in carbon source (Brown *et al.*, 2000). Here again, the pathway shows dependency on Ssa2. Whether this pathway is somehow connected to [URE3] propagation is not yet known. Because of the high structural similarity between Ssa1 and Ssa2, it has been proposed that the functional differences between them with regard to prions are more likely due to differences in physical or functional interactions with co-chaperones, or to differential influences on specific processes the prions might rely on for their propagation, than due to differences in intrinsic Hsp70 activities (Masison *et al.*, 2009; Sharma and Masison, 2008).

Whether there is an absolute requirement of Hsp70 for yeast prion propagation cannot be tested because abundant expression of at least one of the four Ssa proteins is essential for cell growth (Werner-Washburne *et al.*, 1987). Additionally, the ability to interpret data from experiments in which individual *SSA* genes are deleted is limited because Ssa3 and Ssa4 are expressed only under nonoptimal conditions and deleting both Ssa1 and Ssa2 strongly induces expression of Ssa3/4 and other stress-response proteins including Hsp104. To overcome some of these limitations and compare functions of the different Ssa isoforms directly, a system was developed to express plasmid-borne Hsp70 in cells lacking all the chromosomal *SSA* genes. In this system, any Hsp70, regardless of origin, can be expressed and evaluated for its ability to complement Ssa functions in yeast growth and prion propagation (Sharma and Masison, 2008; Tutar *et al.*, 2006). Constitutive (Hsc) isoforms of primate and plant Hsp70 support yeast growth and [*PSI*$^+$] prion propagation, although not as well as Ssa1, whereas their inducible (Hsp) counterparts do not support viability. This work points to important differences between constitutive and inducible Hsp70s and demonstrates the utility of this system for comparative functional analysis of chaperones.

Depending on the isoform encoded, cells engineered to express only one of the Ssa1–Ssa4 isoforms from the strong constitutive Ssa2 promoter display a range of growth rates and a wide variation in the strength and stability of [*PSI*$^+$] and [URE3] (Sharma and Masison, 2008). These observations uncover additional functional differences among these four Hsp70s. The variations in the ability of each Hsp70 to support growth likely reflect important distinctions in the way the different isoforms perform in essential cellular processes. Although cells expressing Ssa1 or Ssa2 grow like wild type, cells expressing the inducible Ssa3 or Ssa4 isoforms grow more slowly, which is reminiscent of the differences between constitutive and inducible isoforms of primate and plant Hsp70s with regard to supporting yeast viability (Tutar *et al.*, 2006). This difference in function between constitutive and inducible Hsp70s within and across species suggests that reliance on the housekeeping responsibilities of the inducible Hsp70s lessened as they diverged for improved stress protection or for increased efficiency of cooperation with other inducible chaperones or co-chaperones.

With regard to [*PSI*$^+$] and [URE3], effects of an Hsp70 isoform on one prion generally are opposite to its effects on the other. Ssa3 enhances propagation of

[*PSI*$^+$] but antagonizes propagation of [URE3]. Ssa1 is similar but with less pronounced effects. In contrast, Ssa2 slightly weakens [*PSI*$^+$] but is best among the Ssas at promoting [URE3] propagation. Ssa4 is the only Ssa that affects the two prions similarly, inhibiting both of them. Therefore, highly homologous Hsp70s affect propagation of the same prions differently, and whether their influence on propagation is positive or negative depends on the prion. The opposite ways that individual Hsp70 isoforms affect [*PSI*$^+$] and [URE3] define additional functional distinctions among the Hsp70s and show that these prions have different requirements of Hsp70 function for their propagation. Because the functions of Ssa1–Ssa4 are highly redundant, these prions represent an exquisitely sensitive system for monitoring differences in Hsp70 function and identifying the molecular bases of these differences.

The mechanisms by which the different Hsp70s affect prion propagation are not yet known, although there are clues. Each of the different Hsp70s could affect prions to different degrees or in different directions by influencing rates of growth or fragmentation, or both, of prion polymers. The Hsp70s might also directly or indirectly affect transmission of prions to daughter cells during cell division. [URE3] has a strong but unstable phenotype in cells expressing Ssa1 as the only Ssa protein (Sharma and Masison, 2008). A possible explanation for this phenotype is that Ssa1 inhibits the replication but not growth of [URE3] polymers, which would cause a reduction in the number of prions and reduce the chances of daughter cells inheriting them, but not the incorporation of monomer into polymer. Alternatively, Ssa1 could interfere with a process needed for efficient transmission of [URE3] prions. In contrast to cells expressing only Ssa1, those expressing only Ssa4 have a weak but stable [URE3] phenotype, so Ssa4 could reduce the growth but not replication of polymers. In this situation, the depletion of soluble [URE3] by the growth of polymers is not as efficient, causing a weaker phenotype, but new prions are produced at a normal rate, so that [URE3] is mitotically stable. Similarly, the weak unstable phenotype of cells expressing Ssa3 could indicate that both the processes are inhibited, whereas the strong and stable phenotype of cells expressing Ssa2 suggests that both growth and replication of prions proceed optimally. Therefore, if one assumes that Hsp70s affect only growth and/or replication, then the different ways [URE3] propagates in cells expressing each of the four Ssas would reflect all four of the ways that these processes could be affected (Sharma and Masison, 2008) (Table 9.2). Again, specificities of interactions with co-chaperones probably contribute to the differences.

The four Ssa protein orthologs of the unrelated yeast *Yarrowia lipolytica* that are roughly 95% identical to each other and 80% identical to each of the *Saccharomyces* Ssa proteins, designated Ssa5–Ssa8, were also analyzed for their ability to support growth and propagation of [*PSI*$^+$] and [URE3] (Sharma *et al.*, 2009a). Because of their high similarity, assignment of constitutive or inducible status could not be made until after assessing the transcription profiles in their native context. As anticipated, although Ssa5–Ssa8 each supported growth, the stress-responsive isoforms again were not as effective. Additionally, each Hsp70

Table 9.2 Model of how Ssa proteins could produce different phenotypes by affecting prion growth and replication

	Process Affected[a]		Ssa[b]
Prion Phenotype	Growth	Replication	*[URE3]*
Strong, stable	+	+	Ssa2
Strong, unstable	+	−	Ssa1
Weak, stable	−	+	Ssa4
Weak, unstable	−	−	Ssa3

Sharma and Masison (2008).
[a]+ process enhanced or no affect; −, process inhibited.
[b]indicated Hsp70 is the only Ssap in the cell.

influenced propagation of the two prions differently and overall there was little correlation between the ability to support growth and the effects on the prions. The variations in the ability of Ssa5–Ssa8 to support growth likely reflect differences in the ability to function with interspecies factors in essential cellular processes. Similarly, the differences in the effects on prion phenotype likely reflect differences in the efficiency of interactions with cellular factors important for prion propagation.

In an effort to understand how the different Hsp70s influence the physical properties of Sup35, the aggregation state of a coexpressed Sup35NM-GFP fusion (GFP in place of the functional Sup35 C-terminal domain (CTD)) was monitored by fluorescence in *[PSI⁺]* cells expressing each isoform individually. The size of the prion polymers in lysates of these same cells expressing the fusion protein was monitored by agarose gel electrophoresis (Kryndushkin *et al.*, 2003; Sharma *et al.*, 2009a). The latter technique has been used widely to estimate the abundance and size range of SDS-resistant prion polymers. Although it can readily detect differences in the average size of polymers between strains, whether the differences are due to effects on polymer growth or fragmentation cannot be distinguished because it shows steady-state conditions. Although several differences and similarities of both aggregation state and polymer size were observed, there was no correlation of fluorescent aggregation patterns or polymer sizes with the prion phenotypes. This inability to ascribe characteristic changes in prion phenotypes to distinctions in the physical properties of prion proteins shows that the dynamics of prion protein aggregation or polymer growth and fragmentation do not alone determine the efficiency of prion propagation.

9.3.4 Hsp70 Mutants

Aside from overexpression studies using candidate chaperones, early work using an unbiased screen for factors other than Sup35 that were important for *[PSI⁺]* propagation identified a mutant of Ssa1, designated Ssa1-21, which contains a substitution of tryptophan for leucine at residue 483 (L483W) in the SBD (Jung *et al.*, 2000). Ssa1-21 impairs *[PSI⁺]* propagation, reducing both the amount of insoluble prion form of Sup35 and the apparent number of [*PSI*$^+$] prions per cell. The effects of Ssa1-21 are dominant in the presence of the functionally redundant Ssa2 or when Ssa1 is expressed in place of Ssa2. When Ssa1-21 is the only Ssa present in the cell, [*PSI*$^+$] cannot propagate at all. Further study showed that while prion aggregates in mutant cells are bigger, the size range of individual prion polymers that come together to form these aggregates are the same as in wild-type cells (Song *et al.*, 2005). These data suggest that the rates of polymer growth and fragmentation are either unaffected or are affected to a precisely complementary degree in opposite directions. Assuming the simpler explanation that these processes are unaffected to be more probable, it was proposed that Ssa1-21p either promotes aggregation of individual prion polymers or interferes with recovery of prions from the large aggregates.

To gain more insight into how Ssa1-21 function is altered, Ssa1-21 was mutagenized to identify second-site mutations that restore normal [*PSI*$^+$] propagation (Jones and Masison, 2003). Such mutations reduce the affinity of Ssa1-21 for substrate or for Hsp40 and other co-chaperones that promote substrate binding, suggesting that increased substrate binding by Ssa1-21 inhibits prion propagation (Figure 9.3). This conclusion was supported by work showing that [*PSI*$^+$] propagation can be restored in *SSA1-21* cells by deleting the Hsp70/Hsp90 co-chaperones Sti1 and Cns1 (Jones *et al.*, 2004), which stimulate Ssa1 ATPase and promote substrate binding, or by overexpressing the NEF Fes1, which promotes substrate release. Additionally, altering the expression of these co-chaperones in wild-type cells in ways expected to prolong or to facilitate Hsp70 substrate release inhibits or improves prion propagation, respectively. Therefore, altering the Hsp70 substrate binding/release cycle in the same direction, either by mutating Hsp70 or by altering its regulators causes the same effects on prions. How an increased avidity of Hsp70 for substrate leads to increased aggregation of prion polymers *in vivo* remains to be worked out, and whether the substrate in question is the prion itself remains unclear.

Wild-type Ssa1 was also mutagenized to identify additional alterations of Hsp70 that impaired [*PSI*$^+$]. Of over 20 additional mutations in Ssa1 that cause prion-inhibiting effects such as L483W, all but one are located in the ATPase domain and the other is in the nearby residue 481 (Jones and Masison, 2003; Loovers *et al.*, 2007). Nevertheless, the new Ssa1 mutants that were tested are suppressed by the same second-site mutations and co-chaperone alterations that suppress Ssa1-21, implying that the prion-inhibiting mutations, including those in different structural domains, affect a similar Hsp70 function. The findings also imply that fine-tuning of Hsp70 activity is more important than substrate binding per se, and that the defect in Ssa-21 and the other mutants is rooted in the inability to be properly regulated.

Conservation of the Hsp70 function altered by these mutations was demonstrated by a screen that identified similar mutations in Ssa2 that alter [*PSI*$^+$] in similar ways (Loovers *et al.*, 2007). Nevertheless, when the L483W mutation made in Ssa2 was used to test the effects on prions, the resulting inhibition of [*PSI*$^+$] and [URE3] was similar but not identical, and the effects on [URE3] were much more detrimental (Sharma and Masison, 2008). Together, these findings uncover other similarities and distinctions between Ssa1 and Ssa2.

Ssa1-21 and two other Ssa1 mutants that similarly interfered with [*PSI*$^+$] were analyzed biochemically to determine how the mutations affected Hsp70 enzymatic activities. Consistent with the genetic data indicating that [*PSI*$^+$] propagation can be inhibited by increasing stimulation of Ssa1-21 ATPase by cofactors, the intrinsic ATPase activity of Ssa1-21 was found to be elevated 10-fold (Needham and Masison, 2008). Surprisingly, however, intrinsic ATPase activity of the mutants in the ATPase domain was not much different than that of the wild type. Although these mutations increased the ability of Ssa1 to refold denatured protein *in vitro*, they had different effects on the synergistic stimulation of ATPase by the substrate and Hsp40. Moreover, when the influence of combinations of substrates and co-chaperones on ATP hydrolysis, ADP release, substrate-binding affinity, and protein refolding was compared, each of the mutant proteins was different. Overall, the effects of the different Ssa1 mutations on intrinsic and cofactor-stimulated Hsp70 activities do not identify a specific alteration in Hsp70 activity that can lead to anti-prion effects, but it is clear that regulation of each of the mutant protein's activities is disrupted, perhaps through effects on communication between functional domains.

An intriguing feature of the L483W substitution and most of the other prion-impairing mutations in Ssa1 and Ssa2 is that although they are in highly conserved residues and alter Hsp70 function considerably, they have little or no effects on growth under a variety of conditions tested, even when they are expressed as the only Ssa in the cell. The absence of a growth phenotype suggests that the mutations affect a specific and nonessential Hsp70 function. Although the mutations alter specific Hsp70 activities in different ways, which is consistent with the effects on prions being indirect, they likely produce a common perturbation of the regulation of Hsp70 function with certain cofactors or in a certain cellular process that is important for prion propagation but not cell growth. The dominance of the mutations implies a gain of function, which might be tolerated better than the loss of Hsp70 activity. Because Hsp70 acts in many important cellular processes, the surprising fact that such a dramatically defective Hsp70 supports growth at all, nevertheless so well, implies that some aspect of the regulation of Hsp70 activity in the cell must buffer this defect. That there are many ways of altering Hsp70 or its co-chaperones to produce a similar prion-inhibitory effect with minimal effects on cell growth points to Hsp70 and its regulators as potential targets for therapy of amyloidoses. Research in this area is increasing (Patury *et al.*, 2009).

The *SSA1-21* mutant also provides insight into the mechanism of curing of [*PSI*$^+$] by Hsp104. A screen for mutations in Hsp104 that restore prion propagation in *SSA1-21* cells identified residues in the Hsp104 N-terminal domain (NTD) (Hung and Masison, 2006). Mutations in these residues not only suppress the effects of Ssa1-21 but also render Hsp104 unable to cure [*PSI*$^+$] when overexpressed. Despite their lack of curing activity, these Hsp104 mutants, as well as Hsp104 lacking the entire NTD, are fully capable of supporting [*PSI*$^+$]

propagation, disaggregating proteins from aggregates, and protecting cells from exposure to lethal heat, which shows that the NTD is dispensable for Hsp104 to recognize and to act on both highly structured Sup35 prions and amorphous aggregates of thermally denatured protein. Therefore, elevating protein disaggregation capability by overexpressing Hsp104 apparently is not sufficient to cure cells of [*PSI*$^+$], which suggests that the mechanism of Hsp104 curing is distinct from or involves more than a direct action on prions by Hsp104 disaggregation activity.

9.3.5 Hsp40s/J-Proteins

The influence of Hsp40s on prion propagation and toxicity is the subject of another chapter in this volume (Chapter 8) and here the discussion focuses on how such influences could be due to their effects on Hsp70 regulation.

Hsp40s are obligate Hsp70 co-chaperones that cooperate with Hsp70 in all its roles (Craig *et al.*, 2006; Walsh *et al.*, 2004). They are the largest family within the more broadly classified Hsp70 co-chaperones termed *J-proteins*, which are defined by a conserved structural region called a *J-domain*. The J-domain, named for the originally described Hsp40 of *E. coli* DnaJ, directs interaction with a site on the ATPase domain of Hsp70. A conserved tripeptide HPD (histidine–proline–aspartate) within the J-domain is required for Hsp40 stimulation of Hsp70 ATP hydrolysis. Hsp40s typically bind substrates with specificity overlapping that of Hsp70 and can present substrates to Hsp70. Hsp40s thus coordinate substrate trapping by Hsp70 with ATP hydrolysis. Hsp40s can also act independently as chaperones by binding unfolded proteins and preventing aggregation (Cyr, 1995). This binding can also "hold" proteins in a conformation that facilitates refolding by Hsp70. For example, heating purified luciferase causes it to aggregate, and subsequent addition of Hsp70 does little to refold the protein to its active state. If Hsp40 is included when luciferase is heated, then it binds the unfolded enzyme to prevent aggregation and holds it in a conformation that can be acted on by Hsp70 to restore activity (Lu and Cyr, 1998). The term *holdase* is used to describe this activity.

As in other eukaryotes, *S. cerevisiae* encodes three subclasses of J-proteins with multiple members in each group (Craig *et al.*, 2006; Walsh *et al.*, 2004). Class III proteins share only the J-domain and do not bind nonnative proteins, whereas class I and II proteins, which represent canonical Hsp40s, have a glycine-rich (G) region and C-terminal SBD. Class I proteins also have a zinc-binding motif between the G and CTD regions. Class I and II Hsp40s are not only functionally redundant but also possess nonoverlapping activities (Fan *et al.*, 2004; Johnson and Craig, 2001; Lopez *et al.*, 2003; Lu and Cyr, 1998). The major cytosolic class I and II Hsp40s in yeast are Ydj1 and Sis1, respectively. Organisms typically encode more Hsp40s than Hsp70s, and Hsp70s can interact functionally with more than one Hsp40. This promiscuity not only allows complexity for interactions with a wide range of substrates and other chaperones but also provides specificity and fine-tuning in the regulation of Hsp70 functions and the recruitment of Hsp70 to specific cellular locations (Craig *et al.*, 2006; Higurashi *et al.*, 2008; Walsh *et al.*, 2004; Young *et al.*, 2004).

As with the other chaperones, identification of Hsp40s in connection with prions was first shown in overexpression studies. Elevating Ydj1 expression was found to cure cells of [URE3] (Moriyama *et al.*, 2000), and another class I protein Apj1 was identified in a screen for factors that destabilize the [PSI^{+}_{PS}] prion formed of the *P. methanolica—S. cerevisiae* Sup35 hybrid protein (mentioned above) (Kryndushkin *et al.*, 2002). Although typical [PSI^{+}] prions are not affected by overexpressing any Hsp40, Sis1 also was found to impair [PSI^{+}_{PS}]. What distinguishes sensitivity of these prions to these factors remains to be determined.

The mechanism of how overexpressing Ydj1 cures [URE3] was investigated biochemically and genetically. Using purified proteins, it was shown that Ydj1 binds Ure2 in its soluble and amyloid conformations, and that incubating Ydj1 with soluble Ure2 inhibits the formation of Ure2 amyloid (Lian *et al.*, 2007; Savistchenko *et al.*, 2008). Unlike Ydj1, Sis1 neither binds Ure2 nor inhibits Ure2 amyloid formation, and overexpressing Sis1 has no effect on [URE3]. These and other *in vitro* data led to the sensible conclusion that Ydj1 cured cells of [URE3] by binding Ure2 and blocking assembly of prion polymers (Lian *et al.*, 2007; Savistchenko *et al.*, 2008). However, Ydj1 does not block the continued assembly of Ure2 amyloid when it is added after the amyloid is already present, which leaves open the possibility that Ydj1 acting alone would be ineffective at preventing prion polymer assembly in cells already propagating [URE3].

The genetic approach combined random and systematic mutational analysis to determine the activities of Ydj1 required for the curing (Sharma *et al.*, 2009b). All randomly identified mutations in Ydj1 that prevented it from curing [URE3] were in the J-domain, which implied that the interaction of Ydj1 with Hsp70 was crucial for it to impair [URE3] propagation. Directed mutagenesis showed that single amino acid substitutions in the J-domain that prevent Ydj1 stimulation of Hsp70 ATPase abolished curing, but altering residues required for Ydj1 membrane association, dimerization, substrate binding, and substrate transfer to Hsp70 have little or no effect on the ability of Ydj1 to cure cells of [URE3]. Moreover, overexpressing the J-domain alone was enough to cure prions (Higurashi *et al.*, 2008; Sharma *et al.*, 2009b). Therefore, curing of [URE3] by Ydj1 is indirect, requiring only its ability to stimulate Hsp70 ATPase activity. This conclusion is strengthened by the fact that the J-domains alone of Sis1 or Jjj1, another Hsp40 with a role in ribosome biogenesis, also cure cells of [URE3], although the intact proteins do not (Higurashi *et al.*, 2008; Sharma *et al.*, 2009b). These findings suggest that Hsp40 acts upstream of Hsp70, and that influencing Hsp70 activity in a particular manner perturbs prion propagation.

What makes Ydj1 unique among the Hsp40s in being able to cure [URE3] is unclear, but it could be that the other Hsp40s are recruited to subcellular structures or locations that preclude its ability to act on [URE3], or that Ydj1 interacts more effectively or specifically with Hsp70 isoforms necessary for the effect. Support for the latter explanation was provided in later work showing that

[URE3] is cured efficiently from cells expressing only Ssa1 or Ssa2, but weakly from cells expressing only Ssa3 or Ssa4 (Sharma *et al.*, 2009a). Together these findings show not only that Ydj1 requires Hsp70 to cure [URE3] but also that isoform-specific interactions of Hsp40 and Hsp70 are important for the anti-prion effect.

Sis1 is essential for [*PIN*$^+$]/[*RNQ*$^+$] propagation (Higurashi *et al.*, 2008), and cells expressing Sis1 lacking its G-region (*sis1*Δ*GF*) in place of wild-type Sis1 are unable to support [*RNQ*$^+$], although they grow well (Sondheimer *et al.*, 2001). Overexpression of *sis1*Δ*GF* in wild-type cells also inhibits cell growth and [*RNQ*$^+$] propagation, but only if it is capable of interacting with Hsp70 (Aron *et al.*, 2005). Overexpressing a Ydj1 mutant lacking the same region has no such effects. A binding site for Sis1 on Rnq1 was identified near but outside the defined prion-determining region and alteration of a single amino acid in this site considerably reduces ability of Sis1 to bind Rnq1 and causes loss of [*RNQ*$^+$] (Bardill *et al.*, 2009; Douglas *et al.*, 2008). However, a hybrid prion containing the prion domain of Rnq1 (lacking the Sis1-binding site) in place of the Sup35 prion domain, designated [*RPS*$^+$], also fails to propagate in cells expressing *sis1*Δ*GF* as the only source of Sis1, indicating that a GF-dependent activity of Sis1 that is not involved in direct binding to this site on Rnq1 is required for [*RNQ*$^+$] propagation. Detailed analysis of the Sis1–[*RNQ*$^+$] interaction defined residues within the GF-region as critical to the Sis1 activity involved in prion propagation and led to the suggestion that interaction with Hsp70 is important for Sis1 effects on [*RNQ*$^+$] (Aron *et al.*, 2007; Lopez *et al.*, 2003).

Unlike [*RNQ*$^+$], [*RPS*$^+$] is curable by overexpressed Hsp104, which implies that regions outside the prion-determining domain (i.e., the middle and CTDs of Sup35) are required for chaperone-mediated effects on prion propagation (Sondheimer *et al.*, 2001). These regions might mediate interactions with specific combinations of chaperones or other cofactors or cellular components. Such differences again reflect the extensive complexity in interactions of prions with co-chaperones, and point to the likelihood that some influences of chaperone interactions go beyond simple direct or mass action effects on kinetics of amyloid formation.

Depleting Sis1 causes increases in both the size of [*RNQ*$^+$] prion polymers and the amount of soluble Rnq1, eventually leading to loss of [*RNQ*$^+$], which suggests that Sis1 is important for the production of prion seeds necessary for [*RNQ*$^+$] propagation (Aron *et al.*, 2007). Depleting Hsp104 affects [*RNQ*$^+$] propagation and Rnq1 aggregation in a manner similar to that of depleting Sis1p. Together, these findings led to proposals that Sis1 acts in [*RNQ*$^+$] propagation through its ability to regulate Hsp70 in a process requiring Hsp104—likely to be polymer fragmentation (Aron *et al.*, 2007; Higurashi *et al.*, 2008; Lopez *et al.*, 2003).

A comprehensive deletion analysis to study the importance of the 13 cytosolic yeast Hsp40s identified an essential role for Sis1, but none of the other Hsp40s, in propagation of not only [*RNQ*$^+$] but also [*PSI*$^+$] and [URE3] (Higurashi *et al.*, 2008; Sahi and Craig, 2007). This work suggests that Sis1 possesses an activity that confers a prion-specific Hsp40 function to the chaperone machinery. A detailed analysis of the fate of Sup35 and [*PSI*$^+$] prions and Hsp104 interactions after depleting Sis1 led to a similar conclusion for a role of Sis1 in the regulation of Hsp70 and the disaggregation machinery in [*PSI*$^+$] propagation (Tipton *et al.*, 2008). These data support current models proposing that Hsp104/Ssa/Sis1 machinery extracts or at least displaces monomers from within prion fibers, which destabilizes the polymers causing them to fragment into smaller pieces that each propagate the prion (Figure 9.1b) (Doyle and Wickner, 2009; Grimminger-Marquardt and Lashuel, 2010; Hung and Masison, 2006; Sweeny and Shorter, 2008; Tessarz *et al.*, 2008; Tipton *et al.*, 2008).

9.3.6 Nucleotide Exchange Factors (NEFs)

Snl1, Fes1, and Sse1/2 are NEFs that regulate yeast cytosolic Hsp70 activity. As seen with the other chaperones and co-chaperones, mutation or altered expression of each NEF influences propagation of different prions in different ways (Fan *et al.*, 2007; Jones *et al.*, 2004; Kryndushkin and Wickner, 2007; Park *et al.*, 2006; Sadlish *et al.*, 2008). Snl1 was originally isolated as a high-copy suppressor of a nuclear pore mutant and was later identified as an NEF on the basis of its homology to the mammalian Hsc70 NEF Bag-1M (Ho *et al.*, 1998; Sondermann *et al.*, 2002). It interacts with both Ssa and Ssb subfamily Hsp70s, and mutations that abolish the ability of Snl1 to interact with Hsp70 also abolish its role in nuclear pore function, indicating that its interaction with Hsp70 is critical for its biological activity (Sondermann *et al.*, 2002). With regard to prions, the only data to date show that Snl1 modestly inhibits curing of [*PSI*$^+$] by overexpressed Hsp104, proposed to be through regulation of Ssa1 (Sadlish *et al.*, 2008).

Fes1 was identified in a search for cytosolic exchange factors structurally similar to Sls1, an NEF for the endoplasmic reticulum localized Hsp70 Kar2 (a homolog of mammalian BiP) (Kabani *et al.*, 2002). Fes1 binds Ssa1 preferentially when it is in the ADP-bound state and promotes ADP release. Fes1's ADP exchange activity varies depending on *in vitro* reaction conditions and the Hsp70 tested, and it can promote the release of ATP (Kabani *et al.*, 2002; Needham and Masison, 2008; Raviol *et al.*, 2006). It also inhibits the ability of Ydj1 to stimulate Ssa1 ATPase. Fes1 also functions as an NEF for Ssb and binds Ssb1 with stronger affinity than Ssa1 (Dragovic *et al.*, 2006b). Similar to Snl1, mutations that abolish Fes1 interaction with Hsp70 also abolish its function *in vivo*, indicating that it functions through the regulation of Hsp70 (Shomura *et al.*, 2005). It is weakest among the three NEFs in promoting the release of ADP from Ssa1 but can compensate for depletion of Sse1 in essential NEF function and in [*PSI*$^+$] formation (Raviol *et al.*, 2006; Sadlish *et al.*, 2008).

Fes1 is required for [URE3] propagation, and, although [*PSI*$^+$] propagation does not require Fes1, depleting Fes1 weakens [*PSI*$^+$]. These findings show that Fes1 has a general prion-promoting activity (Jones *et al.*, 2004; Kryndushkin and Wickner, 2007). Consistent with this conclusion, overexpressing Fes1 can improve [*PSI*$^+$] propagation (Jones *et al.*, 2004). These findings support the notion that prolonging Hsp70 substrate binding inhibits [*PSI*$^+$] propagation, whereas enhancing substrate release promotes it (Figure 9.3).

Sse1 and the 75% identical Sse2 are structurally similar to Hsp70s having ATPase and SBDs, and they were originally designated as a new Hsp70 subfamily sharing roughly 30% identity with the Ssa and Ssb Hsp70s (Mukai *et al.*, 1993). Sse1 is more functionally diverse than the other NEFs, acting as an Hsp90 co-chaperone and having a role in endoplasmic reticulum associated protein degradation (ERAD) (Goeckeler *et al.*, 2002; Hrizo *et al.*, 2007). Cells lacking Sse1 grow slowly, and although deleting Sse2 has little effect on growth, deleting both is lethal (Liu *et al.*, 1999). This lethality is suppressed by overexpression of Fes1 or Snl1, but not of Ssa1 or Ssb1, indicating that the essential activity of Sse1 is its NEF function and not an Hsp70 activity (Sadlish *et al.*, 2008). The importance of Sse1 in Hsp70 regulation is further shown by the findings that it was isolated in a screen for suppressors of a temperature-sensitive Ydj1 mutant and that it enhances stimulation of Ssa1 ATPase by Ydj1 (Goeckeler *et al.*, 2002; Shaner *et al.*, 2005). It acts as an NEF for both Ssa and Ssb Hsp70s and increases their refolding efficiency (Dragovic *et al.*, 2006a; Yam *et al.*, 2005). Sse1 also binds unfolded protein substrates, suggesting that it might have direct Hsp70-like effects on substrates. It is unable to promote refolding, however, leading to the suggestion that it has a role in protein folding as a "holdase" as well as an NEF for Hsp70 (Goeckeler *et al.*, 2008; Shaner *et al.*, 2004). ATP binding, but not hydrolysis, is critical for Sse1 NEF function (Andreasson *et al.*, 2008; Raviol *et al.*, 2006; Shaner *et al.*, 2006). These same properties are also required for its essential function *in vivo*, further demonstrating that the requirement of Sse1 *in vivo* is to regulate Hsp70.

As seen with Fes1, deleting Sse1 also caused complete loss of [URE3], which points to a critical role of NEF activity in [URE3] propagation (Kryndushkin and Wickner, 2007). However, Sse1 was also found to destabilize [URE3] when overexpressed. This dependency of [URE3] on an intermediate amount of Sse1 resembles the dependency of [*PSI*$^+$] on an intermediate amount of Hsp104. The ability of Sse1 to bind ATP, but not to hydrolyze it, is important for its role in [URE3] propagation, which lead to the conclusion that, as with its essential function, its action on prions was through regulation of Hsp70. Overexpressing Sse1 increases the size of Ure2 polymers in [URE3] cells. Again, such an effect could be caused by an inhibition of prion polymer fragmentation or an enhancement of polymer growth. Therefore, Sse1 could be important for either or both of these processes through its ability to regulate Hsp70.

Sse1 also influences [*PSI*$^+$] considerably (Fan *et al.*, 2007; Kryndushkin and Wickner, 2007; Sadlish *et al.*, 2008). Although [*PSI*$^+$] propagates in *sse*1Δ cells, it is weak and unstable. Depleting Sse1 also dramatically reduces the ability of [*PSI*$^+$] to be induced by overexpression of Sup35. Sup35 polymers are larger on average in *sse*Δ [*PSI*$^+$] cells, a property shared by weak variants of [*PSI*$^+$] in wild-type cells. All of these findings are consistent with Sse1 having a positive role in [*PSI*$^+$] propagation. As with its other functions, ATP binding but not hydrolysis is required for Sse1 effects on [*PSI*$^+$], which again led to the proposal that NEF activity of Sse1 helps Hsp70 promote normal [*PSI*$^+$] propagation. This conclusion is further supported by the observations that overexpressing Sse1 promotes both *de novo* appearance of [*PSI*$^+$] and the ability of Ssa1 to inhibit curing by Hsp104, and it inhibits both the ability of overexpressed Hsp104 to cure cells of [*PSI*$^+$] and the ability of Ssb to promote curing by overexpressed Hsp104 (Fan *et al.*, 2007; Sadlish *et al.*, 2008). *In Vitro*, Sse1 in the presence of ATP stimulates assembly of Sup35NM into fibers (Shorter and Lindquist, 2008). Whether it acts in the same manner on intact Sup35 was not tested. Increasing both Sse1 and Ssa1 has an additive effect in promoting *de novo* induction of [*PSI*$^+$] by overexpression of Sup35. Sse1 effects were suggested to be through action on Ssa family Hsp70s since they are more consistent with general [*PSI*$^+$] promoting effects of Ssa than with [*PSI*$^+$] inhibiting effects of Ssb (Fan *et al.*, 2007). In line with this conclusion, the weakening of [*PSI*$^+$] phenotype caused by depletion of Sse1 can be suppressed by elevating Fes1, Snl1, or Ssa1, but not Ssb1 (Sadlish *et al.*, 2008; Yam *et al.*, 2005). These latter findings are also consistent with the conclusion that Sse1 acts in prion propagation through the regulation of Hsp70.

9.3.7 TPR Co-chaperones Influence Prions by Regulating Hsp70

The discovery that alterations in the C-terminal TPR-interaction motif of Ssa1-21 suppressed the ability of Ssa1-21 to inhibit [*PSI*$^+$] led to the finding that the TPR co-chaperones Sti1, Cpr7, and Cns1 can influence prion propagation through their ability to regulate Hsp70 (Jones *et al.*, 2004). As described above, all three of these proteins are established Hsp90 co-chaperones. The influence these TPR proteins have on [*PSI*$^+$] propagation apparently is independent of essential functions of Hsp90 (Jones *et al.*, 2004; Song and Masison, 2005). Cpr7 and Cns1 can interact with Ssa1 *in vivo* and *in vitro* (Duina *et al.*, 1998; Marsh *et al.*, 1998; Tesic *et al.*, 2003), and Sti1 and Cns1 both stimulate Ssa1 ATPase activity (Hainzl *et al.*, 2004; Wegele *et al.*, 2003). As deleting Sti1 restores normal [*PSI*$^+$] propagation in *SSA1-21* cells, these activities are consistent with the suggestion mentioned above that promoting conversion to or stabilizing the ADP-bound state of Hsp70 impairs [*PSI*$^+$]. Accordingly, one way Cpr7 might regulate Hsp70 is by stabilizing the ADP-bound state (Figure 9.3) (Jones *et al.*, 2004).

Sti1 and Cpr7 are also involved in the curing of [*PSI*$^+$] by excess Hsp104 (Moosavi *et al.*, 2009; Reidy and Masison, 2010). The finding that Hsp104 mutants that suppress Ssa1-21 inhibition of [*PSI*$^+$] also fail to cure [*PSI*$^+$] when overexpressed suggests that Ssa1-21 inhibits prions in a manner similar to that of increasing Hsp104. In confirmation of this hypothesis, deleting Sti1, which suppresses Ssa1-21, was found to suppress the ability of overexpressed Hsp104 to cure cells of [*PSI*$^+$] (Reidy and Masison, 2010). This study also identified Hsp90 as a critical component of the curing mechanism. Given that Sti1 and Hsp90 have no apparent role in Hsp104-mediated protein disaggregation, the data are consistent with the earlier suggestion (Hung and Masison, 2006) that curing is not due to direct destruction of prions by Hsp104. In agreement with this conclusion, curing is tightly associated with cell division (Reidy and Masison,

2010)—an unnecessary requirement if prions were simply destroyed. Although Sup35 does not appear to be the target of ubiquitin, normal proteasome function is also important for curing by overexpressed Hsp104, and evidence indicates that the Hsp90 machinery is required for the contribution of ubiquitin to curing (Reidy and Masison, 2010). These data suggest the role for Hsp70/Hsp90 machinery in curing might include turnover of an as yet unidentified protein.

9.4 CONCLUSION

Much evidence supports the widely held view that replication of yeast prions both occurs by and depends upon fragmentation of preexisting prions by the Hsp104/Hsp70/Hsp40 disaggregation machinery. Although Hsp70 is a key factor in the many other effects chaperones and co-chaperones have on yeast prion phenotypes, the molecular mechanisms underlying these effects are much less certain. Hsp40s bind misfolded proteins with substrate specificity that overlaps that of Hsp70, but compelling evidence indicates that their influence on prions is primarily indirect through their regulation of Hsp70 activity rather than through direct interactions with prion proteins. Similarly, the cumulative data indicate that Hsp70 NEFs, including those of the Hsp110 family that bind peptide substrates (i.e., Sse), also act on prions by their influence on Hsp70's reaction cycle. TPR co-chaperones can also affect prions through the regulation of Hsp70. Hsp70 is therefore a central chaperone whose activity can be modified in many ways to influence the strength and stability of different yeast prions. Such effects presumably reflect alteration of one or more of the processes of amyloid growth, fragmentation, and transmission.

How Hsp70s cause so many different effects on prion propagation is unknown, but specific Hsp70/co-chaperone complexes that possess certain affinity and/or specificity of interaction with prion proteins are likely to be important for propagation of certain prions (Jones and Tuite, 2005; Sharma *et al.*, 2009a). In addition to direct effects, chaperone combinations might impact prions by virtue of their importance for controlling cellular processes that one prion might rely on more than the others. It is also clear that one prion can be profoundly affected by alterations of chaperone abundance or function that have little or no effect on other prions. This observation has led to the suggestion that prion variants could be distinguished by their responses to various chaperone deficiencies or overproduction (Kryndushkin and Wickner, 2007).

Hsp40s, NEFs, Hsp70s, and Hsp104, all copurify more efficiently with prion proteins from lysates of cells where prions are present than from cells in which the same proteins are in the soluble form (Bagriantsev *et al.*, 2008; Lopez *et al.*, 2003; Nevzglyadova *et al.*, 2009; Sondheimer *et al.*, 2001). These findings suggest that these protein folding factors physically interact with prion forms of the proteins in cells, which indicates that their effects on prions can be direct. When added alone or in certain combinations to purified prion proteins, these chaperones can affect oligomer formation and amyloid assembly in both positive and negative ways (Krzewska and Melki, 2006; Shorter and Lindquist, 2008), indicating that direct interactions could promote or inhibit prion propagation. On the other hand, Hsp70 mutations that alter substrate affinity in opposite directions can have identical inhibitory effects on prion propagation (Jones and Masison, 2003; Needham and Masison, 2008). Therefore, substrate interactions alone cannot explain all the differences in the ways chaperones influence prion phenotype, and whether a direct interaction of a chaperone with a prion will promote or impair prion propagation is hard to predict. Moreover, cellular processes, such as those involved in endosome dynamics, UPS control of damaged proteins, and actin polymerization and localization, can affect prion propagation and toxicity (Allen *et al.*, 2007; Chernova *et al.*, 2003; Ganusova *et al.*, 2006; Meriin *et al.*, 2007). Obviously, these processes cannot be assessed easily for their influence on amyloid formation *in vitro*. The complexity of chaperone interactions and functions in many cellular processes is daunting, and it is clear that a combined approach of genetics, biochemistry, and cytology is needed to understand how they influence prion propagation. It might be enlightening, for example, to test if a mutant chaperone, alone or as a component of chaperone machinery, influences amyloid formation *in vitro* in ways predicted by the way it affects prion propagation. Mutants in other components of the machinery that suppress effects of the original mutant on prion propagation could then be tested to see if they restore function of the machinery *in vitro*. Obtaining predictable results would strengthen our confidence that we understand the system at a molecular level.

Similarities of the amyloidogenic yeast prions are that they require the Hsp104 disaggregation machinery and they are structurally related. Although Hsp70 with its regulators can act on misfolded and aggregated proteins independently of Hsp104, the efficiency that individual Hsp70 isoforms collaborate with co-chaperones to function as components of the Hsp104 machinery probably contributes significantly to the overall effects on prion propagation. Additionally, the amyloid structures of the yeast prions that have been determined so far have a similar parallel in-register β-sheet conformation (Shewmaker *et al.*, 2006, 2008, 2009; Wickner *et al.*, 2008a, 2008b). Therefore, despite differences in the requirements prions have for propagation, it is possible that some feature of this structure, independent of amino acid sequence, is important for efficient chaperone recognition or action.

Despite their overall structural similarities, the basis of distinctions in the ways different prions respond to chaperone alterations could reside in conformational variations of the prion polymers that confer susceptibility to the action of certain chaperones. Even variants of prions formed of the same protein can arise due to differences in the number of residues that form the core of the fibers (Tanaka *et al.*, 2006). Although differences in kinetics of amyloid assembly and physical strength of fibers likely contribute to the phenotypic variation, the differences in structure might also enhance or restrict access to chaperones or alter the efficiency with which chaperones process them. Alternatively, specific amino acid sequences of prion proteins, or the ways the different prion proteins associate with cofactor partners could determine which chaperones most affect their propagation.

Non-amyloid-forming regions of proteins that assemble into amyloid retain normal structure and function (Baxa *et al.*, 2002, 2004; Krzewska *et al.*, 2007), so the proteins in their prion forms can still associate with cofactors that might influence their behavior. For example, Sup35 interacts with Sup45 on ribosomes and Ure2 interacts with the Gln3 transcription factor. Such interactions direct these prion proteins to different cellular locations or processes that might be important for propagation of [PSI^+] and [URE3].

On a cautionary note, feedback regulation of chaperone abundance and activity is complex so care should be taken when interpreting that effects of overproducing chaperones are simply due to mass action effects of increasing an enzymatic activity of the elevated protein. Several examples have been noted above that argue against such specific effects.

Lastly, it should be noted that the term prion defines a proteinaceous infectious element, and, since yeast cells are very different than the diseased mammalian neurons for which the term was originally coined, there are obvious differences in what the term implies for the two systems. One difference is that yeast prion amyloid propagates intracellularly, which is similar to a large number of mammalian amyloid disorders. Another important difference is that fungal prions generally require the Hsp104 machinery, of which there is no mammalian counterpart, for their propagation. Nevertheless, Hsp70 is highly conserved and human Hsc70 has been shown to support both yeast growth and prion propagation, thus providing a system for studying human chaperones, and Hsp104 has potential applications for therapy in protein folding disorders (Tutar *et al.*, 2006; Vashist *et al.*, 2010) (see Chapter 7). As yeast prions are amyloid-based protein folding problems, they will no doubt continue to provide considerable insight into the cellular processes that influence formation, propagation, and elimination of cytoplasmic amyloid.

ACKNOWLEDGMENT

I thank my colleagues at NIH and LBG/NIDDK for their helpful discussions and comments on the manuscript. This work was supported by the intramural program of the National Institutes of Health, National Institute of Diabetes and Digestive and Kidney Diseases.

REFERENCES

Acebron SP, Martin I, del Castillo U, Moro F, Muga A. DnaK-mediated association of ClpB to protein aggregates. A bichaperone network at the aggregate surface. *FEBS Lett* 2009;**583**:2991–2996.

Allen KD, Chernova TA, Tennant EP, Wilkinson KD, Chernoff YO. Effects of ubiquitin system alterations on the formation and loss of a yeast prion. *J Biol Chem* 2007;**282**:3004–3013.

Allen KD, Wegrzyn RD, Chernova TA, Muller S, Newnam GP, Winslett PA, Wittich KB, Wilkinson KD, Chernoff YO. Hsp70 chaperones as modulators of prion life cycle:

novel effects of Ssa and Ssb on the *Saccharomyces cerevisiae* prion [*PSI*+]. *Genetics* 2005;**169**:1227–1242.

Andreasson C, Fiaux J, Rampelt H, Mayer MP, Bukau B. Hsp110 is a nucleotide-activated exchange factor for Hsp70. *J Biol Chem* 2008;**283**:8877–8884.

Aron R, Higurashi T, Sahi C, Craig EA. J-protein co-chaperone Sis1 required for generation of [RNQ+] seeds necessary for prion propagation. *EMBO J* 2007;**26**:3794–3803.

Aron R, Lopez N, Walter W, Craig EA, Johnson J. *In Vivo* bipartite interaction between the Hsp40 Sis1 and Hsp70 in *Saccharomyces cerevisiae*. *Genetics* 2005;**169**:1873–1882.

Bach S, Tribouillard D, Talarek N, Desban N, Gug F, Galons H, Blondel M. A yeast-based assay to isolate drugs active against mammalian prions. *Methods* 2006;**39**:72–77.

Bagriantsev SN, Gracheva EO, Richmond JE, Liebman SW. Variant-specific [*PSI*+] infection is transmitted by Sup35 polymers within [*PSI*+] aggregates with heterogeneous protein composition. *Mol Biol Cell* 2008;**19**:2433–2443.

Bagriantsev S, Liebman S. Modulation of Abeta42 low-n oligomerization using a novel yeast reporter system. *BMC Biol* 2006;**4**:32.

Bardill JP, Dulle JE, Fisher JR, True HL. Requirements of Hsp104p activity and Sis1p binding for propagation of the [RNQ(+)] prion. *Prion* 2009;**3**:151–160.

Baxa U, Ross PD, Wickner RB, Steven AC. The N-terminal prion domain of Ure2p converts from an unfolded to a thermally resistant conformation upon filament formation. *J Mol Biol* 2004;**339**:259–264.

Baxa U, Speransky V, Steven AC, Wickner RB. Mechanism of inactivation on prion conversion of the *Saccharomyces cerevisiae* Ure2 protein. *Proc Natl Acad Sci U S A* 2002;**99**:5253–5260.

Borchsenius AS, Muller S, Newnam GP, Inge-Vechtomov SG, Chernoff YO. Prion variant maintained only at high levels of the Hsp104 disaggregase. *Curr Genet* 2006;**49**:21–29.

Borchsenius AS, Wegrzyn RD, Newnam GP, Inge-Vechtomov SG, Chernoff YO. Yeast prion protein derivative defective in aggregate shearing and production of new 'seeds'. *EMBO J* 2001;**20**:6683–6691.

Brachmann A, Baxa U, Wickner RB. Prion generation *in vitro*: amyloid of Ure2p is infectious. *EMBO J* 2005;**24**:3082–3092.

Brown CR, McCann JA, Chiang HL. The heat shock protein Ssa2p is required for import of fructose-1, 6-bisphosphatase into Vid vesicles. *J Cell Biol* 2000;**150**:65–76.

Byrne LJ, Cole DJ, Cox BS, Ridout MS, Morgan BJ, Tuite MF. The number and transmission of [*PSI*] prion seeds (Propagons) in the yeast *Saccharomyces cerevisiae*. *PLoS One* 2009;**4**:e4670.

Caplan AJ, Douglas MG. Characterization of YDJ1: a yeast homologue of the bacterial dnaJ protein. *J Cell Biol* 1991;**114**:609–621.

Caplan AJ, Tsai J, Casey PJ, Douglas MG. Farnesylation of YDJ1p is required for function at elevated growth temperatures in *Saccharomyces cerevisiae*. *J Biol Chem* 1992;**267**:18890–18895.

Chacinska A, Szczesniak B, Kochneva-Pervukhova NV, Kushnirov VV, Ter-Avanesyan MD, Boguta M. Ssb1 chaperone is a [*PSI*+] prion-curing factor. *Curr Genet* 2001; **39**:62–67.

Chai Y, Koppenhafer SL, Bonini NM, Paulson HL. Analysis of the role of heat shock protein (Hsp) molecular chaperones in polyglutamine disease. *J Neurosci* 1999;**19**: 10338–10347.

Chen S, Smith DF. Hop as an adaptor in the heat shock protein 70 (Hsp70) and hsp90 chaperone machinery. *J Biol Chem* 1998;**273**:35194–35200.

Chernoff YO, Lindquist SL, Ono B, Inge-Vechtomov SG, Liebman SW. Role of the chaperone protein Hsp104 in propagation of the yeast prion-like factor [*PSI*+]. *Science* 1995;**268**:880–884.

Chernoff YO, Newnam GP, Kumar J, Allen K, Zink AD. Evidence for a protein mutator in yeast: role of the Hsp70-related chaperone ssb in formation, stability, and toxicity of the [*PSI*] prion. *Mol Cell Biol* 1999;**19**:8103–8112.

Chernova TA, Allen KD, Wesoloski LM, Shanks JR, Chernoff YO, Wilkinson KD. Pleiotropic effects of Ubp6 loss on drug sensitivities and yeast prion are due to depletion of the free ubiquitin pool. *J Biol Chem* 2003;**278**:52102–52115.

Collins SR, Douglass A, Vale RD, Weissman JS. Mechanism of prion propagation: amyloid growth occurs by monomer addition. *PLoS Biol* 2004;**2**: e321.

Craig EA, Huang P, Aron R, Andrew A. The diverse roles of J-proteins, the obligate Hsp70 co-chaperone. *Rev Physiol Biochem Pharmacol* 2006;**156**:1–21.

Craig E, Ziegelhoffer T, Nelson J, Laloraya S, Halladay J. Complex multigene family of functionally distinct Hsp70s of yeast. *Cold Spring Harb Symp Quant Biol* 1995;**60**:441–449.

Cummings CJ, Sun Y, Opal P, Antalffy B, Mestril R, Orr HT, Dillmann WH, Zoghbi HY. Over-expression of inducible HSP70 chaperone suppresses neuropathology and improves motor function in SCA1 mice. *Hum Mol Genet* 2001;**10**:1511–1518.

Cyr DM. Cooperation of the molecular chaperone Ydj1 with specific Hsp70 homologs to suppress protein aggregation. *FEBS Lett* 1995;**359**:129–132.

Daugaard M, Rohde M, Jaattela M. The heat shock protein 70 family: Highly homologous proteins with overlapping and distinct functions. *FEBS Lett* 2007;**581**:3702–3710.

Derkatch IL, Bradley ME, Zhou P, Chernoff YO, Liebman SW. Genetic and environmental factors affecting the de novo appearance of the [*PSI*+] prion in *Saccharomyces cerevisiae*. *Genetics* 1997;**147**:507–519.

Dezwaan DC, Freeman BC. HSP90: the Rosetta stone for cellular protein dynamics? *Cell Cycle* 2008;**7**:1006–1012.

Douglas PM, Treusch S, Ren HY, Halfmann R, Duennwald ML, Lindquist S, Cyr DM. Chaperone-dependent amyloid assembly protects cells from prion toxicity. *Proc Natl Acad Sci U S A* 2008;**105**:7206–7211.

Doyle SM, Wickner S. Hsp104 and ClpB: protein disaggregating machines. *Trends Biochem Sci* 2009;**34**:40–48.

Dragovic Z, Broadley SA, Shomura Y, Bracher A, Hartl FU. Molecular chaperones of the Hsp110 family act as nucleotide exchange factors of Hsp70s. *EMBO J* 2006a;**25**: 2519–2528.

Dragovic Z, Shomura Y, Tzvetkov N, Hartl FU, Bracher A. Fes1p acts as a nucleotide exchange factor for the ribosome-associated molecular chaperone Ssb1p. *Biol Chem* 2006b;**387**:1593–1600.

Duennwald ML, Jagadish S, Muchowski PJ, Lindquist S. Flanking sequences profoundly alter polyglutamine toxicity in yeast. *Proc Natl Acad Sci U S A* 2006;**103**: 11045–11050.

Duina AA, Marsh JA, Kurtz RB, Chang HC, Lindquist S, Gaber RF. The peptidyl-prolyl isomerase domain of the CyP-40 cyclophilin homolog Cpr7 is not required to support growth or glucocorticoid receptor activity in *Saccharomyces cerevisiae*. *J Biol Chem* 1998;**273**:10819–10822.

Eaglestone SS, Cox BS, Tuite MF. Translation termination efficiency can be regulated in *Saccharomyces cerevisiae* by environmental stress through a prion-mediated mechanism. *EMBO J* 1999;**18**:1974–1981.

Eaglestone SS, Ruddock LW, Cox BS, Tuite MF. Guanidine hydrochloride blocks a critical step in the propagation of the prion-like determinant [*PSI*+] of *Saccharomyces cerevisiae*. *Proc Natl Acad Sci U S A* 2000;**97**:240–244.

Fan CY, Lee S, Ren HY, Cyr DM. Exchangeable chaperone modules contribute to specification of type I and type II Hsp40 cellular function. *Mol Biol Cell* 2004;**15**:761–773.

Fan Q, Park KW, Du Z, Morano KA, Li L. The role of Sse1 in the de novo formation and variant determination of the [*PSI*+] prion. *Genetics* 2007;**177**:1583–1593.

Ferreira PC, Ness F, Edwards SR, Cox BS, Tuite MF. The elimination of the yeast [*PSI*+] prion by guanidine hydrochloride is the result of Hsp104 inactivation. *Mol Microbiol* 2001;**40**:1357–1369.

Ferreira ST, Vieira MN, De Felice FG. Soluble protein oligomers as emerging toxins in Alzheimer's and other amyloid diseases. *IUBMB Life* 2007;**59**:332–345.

Flaherty KM, DeLuca-Flaherty C, McKay DB. Three-dimensional structure of the ATPase fragment of a 70K heat-shock cognate protein. *Nature* 1990;**346**:623–628.

Ganusova EE, Ozolins LN, Bhagat S, Newnam GP, Wegrzyn RD, Sherman MY, Chernoff YO. Modulation of prion formation, aggregation, and toxicity by the actin cytoskeleton in yeast. *Mol Cell Biol* 2006;**26**:617–629.

Glover JR, Lindquist S. Hsp104, Hsp70, and Hsp40: a novel chaperone system that rescues previously aggregated proteins. *Cell* 1998;**94**:73–82.

Goeckeler JL, Petruso AP, Aguirre J, Clement CC, Chiosis G, Brodsky JL. The yeast Hsp110, Sse1p, exhibits high-affinity peptide binding. *FEBS Lett* 2008;**582**: 2393–2396.

Goeckeler JL, Stephens A, Lee P, Caplan AJ, Brodsky JL. Overexpression of yeast Hsp110 homolog Sse1p suppresses ydj1-151 thermosensitivity and restores Hsp90-dependent activity. *Mol Biol Cell* 2002;**13**:2760–2770.

Gokhale KC, Newnam GP, Sherman MY, Chernoff YO. Modulation of prion-dependent polyglutamine aggregation and toxicity by chaperone proteins in the yeast model. *J Biol Chem* 2005;**280**:22809–22818.

Goloubinoff P, Mogk A, Zvi AP, Tomoyasu T, Bukau B. Sequential mechanism of solubilization and refolding of stable protein aggregates by a bichaperone network. *Proc Natl Acad Sci U S A* 1999;**96**:13732–13737.

Gragerov A, Zeng L, Zhao X, Burkholder W, Gottesman ME. Specificity of DnaK-peptide binding. *J Mol Biol* 1994;**235**:848–854.

Grimminger V, Richter K, Imhof A, Buchner J, Walter S. The prion curing agent guanidinium chloride specifically inhibits ATP hydrolysis by Hsp104. *J Biol Chem* 2004;**279**:7378–7383.

Grimminger-Marquardt V, Lashuel HA. Structure and function of the molecular chaperone Hsp104 from yeast. *Biopolymers* 2010;**93**:252–276.

Ha J-H, Johnson ER, McKay DB, Sousa MC, Takeda S, Wilbanks SM. Structure and mechanism of Hsp70 Proteins. In: Bukau B, editor. *Molecular chaperones and folding catalysts regulation, cellular function and mechanism*. Amsterdam: Harwood Academic Publishers; 1999. pp. 573–607.

Hainzl O, Wegele H, Richter K, Buchner J. Cns1 is an activator of the Ssa1 ATPase activity. *J Biol Chem* 2004;**279**:23267–23273.

Harrison CJ, Hayer-Hartl M, Di Liberto M, Hartl F, Kuriyan J. Crystal structure of the nucleotide exchange factor GrpE bound to the ATPase domain of the molecular chaperone DnaK. *Science* 1997;**276**:431–435.

Higurashi T, Hines JK, Sahi C, Aron R, Craig EA. Specificity of the J-protein Sis1 in the propagation of 3 yeast prions. *Proc Natl Acad Sci U S A* 2008;**105**:16596–16601.

Ho AK, Raczniak GA, Ives EB, Wente SR. The integral membrane protein snl1p is genetically linked to yeast nuclear pore complex function. *Mol Biol Cell* 1998;**9**:355–373.

Hrizo SL, Gusarova V, Habiel DM, Goeckeler JL, Fisher EA, Brodsky JL. The Hsp110 molecular chaperone stabilizes apolipoprotein B from endoplasmic reticulum-associated degradation (ERAD). *J Biol Chem* 2007;**282**:32665–32675.

Hung GC, Masison DC. N-terminal domain of yeast Hsp104 chaperone is dispensable for thermotolerance and prion propagation but necessary for curing prions by Hsp104 overexpression. *Genetics* 2006;**173**:611–620.

Inoue Y, Kishimoto A, Hirao J, Yoshida M, Taguchi H. Strong growth polarity of yeast prion fiber revealed by single fiber imaging. *J Biol Chem* 2001;**276**:35227–35230.

Inoue Y, Taguchi H, Kishimoto A, Yoshida M. Hsp104 binds to yeast Sup35 prion fiber but needs other factor(s) to sever it. *J Biol Chem* 2004;**279**:52319–52323.

James P, Pfund C, Craig EA. Functional specificity among Hsp70 molecular chaperones. *Science* 1997;**275**:387–389.

Johnson JL, Craig EA. An essential role for the substrate-binding region of Hsp40s in *Saccharomyces cerevisiae*. *J Cell Biol* 2001;**152**:851–856.

Jones GW, Masison DC. *Saccharomyces cerevisiae* Hsp70 mutations affect [*PSI+*] prion propagation and cell growth differently and implicate Hsp40 and tetratricopeptide repeat cochaperones in impairment of [*PSI+*]. *Genetics* 2003;**163**:495–506.

Jones G, Song Y, Chung S, Masison DC. Propagation of *Saccharomyces cerevisiae* [*PSI+*] prion is impaired by factors that regulate Hsp70 substrate binding. *Mol Cell Biol* 2004;**24**:3928–3937.

Jones GW, Tuite MF. Chaperoning prions: the cellular machinery for propagating an infectious protein? *Bioessays* 2005;**27**:823–832.

Jung G, Jones G, Wegrzyn RD, Masison DC. A role for cytosolic hsp70 in yeast [*PSI+*] prion propagation and [*PSI+*] as a cellular stress. *Genetics* 2000;**156**:559–570.

Jung G, Masison DC. Guanidine hydrochloride inhibits Hsp104 activity *in vivo*: a possible explanation for its effect in curing yeast prions. *Curr Microbiol* 2001;**43**:7–10.

Kabani M, Beckerich JM, Brodsky JL. Nucleotide exchange factor for the yeast Hsp70 molecular chaperone Ssa1p. *Mol Cell Biol* 2002;**22**:4677–4689.

King CY, Diaz-Avalos R. Protein-only transmission of three yeast prion strains. *Nature* 2004;**428**:319–323.

Klucken J, Shin Y, Masliah E, Hyman BT, McLean PJ. Hsp70 reduces alpha-synuclein aggregation and toxicity. *J Biol Chem* 2004;**279**:25497–25502.

Krobitsch S, Lindquist S. Aggregation of huntingtin in yeast varies with the length of the polyglutamine expansion and the expression of chaperone proteins. *Proc Natl Acad Sci U S A* 2000;**97**:1589–1594.

Kryndushkin DS, Alexandrov IM, Ter-Avanesyan MD, Kushnirov VV. Yeast [*PSI*+] prion aggregates are formed by small Sup35 polymers fragmented by Hsp104. *J Biol Chem* 2003;**278**:49636–49643.

Kryndushkin DS, Shewmaker F, Wickner RB. Curing of the [URE3] prion by Btn2p, a Batten disease-related protein. *EMBO J* 2008;**27**:2725–2735.

Kryndushkin DS, Smirnov VN, Ter-Avanesyan MD, Kushnirov VV. Increased expression of Hsp40 chaperones, transcriptional factors, and ribosomal protein Rpp0 can cure yeast prions. *J Biol Chem* 2002;**277**:23702–23708.

Kryndushkin D, Wickner RB. Nucleotide exchange factors for Hsp70s are required for [URE3] prion propagation in *Saccharomyces cerevisiae*. *Mol Biol Cell* 2007;**18**: 2149–2154.

Krzewska J, Melki R. Molecular chaperones and the assembly of the prion Sup35p, an *in vitro* study. *EMBO J* 2006;**25**:822–833.

Krzewska J, Tanaka M, Burston SG, Melki R. Biochemical and functional analysis of the assembly of full-length Sup35p and its prion-forming domain. *J Biol Chem* 2007;**282**:1679–1686.

Kushnirov VV, Kryndushkin DS, Boguta M, Smirnov VN, Ter-Avanesyan MD. Chaperones that cure yeast artificial [*PSI*+] and their prion-specific effects. *Curr Biol* 2000;**10**:1443–1446.

Li A, Harris DA. Mammalian prion protein suppresses Bax-induced cell death in yeast. *J Biol Chem* 2005;**280**:17430–17434.

Lian HY, Zhang H, Zhang ZR, Loovers HM, Jones GW, Rowling PJ, Itzhaki LS, Zhou JM, Perrett S. Hsp40 interacts directly with the native state of the yeast prion protein Ure2 and inhibits formation of amyloid-like fibrils. *J Biol Chem* 2007;**282**:11931–11940.

Liang J, Clark-Dixon C, Wang S, Flower TR, Williams-Hart T, Zweig R, Robinson LC, Tatchell K, Witt SN. Novel suppressors of alpha-synuclein toxicity identified using yeast. *Hum Mol Genet* 2008;**17**:3784–3795.

Liu XD, Morano KA, Thiele DJ. The yeast Hsp110 family member, Sse1, is an Hsp90 cochaperone. *J Biol Chem* 1999;**274**:26654–26660.

Loovers HM, Guinan E, Jones GW. Importance of the Hsp70 ATPase domain in yeast prion propagation. *Genetics* 2007;**175**:621–630.

Lopez N, Aron R, Craig EA. Specificity of class II Hsp40 Sis1 in maintenance of yeast prion [RNQ+]. *Mol Biol Cell* 2003;**14**:1172–1181.

Lu Z, Cyr DM. Protein folding activity of Hsp70 is modified differentially by the hsp40 co-chaperones Sis1 and Ydj1. *J Biol Chem* 1998;**273**:27824–27830.

Luke MM, Sutton A, Arndt KT. Characterization of SIS1, a *Saccharomyces cerevisiae* homologue of bacterial dnaJ proteins. *J Cell Biol* 1991;**114**:623–638.

Lum R, Tkach JM, Vierling E, Glover JR. Evidence for an unfolding/threading mechanism for protein disaggregation by *Saccharomyces cerevisiae* Hsp104. *J Biol Chem* 2004;**279**:29139–29146.

Marsh JA, Kalton HM, Gaber RF. Cns1 is an essential protein associated with the hsp90 chaperone complex in *Saccharomyces cerevisiae* that can restore cyclophilin 40-dependent functions in cpr7Delta cells. *Mol Cell Biol* 1998;**18**:7353–7359.

Masison DC, Kirkland PA, Sharma D. Influence of Hsp70s and their regulators on yeast prion propagation. *Prion* 2009;**3**:65–73.

Mathur V, Hong JY, Liebman SW. Ssa1 overexpression and [PIN(+)] variants cure [*PSI*(+)] by dilution of aggregates. *J Mol Biol* 2009;**390**:155–167.

Meimaridou E, Gooljar SB, Chapple JP. From hatching to dispatching: the multiple cellular roles of the Hsp70 molecular chaperone machinery. *J Mol Endocrinol* 2009;**42**:1–9.

Meriin AB, Zhang X, Alexandrov IM, Salnikova AB, Ter-Avanesian MD, Chernoff YO, Sherman MY. Endocytosis machinery is involved in aggregation of proteins with expanded polyglutamine domains. *FASEB J* 2007;**21**:1915–1925.

Meriin AB, Zhang X, He X, Newnam GP, Chernoff YO, Sherman MY. Huntington toxicity in yeast model depends on polyglutamine aggregation mediated by a prion-like protein Rnq1. *J Cell Biol* 2002;**157**:997–1004.

Moosavi B, Wongwigkarn J, Tuite MF. Hsp70/Hsp90 co-chaperones are required for efficient Hsp104-mediated elimination of the yeast [*PSI*(+)] prion but not for prion propagation. *Yeast* 2009;**27**:167–179.

Moriyama H, Edskes HK, Wickner RB. [URE3] prion propagation in *Saccharomyces cerevisiae*: requirement for chaperone Hsp104 and curing by overexpressed chaperone Ydj1p. *Mol Cell Biol* 2000;**20**:8916–8922.

Muchowski PJ, Schaffar G, Sittler A, Wanker EE, Hayer-Hartl MK, Hartl FU. Hsp70 and hsp40 chaperones can inhibit self-assembly of polyglutamine proteins into amyloid-like fibrils. *Proc Natl Acad Sci U S A* 2000;**97**:7841–7846.

Mukai H, Kuno T, Tanaka H, Hirata D, Miyakawa T, Tanaka C. Isolation and characterization of SSE1 and SSE2, new members of the yeast HSP70 multigene family. *Gene* 1993;**132**:57–66.

Needham PG, Masison DC. Prion-impairing mutations in Hsp70 chaperone Ssa1: effects on ATPase and chaperone activities. *Arch Biochem Biophys* 2008;**478**:167–174.

Nelson RJ, Ziegelhoffer T, Nicolet C, Werner-Washburne M, Craig EA. The translation machinery and 70kd heat shock protein cooperate in protein synthesis. *Cell* 1992;**71**:97–105.

Nemecek J, Nakayashiki T, Wickner RB. A prion of yeast metacaspase homolog (Mca1p) detected by a genetic screen. *Proc Natl Acad Sci U S A* 2009;**106**:1892–1896.

Ness F, Ferreira P, Cox BS, Tuite MF. Guanidine hydrochloride inhibits the generation of prion "seeds" but not prion protein aggregation in yeast. *Mol Cell Biol* 2002;**22**: 5593–5605.

Nevzglyadova OV, Artemov AV, Mittenberg AG, Solovyov KV, Kostyleva EI, Mikhailova EV, Kuznetsova IM, Turoverov KK, Soidla TR. Prion-associated proteins in yeast: comparative analysis of isogenic [*PSI*(+)] and [*psi*(-)] strains. *Yeast* 2009;**26**:611–631.

Newnam GP, Wegrzyn RD, Lindquist SL, Chernoff YO. Antagonistic interactions between yeast chaperones Hsp104 and Hsp70 in prion curing. *Mol Cell Biol* 1999;**19**: 1325–1333.

Ohba M. Modulation of intracellular protein degradation by SSB1-SIS1 chaperone system in yeast *S. cerevisiae*. *FEBS Lett* 1997;**409**:307–311.

Outeiro TF, Muchowski PJ. Molecular genetics approaches in yeast to study amyloid diseases. *J Mol Neurosci* 2004;**23**:49–60.

Palhano FL, Rocha CB, Bernardino A, Weissmuller G, Masuda CA, Montero-Lomeli M, Gomes AM, Chien P, Fernandes PM, Foguel D. A fluorescent mutant of the NM domain of the yeast prion Sup35 provides insight into fibril formation and stability. *Biochemistry* 2009;**48**:6811–6823.

Park KW, Hahn JS, Fan Q, Thiele DJ, Li L. De novo appearance and "strain" formation of yeast prion [*PSI*+] are regulated by the heat-shock transcription factor. *Genetics* 2006;**173**:35–47.

Patel BK, Liebman SW. "Prion-proof" for [PIN+]: infection with *in vitro*-made amyloid aggregates of Rnq1p-(132–405) induces [PIN+]. *J Mol Biol* 2007;**365**:773–782.

Patury S, Miyata Y, Gestwicki JE. Pharmacological targeting of the Hsp70 chaperone. *Curr Top Med Chem* 2009;**9**:1337–1351.

Paushkin SV, Kushnirov VV, Smirnov VN, Ter-Avanesyan MD. Propagation of the yeast prion-like [*PSI*+] determinant is mediated by oligomerization of the SUP35-encoded polypeptide chain release factor. *EMBO J* 1996;**15**:3127–3134.

Pelham H, Lewis M, Lindquist S. Expression of a *Drosophila* heat shock protein in mammalian cells: transient association with nucleoli after heat shock. *Philos Trans R Soc Lond B Biol Sci* 1984;**307**:301–307.

Prusiner S. Biology of Prion diseases - lessons for other neurodegenerative disorders. *J Neurol Sci* 2010;**285**:S1–S1.

Raviol H, Sadlish H, Rodriguez F, Mayer MP, Bukau B. Chaperone network in the yeast cytosol: Hsp110 is revealed as an Hsp70 nucleotide exchange factor. *EMBO J* 2006;**25**:2510–2518.

Reidy M, Masison DC. Sti1 regulation of Hsp70 and Hsp90 is critical for curing of *S. cerevisiae* [*PSI*+] prions by Hsp104. *Mol Cell Biol* 2010;**30**:3542–3552.

Richarme G, Kohiyama M. Specificity of the *Escherichia coli* chaperone DnaK (70-kDa heat shock protein) for hydrophobic amino acids. *J Biol Chem* 1993;**268**: 24074–24077.

Roberts BT, Moriyama H, Wickner RB. [URE3] prion propagation is abolished by a mutation of the primary cytosolic Hsp70 of budding yeast. *Yeast* 2004;**21**:107–117.

Rohde M, Daugaard M, Jensen MH, Helin K, Nylandsted J, Jaattela M. Members of the heat-shock protein 70 family promote cancer cell growth by distinct mechanisms. *Genes Dev* 2005;**19**:570–582.

Sadlish H, Rampelt H, Shorter J, Wegrzyn RD, Andreasson C, Lindquist S, Bukau B. Hsp110 chaperones regulate prion formation and propagation in *S. cerevisiae* by two discrete activities. *PLoS One* 2008;**3**:e1763.

Sahi C, Craig EA. Network of general and specialty J protein chaperones of the yeast cytosol. *Proc Natl Acad Sci U S A* 2007;**104**:7163–7168.

Sanchez Y, Lindquist SL. HSP104 required for induced thermotolerance. *Science* 1990;**248**:1112–1115.

Satpute-Krishnan P, Serio TR. Prion protein remodelling confers an immediate phenotypic switch. *Nature* 2005;**437**:262–265.

Savistchenko J, Krzewska J, Fay N, Melki R. Molecular chaperones and the assembly of the prion Ure2p *in vitro*. *J Biol Chem* 2008;**283**:15732–15739.

Scheibel T, Kowal AS, Bloom JD, Lindquist SL. Bidirectional amyloid fiber growth for a yeast prion determinant. *Curr Biol* 2001;**11**:366–369.

Scheufler C, Brinker A, Bourenkov G, Pegoraro S, Moroder L, Bartunik H, Hartl FU, Moarefi I. Structure of TPR domain-peptide complexes: critical elements in the assembly of the Hsp70-Hsp90 multichaperone machine. *Cell* 2000;**101**:199–210.

Schlegel T, Mirus O, von Haeseler A, Schleiff E. The tetratricopeptide repeats of receptors involved in protein translocation across membranes. *Mol Biol Evol* 2007;**24**:2763–2774.

Schlieker C, Tews I, Bukau B, Mogk A. Solubilization of aggregated proteins by ClpB/DnaK relies on the continuous extraction of unfolded polypeptides. *FEBS Lett* 2004;**578**:351–356.

Schroder H, Langer T, Hartl FU, Bukau B. DnaK, DnaJ and GrpE form a cellular chaperone machinery capable of repairing heat-induced protein damage. *EMBO J* 1993;**12**:4137–4144.

Schwimmer C, Masison DC. Antagonistic interactions between yeast [*PSI*(+)] and [URE3] prions and curing of [URE3] by Hsp70 protein chaperone Ssa1p but not by Ssa2p. *Mol Cell Biol* 2002;**22**:3590–3598.

Shaner L, Sousa R, Morano KA. Characterization of Hsp70 binding and nucleotide exchange by the yeast Hsp110 chaperone Sse1. *Biochemistry* 2006;**45**:15075–15084.

Shaner L, Trott A, Goeckeler JL, Brodsky JL, Morano KA. The function of the yeast molecular chaperone Sse1 is mechanistically distinct from the closely related hsp70 family. *J Biol Chem* 2004;**279**:21992–22001.

Shaner L, Wegele H, Buchner J, Morano KA. The yeast Hsp110 Sse1 functionally interacts with the Hsp70 chaperones Ssa and Ssb. *J Biol Chem* 2005;**280**:41262–41269.

Sharma D, Martineau CN, Le Dall MT, Reidy M, Masison DC, Kabani M. Function of SSA subfamily of Hsp70 within and across species varies widely in complementing *Saccharomyces cerevisiae* cell growth and prion propagation. *PLoS One* 2009a;**4**: e6644.

Sharma D, Masison DC. Functionally redundant isoforms of a yeast Hsp70 chaperone subfamily have different antiprion effects. *Genetics* 2008;**179**:1301–1311.

Sharma D, Stanley RF, Masison DC. Curing of yeast [URE3] prion by the Hsp40 cochaperone Ydj1p is mediated by Hsp70. *Genetics* 2009b;**181**:129–137.

Shewmaker F, Kryndushkin D, Chen B, Tycko R, Wickner RB. Two prion variants of Sup35p have in-register parallel beta-sheet structures, independent of hydration. *Biochemistry* 2009;**48**:5074–5082.

Shewmaker F, Ross ED, Tycko R, Wickner RB. Amyloids of shuffled prion domains that form prions have a parallel in-register beta-sheet structure. *Biochemistry* 2008;**47**:4000–4007.

Shewmaker F, Wickner RB, Tycko R. Amyloid of the prion domain of Sup35p has an in-register parallel beta-sheet structure. *Proc Natl Acad Sci U S A* 2006;**103**: 19754–19759.

Shomura Y, Dragovic Z, Chang HC, Tzvetkov N, Young JC, Brodsky JL, Guerriero V, Hartl FU, Bracher A. Regulation of Hsp70 function by HspBP1: structural analysis reveals an alternate mechanism for Hsp70 nucleotide exchange. *Mol Cell* 2005;**17**:367–379.

Shorter J, Lindquist S. Hsp104 catalyzes formation and elimination of self-replicating Sup35 prion conformers. *Science* 2004;**304**:1793–1797.

Shorter J, Lindquist S. Destruction or potentiation of different prions catalyzed by similar Hsp104 remodeling activities. *Mol Cell* 2006;**23**:425–438.

Shorter J, Lindquist S. Hsp104, Hsp70 and Hsp40 interplay regulates formation, growth and elimination of Sup35 prions. *EMBO J* 2008;**27**:2712–2724.

Sikorski RS, Boguski MS, Goebl M, Hieter P. A repeating amino acid motif in CDC23 defines a family of proteins and a new relationship among genes required for mitosis and RNA synthesis. *Cell* 1990;**60**:307–317.

Singh A, Helms C, Sherman F. Mutation of the non-Mendelian suppressor, *PSI*, in yeast by hypertonic media. *Proc Natl Acad Sci U S A* 1979;**76**:1952–1956.

Smith DF. Tetratricopeptide repeat cochaperones in steroid receptor complexes. *Cell Stress Chaperones* 2004;**9**:109–121.

Sondermann H, Ho AK, Listenberger LL, Siegers K, Moarefi I, Wente SR, Hartl FU, Young JC. Prediction of novel Bag-1 homologs based on structure/function analysis identifies Snl1p as an Hsp70 co-chaperone in *Saccharomyces cerevisiae*. *J Biol Chem* 2002;**277**:33220–33227.

Sondheimer N, Lindquist S. Rnq1: an epigenetic modifier of protein function in yeast. *Mol Cell* 2000;**5**:163–172.

Sondheimer N, Lopez N, Craig EA, Lindquist S. The role of Sis1 in the maintenance of the [RNQ+] prion. *EMBO J* 2001;**20**:2435–2442.

Song Y, Masison DC. Independent regulation of Hsp70 and Hsp90 chaperones by Hsp70/Hsp90-organizing protein Sti1 (Hop1). *J Biol Chem* 2005;**280**:34178–34185.

Song Y, Wu YX, Jung G, Tutar Y, Eisenberg E, Greene LE, Masison DC. Role for Hsp70 chaperone in *Saccharomyces cerevisiae* prion seed replication. *Eukaryot Cell* 2005;**4**:289–297.

Soper JH, Roy S, Stieber A, Lee E, Wilson RB, Trojanowski JQ, Burd CG, Lee VM. Alpha-synuclein-induced aggregation of cytoplasmic vesicles in *Saccharomyces cerevisiae*. *Mol Biol Cell* 2008;**19**:1093–1103.

Sousa R, Lafer EM. Keep the traffic moving: mechanism of the Hsp70 motor. *Traffic* 2006;**7**:1596–1603.

Sweeny EA, Shorter J. Prion proteostasis: Hsp104 meets its supporting cast. *Prion* 2008;**2**:135–140.

Szabo A, Langer T, Schroder H, Flanagan J, Bukau B, Hartl FU. The ATP hydrolysis-dependent reaction cycle of the *Escherichia coli* Hsp70 system DnaK, DnaJ, and GrpE. *Proc Natl Acad Sci U S A* 1994;**91**:10345–10349.

Tanaka M, Chien P, Naber N, Cooke R, Weissman JS. Conformational variations in an infectious protein determine prion strain differences. *Nature* 2004;**428**:323–328.

Tanaka M, Collins SR, Toyama BH, Weissman JS. The physical basis of how prion conformations determine strain phenotypes. *Nature* 2006;**442**:585–589.

Tesic M, Marsh JA, Cullinan SB, Gaber RF. Functional interactions between Hsp90 and the co-chaperones Cns1 and Cpr7 in *Saccharomyces cerevisiae*. *J Biol Chem* 2003;**278**: 32692–32701.

Tessarz P, Mogk A, Bukau B. Substrate threading through the central pore of the Hsp104 chaperone as a common mechanism for protein disaggregation and prion propagation. *Mol Microbiol* 2008;**68**:87–97.

Thomas PJ, Qu BH, Pedersen PL. Defective protein folding as a basis of human disease. *Trends Biochem Sci* 1995;**20**:456–459.

Tipton KA, Verges KJ, Weissman JS. *In vivo* monitoring of the prion replication cycle reveals a critical role for Sis1 in delivering substrates to Hsp104. *Mol Cell* 2008;**32**:584–591.

Tribouillard-Tanvier D, Beringue V, Desban N, Gug F, Bach S, Voisset C, Galons H, Laude H, Vilette D, Blondel M. Antihypertensive drug guanabenz is active *in vivo* against both yeast and mammalian prions. *PLoS One* 2008;**3**:e1981.

Tuite MF, Mundy CR, Cox BS. Agents that cause a high frequency of genetic change from [*PSI*+] to [*PSI*-] in *Saccharomyces cerevisiae*. *Genetics* 1981;**98**:691–711.

Tutar Y, Song Y, Masison DC. Primate chaperones Hsc70 (constitutive) and Hsp70 (induced) differ functionally in supporting growth and prion propagation in *Saccharomyces cerevisiae*. *Genetics* 2006;**172**:851–861.

Vashist S, Cushman M, Shorter J. Applying Hsp104 to protein-misfolding disorders. *Biochem Cell Biol* 2010;**88**:1–13.

Wacker JL, Huang SY, Steele AD, Aron R, Lotz GP, Nguyen Q, Giorgini F, Roberson ED, Lindquist S, Masliah E, *et al.* Loss of Hsp70 exacerbates pathogenesis but not levels of fibrillar aggregates in a mouse model of Huntington's disease. *J Neurosci* 2009;**29**:9104–9114.

Walsh P, Bursac D, Law YC, Cyr D, Lithgow T. The J-protein family: modulating protein assembly, disassembly and translocation. *EMBO Rep* 2004;**5**:567–571.

Wandinger SK, Richter K, Buchner J. The Hsp90 chaperone machinery. *J Biol Chem* 2008;**283**:18473–18477.

Warrick JM, Chan HY, Gray-Board GL, Chai Y, Paulson HL, Bonini NM. Suppression of polyglutamine-mediated neurodegeneration in *Drosophila* by the molecular chaperone HSP70. *Nat Genet* 1999;**23**:425–428.

Wegele H, Haslbeck M, Reinstein J, Buchner J. Sti1 is a novel activator of the Ssa proteins. *J Biol Chem* 2003;**278**:25970–25976.

Wegrzyn RD, Bapat K, Newnam GP, Zink AD, Chernoff YO. Mechanism of prion loss after Hsp104 inactivation in yeast. *Mol Cell Biol* 2001;**21**:4656–4669.

Wegrzyn RD, Deuerling E. Molecular guardians for newborn proteins: ribosome-associated chaperones and their role in protein folding. *Cell Mol Life Sci* 2005;**62**: 2727–2738.

Weibezahn J, Tessarz P, Schlieker C, Zahn R, Maglica Z, Lee S, Zentgraf H, Weber-Ban EU, Dougan DA, Tsai FT, *et al.* Thermotolerance requires refolding of aggregated proteins by substrate translocation through the central pore of ClpB. *Cell* 2004;**119**: 653–665.

Werner-Washburne M, Stone DE, Craig EA. Complex interactions among members of an essential subfamily of hsp70 genes in *Saccharomyces cerevisiae*. *Mol Cell Biol* 1987;**7**:2568–2577.

Wickner RB. [URE3] as an altered URE2 protein: evidence for a prion analog in *Saccharomyces cerevisiae*. *Science* 1994;**264**:566–569.

Wickner RB, Dyda F, Tycko R. Amyloid of Rnq1p, the basis of the [PIN+] prion, has a parallel in-register beta-sheet structure. *Proc Natl Acad Sci U S A* 2008a;**105**: 2403–2408.

Wickner RB, Shewmaker F, Kryndushkin D, Edskes HK. Protein inheritance (prions) based on parallel in-register beta-sheet amyloid structures. *Bioessays* 2008b;**30**: 955–964.

Willingham S, Outeiro TF, DeVit MJ, Lindquist SL, Muchowski PJ. Yeast genes that enhance the toxicity of a mutant huntingtin fragment or alpha-synuclein. *Science* 2003;**302**:1769–1772.

Winderickx J, Delay C, De Vos A, Klinger H, Pellens K, Vanhelmont T, Van Leuven F, Zabrocki P. Protein folding diseases and neurodegeneration: lessons learned from yeast. *Biochim Biophys Acta* 2008;**1783**:1381–1395.

Wu Y, Sha B. Crystal structure of yeast mitochondrial outer membrane translocon member Tom70p. *Nat Struct Mol Biol* 2006;**13**:589–593.

Yam AY, Albanese V, Lin HT, Frydman J. Hsp110 cooperates with different cytosolic HSP70 systems in a pathway for de novo folding. *J Biol Chem* 2005;**280**: 41252–41261.

Young JC, Agashe VR, Siegers K, Hartl FU. Pathways of chaperone-mediated protein folding in the cytosol. *Nat Rev Mol Cell Biol* 2004;**5**:781–791.

Young JC, Hoogenraad NJ, Hartl FU. Molecular chaperones Hsp90 and Hsp70 deliver preproteins to the mitochondrial import receptor Tom70. *Cell* 2003;**112**:41–50.

Zhu X, Zhao X, Burkholder WF, Gragerov A, Ogata CM, Gottesman ME, Hendrickson WA. Structural analysis of substrate binding by the molecular chaperone DnaK. *Science* 1996;**272**:1606–1614.

Zietkiewicz S, Krzewska J, Liberek K. Successive and synergistic action of the Hsp70 and Hsp100 chaperones in protein disaggregation. *J Biol Chem* 2004;**279**:44376–44383.

Zietkiewicz S, Lewandowska A, Stocki P, Liberek K. Hsp70 chaperone machine remodels protein aggregates at the initial step of Hsp70-Hsp100-dependent disaggregation. *J Biol Chem* 2006;**281**:7022–7029.

10

ALS AND THE COPPER CHAPERONE CCS

Marjatta Son and Jeffrey L. Elliott

Department of Neurology, University of Texas, Southwestern Medical Center, Dallas, TX, USA

10.1 INTRODUCTION

Proteins that rely on metal ions for activity can be found in multiple subcellular compartments, including the cytosol, nucleus, endoplasmic reticulum, peroxisomes, and mitochondria. Metal ions such as copper pose a distinct problem to living cells in that these ions are not only needed cofactors for metalloproteins and essential for viability but are also potentially toxic. The intracellular free copper ion concentration is very low; it is less than one free copper atom per cell in yeast (Rae *et al.*, 1999). This pool of free copper ions is not used in physiological activation of metalloenzymes. Therefore, copper-containing proteins acquire their copper from specific copper transporters (CTRs) or chaperones. CTRs, like metal transporters in general, must have specific structural and functional characteristics that allow the correct metal ion to be transported and presented to the appropriate metalloprotein in such a manner as to avoid toxic side reactions of the bound metal ions. Therefore, cells have evolved specific trafficking as well as utilization pathways to ensure proper delivery of copper to targets and detoxification pathways to prevent the accumulation of free copper ions (Luk *et al.*, 2003; Rae *et al.*, 2001).

Protein Chaperones and Protection from Neurodegenerative Diseases, First Edition.
Edited by Stephan N. Witt.

Intracellular copper chaperones share similar characteristics. The action of the copper chaperones appears highly specific; each chaperone molecule preferentially interacts with one copper-requiring target. A structural hallmark of the copper chaperones is the similarity of the protein fold between the chaperone and its target protein. Metal insertion into the target protein appears to involve a docking of the metal donor and acceptor sites in close proximity to one another. The copper chaperones may function by lowering the activation barrier for metal transfer into specific protein-binding sites (O'Halloran and Culotta, 2000).

10.2 COPPER TRAFFICKING PATHWAYS

Specific pathways for copper trafficking exist in mammalian cells to maintain proper intracellular copper homeostasis. Copper trafficking depends on the redox state of copper ions with only reduced copper (Cu^{+1}) being able to be transported into cells. Since environmental copper exists primarily in an oxidized form (Cu^{+2}), copper reduction is the first step in copper trafficking pathways (Bartnikas and Gitlin, 2001). The STEAP family of oxido-reductases at the cell surface serves as key reductases in this process (Ohgami *et al.*, 2006). The CTR1, an integral membrane protein, internalizes reduced copper into cells functioning as a major copper importer at the plasma membrane (Lee *et al.*, 2000). Subsequently, copper is delivered to specific destinations by copper chaperones. A soluble copper carrier, ATOX1, delivers copper to ATP7A and ATP7B in the secretory compartment. ATOX1 has an N-terminal copper-binding motif (CXXC) that directly interacts with the N-terminal copper-binding domain of ATP7A and ATP7B (MTCXXC). ATP7A and ATP7B transport copper into the lumen of the Golgi, where the metal can then be incorporated into the copper-dependent proteins. They also export copper out of the cell by translocating to the plasma membrane when the intracellular copper levels are high (Prohaska, 2008; Turski and Thiele, 2009).

Copper is also required within the mitochondrion for the maturation of cytochrome *c* oxidase (complex IV). The two proteins implicated in copper ion translocation in the mitochondrion are COX17 and COX19. Both proteins are conserved in eukaryotic cells and exhibit a dual localization in both the cytosol and the mitochondrial intermembrane space (IMS). Once inside the IMS, copper is passed from COX17 to SCO2 and SCO1 or to COX11. SCO1 and SCO2 transfer copper to the COX2 subunit of complex IV, whereas COX11 transfers copper to the COX1 subunit of complex IV (Cobine *et al.*, 2006). Copper may also be available for mitochondrial metalloenzymes from a non-proteinaceous pool localized in the mtiochondrial matrix (Cobine *et al.*, 2006).

The delivery of copper to SOD1 is mediated through a soluble factor copper chaperone for SOD1 (CCS) (Culotta *et al.*, 1997). CCS appears to be specific for SOD1 although it has also been found to interact with some other proteins, BACE1 (Angeletti *et al.*, 2005; Dingwall, 2007), and neuronal adaptor protein XII alpha (McLoughlin *et al.*, 2001).

10.3 SUPEROXIDE DISMUTASES

The principal function of the superoxide dismutase (SOD) family is the elimination of superoxide anion radicals. Three distinct forms of SODs have been identified and characterized in mammals: copper/zinc superoxide dismutase (SOD1) (McCord and Fridovich, 1969), manganese superoxide dismutase (SOD2 or MnSOD) (Weisiger and Fridovich, 1973), and extracellular superoxide dismutase (SOD3) (Marklund, 1982). These forms of SOD possess similar enzymatic functions but differ in protein structure, chromosome localization, cellular compartmentalization, and metal cofactor requirements (Culotta *et al.*, 2006). SOD1 and SOD3 require copper as a metal cofactor, whereas SOD2 requires manganese.

All known SODs require a redox active transition metal in the active site to accomplish the catalytic breakdown of superoxide anions. SOD1 is a homodimeric copper and zinc-containing enzyme, which catalyzes the disproportionation of superoxide anion to hydrogen peroxide and oxygen in a reaction mediated through the cyclic reduction and oxidation of the bound copper ion:

$$Cu^{2+}, Zn^{2+} - SOD1 + O_2{}^- \leftrightarrow Cu^{1+}, Zn^{2+} - SOD1 + O_2$$

$$Cu^{1+}, Zn^{2+} - SOD1 + O_2{}^- + 2H^+ \leftrightarrow Cu^{2+}, Zn^{2+} - SOD1 + H_2O_2$$

The copper cofactor catalyzes both a one-electron oxidation (first step) and a one-electron reduction (second step) of separate superoxide anions to give the overall disproportionation reaction (Furukawa and O'Halloran, 2006; McCord and Fridovich, 1969). These reactions typically require no external source of redox equivalents and are thus self-contained components of the antioxidant machinery. This allows the copper- and zinc-containing SODs to function in a variety of intracellular and extracellular environments (Culotta *et al.*, 2006).

SOD3 also depends on copper for its activity and is typically made in vascular smooth muscle cells and secreted into the extracellular environment. The central core of the SOD3 polypeptide is homologous to SOD1 but possesses extensions at the N- and C-termini. Unlike SOD1, it forms stable tetramers with interchain disulfides that stabilize the quaternary structure. The copper-loading pathways for SOD3 are also quite different from those observed for SOD1. The secreted SOD3 protein appears to be loaded with copper in CCS-independent intracellular steps that occur within the secretory compartments by utilizing ATOX1 copper chaperone pathway (Jeney *et al.*, 2005).

In eukaryotes, SOD2 generally resides in the matrix of the mitochondria. The SOD2 polypeptide is encoded by a nuclear gene and is imported into the mitochondrial matrix. Studies in yeast have shown that manganese is not incorporated into the preexisting pool of apo-SOD2. Instead, metal activation requires protein translation with manganese insertion occurring only with newly synthesized SOD2 molecules that are subsequently imported into mitochondria. Thus, the metallation of SOD2 differs from that of SOD1 where copper can be incorporated into the apoenzyme without new protein synthesis (Luk *et al.*, 2005).

The biological significance of SOD1 initially depended on its dismutase function and the elimination of toxic oxygen species. However, because mutations in human SOD1 have been linked to familial amyotrophic lateral sclerosis (fALS), the need to understand the biology of SOD1 has taken on added significance. Because of the important interactions between CCS and SOD1, much research has focused on CCS and its potential role in disease mechanisms.

10.4 STRUCTURE OF Cu, Zn SUPEROXIDE DISMUTASE

Human SOD1 is a 153 amino acid, 32-kDa homodimeric enzyme in which each subunit folds as an eight-stranded Greek key β-barrel. Each SOD1 monomer binds one copper and one zinc ion and contains one intrasubunit disulfide bond. Two loop elements project from the β-barrel that are important in metal ion binding and in the formation of the active site. These are termed the *zinc loop* (loop IV) and the *electrostatic loop* (loop VII), which form a large part of the active site of the enzyme. The electrostatic loop contains several charged residues producing an electrostatic field that is suitable for guiding the superoxide anion to the copper ion. In the mature enzyme, the disulfide loop, a substructure of loop IV, is covalently linked to the β-barrel through a disulfide bond between C57 and C146. The disulfide bond between C57 and C146 (human SOD1 numbering) is conserved in all SOD1 structures, and it stabilizes loop IV, which plays an important role in SOD1 dimerization. The other two-cysteine residues, C6 and C111, do not normally form disulfide bonds. Interestingly, mouse SOD1 contains a serine residue rather than a cysteine at codon 111. Copper binding is coordinated by four histidine residues (H46, H48, H63, and H120), whereas zinc binding is coordinated by three histidines and one aspartic acid (H63, H71, H80, and D83). Thus, one histidine residue, H63, simultaneously interacts with both zinc and copper metal ions, forming a histidine bridge that is unique to SOD1 (Banci *et al.*, 2002, 2006; Hart, 2006; Seetharaman *et al.*, 2009; Strange *et al.*, 2003). The mechanism of zinc incorporation into SOD1 is unknown although it might be incorporated only into the newly synthesized SOD1 molecules.

10.5 CLONING AND IDENTIFICATION OF A COPPER CHAPERONE FOR SOD1

The *Saccharomyces cerevisiae LYS7* gene has been cloned based on its genetic map position and by a complementation analysis in yeast (Horecka *et al.*, 1995). The 1453 base pair sequence contains an open reading frame that predicted a unique 249 amino acid protein. A mammalian homolog for LYS7 was termed *CCS* (Bartnikas and Gitlin, 2001; Culotta *et al.*, 1997). The human CCS cDNA sequence contains an 825-base pair open reading frame and encodes a protein product of 274 amino acids (Moore *et al.*, 2000). Overall, human CCS and yeast

Table 10.1 Alignment of human, mouse, and yeast CCS protein sequences

```
                                  •  •
Human   1  MASDSGNQGT LCTLEFAVQM TCQSCVDAVR KSLQGVAGVQ DVEVHLEDQM
Mouse   1  MASKSGDGGT VCALEFAVQM SCQSCVDAVH KTLKGVAGVQ NVDVQLENQM
                                  •  •
Yeast   1  MTTNDTYEAT -----YAIPM HCENCVNDIK ACLKNVPGIN SLNFDIEQQI

Human  51  VLVHTTLPSQ EVQALLEGTG RQAVLKGMGS GQLQNLGAAV AIL-------
Mouse  51  VLVQTTLPSQ EVQALLESTG RQAVLKGMGS SQLQNLGAAV AIL-------
Yeast  46  MSVESSVAPS TIINTLRNCG KDAIIRGAGK PN----SSAV AILETFQKYT
                                                       * *
Human  94  ---GGPGTVQ GVVRFLQLTP ERCLIEGTID GL-EPGLHGL HVHQYGDLTN
Mouse  94  ---EGCGSIQ GVVRFLQLSS ELCLIEGTID GL-EPGLHGL HVHQYGDLTR
                                                    *    *
Yeast  92  IDQKKDTAVR GLARIVQVGE NKTLFDITVN GVPEAGNYHA SIHEKGDVSK

                  *#       #         #  #
Human 140  NCNSCGNHFN PDGASHGGPQ DSDRHRGDLG NVRADADGRA IFRMEDEQLK
Mouse 140  DCNSCGDHFN PDGASHGGPQ DTDRHRGDLG NVRAEAGGRA TFRIEDKQLK
                    *
Yeast 142  GVESTGKVWH ----KFDEPI EC-FNESDLG --KNLYSGKT FLSA---PLP

                       *
Human 190  VWDVIGRSLI IDEGEDDLGR GGHPLSKITG NSGERLACGI IARSAGLFQN
Mouse 190  VWDVIGRSLV IDEGEDDLGR GGHPLSKITG NSGKRLACGI IARSAGLFQN
                                   *
Yeast 182  TWQLIGRSFV ISKSLN---- --HPENEPSS VKDYSF-LGV IARSAGVWEN

               • •
Human 240  PKQICSCDGL TIWEERGRPI AGKGRKESAQ PPAHL
Mouse 240  PKQICSCDGL TIWEERGRPI AGQGRKDSAQ PPAHL
               • •
Yeast 225  NKQVCACTGK TVWEERKDAL ANNIK
```

• = copper-binding sites.
∗ = putative copper-binding sites.
= zinc-binding sites.

LYS7 share 28% amino acid identity (Table 10.1). One significant difference in the properties of the human and yeast CCS proteins is that hCCS is a dimer, regardless of copper content or protein concentration. The apo-form of yeast CCS protein is monomeric but forms a mixture of monomers and dimers upon binding the copper ion (Schmidt *et al.*, 1999a). Despite the difference in oligomerization state of the metallochaperone itself, human CCS is able to substitute for yeast CCS in the activation of yeast SOD1 *in vitro* assays (Rae *et al.*, 2001).

The chromosomal location of human *CCS* gene is in the human chromosome 11q13. The human *CCS* gene is organized in eight exons, and from the translation start site to the translation termination site, it spans 12,798 base pairs of genomic DNA (Silahtaroglu *et al.*, 2002). The genomic organization of mouse CCS is very similar to that of human CCS, as it also has eight exons. Mouse CCS sequence is in the proximal or centromeric end of murine chromosome 19. Mouse CCS cDNA contains an 825 base pair open reading frame, corresponding to the 274 amino acid protein that is over 80% identical to human CCS protein (Moore *et al.*, 2000). Human chromosome 11q13 and the proximal end of murine chromosome 19 are corresponding regions within their respective genomes. Rat CCS is also a 274 amino acid protein and very similar to human CCS and mouse CCS (Hiromura *et al.*, 2000).

10.6 CCS STRUCTURE

Human CCS is a 54-kDa homodimeric protein that carries out the final post-translational processing of SOD1. Structural and functional properties of the individual domains of both the yeast (yCCS) and human (hCCS) proteins have been determined from biochemical and crystallographic studies. Both hCCS and yCCS consist of three distinct domains, which are all involved in the activation of SOD1 (Table 10.2) (Barry and Blackburn, 2008; Culotta *et al.*, 2006; Furukawa and O'Halloran, 2006; Lamb *et al.*, 1999; Stasser *et al.*, 2007).

The 8-kD amino terminal domain I (residues 1–85) is important in scavenging copper from the cells and accepting it from the transporters. Domain I is the copper-binding site with the consensus motif MXCXXC (MTCQSC in hCCS), which is conserved in several other copper-binding proteins including ATOX1. In the MTCQSC motif, both cysteine residues, C22 and C25, are involved in copper binding. Domain I itself is required for acquisition of copper, although the MXCXXC motif is not essential to SOD1 activation in the cell under normal conditions. A CCS mutant in which domain I is truncated can still activate SOD1,

Table 10.2 Alignment of human CCS and SODI protein sequences

				• •		
CCS	1	MASDSGNQGT	LCTLEFAVQM	TCQSCVDAVR	KSLQGVAGVQ	DVEVHLEDQM
CCS	51	VLVHTTLPSQ	EVQALLEGTG	RQAVLKGMGS	GQLQNLGAAV	AILGGPGTVQ
SOD1	1				MATKAV	CVLKGDGPVQ
					* *	*#
CCS	101	GVVRFLQLTP	E-RCLIEGTI	DGLEPGLHGL	HVHQYGDLTN	NCNSCGNHFN
SOD1	17	GIINFEQKES	NGPVKVWGSI	KGLTEGLHGF	HVHEFGDNTA	GCTSAGPHFN
		#	# #			
CCS	150	PDGASHGGPQ	DSDRHRGDLG	NVRADADGRA	IFRMEDEQLK	VWD---VIGR
SOD1	67	PLSRKHGGPK	DEERHVGDLG	NVTADKDGVA	DVSIEDSVIS	LSGDHCIIGR
		*				• •
CCS	197	SLIIDEGEDD	LGRGGHPLSK	ITGNSGERLA	CGIIARSAGL	FQNPKQICSC
SOD1	117	TLVVHEKADD	LGKGGNEEST	KTGNAGSRLA	CGVIGIAQ	
CCS	247	DGLTIWEERG	RPIAGKGRKE	SAQPPAHL		

• = copper-binding sites in CSS.
∗ = copper-binding sites in SODI and putative copper-binding sites in CCS.
= zinc-binding sites in SODI and in CCS.

although at a much reduced rate. In addition, hCCS with C22S, C25S mutations still exhibit 70% metallochaperone activity compared to the wild-type protein *in vitro* (Stasser *et al.*, 2005).

In mammals, CCS domain II has remarkable sequence and structural homology to SOD1: hCCS 16-kD domain II (residues 86–234) is 47% identical to hSOD1. In fact, a partial hCCS sequence was previously identified as hSOD4 (Lamb *et al.*, 2000b). Domain II of hCCS consists of an eight-stranded β-barrel and a zinc-binding site formed by two extended loops. The first of these loops provides the ligands to a bound zinc ion and is analogous to the zinc subloop in SOD1. The second loop structurally resembles the SOD1 electrostatic channel loop, but lacks many of the residues important for catalysis. The hCCS domain II, proximal to the dimer interface, also has a single loop that is unique to the chaperones and not found in SOD1 (Furukawa and O'Halloran, 2006; Lamb *et al.*, 2000b).

Human CCS domain II includes a highly conserved dimerization interface with retention of all of the zinc binding and three of the four copper-binding residues corresponding to those found in SOD1. The zinc-binding site in CCS is comparable to that in SOD1: the zinc ion in hCCS is coordinated by H147, H155, H164, and D167 in a distorted tetrahedral geometry. These residues are analogous to the SOD1 zinc-binding ligands H63, H71, H80, and D83, respectively. Mutations of H147 or D167 to alanine in the zinc site affect hCCS activity and suggest a pivotal role of zinc binding for the function of hCCS. Since zinc participates in the stabilization of the tertiary structure of SOD1, these residues in hCCS are also expected to be responsible for stabilizing hCCS (Endo *et al.*, 2000).

A second potential metal-binding site corresponding to the SOD1 copper-binding site is located adjacent to the zinc site in hCCS. In contrast to SOD1, in hCCS, domain II does not contain a copper ion bound in the catalytic site. In SOD1, the copper is coordinated by four histidine residues, H46, H48, H63, and H120. The first three histidine residues of SOD1 are conserved in hCCS with the corresponding residues being H130, H132, and H147, respectively. The hCCS H147 residue is homologous to SOD1 H63 and bridges the copper and zinc ions. Instead of the fourth histidine residue, analogous to H120 of SOD1, hCCS has an aspartic acid, D200 (D201 if methionine is counted). In hCCS, C141 and C227 form a disulfide bridge that is homologous to the essential C57–C146 disulfide bridge in SOD1 (Barry *et al.*, 2008). Given the similarity of the SOD1 and CCS domain II dimer interfaces, CCS has been proposed to interact with SOD1 through SOD1-like domain II by mimicking SOD1 dimerization. One probable function of CCS domain II is recognition of the target enzyme SOD1 and orientation of the copper donor and acceptor sites in the two proteins (Lamb *et al.*, 2000a).

The overall structures of domain II in human CCS and yeast CCS are similar. However, yCCS does not contain the two loops analogous to the electrostatic channel loop and the zinc-binding loop. In the yCCS domain II, there are no zinc-binding residues; however, yCCS has four histidine residues available for copper binding, namely, H130, H134, H151, and H198. Yeast CCS lacks the histidine residue corresponding to H147 of hCCS and the aspartate residue corresponding to D167 of hCCS. SOD1 activation by yCCS is still observed after mutations of the histidine residues to alanine. Thus, the potential copper-binding site in domain II is not required for copper activation of SOD1 (Lamb *et al.*, 1999; Schmidt *et al.*, 1999a).

Despite the conservation of sequences between SOD1 and CCS, CCS exhibits no detectable dismutase activity (Schmidt *et al.*, 1999b). However, if the fourth potential copper-binding ligand (D200) is mutated to histidine in human CCS, D200H hCCS gains significant superoxide scavenging activity when expressed in yeast. Because D200H hCCS retains its metallochaperone capacity, it consequently can be considered as a bifunctional SOD that can activate itself with copper. The aspartic acid at hCCS position 200 is well conserved among mammalian CCS proteins, and it is possible that this residue has evolved to prevent deleterious reactions involving copper bound to CCS. The aspartic acid 200 in hCCS is in analogous position to histidine 120 in hSOD1, which is highly susceptible to oxidation, whereas the remaining three histidine copper ligands are not (Uchida and Kawakishi, 1994). Hence, the presence of D200 in hCCS prohibits redox damage both to CCS itself and to its target, SOD1. The D200H mutation in hCCS does not appear to affect its stability based on *in vitro* aggregation assays (Son *et al.*, 2003). Therefore, D200H hCCS represents a self-sufficient SOD molecule that can charge itself with copper. Having two separate molecules, one for copper incorporation and the other for superoxide scavenging, may allow for a better regulation of SOD1 activity at the posttranslational level, suitable for various intracellular environments in a multicellular organism (Schmidt *et al.*, 1999b).

Carboxyl-terminal domain III (residues 235–274) of hCCS is a short polypeptide (30–40 amino acids) without any preference to form secondary structures. Domain III is essential to CCS function and for copper transfer to SOD1. Domain III has a CXC motif (C244 and C246 in human CCS) that is highly conserved among all species. The domain III polypeptide alone is sufficient to bind copper, and the cysteine-to-serine mutations in the CXC motif dramatically decrease the metallochaperone activity *in vivo* and *in vitro*. Domain III does not possess any homology to SOD1 or other known proteins (Barry and Blackburn, 2008; Furukawa and O'Halloran, 2006; Schmidt *et al.*, 1999a; Stasser *et al.*, 2007). It has a putative peroxisomal localization sequence (AHL) at the COOH terminus (Casareno *et al.*, 1998) and shows a weak homology with a part of prolyl cis–trans isomerase (Furukawa and O'Halloran, 2006).

10.7 CCS–SOD1 INTERACTIONS

SOD1 provides the primary defense against superoxide damage by catalyzing the dismutation of superoxide radicals to oxygen and hydrogen peroxide. SOD1 is normally quite stable in large part because of copper and zinc binding and oxidation of an intramolecular disulfide that is relatively rare in cytosolic proteins because of the highly reducing environment. The immature apo-SOD1 exists in a monomeric form where the critical disulfide between C57 and C146 is reduced (Arnesano *et al.*, 2004). SOD1 requires several posttranslational modifications to form an active molecule. Given the similarity of the SOD1 and CCS domain II dimer interfaces, CCS has been proposed to interact with SOD1 through its SOD1-like domain II (Casareno *et al.*, 1998). CCS recognizes and binds to a disulfide-reduced, copper-deficient form of SOD1, inserts the catalytic copper ion through direct insertion, and apparently functions to protect the copper ion from binding to other intracellular copper-binding proteins. CCS also catalyzes the oxidation of the intrasubunit disulfide bond in each SOD1 monomer in an oxygen-dependent redox process. The C57 and C146 residues of SOD1 can be oxidized by oxygen alone but slowly; disulfide formation is greatly accelerated by copper-bound CCS. CCS exhibits a strong dependence on copper for oxidizing the SOD1 disulfide. Copper-bound CCS facilitates disulfide oxidation and isomerization in a stepwise conversion of the immature form of the wild-type SOD1 enzyme to the active state (Brown *et al.*, 2004; Furukawa *et al.*, 2004; Lamb *et al.*, 2001; Rae *et al.*, 2001; Schmidt *et al.*, 2000). Hindering these posttranslational modifications results in immature, metal-deficient, and/or disulfide-reduced forms of SOD1 that could form higher order oligomers, which are believed to be relevant in fALS toxicity (Karch *et al.*, 2009; Zetterstrom *et al.*, 2007).

Although the zinc and disulfide bonds are not directly involved in the enzymatic activity, these modifications are required for proper stability and formation of the active site of SOD1 (Leitch *et al.*, 2009b). Despite important roles of zinc in the SOD1 folding and structure, little is known on how SOD1 acquires zinc in the cell. According to *in vitro* studies, hCCS can bind zinc in domain II and possibly in domain I; therefore, hCCS may also function as a “zinc chaperone” protein (Furukawa and O’Halloran, 2006).

Furukawa and O’Halloran (2006) have proposed the following mechanism of CCS-dependent SOD1 activation in yeast. Yeast CCS is able to rapidly catalyze the formation of the active SOD1 using a copper, disulfide-reduced E,Zn-SOD1 (E stands for empty metal bonding site) and molecular oxygen as substrates. The first step is the incorporation of a zinc ion into the disulfide-reduced E,E-SOD1. In the next step, the disulfide-reduced E,Zn-SOD1 forms a heterodimeric complex with the copper chaperone. In the subsequent reaction, the CCS-SOD1 heterodimer forms an intermolecular disulfide bond between SOD1 C57 and yCCS C229 in domain III, in an oxygen-dependent reaction. The formation of the heterodimer is followed by insertion of the copper-loaded CXC domain III polypeptide into the catalytic center of the E,Zn-SOD1. Copper insertion in the heterodimeric complex and binding in SOD1 could induce conformational changes, especially around the cysteine residues, promoting the protein disulfide isomerase-like reaction from intermolecular CCS-SOD1 disulfide to intramolecular SOD1 C57-C146 disulfide bond. Disulfide exchange within the heterodimeric intermediate allows the release of fully metallated, disulfide-oxidized Cu,Zn-SOD1 from the complex, leaving CCS domain III in the apoform and ready for the next cycle. CCS-mediated copper delivery and disulfide bond oxidation may be linked events because CCS itself must be loaded with copper to catalyze the thiol-disulfide oxidation (Winkler *et al.*, 2009). To assess mechanistic aspects of CCS interaction with SOD1 in mammalian systems, increasing amounts of copper was added to hCCS *in vitro* (Seetharaman *et al.*, 2009; Stasser *et al.*, 2007). Upon addition of a single equivalent of copper per CCS molecule, the "canonical" CCS dimer mediated by domain II dissociated into monomers. Addition of more copper resulted in the appearance of a "noncanonical" dimer mediated by Cu4S6 copper cluster formed in part by the CXC motifs of CCS domain III. This canonical-to-noncanonical CCS dimer transition could possibly function as a copper-sensing switch to make CCS domain II available to nascent SOD1 binding only when sufficient copper is present (Seetharaman *et al.*, 2009). On the basis of these studies, a second model of SOD1 maturation was proposed for mammalian systems: the "canonical" CCS dimer is loaded with copper to generate the "noncanonical" CCS dimer mediated by Cu4S6 cluster. SOD1 is translated and zinc is loaded at some point. Nascent SOD1 then binds to domain II of the "noncanonical," copper-loaded CCS dimer. Copper from the Cu4S6 cluster is transferred to nascent SOD1 and the intrasubunit disulfide bond in SOD1 is formed. Upon being depleted from copper, CCS reforms the "canonical" CCS domain II mediated copper-free dimer and the cycle repeats. The pathogenic SOD1 mutations may interfere with CCS-mediated SOD1 maturation at the various points in the cycle (Seetharaman *et al.*, 2009).

CCS-dependent metallation and formation of the intramolecular disulfide bond in SOD1 are dependent on oxygen. Studies in yeast have shown that oxidizing equivalents, ultimately derived from oxygen, are employed in the formation of disulfide bonds. First, a heterodimeric disulfide bond is formed linking domain III of CCS with C57 of SOD1. In a subsequent step, the intermolecular disulfide undergoes an exchange reaction in which SOD1 C57 replaces the cysteine of CCS domain III to form the C57–C146 SOD1 intramolecular disulfide (Culotta *et al.*, 2006). In agreement, the *in vitro* studies on human CCS and SOD1 have shown that copper-bound CCS cannot activate the apoform of SOD1 in the absence of oxygen. Under anaerobic conditions, the cysteine residues in SOD1 remain reduced even after incubation with copper-bound CCS, and SOD1 remains inactive. On the other hand, one of the earliest cellular responses to oxidative stress is the immediate posttranslational activation of the available apo-SOD1 protein pool by copper-bound CCS. These results indicate that CCS directly mediates the oxygen-dependent activation of SOD1 and suggest an additional physiological function for CCS (Brown *et al.*, 2004; Furukawa and O'Halloran, 2006; Furukawa *et al.*, 2004).

Yeast SOD1 is completely dependent on CCS for insertion of the catalytic copper and oxidation of the disulfide bond. Although ySOD1 is dependent on CCS for activity, the SOD1 molecules of other eukaryotes are not. Mouse SOD1 was found to be only partially dependent on CCS, as fibroblasts derived from CCS knockout mice still retained some residual superoxide activity (Wong *et al.*, 2000). Similarly, human SOD1 transfected into mutant yeast cells lacking CCS also maintains some enzymatic activity (Corson *et al.*, 1998). As another extreme,

C. elegans does not have CCS, and its SOD1 can be fully activated independently of CCS (Jensen and Culotta, 2005; Leitch *et al.*, 2009a). Thus, there exist alternative pathways for delivery of copper to SOD1 and for SOD1 disulfide bond maturation that are CCS-independent.

Both CCS-dependent and CCS-independent pathways compete for the same intercellular sources of copper. Furthermore, CCS-dependent as well as CCS-independent activation occurs rapidly with a preexisting pool of apo-SOD1 without the need for new protein synthesis. The dependency of SOD1 on CCS for maturation correlates with the propensity of the SOD1 disulfide bond to be formed (Leitch *et al.*, 2009b). First, the CCS-dependent yeast SOD1 is totally reliant on copper and CCS for disulfide oxidation, whereas the partially CCS-independent human SOD1 and fully CCS-independent *C. elegans* SOD1 show correspondingly lower requirements for copper in disulfide oxidation. Second, the CCS-independent activation of SOD1 does not exhibit the same requirement for molecular oxygen as does CCS-dependent activation. CCS-independent activation of hSOD1 can be observed even under hypoxic and anoxic conditions. Therefore, SOD1 molecules that are activated in both CCS-dependent and -independent manners can retain activity over a wide range of oxygen tensions, which is critical in tissues of multicellular organisms having wide variations in oxygen tension (Furukawa and O'Halloran, 2006; Leitch *et al.*, 2009b).

The compensatory pathways for SOD1 maturation in the absence of CCS have been studied. CCS-independent activation of mammalian SOD1 involves glutathione, particularly the reduced form (GSH). The GSH-dependent pathway can act on both wild-type and fALS mutant variants of hSOD1, indicating that mutations in SOD1 do not interfere with the CCS-independent pathway, at least in the case of mutants that retain their enzymatic activity (Carroll *et al.*, 2004). According to *in vitro* studies on human and yeast SOD1, the side chain of C57 could play a role for the incorporation of copper into the protein, regardless of whether this occurs through the interaction with CCS or with other copper-containing species, for example, copper–glutathione complex (Banci *et al.*, 2006). CCS-independent activation has a strict dependence on GSH concentration and the redox potential in the cell. By comparison, CCS is reactive over a range of intracellular GSH concentrations. Studies *in vitro* have shown that the formation of disulfide bond and SOD1 maturation happen much slower in CCS-independent pathways (Furukawa *et al.*, 2004). The CCS-independent pathway is also sensitive to certain structural perturbations in SOD1 structure. Specifically, proline residues at SOD1 positions 142 and 144 prohibit CCS-independent activation but have no effect on CCS-dependent activation (Carroll *et al.*, 2004; Jensen and Culotta, 2005). Yeast SOD1 naturally contains these prolines and is totally dependent on CCS for acquiring copper *in vivo*. However, SOD1 prolines at codons 142 and 144 might not be the only residues to influence CCS-independent activation and disulfide formation. Interestingly, CCS domain III shows a weak homology with a part of prolyl cis–trans isomerase (Furukawa and O'Halloran, 2006). A secondary CCS-independent pathway for introducing the disulfide bond into the

SOD1 polypeptide might involve protein disulfide isomerase, mainly an endoplasmic reticulum protein, which catalyzes the formation and rearrangement of disulfide bonds (Atkin *et al.*, 2008). Protein disulfide isomerase expression has been shown to be elevated in mutant SOD1 transgenic rats (Ahtoniemi *et al.*, 2008). However, there is very little evidence that protein disulfide isomerase normally plays a role in SOD1 disulfide oxidation. In addition to CCS, there may be other proteins that could supply SOD1 with copper. For instance, the apoform of mammalian SOD1 has been proposed to acquire copper from metallothioneins (Suzuki and Kuroda, 1995).

Mutations in either CCS or SOD1, which lead to structural alterations affecting the interaction of CCS with SOD1, might alter rates in one or several steps in the maturation cycle resulting in an accumulation of immature forms of SOD1 (Seetharaman *et al.*, 2009). Alterations in the dimerization interfaces of either SOD1 or CCS molecules interfere with their interaction. When the dimerization of yeast SOD1 is disrupted by mutations in SOD1 at the subunit interface (F50E/G51E), the mutated SOD1 is not activated by yCCS. Similarly, when the corresponding amino acid residues in yCCS domain II are mutated (K136E/G137E), activation of ySOD1 is not observed. The K136E, G137E mutation in yCCS prevents it from homodimerizing and also from forming heterodimers with SOD1. Likewise, hCCS having the comparable mutations (Y134E, G135E) cannot activate SOD1. These findings support the idea that CCS domain II interacts with the SOD1 subunit interface and that this interaction is dependent on particular sequences at the dimer interfaces of both SOD1 and CCS domain II (Furukawa and O'Halloran, 2006; Schmidt *et al.*, 2000). The SOD1-like domain II of CCS is necessary but is not sufficient to facilitate the interactions with SOD1. Normal functions of other domains of CCS are also needed for SOD1 activation. In hCCS, domain I includes the copper-binding motif, MTCQSC, thought to be critical for copper acquisition under the copper-limiting conditions within the cell. Mutated hCCS protein molecules, either with deleted domain I or with mutated cysteine residues in the copper-binding motif (C22S, C25S), were not able to bind copper; therefore, they could not activate SOD1 (Caruano-Yzermans *et al.*, 2006).

Studies in yeast have found that the domain III mutant C229S, C231S of yCCS can bind to SOD1 but cannot activate SOD1. The C229S, C231S mutant of yCCS cannot form the heterodimeric disulfide bridge between C57 in ySOD1 and C229 in yCCS, which is a critical step leading to the formation of the homodimeric C57–C146 disulfide bridge in mature SOD1. Likewise, in hCCS, the corresponding C244S, C246S mutant in domain III still harbors all the domain II sequences required for interacting with hSOD1. Therefore, the mutant C244S, C246S hCCS has the capacity to interact with SOD1; however, it can neither insert copper nor oxidize the disulfide (Caruano-Yzermans *et al.*, 2006). Thus, the interaction between mutant 244S, 246S hCCS and hSOD1 is considered to be "nonproductive" because it cannot promote the disulfide oxidation in SOD1. On the contrary, it can lead to the accumulation of the disulfide-reduced form of SOD1 (Proescher *et al.*, 2008).

Besides donating copper to SOD1 and oxidizing its intramolecular disulfide bridge, CCS can also protect mutant SOD1 from misfolding. Mammalian cultures were cotransfected with mutant A4V SOD1 and either wild-type CCS or mutant C244S, C246S hCCS. Both wild-type and mutant CCS could help prevent A4V SOD1 from aggregating and decrease the amount of the detergent-insoluble form of SOD1. These results indicate that CCS protection against SOD1 misfolding does not require disulfide oxidation or copper loading. However, physical interaction with SOD1 and CCS is required for this chaperoning function, as the mutation Y134E,G135E in domain II of hCCS, which impairs the interaction with SOD1, could not decrease the level of misfolded A4V SOD1 (Proescher *et al.*, 2008). Human CCS with a G168R mutation in domain II, homologous to the hSOD1 G85R mutation, forms aggregates and high molecular mass protein complexes when expressed in mammalian tissue cultures (Son *et al.*, 2003). However, the *in vivo* significance of these findings remains to be determined.

All mutant SOD1 proteins, derived from cell cultures or from transgenic mice, may undergo incomplete posttranslational modifications of nascent SOD1 polypeptides because of their defective interaction with CCS. The ability of CCS to activate the newly translated mutant SOD1 proteins is dependent on the kinetics of copper loading into CCS and the biophysical properties of the different SOD1 proteins, including their overall stability (Karch *et al.*, 2009).

The truncation mutants, including L126Z, are lacking the C146 residue that is necessary for wild-type SOD1 heterodimerization with hCCS domain II. Therefore, CCS probably cannot interact and stabilize the nascent L126Z SOD1 molecules (Jonsson *et al.*, 2004; Wang *et al.*, 2005; Watanabe *et al.*, 2005). Likewise, CCS could not stabilize the pathogenic SOD1 mutants, C57R and C146R, which are unable to form the intra-subunit disulfide bond. The "β-barrel" mutants or "wild-type like" mutants, such as G37R and G93A, can interact with CCS. However, because the SOD1:CCS ratio is normally between 15:1 and 30:1, CCS might not be able to cycle through the entire pool of these SOD1 mutants before they are oligomerized, insolubilized, or degraded (Seetharaman *et al.*, 2009). Interestingly, recent structural studies of the metallated G93A SOD1 mutant revealed a destabilization in the remote metal-binding region, indicating that even in the G93A SOD1 mutant, the metal-binding region is altered that may affect the intermolecular CCS–SOD1 interactions (Museth *et al.*, 2009). The "metal-binding site" mutants are not as unstable as the "β-barrel" mutants. However, CCS cannot normally interact with these mutants because their structure prevents metal binding and/or interferes with SOD1–CCS interactions. Mutant G85R SOD1 has been found to be metal deficient, and this metal deficiency is linked to an increased propensity for G85R SOD1 to exist as monomeric disulfide-reduced species (Cao *et al.*, 2008; Furukawa *et al.*, 2008; Seetharaman *et al.*, 2009; Winkler *et al.*, 2009). Thus, due to the inability to interact normally with CCS molecules, mutant SOD1 can form abnormal folding intermediates prone to aggregate and/or to interfere with other proteins and cellular processes.

10.8 CCS REGULATION

The activity of many cupro-enzymes decreases with copper deficiency because of the lack of metal. In copper-deficient cells, there is likely to be a competition between chaperones and other copper-binding proteins for the limited supply of copper. Nutritional studies have shown that the abundance of CCS is inversely proportional to the amount of copper in tissues (Prohaska *et al.*, 2003). Cells respond to low copper levels by increasing the CCS expression. Conversely, when copper is abundant, CCS is expressed at low concentrations, compatible with its catalytic role in activating SOD1 (Bertinato and L'Abbe, 2003). Evident elevation in CCS levels is one of the most dependable changes in copper-binding proteins due to copper deficiency, and therefore, it may be useful for the evaluation of copper status in animals (Harvey and McArdle, 2008). Increased levels of CCS likely increase the efficiency of copper transfer to SOD1, because CCS functions by scavenging for copper and delivering the metal specifically to SOD1. Increased levels of CCS in copper-deficient cells make it possible to rapidly deliver copper into apo-SOD1 as soon as copper becomes available (Bertinato and L'Abbe, 2003).

There are several possible mechanisms for the increased expression of CCS induced by copper deficiency. CCS may be regulated at the transcriptional level, possibly by a transcription factor that is sensitive to changes in copper levels. Alternatively, CCS may be regulated by posttranslational mechanisms, for instance, copper-loaded CCS may be degraded faster than apo-CCS. Several studies indicate that copper-dependent regulation of CCS is posttranslational. CCS protein but not mRNA is higher in organs from copper-deficient mice and rats (Prohaska *et al.*, 2003). CCS stability and function were determined in fibroblasts derived from CCS knockout mice and transfected with cDNA coding for human wild-type CCS. The CCS, expressed in the transfected fibroblasts, was able to incorporate copper into SOD1, as measured by SOD1 enzymatic activity. The abundance of CCS protein was inversely proportional to the copper content in the culture media suggesting that copper-dependent regulation of CCS is posttranslational (Caruano-Yzermans *et al.*, 2006).

Many cytosolic proteins, whose function is dependent on their rapid regulation in expression, are degraded by the 26S proteosome, a large multisubunit protease (Lehman, 2009). Switching rodent and human cell lines from copper-deficient to copper-rich medium promoted the rapid degradation of their endogenous CCS, which could be blocked by the proteosome inhibitors. CCS degradation was slower in copper-deficient cells than in cells cultured in copper-rich medium. These data indicate that copper regulates CCS expression by modulating its degradation by the 26S proteosome and suggest a novel role for CCS in prioritizing the utilization of copper when it is scarce (Bertinato and L'Abbe, 2003).

CCS consists of three domains that all take part in the copper acquisition and delivery to SOD1. To study the mechanism of CCS regulation, fibroblasts from $CCS^{-/-}$ mice were transfected with cDNAs of either wild-type or mutant human CCS and the effects of copper on CCS levels were measured. Although domain I of hCCS is required for CCS-dependent copper incorporation into SOD1, domain I plays no role in the copper-dependent regulation of CCS. Instead, cysteine residues in the CXC motif of domain III have a critical role in mediating copper-dependent turnover of hCCS. When C244 and C246 in domain III of hCCS were mutated to serine, the copper-dependent regulation was lost. These same C244 and C246 residues, critical for copper-dependent degradation of CCS, are also needed for copper incorporation to SOD1; however, these two processes are mechanistically distinct (Caruano-Yzermans *et al.*, 2006).

The degradation of CCS is also dependent on the interaction with SOD1, as tissue lysates derived from SOD1 knockout mice have significantly decreased levels of CCS when compared with lysates from wild-type littermates. This difference in the steady-state levels of CCS is due to an increased turnover of CCS in SOD1 knockout mice. In fact, the domain II mutant CCS (Y134E, G135E) that is unable to hetero-dimerize with SOD1 has a higher turnover rate than the wild-type CCS protein. The Y134E, G135E mutant CCS is also unable to homodimerize, which indicates that the process of copper-dependent turnover likely involves the recognition of the CCS monomer (Caruano-Yzermans *et al.*, 2006). Binding of increasing amounts of copper to CCS leads to allosteric conformational changes in the CCS molecule, that could allow CCS to be targeted for proteosomal degradation (Schmidt *et al.*, 1999a). Binding of excess copper by the CXC motif might lead to disruption of the CCS domain II dimer interface, allowing access for either SOD1 association or proteasomal targeting (Caruano-Yzermans *et al.*, 2006). Mutations in SOD1, leading to impaired CCS–SOD1 interactions, might also promote abnormal CCS degradation and dysregulation of copper trafficking pathways. In fact, G93A mutant, but not wild-type SOD1 shifted intracellular copper homeostasis toward copper accumulation in the spinal cord of SOD1 transgenic mice during disease progression (Tokuda *et al.*, 2009).

Gitlin and coworkers have presented a model of intracellular cytoplasmic copper homeostasis with specific relevance to SOD1 activity (Caruano-Yzermans *et al.*, 2006). According to this model, at the steady state, the amount of copper bound to CCS may be minimal, as copper is rapidly transferred from CCS to SOD1. When the amount of copper increases, SOD1 becomes saturated with copper and the pool of holo-CCS, defined as CCS with copper bound at the CXC site, starts to accumulate. This results in conformational changes in the CCS molecule that may lead to CCS ubiquitination and subsequent degradation. Such a hypothesis is consistent with the decrease in CCS observed in the absence of SOD1 or impaired interaction with SOD1, each of which would also increase the abundance of holo-CCS.

10.9 CCS DISTRIBUTION

The tissue distribution and expression of CCS in mammalian systems appear to parallel those of SOD1. SOD1 is an extremely abundant cellular protein and is particularly plentiful in the spinal cord and brain where it has been estimated to comprise between 0.1% and 2.0% of the detergent-soluble protein. Quantitatively, CCS is about 15- to 30-fold less abundant than SOD1. CCS is expressed by multiple tissues and is present in high concentrations in kidney and liver. Within the central nervous system (CNS), CCS is found in neurons and in some astrocytes. CCS is present in many neuronal populations, although it appears to be selectively enriched in a subset of neurons, particularly in motor neurons in the spinal cord, pyramidal neurons in the cortex, deep cerebellar nuclei, and Purkinje cells. Ependymal cells of the spinal cord central canal are also intensely immune-reactive for CCS. Both SOD1 and CCS are present together in cells that degenerate in amyotrophic lateral sclerosis (ALS), which emphasizes a potential role of CCS in mutant SOD1-mediated toxicity (Culotta *et al.*, 1997; Pardo *et al.*, 1995; Rothstein *et al.*, 1999).

The subcellular distributions of CCS and SOD1 are very similar. Both CCS and SOD1 are predominantly cytosolic proteins, but they are also found in other cellular compartments including nucleus (Crapo *et al.*, 1992; Son *et al.*, 2007), endoplasmic reticulum (Kikuchi *et al.*, 2006; Son *et al.*, 2007), perixosomes (Islinger *et al.*, 2009), and mitochondria (Okado-Matsumoto and Fridovich, 2001; Son *et al.*, 2007; Sturtz *et al.*, 2001).

10.10 MITOCHONDRIAL LOCALIZATION

In yeast, both CCS and SOD1 appear to play important roles in mitochondrial function. The effects of mitochondrial antioxidant genes on cell survival were studied in yeast cultures. It was found that among the known mitochondrial antioxidant genes (TTR1, CCD1, SOD1, GLO4, TRR2, TRX3, CCS1, SOD2, GRX5, and PRX1), deletion of only three genes, namely, *SOD1, SOD2*, and *CCS* impacted survival, whereas deletion of the other genes had little or no effect on yeast survival (Unlu and Koc, 2007).

The SOD enzymes represent a first line of defense against reactive superoxide species. One major source of superoxide is the electron transport chain complexes located in the inner membrane of mitochondria; 1–2% of the oxygen consumed during respiration is incompletely reduced to superoxide anions. Since superoxide anions do not readily diffuse across the mitochondrial membranes, SODs need to be present in high concentrations on both sides of the inner membrane (Okado-Matsumoto and Fridovich, 2001). SOD1 represents 90% of the total SOD activity in the cell. In the CNS, mitochondria of transgenic mice expressing either wild-type or mutant SOD1, SOD1 was detected in the IMS (Higgins *et al.*, 2002; Jaarsma *et al.*, 2001; Mattiazzi *et al.*, 2002; Sturtz *et al.*, 2001; Vande Velde *et al.*, 2008; Vijayvergiya *et al.*, 2005), in the matrix (Vijayvergiya *et al.*, 2005), and on the outer membrane (Higgins *et al.*, 2003; Liu *et al.*, 2004; Pasinelli *et al.*, 2004; Vande Velde *et al.*, 2008). SOD1 also associated with mitochondrial fractions from spinal cord of fALS patient with G127X SOD1 truncation mutation (Liu *et al.*, 2004).

The mitochondrial distributions of SOD1 and CCS have been studied in yeast. A fraction of CCS together with SOD1 has been found in the IMS of mitochondria, even if neither CCS nor SOD1 contains any recognizable mitochondrial targeting signal. CCS has been shown to strongly influence the mitochondrial accumulation of SOD1 in yeast. When the amount of CCS that localizes to the mitochondria was decreased, the mitochondrial SOD1 levels were also decreased. Conversely, when the bulk of CCS was targeted to the mitochondrial IMS, SOD1 protein levels in the IMS were greatly increased (Sturtz *et al.*, 2001). *In vitro* mitochondrial import assays were used to study how CCS-mediated posttranslational modification of SOD1 controls its partitioning between the mitochondria and cytosol in yeast (Field *et al.*, 2003). Only a very immature form of the SOD1 polypeptide that is apo for both copper and zinc can efficiently enter the mitochondria. Moreover, a conserved disulfide in SOD1 that is essential for activity must be reduced to facilitate mitochondrial uptake of SOD1. In agreement with the studies in yeast, only the apoform of hSOD1 was imported into mouse liver mitochondria (Okado-Matsumoto and Fridovich, 2001). Once inside the mitochondria, SOD1 is converted into an active holo-enzyme through copper acquisition and oxidation of disulfide bridges. The presence of high levels of yCCS in the mitochondrial IMS resulted in enhanced mitochondrial accumulation and retention of SOD1. The CCS-mediated retention of SOD1 is not dependent on copper loading of the enzyme but does require protein–protein interactions at the heterodimerization interface of SOD1 and yCCS as well as all conserved cysteine residues in both molecules. The mitochondrial retention of SOD1 with F50E and G51E mutations, which disrupt the CCS–SOD1 heterodimer formation, was not dependent on CCS. The effect of CCS on mitochondrial accumulation of SOD1 could also be abolished by mutations in C57 residue of SOD1 and C229 residue of yCCS that normally form the intermolecular disulfide bridge. In addition, mutations C146S of SOD1 and C231S of CCS prevented CCS-mediated accumulation of SOD1 inside mitochondria (Field *et al.*, 2003). These cysteines play a critical role in the maturation of SOD1 to the fully metallated and oxidized state. Through its direct involvement in the maturation of SOD1, CCS can greatly influence the partitioning of SOD1 between cytosolic and mitochondrial pools.

The molecular mechanism underlying the mitochondrial import of the nuclear-encoded IMS proteins, including CCS and SOD1, have been elucidated in yeast (Bihlmaier *et al.*, 2008; Reddehase *et al.*, 2009) and in mammalian cells (Kawamata and Manfredi, 2008). The import of proteins into the IMS is dependent on coupled oxidation–reduction reactions (Mesecke *et al.*, 2005). The disulfide relay system, involving Mia40 and Erv1, drives the import of these cysteine-rich proteins into the IMS of mitochondria by an oxidative folding mechanism. Mia40 is an inner membrane protein that forms an intermolecular disulfide bridge with the imported proteins. Mia40 has a twin Cx9C motif, although MIA40 itself is imported through the usual mitochondrial presequence pathway (Khalimonchuk and Winge, 2008). The substrates of Mia40 are a twin CX_2C motif containing Erv1 and a series of proteins with conserved cysteine twin CX_3C or CX_9C motifs, which are critical for their accumulation in the mitochondrial IMS. Herrmann and coworkers have studied the mechanism of disulfide relay system in mitochondria (Herrmann and Hell, 2005). The newly synthesized IMS proteins traverse the TOM complex of the outer membrane in a reduced, unfolded conformation (Lutz *et al.*, 2003), and in the inner membrane, they are transiently trapped by Mia40 via intermolecular disulfide bonds. Upon reorganizing these bonds, intramolecular disulfide bonds are formed in the imported proteins, which are subsequently released into the IMS. Once the proteins are folded, they are not able to cross the outer membrane, and therefore, they are trapped within the mitochondria. During this reaction, Mia40 becomes reduced and needs to be reoxidized to regain its ability to import more proteins. Oxidation of Mia40 is carried out by Erv1, a sulfhydryl oxidase. Erv1 directly interacts with Mia40 and shuttles electrons from reduced Mia40 to oxidized cytochrome *c*, from where they are shuttled through cytochrome *c* oxidase (complex IV) to molecular oxygen, or to cytochrome *c* peroxidase. Thus, Erv1 provides a molecular link between disulfide relay system and the respiratory chain. The oxidative activity of Erv1 strongly depends on the oxygen concentration in mitochondria. Therefore, Erv1 may function as a molecular switch that adapts mitochondrial activities to the oxygen levels in the cell (Bihlmaier *et al.*, 2008; Hell, 2008; Reddehase *et al.*, 2009). The import of CCS and SOD1 into mitochondria has been shown to depend on the function of both Mia40 and Erv1 in yeast. However, Mia40 is the limiting factor, as depletion of Mia40 results in the decreased levels of CCS and SOD1. Likewise, overexpression of Mia40 increased the mitochondrial fractions of both proteins. These findings are consistent with the observation that depletion of Erv1 greatly decreased the levels of SOD1 and CCS in mitochondria. On the other hand, overexpression of Erv1 did not increase the levels of SOD1 and CCS indicating that Erv1, in contrast to Mia40, is not a limiting factor in the import of these proteins into mitochondria (Mesecke *et al.*, 2005; Reddehase *et al.*, 2009).

The disulfide relay system of Mia40 and Erv1 also drives the transport of CCS into the IMS. In yeast, Mia40 was shown to form disulfide intermediates with CCS but not with SOD1, suggesting a role for Mia40 in the generation of disulfide bonds in yCCS, which contains CX_2C and CXC motifs (Reddehase *et al.*, 2009). In agreement, in mammalian culture systems, Mia40 was found to coimmunoprecipitate with hCCS, which contains two CX_2C motifs and a single CxC motif (Kawamata and Manfredi, 2008). These studies indicate that CCS is a substrate of Mia40 and that Mia40 directly interacts with CCS via a disulfide intermediate, which subsequently mediates the import of SOD1 into the IMS. Disulfide bonds are transferred from Erv1 via Mia40 to CCS, which then introduces the disulfide bond and the copper into SOD1, thereby triggering folding and trapping of SOD1 in the IMS. Although human SOD1 does not contain the typical cysteine repeat motifs, it has four cysteines, all of which are found to play a role for its retention in mitochondria (Kawamata and Manfredi, 2008). Rodent SOD1s do not have C111; instead they have serine in that position; however, they localize in mitochondria (Okado-Matsumoto and Fridovich, 2001). In the SOD1 import reaction, the function of CCS is comparable to that of Mia40.

Both Mia40 and CCS are found in the IMS in an active, oxidized form, and they interact with their substrates via intermolecular disulfide bonds. CCS and Mia40 release their substrates as oxidized proteins and thereby are themselves reduced, and therefore, CCS as well as Mia40 need to be reoxidized to be reactivated. In addition, both Mia40 and CCS are found to be rate-limiting for their import reactions. Thus, CCS may act as an import facilitator protein, similar to Mia40, in the import of SOD1 into mitochondria (Bihlmaier *et al.*, 2008).

Because molecular oxygen drives the oxidative protein import system, the oxygen concentration is potentially critical for import efficiency (Bihlmaier *et al.*, 2008). At low oxygen concentrations, Mia40 is largely reduced and inactive, whereas at oxygen-saturated conditions, Mia40 is highly active. This suggests that the Erv1-mediated import system is particularly active at elevated oxygen concentrations, but is less active when oxygen concentrations are low. Since the activity of the disulfide relay system is coupled to the respiratory chain by cytochrome *c*, regulation of the respiratory chain might modulate the import and levels of CCS and SOD1 in mitochondria to meet the physiological requirements in mitochondria (Reddehase *et al.*, 2009). In mammalian tissue cultures, CCS mitochondrial import is regulated by oxygen concentration: physiological (6%) oxygen promotes import, whereas very high (20%) oxygen prevents it. As the content of CCS in mammalian cells is 15- to 30-fold less than that of SOD1, it is likely that the localization of CCS is the limiting factor, which dictates the subcellular distribution of enzymatically active SOD1. When the cytosol is exposed to an increasingly oxidative environment, such as the presence of high oxygen, CCS localizes primarily in the cytosol and retains SOD1 in this compartment. A shift toward a more oxidative environment in the IMS, for example, during mitochondrial oxidative stress, will result in a more efficient disulfide relay import system, enhancing CCS and SOD1 import into mitochondria to scavenge mitochondrial superoxide. Therefore, SOD1 localization responds to changes in environmental conditions following redistribution of CCS, which operates as an oxygen sensor (Bihlmaier *et al.*, 2008; Kawamata and Manfredi, 2008).

The relative amounts of SOD1 accumulating in brain versus liver mitochondria were compared in transgenic mice expressing either wild-type or mutant hSOD1. The proportion of SOD1, either wild type or mutant, accumulating in mitochondria was found to be higher in brain than that in the liver, while the total cellular amount of SOD1 was comparable. Perhaps in response to increased oxidative stress in post-mitotic tissues that have a higher oxidative metabolism, more SOD1 is accumulated into neuronal mitochondria (Vijayvergiya *et al.*, 2005).

10.11 PEROXISOMAL LOCALIZATION

SOD1 metabolizes superoxide into oxygen and hydrogen peroxide, the substrate of the major peroxisomal enzyme catalase. Therefore, it has been proposed that SOD1 is also a peroxisomal enzyme (Moreno *et al.*, 1997). Recently, both SOD1 and CCS have been found in peroxisomes (Islinger *et al.*, 2009). Most newly synthesized peroxisomal proteins are imported in a receptor-mediated fashion, depending on the interaction of a peroxisomal targeting signal (PTS) with its cognate targeting receptor located in the cytoplasm. Heterologous protein complexes might also be imported into peroxisomes. The distributions of CCS and SOD1 between cytosolic and peroxisomal compartments differ from each other. Although SOD1 is mostly cytosolic with only a small fraction found in perixosomes, the relative amount of CCS in peroxisomes is much higher. CCS has a peroxisomal targeting sequence in domain III (Casareno *et al.*, 1998). CCS-mediated SOD1 import into peroxisomes was abolished by deletion of a part of CCS domain II, which is required for CCS–SOD1 heterodimer formation. SOD1/CCS co-import is the first demonstration of a physiologically relevant "shuttle" import into mammalian peroxisomes. These data support the idea that proteins with physiological functions in several subcellular compartments can be directed to their respective locations in a chaperone-mediated fashion (Islinger *et al.*, 2009).

10.12 CCS KNOCKOUT MICE

Although *in vitro* experiments have yielded much information about the important role of CCS in the delivery of copper and maturation of SOD1, the test of these observations in a mammalian system requires the use of genetically altered animals. To determine the role of CCS in mammalian copper homeostasis and in SOD1 activation, CCS knockout mice (mice with targeted disruption of CCS alleles) were generated (Wong *et al.*, 2000). According to these studies, CCS knockout mice are viable and possess normal levels of SOD1 protein within the nervous system, but they have marked reductions in SOD1 activity in several tissues, including kidney, liver, and brain, when compared with control littermates. Metabolic labeling studies demonstrated that the decrease in SOD1 activity in CCS knockout mice is the direct result of impaired copper incorporation into SOD1, and that this effect was specific, because no abnormalities were observed in copper uptake, distribution, or incorporation into other cupro-proteins (Subramaniam *et al.*, 2002; Wong *et al.*, 2000). In CCS knockout mice, the endogenous mouse SOD1 was not totally abolished; about 15% of SOD1 activity remained (Beckman *et al.*, 2002), indicating that CCS is important but not essential for mammalian SOD1 to acquire copper and have activity. This active fraction of SOD1 in the CCS knockouts is likely copper-metallated and matured via CCS-independent pathways. It is unclear whether the small amount of enzymatically active SOD1 occurs because of a relative general inefficiency of CCS-independent pathways in cells or whether SOD1 has differential requirements for CCS-dependent and independent activation in various subcellular compartments. For instance, in mitochondria it is possible that SOD1 may be preferentially activated via CCS-independent mechanism, while CCS-dependent mechanism may provide most of the activation in cytosol.

The phenotype of the CCS knockout mice is relatively unremarkable since there are no distinct neurological or other physical abnormalities. On a microscopic level, CCS knockout mice do not exhibit significant neuronal pathology. However, CCS-deficient mice are hypersensitive to axonal injury and paraquat exposure showing decreased neuronal survival, and they also reveal decreased female fertility (Wong *et al.*, 2000). This phenotype is similar to SOD1-deficient mice and indicates that the loss of CCS in mammals physiologically and functionally mimics the loss of SOD1. CCS, like SOD1, is not required for normal neuronal homeostasis but is needed for a survival response after certain stresses or injuries.

10.13 CCS OVEREXPRESSION IN TRANSGENIC MICE

In order to study the effect of CCS overexpression on neuronal function, transgenic mice, expressing wild-type human CCS, were generated (Son *et al.*, 2007). The CCS construct was made using the mouse prion protein promoter shown to express transgenes at high levels within the CNS. Several of these lines expressed the hCCS protein at significantly higher levels than nontransgenic mice. The highest levels of hCCS were expressed within the CNS, particularly in the cerebellum, brainstem, and spinal cord. hCCS was also found in some organs outside CNS but at lower levels, mainly in kidney and heart, with lesser amount in liver. Within the spinal cord, hCCS was primarily localized within neuronal cell bodies, including ventral horn motor neurons, but there was also hCCS in the neuropil and white matter tracts. Immuno-histological analysis indicated that hCCS was mainly found in cytosol but it also localized to mitochondria and nuclei. Endogenous SOD1 levels in the spinal cord were similar in CCS transgenic and nontransgenic littermates, indicating that CCS overexpression does not alter total SOD1 levels *in vivo*.

CCS transgenic mice develop normally and demonstrate no overt neurological phenotype. The survival rate and all motor functions of CCS transgenic mice are comparable to that of nontransgenic mice, with survival extending to about 2.5 years. Histologic examination of the highest expressing line CCS adult mice revealed no evidence of neuronal loss or gliosis. In addition, neuronal mitochondrial morphology and function appear entirely normal. These results indicated that the overexpression of wild-type human CCS in mice does not produce an abnormal motor or neurological phenotype. The availability of CCS knockout and CCS transgenic mice allows for the definitive testing of the role of CCS in ALS related to SOD1 mutations.

10.14 SOD1-RELATED FAMILIAL AMYOTROPHIC LATERAL SCLEROSIS

ALS is a fatal neurodegenerative disease of unknown cause, characterized by the selective and progressive death of both upper and lower motor neurons, leading to progressive paralysis. The disease occurs in sporadic and familial forms with similar clinical courses and common pathological features including neuronal loss, gliosis, the presence of abnormal accumulations of cytoplasmic ubiquitin positive inclusions within surviving neurons, and mitochondrial abnormalities (Rothstein, 2009; Shi *et al.*, 2010). The familial form of ALS (fALS) accounts for 5–10% of cases and has typically an autosomal dominant pattern of inheritance. Among the fALS cases, the largest number (about 20%) are caused by missense or truncation mutations in the *SOD1* gene (ALS1) (Bruijn *et al.*, 2004; Rosen *et al.*, 1993). More recently, mutations in Tar DNA-binding protein (TDP-43; ALS10) (Kabashi *et al.*, 2008; Sreedharan *et al.*, 2008) and FUS/TLS (ALS6) (Kwiatkowski *et al.*, 2009; Vance *et al.*, 2009) also have been found to cause a smaller number of familial ALS cases.

Across species, the amino acid sequence of SOD1 is highly conserved; 112 of 153 residues are conserved in mammals. Most of the pathogenic mutations occur at residues conserved in mammals. The number of SOD1 mutations linked to ALS is about 150 at present (Prudencio *et al.*, 2009), with the majority being missense mutations. The various mutations of SOD1 are scattered throughout the protein including key areas of the dimer interface, active site, and the β-barrel. Many of the mutations affect side chains facing into the hydrophobic interior of the protein or in the dimer interface. In a few ALS-affected families, about 30 residues on the carboxyl-terminal end are deleted by truncation mutations. These deletions will seriously disrupt the dimer interface and the stability of the zinc-binding loop, but will leave the copper-binding residues intact (Trumbull and Beckman, 2009). Two fALS mutations in humans directly affect the copper-binding ligands of SOD1 (H46R and H48Q) (Arisato *et al.*, 2003). De novo zinc-binding site mutation (H80A) has been reported in one patient who is considered to have sporadic ALS (Alexander *et al.*, 2002).

10.15 SOD1 KNOCKOUT MICE

Much of our current understanding of mechanisms underlying mutant SOD1-induced ALS arises from the generation of mouse models of disease. In order to find out whether ALS, due to mutations in SOD1, was caused by lack of function, mice deficient for SOD1 were generated by targeted gene deletion. Homozygote SOD1 knockout mice were viable and appeared to develop without obvious motor abnormalities (Ho *et al.*, 1998; Reaume *et al.*, 1996). However, adult mice were hypersensitive to axotomy, paraquat-induced toxicity, and focal or global cerebral ischemia. Hence, disruption of SOD1 alone appeared to be insufficient to cause spontaneous motor neuron degeneration in mice without injury or challenge. However, SOD1 null mice were somewhat vulnerable to oxidative damage, as they aged prematurely and died at about 20 months. Experiments using the SOD1 knockout mice indicated that the loss of SOD1 function is not sufficient to cause motor neuron disease *in vivo* and provided evidence that disease does not result from a loss of SOD1 normal function.

10.16 SOD1 TRANSGENIC MICE

Transgenic mice that overexpress human SOD1-harboring mutations linked to fALS develop progressive muscle loss and paralysis characteristic of human ALS. Because these mice contain normal levels of endogenous murine SOD1, disease likely results from a toxic gain-of-function paradigm. Several different human SOD1 mutations have been expressed in transgenic mice. These include nine missense mutations and three carboxy terminal truncated variants (Bruijn *et al.*, 1997; Deng *et al.*, 2006; Gurney *et al.*, 1994; Jonsson *et al.*, 2004; Ripps *et al.*, 1995; Wang *et al.*, 2005; Watanabe *et al.*, 2005; Wong *et al.*, 1995). Transgenic lines with artificial mutations of multiple copper-binding domain were also generated, expressing single H46R, double H46R/H48Q, or quadruple H46R/H48Q/H63G/H120G SOD1 (Chang-Hong *et al.*, 2005; Wang *et al.*, 2002, 2003). Despite differences in transgene copy number, steady-state transcript and protein levels, dismutase activity, and neuropathology, the mutations induce a similar motor phenotype with fatal paralysis consistent with the human ALS motor phenotype.

The interpretation of these models takes into consideration wild-type SOD1 overexpressing transgenic mice to exclude the possibility that disease is related to a nonspecific transgenic protein load. Transgenic mice overexpressing wild-type human SOD1 at levels comparable to the highest mutant SOD1 levels are phenotypically normal and do not develop paralysis (Dal Canto and Gurney, 1995). However, when levels of wild-type human SOD1 were expressed many folds above normal, the mice develop a late-onset vacuolar pathology, axonal loss, and motor neuron degeneration (Jaarsma *et al.*, 2000). In contrast to mouse lines expressing mutant human SOD1, no lines of transgenic wild-type SOD1 mice have succumbed to ALS symptoms to date.

The pathogenic SOD1 mutations have been grouped based on their positions in the structure. One group, the "wild-type like" mutants, includes G93A SOD1 and G37R SOD1. Their metal-binding sites are intact, and they can form intramolecular disulfide bonds. Accordingly, they are active with bound copper in a coordination environment similar to that of wild-type SOD1. However, even in these "wild-type like" mutants, the metal-binding region is not totally stabilized (Museth *et al.*, 2009).

Transgenic mice expressing the G93A SOD1 mutation are the most common and best characterized. Several differing G93A SOD1 lines have been generated although all manifest initial hind limb weakness, progressing paralysis, and ultimately death. These G93A SOD1 lines express differing levels of the mutant SOD1 protein due to variable transgene copy number that determines the rate of disease onset and progression. The higher copy number subline dies at about four months of age (Dal Canto and Gurney, 1995; Gurney *et al.*, 1994), whereas a lower copy number subline has a more delayed phenotype and dies at about eight months of age (Elliott, 2001). The pathological features of spinal motor neurons in these mice include SOD1 positive, ubiquinated hyaline inclusions (Shibata *et al.*, 1998), cytoplasmic SOD1-immunoreactive aggregates (Johnston *et al.*, 2000), and neurofilament-positive inclusions (Tu *et al.*, 1996) that are characteristic of the pathology found in humans with SOD1-related disease. The highest copy number G93A SOD1 line displays extensive progressive mitochondrial vacuoles, whereas mitochondrial vacuolar pathology is minimal in the lower copy number G93A SOD1 mice (Dal Canto and Gurney, 1997). In contrast, the inclusion-based pathology is more pronounced in the lower copy number G93A SOD1 line than in the higher copy number line at the end-stage of the disease.

G37R SOD1 transgenic mice express high levels of G37R SOD1 protein, and their symptoms are very similar to those of higher copy number G93A SOD1 line (Borchelt *et al.*, 1994; Wong *et al.*, 1995). Mice succumb to paralysis in mid-life characterized by extensive spinal mitochondrial and multisystem degeneration. This pronounced vacuolar pathology typical of G37R SOD1 and G93A SOD1 transgenics is not observed in human motor neuron disease (Dal Canto and Gurney, 1995). However, more subtle defects of mitochondria morphology and function have been detected in the nervous system of ALS patients (Shi *et al.*, 2010).

The second group, "metal-binding region" mutants, includes mutations in the metal-binding ligands themselves or in the electrostatic and zinc loop elements that are intimately associated with metal binding, and therefore, they tend to be deficient in copper and/or zinc (Zetterstrom *et al.*, 2007). G85R SOD1 and its murine equivalent G86R SOD1 belong to "metal-binding site" mutations. In contrast to the G93A SOD1 and G37R SOD1 mice, transgenic mice expressing G85R SOD1 or G86R SOD1 have low steady-state levels of SOD1 protein. Both G85R SOD1 and G86R SOD1 cause an aggressive disease with short duration (Bruijn *et al.*, 1997; Morrison *et al.*, 1998; Ripps *et al.*, 1995). Any SOD activity above endogenous SOD1 levels cannot be detected in either G85R SOD1 or G86R SOD1 tissues. The spinal cords of G85R SOD1 and G86R SOD1 mice display SOD1 positive inclusions and ubiquination but not vacuolated mitochondria (Bruijn *et al.*, 1997; Son *et al.*, 2009). The spinal cords of G85R SOD1 and G86R SOD1 transgenic mice have been shown to contain disulfide-reduced forms of the mutant (Jonsson *et al.*, 2006; Son *et al.*, 2009).

To further test the role of copper in mutant SOD1 toxicity, mice expressing the H46R single mutant, H46R/H48Q double mutant, and mice expressing a quadruple SOD1 mutant that disrupts all four copper ligands (H46R/H48Q/H63G/H120G) were developed (Chang-Hong *et al.*, 2005; Wang *et al.*, 2002, 2003). Since H63 is also a ligand for zinc, the zinc-binding site would also be affected in the quadruple mice. Despite the lack of copper, both lines develop motor neuron disease characterized by the appearance of fibrillar SOD1-containing aggregates, ubiquinated inclusions but no mitochondrial vacuoles. However, this does not exclude the role of copper-mediated toxicity in ALS; in this class of metal-deficient SOD1 mutants, the amino acid substitutions introduced alone may alter SOD1 conformation and promote misfolding irrespective of metallation.

The third group is "truncation" mutants, where about 30 base pairs of the N-terminal end are missing. Mice expressing L126X, L126Z, or L126delTT truncation mutants of SOD1 develop late-onset motor neuron disease with rapid progression to death (Jonsson *et al.*, 2004; Wang *et al.*, 2005; Watanabe *et al.*, 2005). The steady-state expression levels of these L126Z and other truncated SOD1 proteins in young asymptomatic mice were lower than endogenous SOD1 protein. However, spinal cords of paralyzed mice accumulated relatively high levels of detergent-insoluble forms of the L126Z variant. Spinal cords of symptomatic L126Z SOD1 mice display ubiquitin immune-reactive inclusions but no mitochondrial vacuolization. Similar to transgenic mice harboring "metal-binding site" mutations, these truncated SOD1 mutants are inactive (Watanabe *et al.*, 2005). Notably, the intramolecular disulfide bond between C57 and C146 is destroyed in these truncation mutants, suggesting that disulfide-reduced SOD1 monomers may be important in pathogenesis (Jonsson *et al.*, 2006; Son *et al.*, 2009).

The dominant phenotype of transgenic ALS mice strongly supports a neurotoxic gain-of-function property of mutant SOD1 that accumulates damage overtime. These primary gain-of-functions likely result from mutation-induced protein conformational changes and alterations to SOD1 structural and functional properties. However, no biochemical feature appears to be common to all mutants. Rather, subsets of SOD1 variant proteins share decreased net charge, disulfide oxidation, stability, and heightened mitochondrial association (Valentine *et al.*, 2005). Hence, most investigations in transgenic mice have focused on putative disease mechanisms downstream of mutant SOD1-induced toxicity.

The mechanisms underlying mutant SOD1-related ALS are still unknown and several hypotheses have been proposed to account for the selective death of upper and lower motor neurons. These include oxidative damage, toxicity from intracellular aggregates, mitochondrial dysfunction, apoptosis, copper-catalyzed oxidative stress, glutamate-mediated excitotoxicity, axonal transport defects, and inflammation. Because the clinical course of the disease is highly variable, the mechanism of motor neuron death may arise from the unfortunate convergence of multiple factors rather than from a single cause (Cozzolino *et al.*, 2009; Rothstein, 2009; Tovar *et al.*, 2009).

10.17 ROLE OF CCS IN SOD1-LINKED FALS

Because CCS plays a critical role in the biochemical maturation and subcellular localization of wild-type SOD1, the possible impact of CCS on mutant SOD1 and on ALS has been investigated. In considering sporadic ALS, human genetics studies have potentially excluded CCS as a candidate gene or modifier (Silahtaroglu *et al.*, 2002). Sequencing of the *CCS* gene in a small cohort of 20 sporadic ALS patients found no polymorphisms or mutations associated with disease to suggest a link between CCS and sporadic ALS. This finding is not surprising, given that the role of SOD1 in sporadic ALS is not probable. Rather, there has been much exciting *in vivo* work using mice with genetic alterations in CCS levels to prove a pivotal role for CCS in mutant SOD1-related familial ALS.

In order to determine whether CCS is required for the development or progression of mutant SOD-related motor neuron disease, CCS knockout mice were crossed to three differing mutant SOD1 transgenic mouse lines (Subramaniam *et al.*, 2002). The influence of CCS on the motor neuron disease was studied by comparing the time of disease onset, the length of survival, and the pathology of G37R, G93A, or G85R SOD1 mice to the respective SOD1 mutant mice lacking CCS (Table 10.3). Loss of CCS did not modify the onset and progression of motor neuron disease in SOD1-mutant mice. Mutant SOD1 mice lacking CCS did not show alterations in the basic neuropathology associated with each particular SOD1 mutation. The degree of mitochondrial vacuolation appeared unchanged in G37R or G93A SOD1 mice lacking CCS compared to the mutant mice with full CCS levels. Similarly, ubiquitin and neurofilament neuronal inclusions were just as prominent in G37R, G85R, or G93A SOD1 with or without CCS. The presence of SOD1-positive protein aggregates was also unchanged in G37R SOD1 mice with or without CCS.

Since CCS regulates the mitochondrial localization of SOD1 *in vitro*, Cleveland and colleagues assessed whether the mitochondrial localization of mutant SOD1 *in vivo* was perturbed by the loss of CCS protein (Liu *et al.*, 2004). G37R SOD1 mice with and without CCS showed no difference in the levels of mutant SOD1 protein associated with the mitochondrial fraction isolated from spinal cords. Although total mitochondrial levels of G37R SOD1 were unchanged in the absence of CCS, there may have been shifts in mutant SOD1 within mitochondrial subcompartments that were not recognized. These *in vivo* experiments definitively prove that mutant SOD1-related disease and mitochondrial localization can occur in the absence of CCS. CCS is not required for the mitochondrial localization of mutant SOD1; however, it is not known whether CCS-independent

Table 10.3 The effect of CCS expression on the life span of wild-type SOD1 or mutant SOD1 transgenic mouse lines

LINE		CCS transgene	CCS endogeneous	CCS hemizygote	CCS knockout
WTSOD1	(1)	2.5y	2.5y		2.5y
G93ASOD1	(1)	36d	242d		
G37RSOD1	(1)	32d	270d		
G86RSOD1	(2)	115d	113d		
L126ZSOD1	(2)	230d	230d		
G93ASOD1	(3)		142d	136d	134d
G37RSOD1	(3)		163d	155d	146d
G85RSOD1	(3)		310d	324d	353d

The lines have been described in the following references:
(1) Son *et al.* (2007).
(2) Son *et al.* (2009).
(3) Subramaniam *et al.* (2002).

pathways substitute and regulate mutant SOD1 levels in mitochondria. Although the loss of CCS did not impact the disease phenotype, the loss of CCS did change mutant SOD1 biochemical parameters *in vivo*. Although SOD1 steady-state levels and total cellular copper levels were unchanged in spinal cords of mutant SOD1 mice lacking CCS, there was a 90% reduction in the incorporation of radio-labeled copper into SOD1 in G37R SOD1 mice in the absence of CCS when compared to G37R SOD1 $CCS^{+/+}$ or G37R SOD $CCS1^{+/-}$ mice (Subramaniam *et al.*, 2002). These experiments confirm that efficient incorporation of copper into mutant SOD1 depends on CCS *in vivo* although the residual copper loading indicates the existence of CCS-independent pathways. In agreement, SOD1 activity from spinal cord lysates was markedly decreased, but not absent (about 20% remaining), in both G37R and G93A mice lacking CCS, whereas there was no G85R-associated enzyme activity either in the presence or in the absence of CCS (Subramaniam *et al.*, 2002). Interestingly, the loss of dismutase activity in G37R SOD1 mice lacking CCS was considerable (about 80% reduced) in the cytoplasmic pool but was minimal in the mitochondrial compartment (Liu *et al.*, 2004). One possible explanation for this observation is that CCS-independent pathways of SOD1 activation and changes in redox state are more prominent in mitochondria than in cytosol. If these parameters (SOD1 maturation and redox state) are indeed critical components of the disease pathway, the importance of the CCS-independent pathways in mitochondria may explain why the loss of CCS does not impact disease onset or progression.

10.18 THE EFFECT OF CCS OVEREXPRESSION ON SOD1 LINKED FALS

Although the experiments with mutant SOD1 transgenic lines crossed with CCS knockout mice indicated that CCS is not required for disease, they could not exclude an important role for CCS in altering SOD1 biochemistry and thereby impacting the disease course. In order to better understand whether CCS could influence mutant SOD1-induced disease, transgenic lines expressing high levels of wild-type human CCS were crossed with SOD1 transgenic lines harboring different mutations (human G93A, human G37R, human L126Z, or mouse G86R) or, as a control, with a wild-type human SOD1 transgenic line (Table 10.3) (Son *et al.*, 2007, 2009). Double transgenic mice expressing both CCS and G93A SOD1 proteins develop a markedly accelerated disease course with neurological symptoms beginning by day 8 with a mean survival of 36 days. In contrast, the parental G93A SOD1 line, a lower copy number subline, develops motor symptoms at about 165 days and survives until 242 days, on average. Crossing CCS transgenic mice with G37R SOD1 transgenic mice yielded similar results. CCS/G37RSOD1 dual transgenic mice develop a progressive neurological disorder by 10 days of age and have a mean survival of 32 days, whereas G37R SOD1 mice live for 270 days. CCS transgenic mice crossed with wild-type SOD1 mice did not manifest any abnormal neurological phenotype and have a normal life span, indicating that the accelerated phenotype in CCS/G93A SOD1 and CCS/G37R SOD1 dual mice was not related to the simultaneous overexpression of CCS and SOD1. Thus, CCS overexpression markedly accelerates disease in both G37R SOD1 and G93A SOD1 mice, and to date represents the most significant acceleration of mutant SOD1 motor neuron disease that has been reported in the literature.

The pathological correlates of this marked phenotypic change in G37R or G93A SOD1 mice provided insights into disease mechanisms. SOD1 and ubiquitin positive aggregates are a pathologic hallmark of disease in mutant SOD1 transgenic mice and believed to be an essential part of mutant SOD1 toxicity (Johnston *et al.*, 2000). However, neither SOD1 protein complexes nor ubiquitin positive cellular inclusions were readily identified in the spinal cords of paralyzed CCS/G93A SOD1 or CCS/G37R SOD1 dual mice. Mutant SOD1-induced neurological dysfunction may therefore occur without prominent signs of SOD1 aggregation or ubiquitin inclusions.

Rather than inclusion-based pathology, CCS/G37R SOD1 or CCS/G93A SOD1 dual mice exhibit a markedly enhanced vacuolar mitochondrial pathology (Figure 10.1). While the parental G93A SOD1 mice manifest some degree of mitochondrial vacuolation within spinal cord neurons later during the disease course, the predominant pathology is largely inclusion based. However, overexpression of CCS in this G93A SOD1 line leads to prominent vacuolar pathology within the first week of life, and by three weeks of age, vacuoles fill spinal motor neuron somata. Electron microscopy revealed that these vacuoles were swollen mitochondria with disruption of the inner mitochondrial membrane but with relative preservation of the outer membrane. An identical enhanced mitochondrial pathology was also seen in young CCS/G37R SOD1 dual mice. Consistent with their normal neurological phenotype, CCS/wild-type SOD1 dual mice have no mitochondrial vacuoles, demonstrating that overexpression of CCS in the context of normal SOD1 did not cause pathological changes in mitochondria. The remarkable acceleration of disease in G37R and G93A SOD1 mice overexpressing CCS is correlated with mitochondrial pathology, not inclusion pathology, and suggests that toxicity of these mutations directly affects the mitochondria.

The ability of CCS to accelerate mutant SOD1-induced disease appears to be SOD1-mutation specific. To determine whether CCS could also impact different groups of SOD1 mutants, CCS transgenic mice were crossed with mice expressing either the truncation mutant, human L126Z SOD1, or the metal-binding region mutant, mouse G86R SOD1. In contrast to β-barrel SOD1 mutants G37R and G93A, CCS overexpression had no impact on the disease course or survival of L126Z or G86R SOD1 mice. The typical inclusion type pathology observed in L126Z SOD1 or G86R SOD1 mice was unchanged in the presence of CCS overexpression, and no vacuolar mitochondrial pathology was observed (Figure 10.1) (Son *et al.*, 2009). The reasons for the mutation specific effect of CCS may be related to basic structural properties of the mutant SOD1 proteins, which affect their subcellular localization, reduction potential, and interactions with CCS.

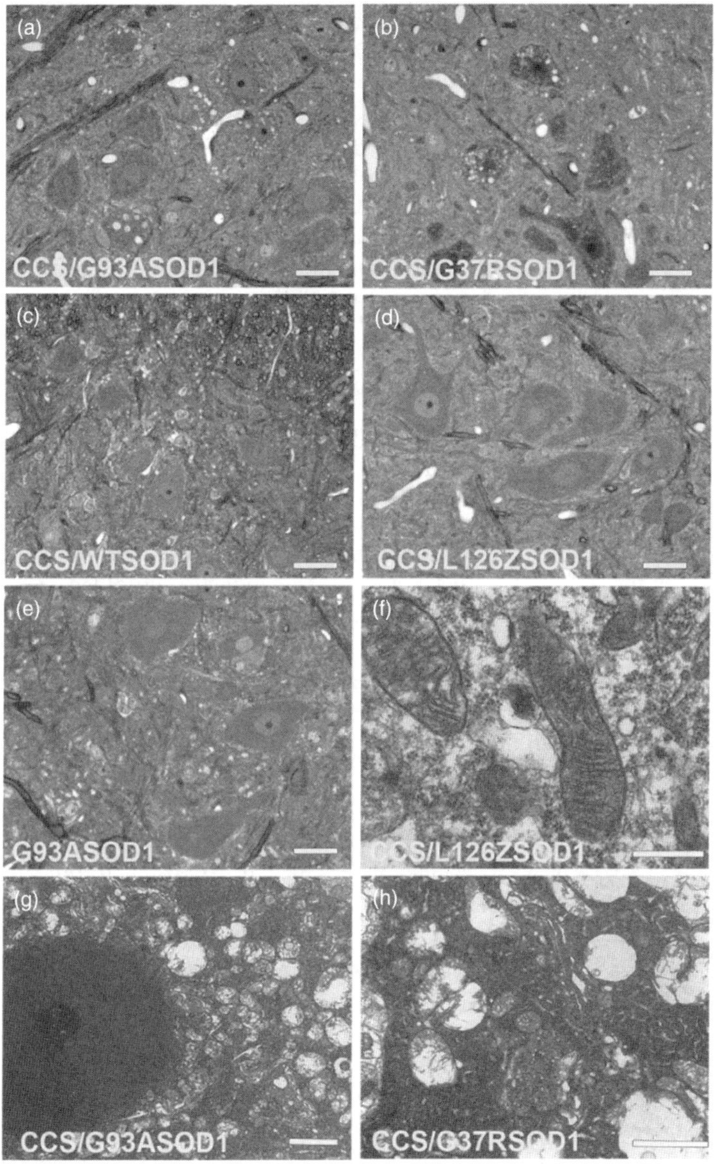

Figure 10.1 Effect of CCS overexpression on spinal cord pathology. (a–e) Toluidine blue-stained plastic sections of lumbar spinal cord ventral horn from three-week old CCS/G93A SOD1 (a), CCS/G37R SOD1 (b), CCS/WT SOD1 (c), CCS/L126Z SOD1 (d), and G93A SOD1 (f) mice. (f–h) Electron micrographs of mitochondria in spinal cord motor neurons from three-week old CCS/L126Z SOD1 (f), CCS/G93A SOD1 (g), and CCS/G37R SOD1 (h) mice. Arrows highlight neurons with vacuoles. Scale bars: a–e = 20 μm, f = 500 nm, g = 2 μm, and h = 2 μm.

The prominent mitochondrial pathology in CCS/G37R and CCS/G93A dual mice coupled with the earlier findings in yeast (Field *et al.*, 2003) regarding the effects of CCS on SOD1 mitochondrial localization indicated that CCS overexpression *in vivo* might affect the subcellular localization of SOD1. Studies on purified mitochondrial fractions as well as immuno-electron microscopy analysis confirmed elevated levels (about 2.5-fold) of SOD1 within spinal cord mitochondria of CCS/G93A SOD1 mice compared to G93A SOD1 mice. CCS overexpression slightly increased the mitochondrial colocalization of SOD1 in CCS/WT-SOD1 dual mice indicating that the increased mitochondrial localization of wild-type SOD1 protein, even in the presence of CCS overexpression, does not cause mitochondrial damage. CCS did not appear to alter total levels of SOD1 within spinal cord of dual mice nor did it alter the subcellular distribution of SOD1 in other compartments such as nuclei or microsomes. These results confirm that CCS overexpression alters the mitochondrial localization of wild type and certain mutant SOD1s and provides a potential explanation for the enhanced mitochondrial pathology in CCS/G93A SOD1 mice.

The import of proteins into the IMS of mitochondria depends on coupled oxidation–reduction reactions of cysteine-rich proteins, including CCS and SOD1 (Mesecke *et al.*, 2005). The disulfide relay system drives the import of these proteins into the IMS of mitochondria in the reduced, unfolded state, where they are trapped after being oxidatively folded. Wild-type SOD1 and the SOD1 mutants that can exist in both redox states, such as G93A SOD1, substantially localize to the mitochondrial IMS, while SOD1 mutants that exist largely in the reduced state, such as G86R or L126Z, would not be expected to participate in this disulfide relay system. Support for this idea comes from the findings that G85R SOD1 localizes mainly to the cytoplasmic surface of the mitochondrial outer membrane (Vande Velde *et al.*, 2008). Furthermore, a frameshift mutant of human SOD1 that is lacking both C111 and C146 accumulates in mitochondria in CCS-independent fashion (Kawamata and Manfredi, 2008). Therefore, while CCS overexpression can increase the mitochondrial sub-localization of some SOD1 mutants, like G93A, CCS would have little impact on altering mitochondrial localization of other mutants, like G85R (Kawamata and Manfredi, 2008; Son *et al.*, 2007).

10.19 SOD1 REDOX STATE AND DISEASE ACCELERATION

CCS overexpression might facilitate the presence of SOD1 in mitochondria and promote its toxicity by altering the redox states of SOD1, with only the reduced form being able to enter mitochondria. Wild-type human SOD1 contains two-cysteine residues (C57 and C146) that form intramolecular disulfide bonds and allow the SOD1 monomer to exist either in oxidized or reduced form. The C57 residue also allows for a transitory disulfide bond interaction between SOD1 and CCS that facilitates SOD1 maturation and oxidation, as detected in yeast (Culotta *et al.*, 2006; Lamb *et al.*, 2001). As predicted, in mouse spinal cords, CCS overexpression results in a marked change of the SOD1 redox state, where virtually all the wild-type SOD1 monomer is shifted to the oxidized form (Proescher *et al.*, 2008). Any wild-type SOD1 oxidized in the mitochondria would be retained there, which might explain the elevation in mitochondrial wild-type SOD1 observed *in vivo*. In contrast to its effects on wild-type SOD1, CCS overexpression increases the proportion of the reduced form of G37R and G93A SOD1. This surprising result suggests that the toxic moiety is the aberrantly reduced form of mutant SOD1, which would disrupt critical redox-dependent pathways in mitochondria. The inability of CCS to accelerate disease in G86R and L126Z mice supports the hypothesis that CCS exerts its effects on disease by promoting the reduced form of mutant SOD1. G86R SOD1 and L126Z SOD1 exist virtually in all the reduced form (Jonsson *et al.*, 2006; Son *et al.*, 2009). CCS overexpression *in vivo* cannot reduce G86R SOD1 and L126Z SOD1 any further and consequently would not be predicted to alter disease course for these mutants. Thus, CCS not only increases levels of SOD1 within mitochondria but also alters its redox state, and both of these correlate with disease acceleration.

The presence of an increased pool of reduced SOD1 within the IMS has deleterious consequences for mitochondrial function. For example, increased amounts of G93A SOD1 within the mitochondrial IMS have been shown to significantly elevate levels of toxic reactive oxygen species (Goldsteins *et al.*, 2008). Mitochondrial function was assessed by measuring the activity and amount of the mitochondrial electron transport chain complexes (Son *et al.*, 2007). Isolated defects in complex IV, cytochrome *c* oxidase, were detected within the spinal cords from both CCS/G37R and CCS/G93A SOD1 mice. Complex IV activity was decreased by over 50% compared to controls, while no changes in complex I, II, or III activities were noted. This loss of complex IV activity was accompanied by significant reductions in the levels of total complex IV as well as in both nuclear and mitochondrial encoded subunits (COX 1 and COX 5b), while levels of complexes I, II, III, and V were not changed. Thus, CCS overexpression in G37R and G93A SOD1 mice does not lead to overall mitochondrial respiratory dysfunction but rather to an isolated deficiency in complex IV. Such reductions in complex IV function and levels were not observed in the spinal cords of CCS/wild-type SOD1, CCS/G86R SOD1, or CCS/L126Z SOD1 mice (Figure 10.2). Thus, isolated complex IV deficiency correlates with mitochondrial pathology and motor neuron defects suggesting a causative role (Son *et al.*, 2008).

The isolated complex IV deficiency in CCS/G37R and CCS/G93A SOD1 spinal cord can potentially be explained by alterations in mutant SOD1 redox state (Figure 10.3). The mitochondrial IMS contains several cysteine-containing redox-sensitive proteins critical for assembly and function of complex IV that could be vulnerable to increased levels of an aberrantly reduced SOD1 species. Complex IV is composed of 3 mitochondrial encoded subunits (COX1, COX2, and COX3) and 10 nuclear-encoded subunits. Mitochondrial encoded subunits COX1 and COX2 form the catalytic core of the complex and require the addition of a number of prosthetic groups and cofactors for functionality. The biogenesis of complex IV also requires several nuclear-encoded protein cofactors which

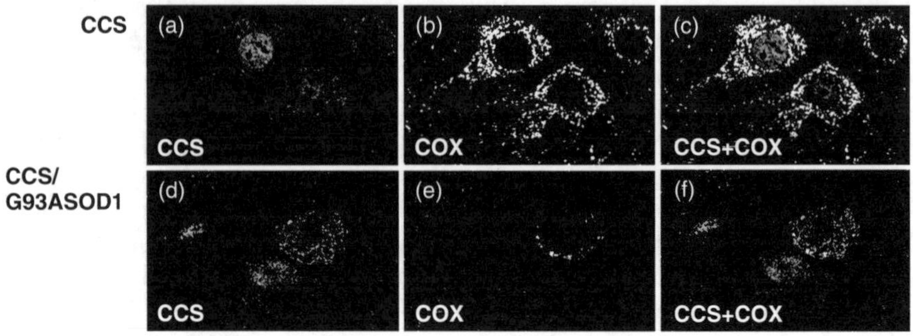

Figure 10.2 COX1 immuno-reactivity in spinal cord motor neurons. Confocal microscopy showing CCS (a and d), COX1 (b and e), and merged images (c and f) from three-week old CCS (top row) and CCS/G93A SOD1 (bottom row) dual transgenic mice. (*A full color version of this figure appears in the color plate section.*)

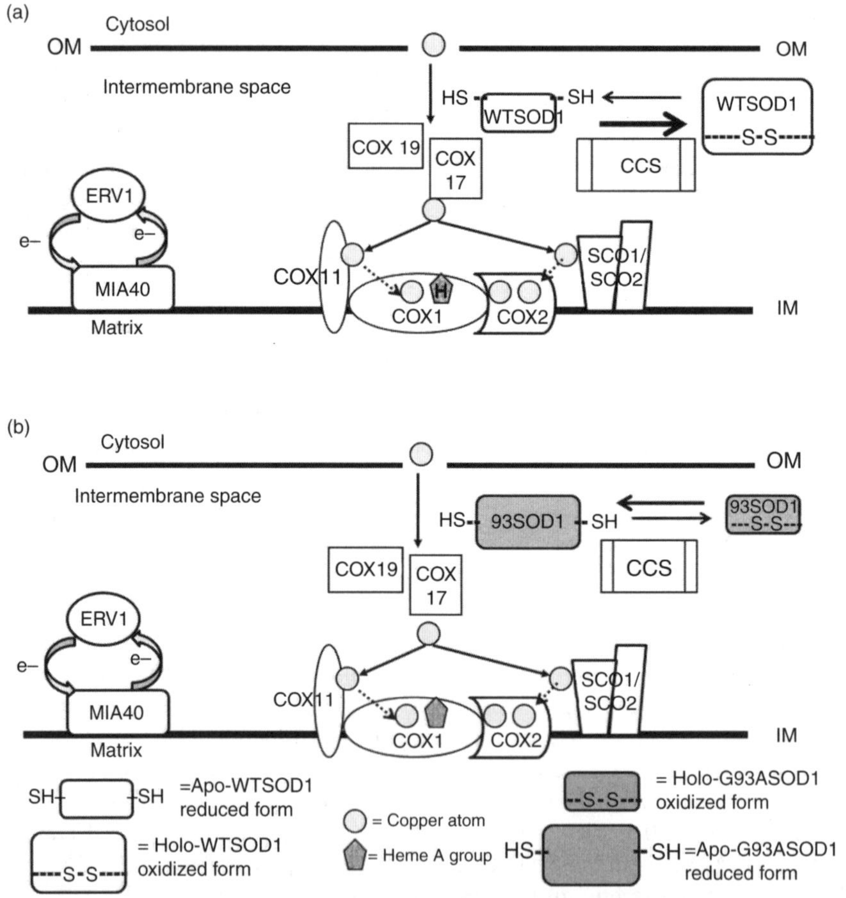

Figure 10.3 Schematic diagram of copper transport to COX subunits 1 and 2 and effect of CCS on redox state of wild-type SOD1 (a) and G93A SOD1 (b).

Table 10.4 Selected cysteine-rich mitochondrial proteins

Protein	Cysteine motifs
CCS	CX_2C, CX_2C, CXC
Mia40	CX_9C, CX_9C, CXC
Erv1	CX_2C, CX_2C
COX10	CX_9C
COX11	CXC
COX15	CX_3C, CX_3C
COX17	CX_9C, CX_9C
COX19	CX_9C, CX_9C
SCO1	CX_3C, CXC
SCO2	CX_3C
COX1	CX_3C

must transit into mitochondria. Mitochondrial encoded COX1 and COX2 subunits both require copper, which normally is imported from the cytoplasm via a complex trafficking system including COX17, COX11, SCO1, SCO2, COX19, and COX 23. COX17 transports copper from the cytosol into the mitochondrial IMS (Cobine *et al.*, 2006). COX 17 then donates copper to COX11 and SCO1, which in turn incorporate copper into COX1 CuB and COX2 CuA sites, respectively. SCO2 is another copper-binding protein required for normal complex IV maturation although its precise role in complex IV metallation is unclear. Human SCO1 and SCO2 mutations produce isolated complex IV deficiency and reductions in subunit levels, in a pattern similar to that in CCS/G93A SOD1 dual mice. How might G93A SOD1 localization within mitochondria interfere with the function of these copper trafficking proteins and lead to isolated complex IV deficiency? COX17, COX19, COX11, SCO1, and SCO2, all of which contain multi-cysteine motifs, are involved in copper transport to complex IV (Table 10.4). Because mutant SOD1 is in an aberrant redox state, it has an increased chance of forming inappropriate interactions with these complex IV accessory proteins containing cysteine disulfide bonds and interfering with their function, yielding an isolated complex IV deficiency. There is extensive current work in progress to determine whether this hypothesis is correct.

10.20 CONCLUSION

Over the past decade, much has been learned about the role CCS plays in the normal physiological function of SOD1 as well as in disease. CCS not only chaperones and presents copper to SOD1 but also promotes the maturation of SOD1 through facilitating its disulfide bond oxidation. In addition, by altering the redox state of SOD1, CCS also significantly impacts the subcellular localization of SOD1. CCS overexpression in G93A SOD1 or G37R SOD1 mice causes the most significant acceleration of mutant SOD1-linked fALS reported in the literature. This effect might be due to alterations in the properties of mutant SOD1 causing the disease, which may be linked to changes in the redox state of SOD1. The work with CCS overexpression in mutant SOD1 mice also provides evidence that a substantial part of the toxicity in G93A and G37R SOD1 mutants is related to effects on mitochondria. SOD1 protein aggregation may be harmful but not necessary for disease onset or progression. The mechanism for SOD1 toxicity in mitochondria is not completely known but may involve alterations in complex IV assembly protein/s that also depend on redox state shifts to deliver copper to complex IV. The specificity of the complex IV deficiency observed in transgenic mice support this idea and may provide a rationale for designing therapeutic agents for the treatment of mutant SOD1-linked fALS.

REFERENCES

Ahtoniemi T, Jaronen M, Keksa-Goldsteine V, Goldsteins G, Koistinaho J. Mutant SOD1 from spinal cord of G93A rats is destabilized and binds to inner mitochondrial membrane. *Neurobiol Dis* 2008;**32**:479–485.

Alexander MD, Traynor BJ, Miller N, Corr B, Frost E, McQuaid S, Brett FM, Green A, Hardiman O. "True" sporadic ALS associated with a novel SOD-1 mutation. *Ann Neurol* 2002;**52**:680–683.

Angeletti B, Waldron KJ, Freeman KB, Bawagan H, Hussain I, Miller CC, Lau KF, Tennant ME, Dennison C, Robinson NJ, *et al.* BACE1 cytoplasmic domain interacts with the copper chaperone for superoxide dismutase-1 and binds copper. *J Biol Chem* 2005;**280**:17930–17937.

Arisato T, Okubo R, Arata H, Abe K, Fukada K, Sakoda S, Shimizu A, Qin XH, Izumo S, Osame M, *et al.* Clinical and pathological studies of familial amyotrophic lateral sclerosis (FALS) with SOD1 H46R mutation in large Japanese families. *Acta Neuropathol* 2003;**106**:561–568.

Arnesano F, Banci L, Bertini I, Martinelli M, Furukawa Y, O'Halloran TV. The unusually stable quaternary structure of human Cu,Zn-superoxide dismutase 1 is controlled by both metal occupancy and disulfide status. *J Biol Chem* 2004;**279**:47998–48003.

Atkin JD, Farg MA, Walker AK, McLean C, Tomas D, Horne MK. Endoplasmic reticulum stress and induction of the unfolded protein response in human sporadic amyotrophic lateral sclerosis. *Neurobiol Dis* 2008;**30**:400–407.

Banci L, Bertini I, Cantini F, D'Amelio N, Gaggelli E. Human SOD1 before harboring the catalytic metal: solution structure of copper-depleted, disulfide-reduced form. *J Biol Chem* 2006;**281**:2333–2337.

Banci L, Bertini I, Cantini F, D'Onofrio M, Viezzoli MS. Structure and dynamics of copper-free SOD: the protein before binding copper. *Protein Sci* 2002;**11**:2479–2492.

Barry AN, Blackburn NJ. A selenocysteine variant of the human copper chaperone for superoxide dismutase. A Se-XAS probe of cluster composition at the domain 3-domain 3 dimer interface. *Biochemistry* 2008;**47**:4916–4928.

Barry AN, Clark KM, Otoikhian A, van der Donk WA, Blackburn NJ. Selenocysteine positional variants reveal contributions to copper binding from cysteine residues in domains 2 and 3 of human copper chaperone for superoxide dismutase. *Biochemistry* 2008;**47**:13074–13083.

Bartnikas TB, Gitlin JD. How to make a metalloprotein. *Nat Struct Biol* 2001;**8**:733–734.

Beckman JS, Esetvez AG, Barbeito L, Crow JP. CCS knockout mice establish an alternative source of copper for SOD in ALS. *Free Radic Biol Med* 2002;**33**:1433–1435.

Bertinato J, L'Abbe MR. Copper modulates the degradation of copper chaperone for Cu,Zn superoxide dismutase by the 26S proteosome. *J Biol Chem* 2003;**278**:35071–35078.

Bihlmaier K, Mesecke N, Kloeppel C, Herrmann JM. The disulfide relay of the intermembrane space of mitochondria: an oxygen-sensing system? *Ann N Y Acad Sci* 2008;**1147**:293–302.

Borchelt DR, Lee MK, Slunt HS, Guarnieri M, Xu ZS, Wong PC, Brown RH Jr, Price DL, Sisodia SS, Cleveland DW. Superoxide dismutase 1 with mutations linked to familial amyotrophic lateral sclerosis possesses significant activity. *Proc Natl Acad Sci U S A* 1994;**91**:8292–8296.

Brown NM, Torres AS, Doan PE, O'Halloran TV. Oxygen and the copper chaperone CCS regulate posttranslational activation of Cu,Zn superoxide dismutase. *Proc Natl Acad Sci U S A* 2004;**101**:5518–5523.

Bruijn LI, Becher MW, Lee MK, Anderson KL, Jenkins NA, Copeland NG, Sisodia SS, Rothstein JD, Borchelt DR, Price DL, *et al.* ALS-linked SOD1 mutant G85R mediates damage to astrocytes and promotes rapidly progressive disease with SOD1-containing inclusions. *Neuron* 1997;**18**:327–338.

Bruijn LI, Miller TM, Cleveland DW. Unraveling the mechanisms involved in motor neuron degeneration in ALS. *Annu Rev Neurosci* 2004;**27**:723–749.

Cao X, Antonyuk SV, Seetharaman SV, Whitson LJ, Taylor AB, Holloway SP, Strange RW, Doucette PA, Valentine JS, Tiwari A, *et al.* Structures of the G85R variant of SOD1 in familial amyotrophic lateral sclerosis. *J Biol Chem* 2008;**283**:16169–16177.

Carroll MC, Girouard JB, Ulloa JL, Subramaniam JR, Wong PC, Valentine JS, Culotta VC. Mechanisms for activating Cu- and Zn-containing superoxide dismutase in the absence of the CCS Cu chaperone. *Proc Natl Acad Sci U S A* 2004;**101**:5964–5969.

Caruano-Yzermans AL, Bartnikas TB, Gitlin JD. Mechanisms of the copper-dependent turnover of the copper chaperone for superoxide dismutase. *J Biol Chem* 2006;**281**:13581–13587.

Casareno RL, Waggoner D, Gitlin JD. The copper chaperone CCS directly interacts with copper/zinc superoxide dismutase. *J Biol Chem* 1998;**273**:23625–23628.

Chang-Hong R, Wada M, Koyama S, Kimura H, Arawaka S, Kawanami T, Kurita K, Kadoya T, Aoki M, Itoyama Y, *et al.* Neuroprotective effect of oxidized galectin-1 in a transgenic mouse model of amyotrophic lateral sclerosis. *Exp Neurol* 2005;**194**:203–211.

Cobine PA, Pierrel F, Winge DR. Copper trafficking to the mitochondrion and assembly of copper metalloenzymes. *Biochim Biophys Acta* 2006;**1763**:759–772.

Corson LB, Strain JJ, Culotta VC, Cleveland DW. Chaperone-facilitated copper binding is a property common to several classes of familial amyotrophic lateral sclerosis-linked superoxide dismutase mutants. *Proc Natl Acad Sci U S A* 1998;**95**:6361–6366.

Cozzolino M, Pesaresi MG, Amori I, Crosio C, Ferri A, Nencini M, Carri MT. Oligomerization of mutant SOD1 in mitochondria of motoneuronal cells drives mitochondrial damage and cell toxicity. *Antioxid Redox Signal* 2009;**11**:1547–1558.

Crapo JD, Oury T, Rabouille C, Slot JW, Chang LY. Copper,zinc superoxide dismutase is primarily a cytosolic protein in human cells. *Proc Natl Acad Sci U S A* 1992;**89**:10405–10409.

Culotta VC, Klomp LW, Strain J, Casareno RL, Krems B, Gitlin JD. The copper chaperone for superoxide dismutase. *J Biol Chem* 1997;**272**:23469–23472.

Culotta VC, Yang M, O'Halloran TV. Activation of superoxide dismutases: putting the metal to the pedal. *Biochim Biophys Acta* 2006;**1763**:747–758.

Dal Canto MC, Gurney ME. Neuropathological changes in two lines of mice carrying a transgene for mutant human Cu,Zn SOD, and in mice overexpressing wild type human SOD: a model of familial amyotrophic lateral sclerosis (FALS). *Brain Res* 1995;**676**:25–40.

Dal Canto MC, Gurney ME. A low expressor line of transgenic mice carrying a mutant human Cu,Zn superoxide dismutase (SOD1) gene develops pathological changes that most closely resemble those in human amyotrophic lateral sclerosis. *Acta Neuropathol* 1997;**93**:537–550.

Deng HX, Shi Y, Furukawa Y, Zhai H, Fu R, Liu E, Gorrie GH, Khan MS, Hung WY, Bigio EH, *et al.* Conversion to the amyotrophic lateral sclerosis phenotype is associated with intermolecular linked insoluble aggregates of SOD1 in mitochondria. *Proc Natl Acad Sci U S A* 2006;**103**:7142–7147.

Dingwall C. A copper-binding site in the cytoplasmic domain of BACE1 identifies a possible link to metal homoeostasis and oxidative stress in Alzheimer's disease. *Biochem Soc Trans* 2007;**35**:571–573.

Elliott JL. Cytokine upregulation in a murine model of familial amyotrophic lateral sclerosis. *Brain Res Mol Brain Res* 2001;**95**:172–178.

Endo T, Fujii T, Sato K, Taniguchi N, Fujii J. A pivotal role of Zn-binding residues in the function of the copper chaperone for SOD1. *Biochem Biophys Res Commun* 2000;**276**:999–1004.

Field LS, Furukawa Y, O'Halloran TV, Culotta VC. Factors controlling the uptake of yeast copper/zinc superoxide dismutase into mitochondria. *J Biol Chem* 2003;**278**:28052–28059.

Furukawa Y, Kaneko K, Yamanaka K, O'Halloran TV, Nukina N. Complete loss of post-translational modifications triggers fibrillar aggregation of SOD1 in the familial form of amyotrophic lateral sclerosis. *J Biol Chem* 2008;**283**:24167–24176.

Furukawa Y, O'Halloran TV. Posttranslational modifications in Cu,Zn-superoxide dismutase and mutations associated with amyotrophic lateral sclerosis. *Antioxid Redox Signal* 2006;**8**:847–867.

Furukawa Y, Torres AS, O'Halloran TV. Oxygen-induced maturation of SOD1: a key role for disulfide formation by the copper chaperone CCS. *EMBO J* 2004;**23**:2872–2881.

Goldsteins G, Keksa-Goldsteine V, Ahtoniemi T, Jaronen M, Arens E, Akerman K, Chan PH, Koistinaho J. Deleterious role of superoxide dismutase in the mitochondrial intermembrane space. *J Biol Chem* 2008;**283**:8446–8452.

Gurney ME, Pu H, Chiu AY, Dal Canto MC, Polchow CY, Alexander DD, Caliendo J, Hentati A, Kwon YW, Deng HX, *et al.* Motor neuron degeneration in mice that express a human Cu,Zn superoxide dismutase mutation. *Science* 1994;**264**:1772–1775.

Hart PJ. Pathogenic superoxide dismutase structure, folding, aggregation and turnover. *Curr Opin Chem Biol* 2006;**10**:131–138.

Harvey LJ, McArdle HJ. Biomarkers of copper status: a brief update. *Br J Nutr* 2008;**99** Suppl 3:S10–S13.

Hell K. The Erv1-Mia40 disulfide relay system in the intermembrane space of mitochondria. *Biochim Biophys Acta* 2008;**1783**:601–609.

Herrmann JM, Hell K. Chopped, trapped or tacked–protein translocation into the IMS of mitochondria. *Trends Biochem Sci* 2005;**30**:205–211.

Higgins CM, Jung C, Ding H, Xu Z. Mutant Cu, Zn superoxide dismutase that causes motoneuron degeneration is present in mitochondria in the CNS. *J Neurosci* 2002;**22**:RC215.

Higgins CM, Jung C, Xu Z. ALS-associated mutant SOD1G93A causes mitochondrial vacuolation by expansion of the intermembrane space and by involvement of SOD1 aggregation and peroxisomes. *BMC Neurosci* 2003;**4**:16.

Hiromura M, Chino H, Sonoda T, Sakurai H. Molecular cloning and characterization of a copper chaperone for copper/zinc superoxide dismutase from the rat. *Biochem Biophys Res Commun* 2000;**275**:394–400.

Ho YS, Gargano M, Cao J, Bronson RT, Heimler I, Hutz RJ. Reduced fertility in female mice lacking copper-zinc superoxide dismutase. *J Biol Chem* 1998;**273**:7765–7769.

Horecka J, Kinsey PT, Sprague GF Jr. Cloning and characterization of the *Saccharomyces cerevisiae* LYS7 gene: evidence for function outside of lysine biosynthesis. *Gene* 1995;**162**:87–92.

Islinger M, Li KW, Seitz J, Volkl A, Luers GH. Hitchhiking of Cu/Zn superoxide dismutase to peroxisomes–evidence for a natural piggyback import mechanism in mammals. *Traffic* 2009;**10**:1711–1721.

Jaarsma D, Haasdijk ED, Grashorn JA, Hawkins R, van Duijn W, Verspaget HW, London J, Holstege JC. Human Cu/Zn superoxide dismutase (SOD1) overexpression in mice causes mitochondrial vacuolization, axonal degeneration, and premature motoneuron death and accelerates motoneuron disease in mice expressing a familial amyotrophic lateral sclerosis mutant SOD1. *Neurobiol Dis* 2000;**7**:623–643.

Jaarsma D, Rognoni F, van Duijn W, Verspaget HW, Haasdijk ED, Holstege JC. CuZn superoxide dismutase (SOD1) accumulates in vacuolated mitochondria in transgenic mice expressing amyotrophic lateral sclerosis-linked SOD1 mutations. *Acta Neuropathol* 2001;**102**:293–305.

Jeney V, Itoh S, Wendt M, Gradek Q, Ushio-Fukai M, Harrison DG, Fukai T. Role of antioxidant-1 in extracellular superoxide dismutase function and expression. *Circ Res* 2005;**96**:723–729.

Jensen LT, Culotta VC. Activation of CuZn superoxide dismutases from *Caenorhabditis elegans* does not require the copper chaperone CCS. *J Biol Chem* 2005;**280**:41373–41379.

Johnston JA, Dalton MJ, Gurney ME, Kopito RR. Formation of high molecular weight complexes of mutant Cu, Zn-superoxide dismutase in a mouse model for familial amyotrophic lateral sclerosis. *Proc Natl Acad Sci U S A* 2000;**97**:12571–12576.

Jonsson PA, Ernhill K, Andersen PM, Bergemalm D, Brannstrom T, Gredal O, Nilsson P, Marklund SL. Minute quantities of misfolded mutant superoxide dismutase-1 cause amyotrophic lateral sclerosis. *Brain* 2004;**127**:73–88.

Jonsson PA, Graffmo KS, Andersen PM, Brannstrom T, Lindberg M, Oliveberg M, Marklund SL. Disulphide-reduced superoxide dismutase-1 in CNS of transgenic amyotrophic lateral sclerosis models. *Brain* 2006;**129**:451–464.

Kabashi E, Valdmanis PN, Dion P, Spiegelman D, McConkey BJ, Vande Velde C, Bouchard JP, Lacomblez L, Pochigaeva K, Salachas F, *et al.* TARDBP mutations in individuals with sporadic and familial amyotrophic lateral sclerosis. *Nat Genet* 2008;**40**:572–574.

Karch CM, Prudencio M, Winkler DD, Hart PJ, Borchelt DR. Role of mutant SOD1 disulfide oxidation and aggregation in the pathogenesis of familial ALS. *Proc Natl Acad Sci U S A* 2009;**106**:7774–7779.

Kawamata H, Manfredi G. Different regulation of wild-type and mutant Cu,Zn superoxide dismutase localization in mammalian mitochondria. *Hum Mol Genet* 2008;**17**:3303–3317.

Khalimonchuk O, Winge DR. Function and redox state of mitochondrial localized cysteine-rich proteins important in the assembly of cytochrome c oxidase. *Biochim Biophys Acta* 2008;**1783**:618–628.

Kikuchi H, Almer G, Yamashita S, Guegan C, Nagai M, Xu Z, Sosunov AA, McKhann GM 2nd, Przedborski S. Spinal cord endoplasmic reticulum stress associated with a microsomal accumulation of mutant superoxide dismutase-1 in an ALS model. *Proc Natl Acad Sci U S A* 2006;**103**:6025–6030.

Kwiatkowski TJ Jr, Bosco DA, Leclerc AL, Tamrazian E, Vanderburg CR, Russ C, Davis A, Gilchrist J, Kasarskis EJ, Munsat T, *et al.* Mutations in the FUS/TLS gene on chromosome 16 cause familial amyotrophic lateral sclerosis. *Science* 2009;**323**:1205–1208.

Lamb AL, Torres AS, O'Halloran TV, Rosenzweig AC. Heterodimer formation between superoxide dismutase and its copper chaperone. *Biochemistry* 2000a;**39**:14720–14727.

Lamb AL, Wernimont AK, Pufahl RA, O'Halloran TV, Rosenzweig AC. Crystal structure of the second domain of the human copper chaperone for superoxide dismutase. *Biochemistry* 2000b;**39**:1589–1595.

Lamb AL, Torres AS, O'Halloran TV, Rosenzweig AC. Heterodimeric structure of superoxide dismutase in complex with its metallochaperone. *Nat Struct Biol* 2001;**8**:751–755.

Lamb AL, Wernimont AK, Pufahl RA, Culotta VC, O'Halloran TV, Rosenzweig AC. Crystal structure of the copper chaperone for superoxide dismutase. *Nat Struct Biol* 1999;**6**:724–729.

Lee J, Prohaska JR, Dagenais SL, Glover TW, Thiele DJ. Isolation of a murine copper transporter gene, tissue specific expression and functional complementation of a yeast copper transport mutant. *Gene* 2000;**254**:87–96.

Lehman NL. The ubiquitin proteasome system in neuropathology. *Acta Neuropathol* 2009;**118**:329–347.

Leitch JM, Jensen LT, Bouldin SD, Outten CE, Hart PJ, Culotta VC. Activation of Cu,Zn-superoxide dismutase in the absence of oxygen and the copper chaperone CCS. *J Biol Chem* 2009a;**284**:21863–21871.

Leitch JM, Yick PJ, Culotta VC. The right to choose: multiple pathways for activating copper,zinc superoxide dismutase. *J Biol Chem* 2009b;**284**:24679–24683.

Liu J, Lillo C, Jonsson PA, Vande Velde C, Ward CM, Miller TM, Subramaniam JR, Rothstein JD, Marklund S, Andersen PM, *et al.* Toxicity of familial ALS-linked SOD1 mutants from selective recruitment to spinal mitochondria. *Neuron* 2004;**43**:5–17.

Luk E, Jensen LT, Culotta VC. The many highways for intracellular trafficking of metals. *J Biol Inorg Chem* 2003;**8**:803–809.

Luk E, Yang M, Jensen LT, Bourbonnais Y, Culotta VC. Manganese activation of superoxide dismutase 2 in the mitochondria of *Saccharomyces cerevisiae. J Biol Chem* 2005;**280**:22715–22720.

Lutz T, Neupert W, Herrmann JM. Import of small Tim proteins into the mitochondrial intermembrane space. *EMBO J* 2003;**22**:4400–4408.

Marklund SL. Human copper-containing superoxide dismutase of high molecular weight. *Proc Natl Acad Sci U S A* 1982;**79**:7634–7638.

Mattiazzi M, D'Aurelio M, Gajewski CD, Martushova K, Kiaei M, Beal MF, Manfredi G. Mutated human SOD1 causes dysfunction of oxidative phosphorylation in mitochondria of transgenic mice. *J Biol Chem* 2002;**277**:29626–29633.

McCord JM, Fridovich I. Superoxide dismutase. An enzymic function for erythrocuprein (hemocuprein). *J Biol Chem* 1969;**244**:6049–6055.

McLoughlin DM, Standen CL, Lau KF, Ackerley S, Bartnikas TP, Gitlin JD, Miller CC. The neuronal adaptor protein X11alpha interacts with the copper chaperone for SOD1 and regulates SOD1 activity. *J Biol Chem* 2001;**276**:9303–9307.

Mesecke N, Terziyska N, Kozany C, Baumann F, Neupert W, Hell K, Herrmann JM. A disulfide relay system in the intermembrane space of mitochondria that mediates protein import. *Cell* 2005;**121**:1059–1069.

Moore SD, Chen MM, Cox DW. Cloning and mapping of murine superoxide dismutase copper chaperone (Ccsd) and mapping of the human ortholog. *Cytogenet Cell Genet* 2000;**88**:35–37.

Moreno S, Nardacci R, Ceru MP. Regional and ultrastructural immunolocalization of copper-zinc superoxide dismutase in rat central nervous system. *J Histochem Cytochem* 1997;**45**:1611–1622.

Morrison BM, Janssen WG, Gordon JW, Morrison JH. Time course of neuropathology in the spinal cord of G86R superoxide dismutase transgenic mice. *J Comp Neurol* 1998;**391**:64–77.

Museth AK, Brorsson AC, Lundqvist M, Tibell LA, Jonsson BH. The ALS-associated mutation G93A in human copper-zinc superoxide dismutase selectively destabilizes the remote metal binding region. *Biochemistry* 2009;**48**:8817–8829.

O'Halloran TV, Culotta VC. Metallochaperones, an intracellular shuttle service for metal ions. *J Biol Chem* 2000;**275**:25057–25060.

Ohgami RS, Campagna DR, McDonald A, Fleming MD. The Steap proteins are metalloreductases. *Blood* 2006;**108**:1388–1394.

Okado-Matsumoto A, Fridovich I. Subcellular distribution of superoxide dismutases (SOD) in rat liver: Cu,Zn-SOD in mitochondria. *J Biol Chem* 2001;**276**:38388–38393.

Pardo CA, Xu Z, Borchelt DR, Price DL, Sisodia SS, Cleveland DW. Superoxide dismutase is an abundant component in cell bodies, dendrites, and axons of motor neurons and in a subset of other neurons. *Proc Natl Acad Sci U S A* 1995;**92**:954–958.

Pasinelli P, Belford ME, Lennon N, Bacskai BJ, Hyman BT, Trotti D, Brown RH Jr. Amyotrophic lateral sclerosis-associated SOD1 mutant proteins bind and aggregate with Bcl-2 in spinal cord mitochondria. *Neuron* 2004;**43**:19–30.

Proescher JB, Son M, Elliott JL, Culotta VC. Biological effects of CCS in the absence of SOD1 enzyme activation: implications for disease in a mouse model for ALS. *Hum Mol Genet* 2008;**17**:1728–1737.

Prohaska JR. Role of copper transporters in copper homeostasis. *Am J Clin Nutr* 2008;**88**:826S–829S.

Prohaska JR, Broderius M, Brokate B. Metallochaperone for Cu,Zn-superoxide dismutase (CCS) protein but not mRNA is higher in organs from copper-deficient mice and rats. *Arch Biochem Biophys* 2003;**417**:227–234.

Prudencio M, Hart PJ, Borchelt DR, Andersen PM. Variation in aggregation propensities among ALS-associated variants of SOD1: correlation to human disease. *Hum Mol Genet* 2009;**18**:3217–3226.

Rae TD, Schmidt PJ, Pufahl RA, Culotta VC, O'Halloran TV. Undetectable intracellular free copper: the requirement of a copper chaperone for superoxide dismutase. *Science* 1999;**284**:805–808.

Rae TD, Torres AS, Pufahl RA, O'Halloran TV. Mechanism of Cu,Zn-superoxide dismutase activation by the human metallochaperone hCCS. *J Biol Chem* 2001;**276**:5166–5176.

Reaume AG, Elliott JL, Hoffman EK, Kowall NW, Ferrante RJ, Siwek DF, Wilcox HM, Flood DG, Beal MF, Brown RH Jr, *et al.* Motor neurons in Cu/Zn superoxide dismutase-deficient mice develop normally but exhibit enhanced cell death after axonal injury. *Nat Genet* 1996;**13**:43–47.

Reddehase S, Grumbt B, Neupert W, Hell K. The disulfide relay system of mitochondria is required for the biogenesis of mitochondrial Ccs1 and Sod1. *J Mol Biol* 2009;**385**:331–338.

Ripps ME, Huntley GW, Hof PR, Morrison JH, Gordon JW. Transgenic mice expressing an altered murine superoxide dismutase gene provide an animal model of amyotrophic lateral sclerosis. *Proc Natl Acad Sci U S A* 1995;**92**:689–693.

Rosen DR, Siddique T, Patterson D, Figlewicz DA, Sapp P, Hentati A, Donaldson D, Goto J, O'Regan JP, Deng HX, *et al.* Mutations in Cu/Zn superoxide dismutase gene are associated with familial amyotrophic lateral sclerosis. *Nature* 1993;**362**:59–62.

Rothstein JD. Current hypotheses for the underlying biology of amyotrophic lateral sclerosis. *Ann Neurol* 2009;**65** Suppl 1:S3–S9.

Rothstein JD, Dykes-Hoberg M, Corson LB, Becker M, Cleveland DW, Price DL, Culotta VC, Wong PC. The copper chaperone CCS is abundant in neurons and astrocytes in human and rodent brain. *J Neurochem* 1999;**72**:422–429.

Schmidt PJ, Kunst C, Culotta VC. Copper activation of superoxide dismutase 1 (SOD1) *in vivo*. Role for protein-protein interactions with the copper chaperone for SOD1. *J Biol Chem* 2000;**275**:33771–33776.

Schmidt PJ, Rae TD, Pufahl RA, Hamma T, Strain J, O'Halloran TV, Culotta VC. Multiple protein domains contribute to the action of the copper chaperone for superoxide dismutase. *J Biol Chem* 1999a;**274**:23719–23725.

Schmidt PJ, Ramos-Gomez M, Culotta VC. A gain of superoxide dismutase (SOD) activity obtained with CCS, the copper metallochaperone for SOD1. *J Biol Chem* 1999b;**274**:36952–36956.

Seetharaman SV, Prudencio M, Karch C, Holloway SP, Borchelt DR, Hart PJ. Immature copper-zinc superoxide dismutase and familial amyotrophic lateral sclerosis. *Exp Biol Med (Maywood)* 2009;**234**:1140–1154.

Shi P, Gal J, Kwinter DM, Liu X, Zhu H. Mitochondrial dysfunction in amyotrophic lateral sclerosis. *Biochim Biophys Acta* 2010;**1802**:45–51.

Shibata N, Hirano A, Kobayashi M, Dal Canto MC, Gurney ME, Komori T, Umahara T, Asayama K. Presence of Cu/Zn superoxide dismutase (SOD) immunoreactivity in neuronal hyaline inclusions in spinal cords from mice carrying a transgene for Gly93Ala mutant human Cu/Zn SOD. *Acta Neuropathol* 1998;**95**:136–142.

Silahtaroglu AN, Brondum-Nielsen K, Gredal O, Werdelin L, Panas M, Petersen MB, Tommerup N, Tumer Z. Human CCS gene: genomic organization and exclusion as a candidate for amyotrophic lateral sclerosis (ALS). *BMC Genet* 2002;**3**:5.

Son M, Cloyd CD, Rothstein JD, Rajendran B, Elliott JL. Aggregate formation in Cu,Zn superoxide dismutase-related proteins. *J Biol Chem* 2003;**278**:14331–14336.

Son M, Fu Q, Puttaparthi K, Matthews CM, Elliott JL. Redox susceptibility of SOD1 mutants is associated with the differential response to CCS over-expression *in vivo*. *Neurobiol Dis* 2009;**34**:155–162.

Son M, Leary SC, Romain N, Pierrel F, Winge DR, Haller RG, Elliott JL. Isolated cytochrome c oxidase deficiency in G93A SOD1 mice overexpressing CCS protein. *J Biol Chem* 2008;**283**:12267–12275.

Son M, Puttaparthi K, Kawamata H, Rajendran B, Boyer PJ, Manfredi G, Elliott JL. Overexpression of CCS in G93A-SOD1 mice leads to accelerated neurological deficits with severe mitochondrial pathology. *Proc Natl Acad Sci U S A* 2007;**104**:6072–6077.

Sreedharan J, Blair IP, Tripathi VB, Hu X, Vance C, Rogelj B, Ackerley S, Durnall JC, Williams KL, Buratti E, *et al.* TDP-43 mutations in familial and sporadic amyotrophic lateral sclerosis. *Science* 2008;**319**:1668–1672.

Stasser JP, Eisses JF, Barry AN, Kaplan JH, Blackburn NJ. Cysteine-to-serine mutants of the human copper chaperone for superoxide dismutase reveal a copper cluster at a domain III dimer interface. *Biochemistry* 2005;**44**:3143–3152.

Stasser JP, Siluvai GS, Barry AN, Blackburn NJ. A multinuclear copper(I) cluster forms the dimerization interface in copper-loaded human copper chaperone for superoxide dismutase. *Biochemistry* 2007;**46**:11845–11856.

Strange RW, Antonyuk S, Hough MA, Doucette PA, Rodriguez JA, Hart PJ, Hayward LJ, Valentine JS, Hasnain SS. The structure of holo and metal-deficient wild-type human Cu, Zn superoxide dismutase and its relevance to familial amyotrophic lateral sclerosis. *J Mol Biol* 2003;**328**:877–891.

Sturtz LA, Diekert K, Jensen LT, Lill R, Culotta VC. A fraction of yeast Cu,Zn-superoxide dismutase and its metallochaperone, CCS, localize to the intermembrane space of mitochondria. A physiological role for SOD1 in guarding against mitochondrial oxidative damage. *J Biol Chem* 2001;**276**:38084–38089.

Subramaniam JR, Lyons WE, Liu J, Bartnikas TB, Rothstein J, Price DL, Cleveland DW, Gitlin JD, Wong PC. Mutant SOD1 causes motor neuron disease independent of copper chaperone-mediated copper loading. *Nat Neurosci* 2002;**5**:301–307.

Suzuki KT, Kuroda T. Transfer of copper and zinc from ionic and metallothionein-bound forms to Cu, Zn–superoxide dismutase. *Res Commun Mol Pathol Pharmacol* 1995;**87**:287–296.

Tokuda E, Okawa E, Ono S. Dysregulation of intracellular copper trafficking pathway in a mouse model of mutant copper/zinc superoxide dismutase-linked familial amyotrophic lateral sclerosis. *J Neurochem* 2009;**111**:181–191.

Tovar YRLB, Santa-Cruz LD, Tapia R. Experimental models for the study of neurodegeneration in amyotrophic lateral sclerosis. *Mol Neurodegener* 2009;**4**:31.

Trumbull KA, Beckman JS. A role for copper in the toxicity of zinc-deficient superoxide dismutase to motor neurons in amyotrophic lateral sclerosis. *Antioxid Redox Signal* 2009;**11**:1627–1639.

Tu PH, Raju P, Robinson KA, Gurney ME, Trojanowski JQ, Lee VM. Transgenic mice carrying a human mutant superoxide dismutase transgene develop neuronal cytoskeletal pathology resembling human amyotrophic lateral sclerosis lesions. *Proc Natl Acad Sci U S A* 1996;**93**:3155–3160.

Turski ML, Thiele DJ. New roles for copper metabolism in cell proliferation, signaling, and disease. *J Biol Chem* 2009;**284**:717–721.

Uchida K, Kawakishi S. Identification of oxidized histidine generated at the active site of Cu,Zn-superoxide dismutase exposed to H2O2. Selective generation of 2-oxo-histidine at the histidine 118. *J Biol Chem* 1994;**269**:2405–2410.

Unlu ES, Koc A. Effects of deleting mitochondrial antioxidant genes on life span. *Ann N Y Acad Sci* 2007;**1100**:505–509.

Valentine JS, Doucette PA, Zittin Potter S. Copper-zinc superoxide dismutase and amyotrophic lateral sclerosis. *Annu Rev Biochem* 2005;**74**:563–593.

Vance C, Rogelj B, Hortobagyi T, De Vos KJ, Nishimura AL, Sreedharan J, Hu X, Smith B, Ruddy D, Wright P, *et al.* Mutations in FUS, an RNA processing protein, cause familial amyotrophic lateral sclerosis type 6. *Science* 2009;**323**:1208–1211.

Vande Velde C, Miller TM, Cashman NR, Cleveland DW. Selective association of misfolded ALS-linked mutant SOD1 with the cytoplasmic face of mitochondria. *Proc Natl Acad Sci U S A* 2008;**105**:4022–4027.

Vijayvergiya C, Beal MF, Buck J, Manfredi G. Mutant superoxide dismutase 1 forms aggregates in the brain mitochondrial matrix of amyotrophic lateral sclerosis mice. *J Neurosci* 2005;**25**:2463–2470.

Wang J, Slunt H, Gonzales V, Fromholt D, Coonfield M, Copeland NG, Jenkins NA, Borchelt DR. Copper-binding-site-null SOD1 causes ALS in transgenic mice: aggregates of non-native SOD1 delineate a common feature. *Hum Mol Genet* 2003;**12**:2753–2764.

Wang J, Xu G, Gonzales V, Coonfield M, Fromholt D, Copeland NG, Jenkins NA, Borchelt DR. Fibrillar inclusions and motor neuron degeneration in transgenic mice expressing superoxide dismutase 1 with a disrupted copper-binding site. *Neurobiol Dis* 2002;**10**:128–138.

Wang J, Xu G, Li H, Gonzales V, Fromholt D, Karch C, Copeland NG, Jenkins NA, Borchelt DR. Somatodendritic accumulation of misfolded SOD1-L126Z in motor neurons mediates degeneration: alphaB-crystallin modulates aggregation. *Hum Mol Genet* 2005;**14**:2335–2347.

Watanabe Y, Yasui K, Nakano T, Doi K, Fukada Y, Kitayama M, Ishimoto M, Kurihara S, Kawashima M, Fukuda H, *et al.* Mouse motor neuron disease caused by truncated SOD1 with or without C-terminal modification. *Brain Res Mol Brain Res* 2005;**135**:12–20.

Weisiger RA, Fridovich I. Superoxide dismutase. Organelle specificity. *J Biol Chem* 1973;**248**:3582–3592.

Winkler DD, Schuermann JP, Cao X, Holloway SP, Borchelt DR, Carroll MC, Proescher JB, Culotta VC, Hart PJ. Structural and biophysical properties of the pathogenic SOD1 variant H46R/H48Q. *Biochemistry* 2009;**48**:3436–3447.

Wong PC, Pardo CA, Borchelt DR, Lee MK, Copeland NG, Jenkins NA, Sisodia SS, Cleveland DW, Price DL. An adverse property of a familial ALS-linked SOD1 mutation causes motor neuron disease characterized by vacuolar degeneration of mitochondria. *Neuron* 1995;**14**:1105–1116.

Wong PC, Waggoner D, Subramaniam JR, Tessarollo L, Bartnikas TB, Culotta VC, Price DL, Rothstein J, Gitlin JD. Copper chaperone for superoxide dismutase is essential to activate mammalian Cu/Zn superoxide dismutase. *Proc Natl Acad Sci U S A* 2000;**97**:2886–2891.

Zetterstrom P, Stewart HG, Bergemalm D, Jonsson PA, Graffmo KS, Andersen PM, Brannstrom T, Oliveberg M, Marklund SL. Soluble misfolded subfractions of mutant superoxide dismutase-1s are enriched in spinal cords throughout life in murine ALS models. *Proc Natl Acad Sci U S A* 2007;**104**:14157–14162.

11

EMERGING AREA: TorsinA, A NOVEL ATP-DEPENDENT FACTOR LINKED TO DYSTONIA

Michal Zolkiewski and Hui-Chuan Wu
Department of Biochemistry, Kansas State University, Manhattan, KS, USA

11.1 TorsinA AND EARLY-ONSET DYSTONIA

Dystonia represents a group of neurological disorders that manifest themselves in involuntary muscle contractions, producing uncontrollable movements and abnormal body postures (Breakefield *et al.*, 2008). Dystonia is one of the most common movement disorders together with essential tremor and Parkinson's disease. Primary dystonia develops in the absence of other apparent neurological traumas or diseases, whereas secondary dystonia may result from brain injury or other disease. Early-onset torsion dystonia (EOTD, also known as *DYT1* or *Oppenheim's dystonia*) is the most common and severe form of primary dystonia with a strong hereditary component (Kamm, 2006). The symptoms of EOTD usually occur first between the ages of 5 and 28, which suggest a developmentally linked etiology. The dystonic symptoms in EOTD typically affect the limbs but may spread to other body parts. There is no apparent loss of neurons in the brains of EOTD patients; their psychological development, intellectual capabilities, and life expectancy are normal, but the progressive

Protein Chaperones and Protection from Neurodegenerative Diseases, First Edition.
Edited by Stephan N. Witt.

nature of the disease often leaves them severely disabled and confined to a wheelchair. No apparent neurodegeneration in EOTD suggests that a dysfunction of neuronal or neuromuscular signaling, rather than cell death, is the cause of the disease. A detailed pathological analysis of the brains from EOTD patients showed some intracellular aggregates or inclusions in neurons (McNaught *et al.*, 2004). The inclusions stained positively for ubiquitin, torsinA (see below), the nuclear envelope (NE) protein lamin A/C, and tau protein. These findings suggest that EOTD may be a protein-folding/aggregation disease. In addition, a potential link between the EOTD pathology and the NE in neurons was established, which later was corroborated in multiple studies (see below).

EOTD is quite rare in the general population (3.4:100,000), but its prevalence among the Ashkenazi Jews is an order of higher magnitude (Kamm, 2006). Studies of the Ashkenazi Jewish families revealed that the high prevalence of EOTD can be linked to a founder mutation that occurred approximately 350 years ago in that ethnic group living in Eastern Europe (Risch *et al.*, 1995). The gene responsible for EOTD in both Jewish and non-Jewish families has been mapped to chromosome 9q34 (Kramer *et al.*, 1994; Ozelius *et al.*, 1992), and in 1997, that gene was identified (Ozelius *et al.*, 1997). The human EOTD-linked gene is termed *DYT1* or *TOR1A* and encodes a 38-kDa protein called *torsinA*. By sequence homology, torsinA is a member of the AAA+ superfamily of *A*TPases *a*ssociated with various cellular *a*ctivities (Neuwald *et al.*, 1999).

Most cases of EOTD are associated with a deletion of a single GAG codon in *DYT1*, which leads to loss of a single glutamic acid residue in the C-terminal region of torsinA (Figure 11.1) (Ozelius *et al.*, 1997). The mutation is autosomal dominant, but the disease manifests itself in only 30–40% of the carriers, which suggests that other genetic or environmental factors may contribute to the onset of EOTD. If a mutation carrier reaches the age of 28 without developing the EOTD symptoms, he or she usually remains disease-free for the rest of life.

There is no effective pharmacological therapy for EOTD. *Botulinum* toxin injections into selected muscle groups produce local suppression of dystonic symptoms in some EOTD patients, but they must be administered repeatedly to maintain the beneficial effect. Recently, surgical deep brain stimulation has been shown to relieve the EOTD symptoms. However, the procedure is very invasive and the understanding of the neurophysiological side effects of this approach is limited (Kamm, 2006). Thus, torsinA appears as an attractive target for developing novel and more conventional EOTD therapies. The remaining part of this chapter describes the current state of our understanding of the biological function of torsinA.

11.2 TorsinA AND OTHER TORSIN PROTEINS IN THE AAA+ SUPERFAMILY

AAA+ ATPases are considered "molecular machines" that use energy from the hydrolysis of adenosine triphosphate (ATP) to induce conformational changes

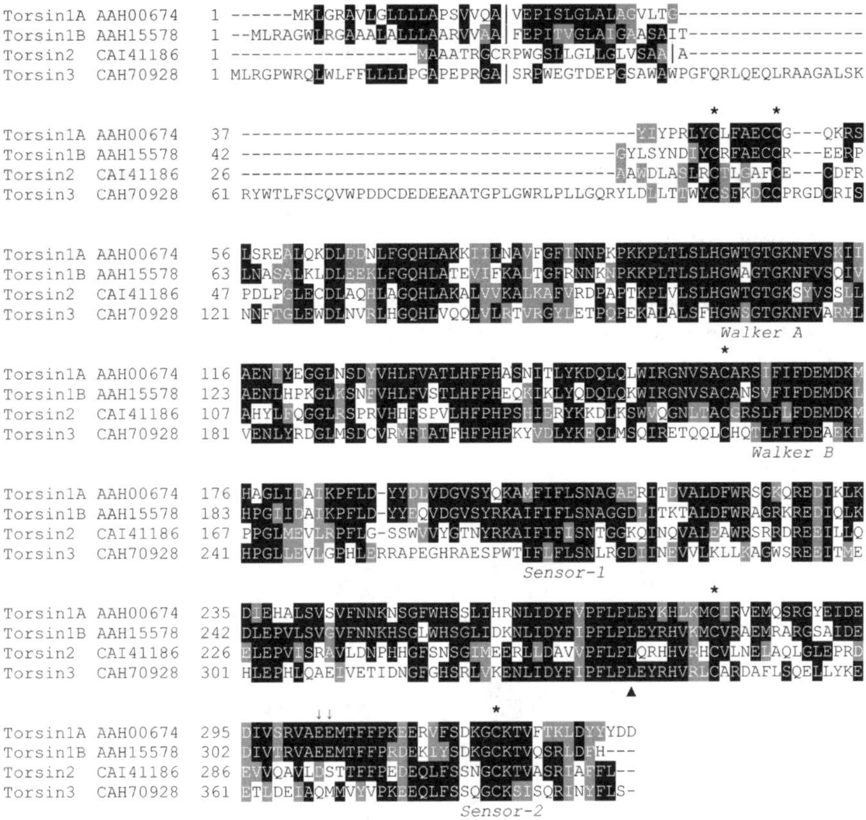

Figure 11.1 Sequence alignment of human torsin sequences: torsin1A, torsin1B, torsin2, and torsin3 (see Figure 11.3). The signal-sequence cleavage sites predicted with SignalP algorithm (Nielsen et al., 1997) are indicated with vertical lines. The proposed boundary between the large and small AAA+ subdomains (Zhu et al., 2008) is marked with a solid arrowhead. The positions of Walker A, Walker B, Sensor-1, and Sensor-2 motifs are indicated below the aligned sequences. Two Glu residues, one of which is deleted in the dystonia-linked mutant torsin1A are indicated by arrows and the conserved Cys residues are indicated by asterisks above the aligned sequences.

in macromolecules or to disassemble macromolecular complexes (Hanson and Whiteheart, 2005; Ogura and Wilkinson, 2001). The known substrates of AAA+ ATPases whose conformation is remodeled by the "machines" are either polypeptides or polynucleotides. Members of the AAA+ family play essential roles in protein disaggregation and degradation, intracellular transport and membrane fusion, DNA replication and repair, and cytoskeletal regulation (Hanson and Whiteheart, 2005). AAA+ ATPases contain one or two conserved ATP-binding sequence modules. The AAA+ modules include the Walker A and B motifs as well as several other sequence motifs that distinguish the AAA+ family from other P-loop containing NTPases (Figure 11.1). In addition to the AAA+

sequence modules, those proteins often contain additional domains that determine their cellular localization, substrate specificity, and function.

Each AAA+ module folds into a two-domain structure with the ATP-binding site located between the subdomains (Figure 11.2). The larger N-terminal Rossman-fold subdomain contains a central β-sheet whose strand tips contain the conserved ATP-binding residues. The smaller C-terminal subdomain is usually α-helical (Erzberger and Berger, 2006; Hanson and Whiteheart, 2005). A characteristic structural feature of AAA+ ATPases is the formation of cylinder-shaped oligomers, usually hexamers. In many cases, the AAA+ oligomers are not stable but assemble and disassemble into monomers dynamically as they process their substrates and turn over ATP (Hartman and Vale, 1999; Haslberger *et al.*, 2008). The association of AAA+ cylinders is stimulated by nucleotides

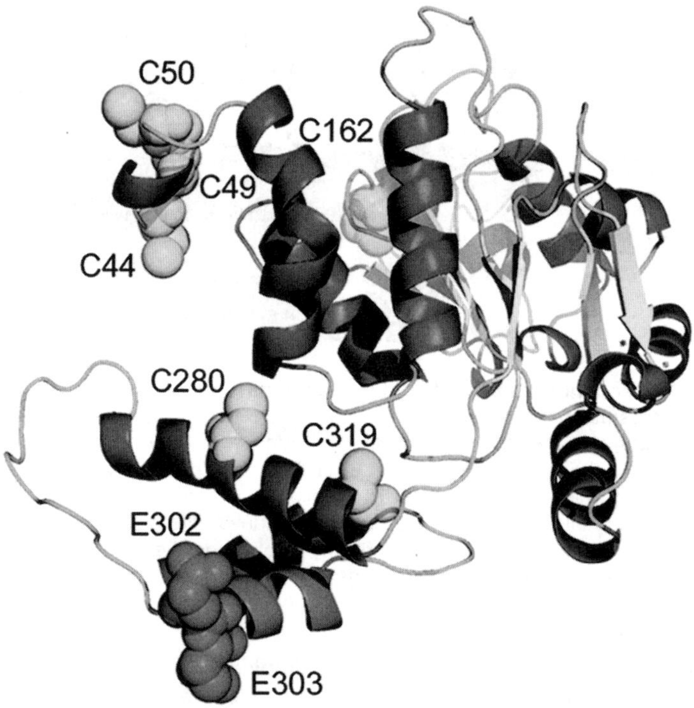

Figure 11.2 Structural model of human torsinA. The model was produced with Swiss Model server (http://swissmodel.expasy.org/) (Arnold et al., 2006; Bordoli et al., 2009) using the structure of the C-terminal AAA+ module of *Thermus thermophilus* ClpB (Lee et al., 2003) as the template. The hydrophobic N-terminal segment of torsinA was not included in the model, that is, the N-terminus shown in this figure corresponds to Arg41 in the sequence of torsinA. The N-terminal subdomain of the AAA+ module is on the top of the figure and the C-terminal subdomain is on the bottom. The Cys residues in torsinA as well as Glu302 and Glu303, one of which residues is deleted in torsinAΔE, are shown with space-fill representation.

but is inhibited by high ionic-strength buffers, which suggests that electrostatic interactions stabilize the oligomers (Barnett and Zolkiewski, 2002). In addition, contacts maintained by the smaller C-terminal subdomain of the AAA+ module contribute to stability of the oligomers (Barnett *et al.*, 2000). Importantly, no biological activity has been found in mutated monomeric variants of AAA+ ATPases, which indicates that the oligomers are required for the AAA+ functionality (Barnett *et al.*, 2000). One reason why monomeric AAA+ ATPases are found biologically inactive is that the nucleotide-binding sites in oligomers are located at the interfaces between subunits (Bochtler *et al.*, 2000; Zhang *et al.*, 2000). As a consequence, AAA+ monomers bind nucleotides with very low affinity and support low rates of ATP hydrolysis (Barnett *et al.*, 2000).

The cylinder-like structure of an AAA+ ATPase in the oligomeric form contains a narrow axial channel open at both ends. For those AAA+ ATPases whose mechanism of function has been studied in detail, the central channel was found to be the substrate-processing site. The conformation of substrate-binding site(s) depends on the nucleotide occupancy, and the substrate-binding affinity is the highest in the presence of ATP (Lee *et al.*, 2007; Rouiller *et al.*, 2002). The high-affinity state of an AAA+ ATPase can be experimentally maintained either by supplying a non-hydrolyzable ATP analog (ATPγS and AMP-PNP) or by introducing a mutation of the conserved glutamate in the Walker B motif, which preserves ATP binding but blocks its hydrolysis (Weibezahn *et al.*, 2003).

After being recognized by the ATPase, a substrate is inserted into the channel, which then either partially or fully translocates across the length of the cylinder (Reid *et al.*, 2001; Weber-Ban *et al.*, 1999). The substrate translocation is driven by ATP hydrolysis. The channel in an AAA+ oligomer is too narrow to accommodate a folded substrate. Thus, insertion into the channel and translocation of the substrate cause gradual unraveling of its structure (Lee *et al.*, 2001). Indeed, many well-studied AAA+ ATPases turned out to be the ATP-driven "unfoldases" of their respective substrates (Zolkiewski, 2006).

Although AAA+ ATPases do manipulate conformation of other macromolecules, they are not usually included in the molecular chaperone family because, with one known exception, they are not involved in maintaining the quality of protein folding and they are not upregulated under stress. The well-studied exception is a bacterial heat shock protein ClpB whose yeast and plant orthologs are known as *Hsp104* and *Hsp101*, respectively (see Chapter 7 for more on Hsp104). ClpB is essential for the survival of bacteria (Squires *et al.*, 1991), fungi (Parsell *et al.*, 1994), and plants (Queitsch *et al.*, 2000) under conditions of severe heat shock, but it is not found in animal proteomes. ClpB specifically recognizes protein aggregates that form under conditions of severe stress and mediates their resolubilization and reactivation (Glover and Lindquist, 1998; Zolkiewski, 1999). The aggregate-reactivation activity of ClpB requires cooperation with the chaperones from Hsp70 and Hsp40 families and involves extraction of single polypeptides from the aggregates by insertion and translocation through the axial channel in the oligomeric ClpB (Weibezahn *et al.*, 2004). Substrates are then released from the ClpB channel and they refold

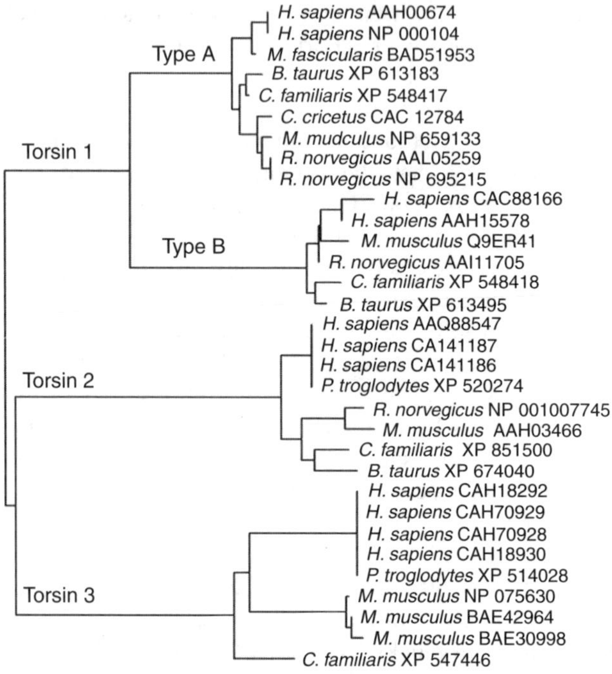

Figure 11.3 Phylogenetic analysis of mammalian torsins. Protein sequence alignments and the phylogenetic analysis of the torsin family were performed using MEGA software (ver. 3.1, www.megasoftware.net) (Kumar et al., 2004). All complete nonidentical torsin protein sequences found in the NCBI database were included in the analysis. The sequence accession numbers are shown in the figure.

with assistance of other chaperones. A related bacterial AAA+ ATPase ClpA uses a similar mechanistic principle as ClpB, but unlike ClpB, ClpA forms a complex with a multimeric peptidase ClpP. Substrates of ClpA are forcefully unfolded during translocation through the channel and then transferred into ClpP for degradation (Reid *et al.*, 2001; Weber-Ban *et al.*, 1999). The ClpA/ClpP complex is a prokaryotic prototype of the eukaryotic proteasome.

There are about 80 *AAA+* genes in the human genome (Ogura and Wilkinson, 2001). Among those, there are four genes of torsins, with one of them, *DYT1*, encoding torsinA. All known mammalian genomes contain four torsin genes, whereas only a single torsin gene is found in insects. Interestingly, no torsin-like genes are found in prokaryotes, fungi, or plants, which suggests that the function of torsins is essential for animal physiology. Phylogenetic analysis showed that in the highly divergent "evolutionary tree" of AAA+ ATPases, torsins are specifically related to the C-terminal AAA+ module of ClpA and ClpB (Iyer *et al.*, 2004), the ATPases involved in protein degradation and aggregate reactivation, respectively (Zolkiewski, 2006). It appears that the AAA+ module of torsins evolved exclusively in the animal lineage from that of ClpA/ClpB.

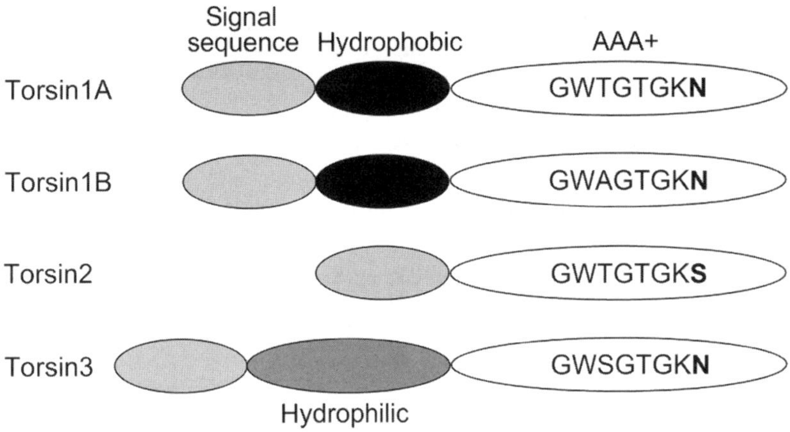

Figure 11.4 Predicted domain structure of torsin subfamilies. Shown are signal sequences, AAA+ modules, and the inserted hydrophobic or hydrophilic domains. The sequence of the Walker A motif is indicated for each subfamily.

This evolutionary link is important because unlike the structures of torsins, those of ClpA and ClpB are available (Guo *et al.*, 2002; Lee *et al.*, 2003) and their mechanisms are better understood (see above).

Multiple sequence alignments and a phylogenetic analysis of mammalian torsin sequences show that all torsins can be divided into three distinct families (torsin1, torsin2, and torsin3) (Figure 11.3). The torsin1 family further diverges into two branches: A and B. The dystonia-linked human torsinA belongs to the torsin1 type A family. All torsins contain a single AAA+ module with a set of conserved nucleotide-binding motifs: Walker A, Walker B, sensor-1, and sensor-2 (Figure 11.1) (Neuwald *et al.*, 1999). The AAA+ module in each torsin sequence is preceded by the N-terminal leader whose length and properties specify the torsin subfamily (Figure 11.4). Each torsin subfamily contains a predicted endoplasmic reticulum (ER)-targeting signal sequence (Figure 11.1), whose cleavage has been confirmed experimentally for human torsinA (Liu *et al.*, 2003). Indeed, torsins are the only known AAA+ ATPases with signal sequences targeting the proteins to the secretory pathway. In the torsin1A and torsin1B subfamilies, the signal sequence is followed by an ~20-amino acid-long hydrophobic segment. The hydrophobic segment is missing in the torsin2 subfamily, and in torsin3, it is replaced by a larger, mostly hydrophilic domain. Thus, all torsins appear to be targeted to the secretory pathway, whereas the hydrophobic domains of torsin1A and torsin1B anchor the proteins in the ER membrane (Liu *et al.*, 2003); torsin2 and torsin3 are likely soluble proteins.

The Walker A motif in the torsin1A, torsin1B, and torsin3 subfamilies contains a non-canonical residue: Asn following the conserved Lys (Figures 11.1 and 11.4). In contrast, the torsin2 subfamily contains a Ser in Walker A, which agrees with the consensus sequence GxxGxGK[T/S]. ClpB and ClpA contain

Thr in the Walker A motifs of their AAA+ modules. There are other important differences between the sequence of the AAA+ module in torsins and in their AAA+ peers. A unique signature of torsins is a network of conserved cysteines, which are not found in ClpA, ClpB, or other AAA+ ATPases (Figure 11.1) (Neuwald *et al.*, 1999). Importantly, one of the cysteines is located within the sensor-2 motif of torsins next to the positively charged residue that interacts with the ATP γ-phosphate. No other AAA+ protein contains a Cys in the sensor-2 motif. The sensor-2 motif resides in the C-terminal subdomain of the AAA+ module and in torsins, it is significantly closer to the proteins' C-terminus than in ClpA and ClpB (Kock *et al.*, 2006). This indicates that the C-terminal domain in torsins may be smaller than in ClpA or ClpB. Since the C-terminal domain is located at the outside of the AAA+ oligomer and "staples" the subunits of the cylinder (Barnett *et al.*, 2000; Bochtler *et al.*, 2000), one might speculate that the oligomers of torsins could be less stable than those of ClpA and ClpB.

Many AAA+ ATPases contain the so-called *channel loop motif* located in their sequence between the Walker A and B motifs. The channel loop corresponds to an unstructured region exposed inside the axial channel that is involved in substrate binding, translocation, and unfolding (Hinnerwisch *et al.*, 2005). The channel loop sequence contains a conserved aromatic residue (usually Tyr) that is essential for substrate binding (Barnett *et al.*, 2005; Schlieker *et al.*, 2004). The loop mobility may be supported by the conserved flanking glycines (Zolkiewski, 2006). The loop motif sequence is GYVG in ClpA and ClpB, the two closest peers of torsins. Intriguingly, no motif similar to the one above can be identified in the torsin sequence between the Walker A and B motifs (Figure 11.1), which is quite puzzling and may indicate that the mechanism of substrate processing by torsins may be distinct from those of ClpA and ClpB.

We investigated the role of the non-canonical Walker A motif found in the torsin1 and torsin3 subfamilies (Figures 11.3 and 11.4). To determine the functional effects due to the introduction of Asn into the Walker A sequence, we replaced the Walker A Thr with Asn in ClpB (Nagy *et al.*, 2009). We found that the T-to-N mutation in Walker A partially inhibited the ATPase activity of ClpB but did not affect the ClpB capability to associate into oligomers. Interestingly, the non-canonical Walker A sequence in ClpB induced preferential binding of adenosine diphosphate (ADP) over ATP as determined by calorimetric titrations. In ClpB, the Walker A Thr is exposed to the nucleotide-binding site in the vicinity of the ATP γ-phosphate (Lee *et al.*, 2003). A replacement of the canonical Thr or Ser with a larger amino acid Asn (as it occurs in torsin1 and torsin3) might interfere with the γ-phosphate binding. This result suggests that the balance between the affinity of torsinA toward ATP and ADP may be shifted in favor of ADP, which might affect the catalytic cycle of the torsinA ATPase.

More importantly, ClpB T-to-N mutants lost the capability to bind substrates with a high affinity even in the presence of saturating ATP (Nagy *et al.*, 2009). As a consequence, ClpB with the torsin-like Walker A sequence showed a low chaperone activity *in vitro* and *in vivo*. These results demonstrate a role of the Walker A motif in maintaining an allosteric linkage between the nucleotide-binding site and the substrate-binding site of ClpB. Thus, those torsins that contain the non-canonical Walker A might utilize distinct mechanisms to couple the ATPase cycle with their substrate-remodeling activity.

Besides the torsin1 and torsin3 families, the Walker A motif with Asn in place of Thr or Ser occurs in the bacterial DNA helicase loader DnaC (Davey *et al.*, 2002). DnaC and torsins are located in two distant branches of the AAA+ phylogenetic tree (Iyer *et al.*, 2004). Thus, it is intriguing that the introduction of Asn into the Walker A motif apparently occurred twice during evolution in two distant forms of life: bacteria and animals. DnaC is not an efficient ATPase and requires other factors to assemble into oligomers (Davey *et al.*, 2002). The non-canonical properties of DnaC and our results for the ClpB T-to-N mutants suggest that torsinA may significantly differ from its AAA+ peers in biochemical functionality.

11.3 BIOLOGICAL MODELS OF THE TorsinA LOSS-OF-FUNCTION

Because of its link to dystonia, torsinA has been a main focus of an intense research since its discovery in 1997, while very limited information has been obtained for the remaining torsin types, as defined in Figures 11.3 and 11.4. The torsinA mRNA and protein are widely expressed in different human tissues (Ozelius *et al.*, 1997; Shashidharan *et al.*, 2000b). In the central nervous system, torsinA is found in dopaminergic neurons of the *substantia nigra pars compacta*, which is consistent with the role of that part of basal ganglia in motor control. Other parts of the brain where torsinA has been detected include neocortex, hippocampus, and cerebellum (Shashidharan *et al.*, 2000b).

TorsinA null mice were first described by Goodchild *et al.* (2005). The heterozygous Tor1A$^{+/-}$ mice did not show any apparent phenotype, but the homozygous Tor1A$^{-/-}$ animals typically failed to feed and died within two days from birth. The Tor1A$^{-/-}$ mice did not show any apparent developmental abnormalities, which could explain the abnormal feeding behavior and the early lethality. The central nervous system appeared normal in the Tor1A$^{-/-}$ mice. The only discernible abnormalities found in those animals were membranous vesicles separated from the NE displayed in electron micrographs of neuronal cells. Interestingly, the NE membrane defects were found in neurons but not in other types of cells of the Tor1A$^{-/-}$ mice. Importantly, an identical early-lethality phenotype and the NE membrane anomalies were also found in the Tor1A$^{\Delta GAG/\Delta GAG}$ knock-in mice that produce only the dystonia-linked mutant torsinA (Goodchild *et al.*, 2005). This result indicates that the dystonia-linked Glu deletion is a loss-of-function mutation, but no clear relationship between the EOTD and the loss of torsinA was found in those animal studies. However, a link between the torsinA loss-of-function and the NE is an important discovery, which has been corroborated in cell culture (see below), and defines a promising direction for studies on cellular pathways affected by torsinA.

In search for an animal model of EOTD, other genetically manipulated animals were produced and characterized. In transgenic mice overexpressing human torsinAΔE, some involuntary movements reminiscent of EOTD were observed (Shashidharan *et al.*, 2005). In addition, immunochemical analysis of neurons showed perinuclear inclusions containing torsinA, ubiquitin, and lamin in striking similarity to the inclusions found in EOTD patients (McNaught *et al.*, 2004). In another study, the heterozygous torsinAΔE knockin mice showed impaired motor coordination and balance as well as intracellular protein aggregates in neurons (Dang *et al.*, 2005). Further work of the same group produced torsinA knockdown mice (Dang *et al.*, 2006) and conditional cerebral cortex-specific torsinA knockout mice (Yokoi *et al.*, 2008), both of which showed motor abnormalities similar to the behaviors observed in the torsinAΔE knockins. The availability of multiple mouse models will be invaluable in testing potential therapies for the EOTD symptoms, once the target pathways and pharmacological leads are identified by the biochemical and cellular studies.

There have been a few studies on the role of torsin proteins in nonmammalian species. Expression of human torsinAΔE in *Drosophila* produced motor defects and morphological abnormalities in neurons (Koh *et al.*, 2004). Downregulation of the *Drosophila* torsin message corresponding to the single torsinA-like gene with RNAi targeted to the eye caused retinal degeneration (Muraro and Moffat, 2006). RNAi-induced suppression of the torsin message in red flour beetle *Tribolium castaneum* at the late-larva stage induced a lethal molting defect in a significant population of the developing adult insects (Z. Liu, M. Zolkiewski, S. Muthukrishnan, unpublished results). In *Caenorhabditis elegans*, mutations in the *ooc-5* gene, an ortholog of *DYT1*, caused abnormal embryonic morphology (Basham and Rose, 2001). All these results support a hypothesis that the torsin function is essential in animal development, but how exactly torsins as ER-targeted AAA+ ATPases might support the physiological processes remains unknown.

11.4 HINTS OF THE CHAPERONE FUNCTION OF TorsinA

Pathogenesis of several neurodegenerative diseases including Parkinson's disease is associated with Lewy bodies, the cytoplasmic inclusions found in the affected neurons. There is a close relationship between Lewy bodies and aggresomes, which are the cellular deposits of misfolded and aggregated proteins (McNaught *et al.*, 2002). Chaperones and the proteasome are targeted to the aggresomes to mediate the refolding and degradation of the accumulated proteins (Johnston *et al.*, 1998). Thus, Lewy bodies may originate from persistent neuronal aggresomes whose clearing does not occur efficiently. A prominent component of Lewy bodies is α-synuclein, a natively disordered aggregation-prone protein with an unknown biological function (Uversky, 2007). Interestingly, torsinA was detected in Lewy bodies in human brain tissue (Shashidharan *et al.*, 2000a) and the close proximity between torsinA and α-synuclein was determined by fluorescence resonance energy transfer (FRET) (Sharma *et al.*, 2001). Moreover, in cells transfected with α-synuclein, overexpression of wild-type (wt) torsinA, but not the dystonia-linked torsinAΔE, inhibited the formation of Lewy-body-like inclusions (McLean *et al.*, 2002). Thus, torsinA might be involved in recognition of aggregates and their suppression.

Another hint of the possible chaperone function of torsinA was found in studies of protein aggregation in *C. elegans*. The cytosolic aggregation of polyglutamine repeats fused to green fluorescent protein (GFP) can be followed *in vivo* by observing the green fluorescence inside the animal bodies (Satyal *et al.*, 2000). Overexpression of human torsinA or its *C. elegans* ortholog was associated with a striking suppression of polyglutamine aggregation (Caldwell *et al.*, 2003). Furthermore, worms treated with a neurotoxin 6-hydroxydopamine underwent neurodegeneration that was partially suppressed in the presence of torsinA (Cao *et al.*, 2005). A similar neuroprotective effect of torsinA was found in *C. elegans* overexpressing α-synuclein.

Altogether, the above results suggest an exciting hypothesis about a possible protein aggregation-suppressing function of torsinA. However, while the phenotypic effects of torsinA in the models of aggregation-linked pathologies have been clearly documented, their mechanism remains a mystery because torsinA and the apparently affected protein aggregates reside on the opposite sides of the ER membrane. Importantly, torsinA is anchored in the ER membrane at the lumen side but does not expose any domains into the cytoplasm (see Kustedjo *et al.*, 2000 and the text below). The observed phenotypes could be trivially explained by the torsinA-induced upregulation of cytosolic chaperones or their activation. Thus, further studies are needed to determine a possible interplay between torsinA and other components of the stress–response machinery. It is also possible that the ER membrane does not separate but indeed connects different components of the aggregation reactions. α-Synuclein is known to interact with phospholipid vesicles and membranes (Uversky, 2007). Moreover, α-synuclein aggregates disrupt the ER–Golgi traffic in yeast, presumably by binding to the transport vesicles (Cooper *et al.*, 2006). Thus, the close association of torsinA and α-synuclein could be explained by the presence of the ER-derived material in Lewy bodies. This hypothetical mechanism implies that beneficial effects of torsinA on the cytosolic aggregates might also involve other ER transmembrane proteins that could associate with torsinA in the lumen and with the aggregates in the cytosol. The identity of such putative "linker" components that could transmit the activity of torsinA across the ER membrane remains unknown so far.

In addition to apparently affecting protein folding in the cytosol, torsinA was found to participate in protein processing within the secretory pathway. Overexpression of torsinA decreased the amount of dopamine transporter and several other polytopic transmembrane proteins present at the cell membrane (Torres *et al.*, 2004). The dystonia-linked torsinAΔE did not affect the intracellular distribution of those proteins. In a different study, torsinAΔE inhibited secretion of a luminescent reporter protein (Hewett *et al.*, 2007), and a suppression of the torsinAΔE message with siRNA reversed the inhibition (Hewett *et al.*, 2008).

All those results, which are consistent with ER-targeting of torsinA, suggest that torsinA might interact with multiple substrates in the ER, which include both soluble and transmembrane proteins and might participate in their processing and transport.

11.5 CELLULAR FATE OF TorsinA

In agreement with the prediction of the ER-targeting of torsinA with the N-terminal signal sequence (Figure 11.1), experiments with transfected mammalian cells identified the torsinA immunoreactivity in the ER (Hewett *et al.*, 2000; Kustedjo *et al.*, 2000). Similarly, the endogenous torsinA co-localized with the lumenal ER proteins in rat PC12 cells (Hewett *et al.*, 2003). The torsinA level did not increase in response to the nerve growth factor-induced PC12 differentiation nor did it respond to heat shock or induction of the unfolded protein response. The ER-targeting was also confirmed for the remaining members of the mammalian torsin family (Figure 11.3) (Hewett *et al.*, 2004; Jungwirth *et al.*, 2010).

From the onset of studies on torsinA, a striking difference was observed between the intracellular localization of wt torsinA and torsinAΔE: unlike the wt, the dystonia-linked mutant protein formed large perinuclear inclusions in the cytoplasm (Hewett *et al.*, 2000; Kustedjo *et al.*, 2000). Subsequent careful studies of the torsinAΔE localization performed by several groups revealed that the apparent inclusions formed at high levels of protein expression whereas at a lower expression level, torsinAΔE localized to the NE (Bragg *et al.*, 2004; Gonzalez-Alegre and Paulson, 2004; Goodchild and Dauer, 2004; Naismith *et al.*, 2004). The apparent "inclusions" of torsinAΔE do not show properties of protein aggregates (Kustedjo *et al.*, 2000); instead, they are a sign of overloading of the tight intermembrane NE space with an overproduced protein. The main question that arose from those important studies is why the dystonia-linked Glu deletion retargets torsinA from the ER to the NE.

It has been demonstrated that wt torsinA can, like torsinAΔE, accumulate in the NE after a mutation of the Walker B motif (Goodchild and Dauer, 2004; Naismith *et al.*, 2004). The Walker B mutation blocks the ATP hydrolysis, which implies that the mutant torsinA in cells may persist in the ATP-bound state. In many AAA+ proteins, ATP-state induces the highest affinity toward substrates, which suggests that the trapping in the NE might be due to a strong interaction with an NE-resident torsinA partner. Indeed, at least two interacting partners of torsinA (Lap1 and nesprin) are located in the NE (see below). Moreover, it has been recently shown that even wt torsinA is quantitatively enriched in the NE of the neuronal cells as compared to the nonneuronal cell types (Giles *et al.*, 2008). This result may help explain why the torsinA knockout mice show neuron-specific NE defects (Goodchild *et al.*, 2005) and suggests that the putative NE-resident partner of torsinA may be enriched in neuronal cells. Altogether, it appears that the dystonia-linked Glu deletion in torsinA does not produce a nonfunctional protein; it rather amplifies the torsinA capability to interact with the component(s) of the NE.

Studies using transfected cell lines suggest that torsinA forms oligomeric complexes, as predicted from its AAA+ lineage (Gonzalez-Alegre and Paulson, 2004; Jungwirth *et al.*, 2010; Torres *et al.*, 2004; Vander Heyden *et al.*, 2009). Interestingly, the apparent oligomerization is enhanced in torsinAΔE but can be inhibited by reduction, which implies intersubunit disulfide links (Gonzalez-Alegre and Paulson, 2004; Gordon and Gonzalez-Alegre, 2008). It should be noted, however, that so far it proved impossible to reconstitute oligomers of torsinA or torsinAΔE *in vitro* using purified proteins (see below), which suggests that torsinA might require an unidentified partner to stimulate its association.

Pulse-chase experiments in neuronal cultured cells determined that the dystonia-linked Glu deletion reduced the torsinA half-life from ~80 h to ~18 h (Giles *et al.*, 2008). The accelerated clearing of torsinAΔE occurs through ER-associated degradation (ERAD) (Giles *et al.*, 2008, 2009b), the pathway that captures defective ER proteins and delivers them to the proteasome (Vembar and Brodsky, 2008). Thus, in spite of being apparently capable of strong interactions in the NE, torsinAΔE becomes an ERAD substrate, which suggests that it is recognized as an abnormal protein by the ER chaperone machinery. Whether torsinAΔE is captured by ERAD within the NE or is first released from the NE to the ER is currently unknown.

11.6 PARTNER PROTEINS

As mentioned above, torsinA has been found in close association with the cytosolic protein α-synuclein in Lewy bodies (Sharma *et al.*, 2001). Tyrosine hydroxylase was also identified as a cytosolic partner of torsinA (Table 11.1) (O'Farrell *et al.*, 2009). Since torsinA resides in the ER lumen, its interaction with cytosolic proteins is likely indirect. Interestingly, tyrosine hydroxylase is an enzyme responsible for catalyzing the biosynthesis of dihydroxyphenylalanine (DOPA), a precursor of the neurotransmitter dopamine. In transfected neurons, tyrosine hydroxylase concentrates at the apparent inclusions containing torsinAΔE (O'Farrell *et al.*, 2009). Whether the recruitment of tyrosine hydroxylase to inclusions affects the dopaminergic neuronal function remains to be determined.

As soon as the distinct localization of torsinAΔE in the NE was established, efforts have been made to discover an interacting component responsible for trapping the dystonia-linked mutant outside the ER. Lamina-associated protein 1 LAP1 was the first identified interacting partner of torsinA in the NE (Goodchild and Dauer, 2005). LAP1 is a transmembrane protein that spans the inner nuclear membrane and interacts with filaments of the nuclear lamina (Kondo *et al.*, 2002). The exact biological function of LAP1 is unknown. The interaction of torsinA with LAP1 is stabilized by the Walker B mutation in torsinA (Goodchild and

Dauer, 2005), which agrees with the AAA+ paradigm and suggests that torsinA recognizes LAP1 as its substrate. Surprisingly, however, the interaction with LAP1 is weaker for torsinAΔE (Naismith *et al.*, 2009). This important result shows that it is not LAP1 that traps torsinAΔE in the NE or is responsible for its mislocalization.

A LAP1-like ER-resident protein LULL1 has also been shown to interact with torsinA (Goodchild and Dauer, 2005). LULL1, like LAP1, is a single-span transmembrane protein that does not contain the lamin-interacting domain. The lumenal domains of LAP1 and LULL1 are similar, and they apparently mediate interactions with torsinA. It is not known what role LULL1 plays in the ER membrane, but its interaction with torsinA is weakened, as it is for LAP1, by the dystonia-linked Glu deletion (Naismith *et al.*, 2009). It was proposed that LAP1 and LULL1 compete for binding to torsinA in the NE and ER, respectively, and the shifts in their binding affinity may control the distribution of torsinA between the two cellular compartments. However, contrary to that model, increasing amounts of LULL1 produce higher concentration of torsinA in the NE (Vander Heyden *et al.*, 2009), suggesting that the interaction with LULL1 in the ER might trigger an event sending torsinA into the NE. The nature of such event and its significance are currently unknown.

The list of identified partners of torsinA does not end with LAP1 and LULL1 (Table 11.1). A novel, torsinA-interacting protein, printor, was recently identified by yeast two-hybrid screen (Giles *et al.*, 2009a). In disagreement with the AAA+

Table 11.1 Identified partners of torsinA

Partner Protein	Partner Localization[a]	Principal Method[b]	Effect of ΔE Mutation on the Interaction Efficiency	References
α-Synuclein	cytosol	FRET	—	Sharma *et al.* (2001)
Kinesin LC 1	cytosol	YTHS, co-IP	Decrease	Kamm *et al.* (2004)
LAP1	NE (INM)	co-IP	Decrease	Goodchild and Dauer (2005); Naismith *et al.* (2009)
LULL1	ER	co-IP	—	Goodchild and Dauer (2005)
Vimentin	cytosol	co-IP	Increase	Hewett *et al.* (2006)
Nesprin	NE (ONM)	co-IP	Increase	Nery *et al.* (2008)
Printor	ER[c]	YTHS, co-IP	Decrease	Giles *et al.* (2009a)
Tyrosine hydroxylase	cytosol	co-IP	Increase	O'Farrell *et al.* (2009)

[a]NE, nuclear envelope; INM, inner nuclear membrane; ONM, outer nuclear membrane; ER, endoplasmic reticulum.

[b]FRET, fluorescence resonance energy transfer; YTHS, yeast two-hybrid system; co-IP, co-immunoprecipitation.

[c]No apparent ER-targeting signal sequence was detected in printor.

paradigm, the Walker B mutation in torsinA partially inhibits the interaction with printor, while the Walker A mutation that suppresses nucleotide binding to torsinA preserves the interaction with printor. These results suggest that printor might not be a substrate of torsinA and the significance of the interaction needs to be further investigated.

Perhaps, the best developed and the most promising hypothesis about the function of torsinA has been based on the identified links with the cytoskeleton. The kinesin light chain was the first identified cellular component that co-immunoprecipitated with torsinA (Kamm *et al.*, 2004). Further experiments revealed association of torsinA with vimentin, a component of intermediate filaments, as well as with actin and tubulin (Hewett *et al.*, 2006). Nery *et al.* recently identified nesprins as the linkers between torsinA and the cytoskeletal components (Nery *et al.*, 2008). Nesprins belong to a group of transmembrane proteins located mainly in the outer nuclear membrane whose N-terminal cytoplasmic domains interact with components of cytoskeleton, while their small C-terminal KASH domains are exposed in the NE (Starr and Fischer, 2005). Within the NE, nesprin KASH domains interact with SUN domains of the inner nuclear membrane proteins. The links maintained by nesprins and other KASH- and SUN-containing proteins between the cytoskeleton and nucleoskeleton are essential for nuclear positioning and migration within cells.

The interaction between torsinA and the nesprin KASH domain is supported by the torsinA C-terminal subdomain (Figure 11.2) that contains the dystonia-linked Glu-deletion site (Nery *et al.*, 2008). Importantly, the interaction is more efficient with torsinAΔE than with wt torsinA. This result suggests that nesprins or, more specifically, KASH domains could be responsible for the apparent trapping of torsinAΔE in the NE. Moreover, fibroblasts lacking torsinA showed slower migration in a wound-healing assay, which was accompanied by a reduced ability to properly position their nuclei (Nery *et al.*, 2008). In another study, the accumulation of torsinA in the NE also affected the localization of SUN-domain containing protein (Vander Heyden *et al.*, 2009). Altogether, these results suggest a role for torsinA in regulating the functionality of linkages spanning both nuclear membranes and supporting proper nuclear positioning and mobility.

11.7 BIOCHEMICAL PROPERTIES OF TorsinA

The identification of torsins as members of the AAA+ superfamily and their apparent similarity to Clp ATPases allow some predictions to be made about the function of torsinA and possible significance of the dystonia-linked mutation. It has been suggested that torsinA, like ClpB, might be a molecular chaperone designed to control the quality of protein folding and to suppress protein aggregation. Unlike other AAA+ ATPases, torsinA would operate in the ER, which brings a question why only animal cells would require the AAA+ functionality in the secretory pathway. Homology modeling of the torsinA structure with the ClpA and ClpB templates showed that the dystonia-linked Glu deletion occurs within the small C-terminal subdomain of the AAA+ module (Figure 11.2 and Kock *et al.*, 2006; Zhu *et al.*, 2008). On the basis of the known biochemical properties of AAA+ ATPases, it could be hypothesized that a loss of a charged residue in the oligomer-supporting domain might inhibit the self-association of torsinA and, as a consequence, its ability to bind nucleotides, hydrolyze ATP, and process its substrates. That attractive hypothesis prompted a number of groups to initiate biochemical studies on torsinA because the idea that an inherited human disease could be due to a simple oligomerization defect seemed very appealing. Moreover, the fact that EOTD is autosomal dominant suggests that the dystonia mutant torsinAΔE may form functionally defective hetero-oligomers with wt torsinA, again focusing attention on the protein's self-association properties. On the other hand, the unique features of torsins described above, as compared to other AAA+ ATPases, suggested that torsinA might employ a less conventional mechanism to perform its function and may not be simply an ER-targeted version of ClpA or ClpB.

Attempts to produce recombinant human torsinA in bacteria proved unsuccessful, but Kustedjo *et al.* (2003) purified torsinA and the dystonia-linked torsinAΔE from Baculovirus-infected Sf9 cells. Both wt torsinA and torsinAΔE were found to be glycoproteins that were soluble only in the presence of a detergent. Low solubility of torsinA is due to the hydrophobic segment that remains at the protein's N-terminus after removal of the signal sequence (Figure 11.1) (Liu *et al.*, 2003). The 20-amino acid-long hydrophobic segment is responsible for association of torsinA with the ER membrane (Liu *et al.*, 2003). Further studies confirmed that torsinA is a peripherally membrane-associated protein (Callan *et al.*, 2007). Deletion of the N-terminal hydrophobic segment prevents membrane association and increases solubility of torsinA in aqueous buffers (Liu *et al.*, 2003). Thus, torsinA consists of two major domains: the N-terminal hydrophobic membrane anchor (amino acids 21–40, see Figure 11.1) and the C-terminal domain exposed in the ER lumen that contains the AAA+ module (amino acids 41–332, see Figures 11.1 and 11.2). The topology of torsinA in the ER is analogous to that of the mitochondrial AAA+ protease Yme1p that is anchored in the inner mitochondrial membrane with an N-terminal hydrophobic domain (Langer, 2000).

Purified torsinA was monomeric in the absence of nucleotides but surprisingly also in the presence of ATP or ADP (Kustedjo *et al.*, 2003). Since detergents may inhibit protein association, we purified a soluble variant of torsinA (torsinAΔ40) without the membrane-anchoring hydrophobic domain. Similar to the full-length torsinA, torsinAΔ40 was monomeric in gel filtration experiments but showed evidence of oligomerization at high concentration in sedimentation equilibrium (Z. Liu, H.-C. Wu, and M. Zolkiewski, unpublished results). Both full-length torsinA and torsinAΔ40 were active ATPases, but their activities were significantly lower than those of Clp ATPases (Kustedjo *et al.*, 2003; Z. Liu, H-C. Wu, and M. Zolkiewski, unpublished results). Altogether, the biochemical studies suggested that not all properties of the purified torsinA may conform to the common AAA+ paradigm.

Importantly, the studies of Kustedjo *et al.* failed to identify a single difference between the biochemical properties of wt torsinA and torsinAΔE (Kustedjo *et al.*, 2003). Both proteins showed the same pattern of glycosylation, the same secondary structure as determined by circular dichroism, the same thermal and chemical stability, the same lack of oligomers, and a very similar ATPase activity. These results indicate that the dystonia-linked Glu deletion does not cause a global destabilization of the torsinA structure. Thus, the intracellular destabilization of torsinAΔE and its susceptibility to ERAD (see above) cannot be simply explained by a loss of structural stability. The homology model (Figure 11.2) suggests that the Glu deletion might affect the local structure of an α-helix within the C-terminal subdomain, but the results of such local structural disruption are impossible to predict in the absence of high-resolution structural data. Importantly, however, the main hypothesis, that is—a possible effect of the ΔE mutation on self-association of torsinA—cannot be tested until the putative torsinA oligomers are reconstituted *in vitro*.

It should be noted that two other studies compared the properties of purified wt torsinA and torsinAΔE and found defects in the torsinAΔE ATPase activity (Konakova and Pulst, 2005; Pham *et al.*, 2006). The significance of those results is uncertain, however, as in both cases the recombinant torsinA variants were fused to other large domains and their folding status after purification from *Escherichia coli* was not determined.

We found that torsinAΔ40 is properly translocated and glycosylated in the ER, but unlike the full-length torsinA, it does not associate with the ER membrane and becomes secreted into the cell culture media that facilitates protein purification (Liu *et al.*, 2003). Interestingly, torsinAΔ40ΔE (the dystonia-linked variant) was apparently trapped inside *Drosophila* S2 and yeast cells and showed a significantly weaker secretion (Liu *et al.*, 2003). In the absence of structural destabilization of torsinAΔE, this result can be explained by "gain-of-function" interactions that occur inside the cells with torsinAΔE but not with wt torsinA. As described above, such trapping interactions that inhibit secretion of torsinAΔ40ΔE are consistent with the targeting of torsinAΔE to the NE (Gonzalez-Alegre and Paulson, 2004; Goodchild and Dauer, 2004; Naismith *et al.*, 2004).

Although secretion of torsinAΔ40 from cells is nonphysiological, it can be considered a rudimentary assay of the intracellular protein processing. We investigated whether the change of the Walker A motif in torsinAΔ40 to the canonical sequence (N-to-T and N-to-S) would affect the protein properties. Unexpectedly, the N109T and N109S mutations of the torsinA Walker A motif inhibited secretion of torsinAΔ40, which prevented us from purifying the protein (H.-C. Wu and M. Zolkiewski, unpublished results). The unexpected "trapping" of torsinAΔ40(N109T) and torsinAΔ40(N109S) inside cells suggests that the introduction of the canonical Walker A into torsinA may affect its nucleotide-dependent functionality and might stimulate interactions with other cellular components. In addition, the inhibition of secretion of the N-to-T(S) Walker A mutants of torsinA is reminiscent of the intracellular retention of the dystonia-linked glutamate-deletion mutant (see above). Since AAA+

ATPases preferentially interact with their substrates in the ATP-bound state, the intracellular trapping would also be expected to arise from the mutation of the Walker B motif, which preserves ATP binding but blocks the hydrolysis of ATP (Hanson and Whiteheart, 2005). Unexpectedly, we found that the Walker B mutation E171Q in torsinAΔ40 strongly stimulated its secretion (H.-C. Wu and M. Zolkiewski, unpublished results). Further biochemical research on torsinA is necessary to explain the apparent "untrapping" of the Walker B mutant, which further emphasizes possible non-canonical properties of torsins.

Zhu *et al.* investigated the biochemical properties of OOC-5, a torsinA homolog from *C. elegans* (Zhu *et al.*, 2008). OOC-5 has been previously found essential in the nematode oogenesis and embryo development (Basham and Rose, 2001). The recombinant soluble lumenal domain of OOC-5 was produced in *E. coli*, and a number of new important protein properties were established (Zhu *et al.*, 2008). It was found that all six conserved cysteines in OOC-5 are involved in disulfide bonds. In particular, the sensor-2 cysteine (Cys319 in human torsinA, see Figures 11.1 and 11.2) is linked with the other Cys within the C-terminal subdomain (Cys280 in human torsinA). Like human torsinA produced in eukaryotic cells, OOC-5 remained monomeric in gel filtration experiments with or without nucleotides. No detectable ATPase activity was observed for OOC-5. However, by studying the thermal stability of OOC-5 in the presence of nucleotides, it was established that OOC-5 does bind ADP and ATP. Interestingly, the binding of ADP induced a conformational change in the reduced but not oxidized OOC-5, as detected by limited proteolysis. The apparent coupling between ADP binding to OOC-5 and the redox state was traced to the disulfide bond of the sensor-2 Cys. Altogether, the results of Zhu *et al.* suggest that in OOC-5 and possibly in other torsins, the nucleotide-dependent AAA+ functionality may be regulated by the redox state. The biological significance of this unique mechanism remains unknown.

11.8 TOWARD THE FUNCTION OF TorsinA AND BEYOND

More than a decade of studies on torsinA and other torsins brought a significant amount of experimental data that were described in this chapter. The field has not converged so far on a definite biological function that could be associated with torsinA and no clear link between the mutation in *DYT1* and EOTD has been established yet. The AAA+ lineage strongly suggests that torsins are ATP-dependent structure-remodeling factors, but no direct evidence of such activity has been so far obtained for any torsin protein. The studies on animal models and in cell culture defined a number of processes potentially requiring the torsinA activity: protein trafficking in the secretory pathway, suppression of protein aggregation, biogenesis of the NE, and control of the nuclear positioning and mobility. The list of identified partners of torsinA (Table 11.1) does not seem to converge upon a single type of recognition mechanism or a single function. For example, LAP1 and LULL1 unquestionably cooperate with torsinA in mammalian cells, but an analogous cooperation is unlikely for the insect torsin, as there are no clear LAP1 and LULL1 orthologs in insects. Thus, it appears that torsinA is quite versatile in its specificity: it may associate with multiple cellular targets in both the ER and the NE. Whether torsinA is a *bona fide* molecular chaperone remains to be determined.

There is an urgent need for more extensive biochemical and structural studies on torsins. None of the interactions of torsinA demonstrated in cell culture studies has been confirmed *in vitro* with purified components. Indeed, most of the expected AAA+ biochemical properties, including ATP-dependent self-association, have not been unequivocally confirmed yet for the recombinant torsinA. Most importantly, only biochemical studies can reveal structural rearrangements that may occur in partner proteins upon their interaction with torsinA.

And what about dystonia? Any one of the biological processes potentially requiring torsinA or indeed all of them could be involved in the etiology of EOTD. An important result on which future models should be built states that the EOTD-linked mutation produces a loss-of-function in torsinA (Goodchild *et al.*, 2005). Yet, with the same mutation, torsinA appears to gain affinity to targets in the NE. Does this imply that the activity of torsinA within the NE is not relevant to dystonia? It is possible that a higher binding affinity of torsinA toward substrates does not translate into its higher activity. It is also possible that the main link to dystonia should be sought in the compartment depleted of torsinAΔE, namely, the ER. Would it be sufficient to bring torsinAΔE back into the ER by a chemical intervention to suppress the symptoms of EOTD? Obviously, many more experiments are needed to solve these dilemmas, and the dystonia research is poised for its most exciting discoveries in this new decade.

REFERENCES

Arnold K, Bordoli L, Kopp J, Schwede T. The SWISS-MODEL workspace: a web-based environment for protein structure homology modelling. *Bioinformatics* 2006;**22**:195–201.

Barnett ME, Nagy M, Kedzierska S, Zolkiewski M. The amino-terminal domain of ClpB supports binding to strongly aggregated proteins. *J Biol Chem* 2005;**280**:34940–34945.

Barnett ME, Zolkiewski M. Site-directed mutagenesis of conserved charged amino acid residues in ClpB from *Escherichia coli*. *Biochemistry* 2002;**41**:11277–11283.

Barnett ME, Zolkiewska A, Zolkiewski M. Structure and activity of ClpB from *Escherichia coli*. Role of the amino- and -carboxyl-terminal domains. *J Biol Chem* 2000;**275**:37565–37571.

Basham SE, Rose LS. The *Caenorhabditis elegans* polarity gene ooc-5 encodes a Torsin-related protein of the AAA ATPase superfamily. *Development* 2001;**128**:4645–4656.

Bochtler M, Hartmann C, Song HK, Bourenkov GP, Bartunik HD, Huber R. The structures of HslU and the ATP-dependent protease HslU-HslV. *Nature* 2000;**403**:800–805.

Bordoli L, Kiefer F, Arnold K, Benkert P, Battey J, Schwede T. Protein structure homology modeling using SWISS-MODEL workspace. *Nat Protoc* 2009;**4**:1–13.

Bragg DC, Camp SM, Kaufman CA, Wilbur JD, Boston H, Schuback DE, Hanson PI, Sena-Esteves M, Breakefield XO. Perinuclear biogenesis of mutant torsin-a inclusions in cultured cells infected with tetracycline-regulated herpes simplex virus type 1 amplicon vectors. *Neuroscience* 2004;**125**:651–661.

Breakefield XO, Blood AJ, Li Y, Hallett M, Hanson PI, Standaert DG. The pathophysiological basis of dystonias. *Nat Rev Neurosci* 2008;**9**:222–234.

Caldwell GA, Cao S, Sexton EG, Gelwix CC, Bevel JP, Caldwell KA. Suppression of polyglutamine-induced protein aggregation in *Caenorhabditis elegans* by torsin proteins. *Hum Mol Genet* 2003;**12**:307–319.

Callan AC, Bunning S, Jones OT, High S, Swanton E. Biosynthesis of the dystonia-associated AAA+ ATPase torsinA at the endoplasmic reticulum. *Biochem J* 2007;**401**:607–612.

Cao S, Gelwix CC, Caldwell KA, Caldwell GA. Torsin-mediated protection from cellular stress in the dopaminergic neurons of *Caenorhabditis elegans*. *J Neurosci* 2005;**25**:3801–3812.

Cooper AA, Gitler AD, Cashikar A, Haynes CM, Hill KJ, Bhullar B, Liu K, Xu K, Strathearn KE, Liu F, Cao S, Caldwell KA, Caldwell GA, Marsischky G, Kolodner RD, Labaer J, Rochet JC, Bonini NM, Lindquist S. Alpha-synuclein blocks ER-Golgi traffic and Rab1 rescues neuron loss in Parkinson's models. *Science* 2006;**313**:324–328.

Dang MT, Yokoi F, McNaught KS, Jengelley TA, Jackson T, Li J, Li Y. Generation and characterization of Dyt1 DeltaGAG knock-in mouse as a model for early-onset dystonia. *Exp Neurol* 2005;**196**:452–463.

Dang MT, Yokoi F, Pence MA, Li Y. Motor deficits and hyperactivity in Dyt1 knockdown mice. *Neurosci Res* 2006;**56**:470–474.

Davey MJ, Fang L, McInerney P, Georgescu RE, O'Donnell M. The DnaC helicase loader is a dual ATP/ADP switch protein. *EMBO J* 2002;**21**:3148–3159.

Erzberger JP, Berger JM. Evolutionary relationships and structural mechanisms of AAA+ proteins. *Annu Rev Biophys Biomol Struct* 2006;**35**:93–114.

Giles LM, Chen J, Li L, Chin LS. Dystonia-associated mutations cause premature degradation of torsinA protein and cell-type-specific mislocalization to the nuclear envelope. *Hum Mol Genet* 2008;**17**:2712–2722.

Giles LM, Li L, Chin LS. Printor, a novel torsinA-interacting protein implicated in dystonia pathogenesis. *J Biol Chem* 2009a;**284**:21765–21775.

Giles LM, Li L, Chin LS. TorsinA protein degradation and autophagy in DYT1 dystonia. *Autophagy* 2009b;**5**:82–84.

Glover JR, Lindquist S. Hsp104, Hsp70, and Hsp40: a novel chaperone system that rescues previously aggregated proteins. *Cell* 1998;**94**:73–82.

Gonzalez-Alegre P, Paulson HL. Aberrant cellular behavior of mutant torsinA implicates nuclear envelope dysfunction in DYT1 dystonia. *J Neurosci* 2004;**24**:2593–2601.

Goodchild RE, Dauer WT. Mislocalization to the nuclear envelope: an effect of the dystonia-causing torsinA mutation. *Proc Natl Acad Sci USA* 2004;**101**: 847–852.

Goodchild RE, Dauer WT. The AAA+ protein torsinA interacts with a conserved domain present in LAP1 and a novel ER protein. *J Cell Biol* 2005;**168**:855–862.

Goodchild RE, Kim CE, Dauer WT. Loss of the dystonia-associated protein torsina selectively disrupts the neuronal nuclear envelope. *Neuron* 2005;**48**:923–932.

Gordon KL, Gonzalez-Alegre P. Consequences of the DYT1 mutation on torsinA oligomerization and degradation. *Neuroscience* 2008;**157**:588–595.

Guo F, Maurizi MR, Esser L, Xia D. Crystal Structure of ClpA, an Hsp100 Chaperone and Regulator of ClpAP Protease. *J Biol Chem* 2002;**277**:46743–46752.

Hanson PI, Whiteheart SW. AAA+ proteins: have engine, will work. *Nat Rev Mol Cell Biol* 2005;**6**:519–529.

Hartman JJ, Vale RD. Microtubule disassembly by ATP-dependent oligomerization of the AAA enzyme katanin. *Science* 1999;**286**:782–785.

Haslberger T, Zdanowicz A, Brand I, Kirstein J, Turgay K, Mogk A, Bukau B. Protein disaggregation by the AAA+ chaperone ClpB involves partial threading of looped polypeptide segments. *Nat Struct Mol Biol* 2008;**15**:641–650.

Hewett J, Gonzalez-Agosti C, Slater D, Ziefer P, Li S, Bergeron D, Jacoby DJ, Ozelius LJ, Ramesh V, Breakefield XO. Mutant torsinA, responsible for early-onset torsion dystonia, forms membrane inclusions in cultured neural cells. *Hum Mol Genet* 2000;**9**:1403–1413.

Hewett JW, Kamm C, Boston H, Beauchamp R, Naismith T, Ozelius L, Hanson PI, Breakefield XO, Ramesh V. TorsinB - perinuclear location and association with torsinA. *J Neurochem* 2004;**89**:1186–1194.

Hewett JW, Nery FC, Niland B, Ge P, Tan P, Hadwiger P, Tannous BA, Sah DW, Breakefield XO. siRNA knock-down of mutant torsinA restores processing through secretory pathway in DYT1 dystonia cells. *Hum Mol Genet* 2008;**17**:1436–1445.

Hewett JW, Tannous B, Niland BP, Nery FC, Zeng J, Li Y, Breakefield XO. Mutant torsinA interferes with protein processing through the secretory pathway in DYT1 dystonia cells. *Proc Natl Acad Sci USA* 2007;**104**:7271–7276.

Hewett JW, Zeng J, Niland BP, Bragg DC, Breakefield XO. Dystonia-causing mutant torsinA inhibits cell adhesion and neurite extension through interference with cytoskeletal dynamics. *Neurobiol Dis* 2006;**22**:98–111.

Hewett J, Ziefer P, Bergeron D, Naismith T, Boston H, Slater D, Wilbur J, Schuback D, Kamm C, Smith N, Camp S, Ozelius LJ, Ramesh V, Hanson PI, Breakefield XO. TorsinA in PC12 cells: localization in the endoplasmic reticulum and response to stress. *J Neurosci Res* 2003;**72**:158–168.

Hinnerwisch J, Fenton WA, Furtak KJ, Farr GW, Horwich AL. Loops in the central channel of ClpA chaperone mediate protein binding, unfolding, and translocation. *Cell* 2005;**121**:1029–1041.

Iyer LM, Leipe DD, Koonin EV, Aravind L. Evolutionary history and higher order classification of AAA+ ATPases. *J Struct Biol* 2004;**146**:11–31.

Johnston JA, Ward CL, Kopito RR. Aggresomes: a cellular response to misfolded proteins. *J Cell Biol* 1998;**143**:1883–1898.

Jungwirth M, Dear ML, Brown P, Holbrook K, Goodchild R. Relative tissue expression of homologous torsinB correlates with the neuronal specific importance of DYT1 dystonia-associated torsinA. *Hum Mol Genet* 2010;**19**:888–900.

Kamm C. Early onset torsion dystonia (Oppenheim's dystonia). *Orphanet J Rare Dis* 2006;**1**:48–48.

Kamm C, Boston H, Hewett J, Wilbur J, Corey DP, Hanson PI, Ramesh V, Breakefield XO. The early onset dystonia protein torsinA interacts with kinesin light chain 1. *J Biol Chem* 2004;**279**:19882–19892.

Kock N, Naismith TV, Boston HE, Ozelius LJ, Corey DP, Breakefield XO, Hanson PI. Effects of genetic variations in the dystonia protein torsinA: identification of polymorphism at residue 216 as protein modifier. *Hum Mol Genet* 2006;**15**:1355–1364.

Koh YH, Rehfeld K, Ganetzky B. A Drosophila model of early onset torsion dystonia suggests impairment in TGF-beta signaling. *Hum Mol Genet* 2004;**13**:2019–2030.

Konakova M, Pulst SM. Dystonia-associated forms of torsinA are deficient in ATPase activity. *J Mol Neurosci* 2005;**25**:105–117.

Kondo Y, Kondoh J, Hayashi D, Ban T, Takagi M, Kamei Y, Tsuji L, Kim J, Yoneda Y. Molecular cloning of one isotype of human lamina-associated polypeptide 1s and a topological analysis using its deletion mutants. *Biochem Biophys Res Commun* 2002;**294**:770–778.

Kramer PL, Heiman GA, Gasser T, Ozelius LJ, de LD, Brin MF, Burke RE, Hewett J, Hunt AL, Moskowitz C. The DYT1 gene on 9q34 is responsible for most cases of early limb-onset idiopathic torsion dystonia in non-Jews. *Am J Hum Genet* 1994;**55**:468–475.

Kumar S, Tamura K, Nei M. MEGA3: Integrated software for Molecular Evolutionary Genetics Analysis and sequence alignment. *Brief Bioinform* 2004;**5**:150–163.

Kustedjo K, Bracey MH, Cravatt BF. Torsin A and its torsion dystonia-associated mutant forms are lumenal glycoproteins that exhibit distinct subcellular localizations. *J Biol Chem* 2000;**275**:27933–27939.

Kustedjo K, Deechongkit S, Kelly JW, Cravatt BF. Recombinant expression, purification, and comparative characterization of torsinA and its torsion dystonia-associated variant Delta E-torsinA. *Biochemistry* 2003;**42**:15333–15341.

Langer T. AAA proteases: cellular machines for degrading membrane proteins. *Trends Biochem Sci* 2000;**25**:247–251.

Lee S, Choi JM, Tsai FT. Visualizing the ATPase cycle in a protein disaggregating machine: structural basis for substrate binding by ClpB. *Mol Cell* 2007;**25**:261–271.

Lee C, Schwartz MP, Prakash S, Iwakura M, Matouschek A. ATP-dependent proteases degrade their substrates by processively unraveling them from the degradation signal. *Mol Cell* 2001;**7**:627–637.

Lee S, Sowa ME, Watanabe YH, Sigler PB, Chiu W, Yoshida M, Tsai FT. The structure of ClpB: a molecular chaperone that rescues proteins from an aggregated state. *Cell* 2003;**115**:229–240.

Liu Z, Zolkiewska A, Zolkiewski M. Characterization of human torsinA and its dystonia-associated mutant form. *Biochem J* 2003;**374**:117–122.

McLean PJ, Kawamata H, Shariff S, Hewett J, Sharma N, Ueda K, Breakefield XO, Hyman BT. TorsinA and heat shock proteins act as molecular chaperones: suppression of alpha-synuclein aggregation. *J Neurochem* 2002;**83**:846–854.

McNaught KS, Kapustin A, Jackson T, Jengelley TA, Jnobaptiste R, Shashidharan P, Perl DP, Pasik P, Olanow CW. Brainstem pathology in DYT1 primary torsion dystonia. *Ann Neurol* 2004;**56**:540–547.

McNaught KS, Shashidharan P, Perl DP, Jenner P, Olanow CW. Aggresome-related biogenesis of Lewy bodies. *Eur J Neurosci* 2002;**16**:2136–2148.

Muraro NI, Moffat KG. Down-regulation of torp4a, encoding the *Drosophila* homologue of torsinA, results in increased neuronal degeneration. *J Neurobiol* 2006;**66**:1338–1353.

Nagy M, Wu HC, Liu Z, Kedzierska-Mieszkowska S, Zolkiewski M. Walker-A threonine couples nucleotide occupancy with the chaperone activity of the AAA+ ATPase ClpB. *Protein Sci* 2009;**18**:287–293.

Naismith TV, Dalal S, Hanson PI. Interaction of torsinA with its major binding partners is impaired by the dystonia-associated DeltaGAG deletion. *J Biol Chem* 2009;**284**:27866–27874.

Naismith TV, Heuser JE, Breakefield XO, Hanson PI. TorsinA in the nuclear envelope. *Proc Natl Acad Sci USA* 2004;**101**:7612–7617.

Nery FC, Zeng J, Niland BP, Hewett J, Farley J, Irimia D, Li Y, Wiche G, Sonnenberg A, Breakefield XO. TorsinA binds the KASH domain of nesprins and participates in linkage between nuclear envelope and cytoskeleton. *J Cell Sci* 2008;**121**:3476–3486.

Neuwald AF, Aravind L, Spouge JL, Koonin EV. AAA+: A class of chaperone-like ATPases associated with the assembly, operation, and disassembly of protein complexes. *Genome Res* 1999;**9**:27–43.

Nielsen H, Engelbrecht J, Brunak S, von Heijne G. Identification of prokaryotic and eukaryotic signal peptides and prediction of their cleavage sites. *Protein Eng* 1997;**10**:1–6.

O'Farrell CA, Martin K, Hutton M, Delatycki MB, Cookson MR, Lockhart PJ. Mutant torsinA interacts with tyrosine hydroxylase in cultured cells. *Neuroscience* 2009;**164**:1127–1137.

Ogura T, Wilkinson AJ. AAA+ superfamily ATPases: common structure—diverse function. *Genes Cells* 2001;**6**:575–597.

Ozelius LJ, Hewett JW, Page CE, Bressman SB, Kramer PL, Shalish C, de Leon D, Brin MF, Raymond D, Corey DP, Fahn S, Risch NJ, Buckler AJ, Gusella JF, Breakefield XO. The early-onset torsion dystonia gene (DYT1) encodes an ATP-binding protein. *Nat Genet* 1997;**17**:40–48.

Ozelius LJ, Kramer PL, de LD, Risch N, Bressman SB, Schuback DE, Brin MF, Kwiatkowski DJ, Burke RE, Gusella JF. Strong allelic association between the torsion dystonia gene (DYT1) andloci on chromosome 9q34 in Ashkenazi Jews. *Am J Hum Genet* 1992;**50**:619–628.

Parsell DA, Kowal AS, Singer MA, Lindquist S. Protein disaggregation mediated by heat-shock protein Hsp104. *Nature* 1994;**372**:475–478.

Pham P, Frei KP, Woo W, Truong DD. Molecular defects of the dystonia-causing torsinA mutation. *Neuroreport* 2006;**17**:1725–1728.

Queitsch C, Hong SW, Vierling E, Lindquist S. Heat shock protein 101 plays a crucial role in thermotolerance in *Arabidopsis*. *Plant Cell* 2000;**12**:479–492.

Reid BG, Fenton WA, Horwich AL, Weber-Ban EU. ClpA mediates directional translocation of substrate proteins into the ClpP protease. *Proc Natl Acad Sci USA* 2001;**98**:3768–3772.

Risch N, de LD, Ozelius L, Kramer P, Almasy L, Singer B, Fahn S, Breakefield X, Bressman S. Genetic analysis of idiopathic torsion dystonia in Ashkenazi Jews and their recent descent from a small founder population. *Nat Genet* 1995;**9**:152–159.

Rouiller I, DeLaBarre B, May AP, Weis WI, Brunger AT, Milligan RA, Wilson-Kubalek EM. Conformational changes of the multifunction p97 AAA ATPase during its ATPase cycle. *Nat Struct Biol* 2002;**9**:950–957.

Satyal SH, Schmidt E, Kitagawa K, Sondheimer N, Lindquist S, Kramer JM, Morimoto RI. Polyglutamine aggregates alter protein folding homeostasis in *Caenorhabditis elegans*. *Proc Natl Acad Sci USA* 2000;**97**:5750–5755.

Schlieker C, Weibezahn J, Patzelt H, Tessarz P, Strub C, Zeth K, Erbse A, Schneider-Mergener J, Chin JW, Schultz PG, Bukau B, Mogk A. Substrate recognition by the AAA+ chaperone ClpB. *Nat Struct Mol Biol* 2004;**11**:607–615.

Sharma N, Hewett J, Ozelius LJ, Ramesh V, McLean PJ, Breakefield XO, Hyman BT. A close association of torsinA and alpha-synuclein in Lewy bodies: a fluorescence resonance energy transfer study. *Am J Pathol* 2001;**159**:339–344.

Shashidharan P, Good PF, Hsu A, Perl DP, Brin MF, Olanow CW. TorsinA accumulation in Lewy bodies in sporadic Parkinson's disease. *Brain Res* 2000a;**877**:379–381.

Shashidharan P, Kramer BC, Walker RH, Olanow CW, Brin MF. Immunohistochemical localization and distribution of torsinA in normal human and rat brain. *Brain Res* 2000b;**853**:197–206.

Shashidharan P, Sandu D, Potla U, Armata IA, Walker RH, McNaught KS, Weisz D, Sreenath T, Brin MF, Olanow CW. Transgenic mouse model of early-onset DYT1 dystonia. *Hum Mol Genet* 2005;**14**:125–133.

Squires CL, Pedersen S, Ross BM, Squires C. ClpB is the *Escherichia coli* heat shock protein F84.1. *J Bacteriol* 1991;**173**:4254–4262.

Starr DA, Fischer JA. KASH 'n Karry: the KASH domain family of cargo-specific cytoskeletal adaptor proteins. *Bioessays* 2005;**27**:1136–1146.

Torres GE, Sweeney AL, Beaulieu JM, Shashidharan P, Caron MG. Effect of torsinA on membrane proteins reveals a loss of function and a dominant-negative phenotype of the dystonia-associated DeltaE-torsinA mutant. *Proc Natl Acad Sci USA* 2004;**101**:15650–15655.

Uversky VN. Neuropathology, biochemistry, and biophysics of alpha-synuclein aggregation. *J Neurochem* 2007;**103**:17–37.

Vander Heyden AB, Naismith TV, Snapp EL, Hodzic D, Hanson PI. LULL1 retargets TorsinA to the nuclear envelope revealing an activity that is impaired by the DYT1 dystonia mutation. *Mol Biol Cell* 2009;**20**:2661–2672.

Vembar SS, Brodsky JL. One step at a time: endoplasmic reticulum-associated degradation. *Nat Rev Mol Cell Biol* 2008;**9**:944–957.

Weber-Ban EU, Reid BG, Miranker AD, Horwich AL. Global unfolding of a substrate protein by the Hsp100 chaperone ClpA. *Nature* 1999;**401**:90–93.

Weibezahn J, Schlieker C, Bukau B, Mogk A. Characterization of a trap mutant of the AAA+ chaperone ClpB. *J Biol Chem* 2003;**278**:32608–32617.

Weibezahn J, Tessarz P, Schlieker C, Zahn R, Maglica Z, Lee S, Zentgraf H, Weber-Ban EU, Dougan DA, Tsai FT, Mogk A, Bukau B. Thermotolerance requires refolding of aggregated proteins by substrate translocation through the central pore of ClpB. *Cell* 2004;**119**:653–665.

Yokoi F, Dang MT, Mitsui S, Li J, Li Y. Motor deficits and hyperactivity in cerebral cortex-specific Dyt1 conditional knockout mice. *J Biochem (Tokyo)* 2008;**143**:39–47.

Zhang X, Shaw A, Bates PA, Newman RH, Gowen B, Orlova E, Gorman MA, Kondo H, Dokurno P, Lally J, Leonard G, Meyer H, van Heel M, Freemont PS. Structure of the AAA ATPase p97. *Mol Cell* 2000;**6**:1473–1484.

Zhu L, Wrabl JO, Hayashi AP, Rose LS, Thomas PJ. The torsin-family AAA+ protein OOC-5 contains a critical disulfide adjacent to Sensor-II that couples redox state to nucleotide binding. *Mol Biol Cell* 2008;**19**:3599–3612.

Zolkiewski M. ClpB cooperates with DnaK, DnaJ, and GrpE in suppressing protein aggregation. A novel multi-chaperone system from *Escherichia coli*. *J Biol Chem* 1999;**274**:28083–28086.

Zolkiewski M. A camel passes through the eye of a needle: protein unfolding activity of Clp ATPases. *Mol Microbiol* 2006;**61**:1094–1100.

12

THERAPEUTICS: HARNESSING THE POWER OF MOLECULAR AND PHARMACOLOGICAL CHAPERONES

David S. Gross

Department of Biochemistry and Molecular Biology, Louisiana State University Health Sciences Center, Shreveport, LA, USA

Ronald L. Klein

Department of Pharmacology, Toxicology and Neuroscience, Louisiana State University Health Sciences Center, Shreveport, LA, USA

Stephan N. Witt

Department of Biochemistry and Molecular Biology, Louisiana State University Health Sciences Center, Shreveport, LA, USA

12.1 MANIPULATING THE HEAT SHOCK RESPONSE TO BLOCK NEURODEGENERATION

12.1.1 Molecular Chaperones – What Are They and What Do They Do?

Molecular chaperones are a diverse set of proteins that facilitate the folding of newly synthesized polypeptide chains, assist protein translocation across intracellular membranes, inhibit and reverse aggregation that may occur after exposure

Protein Chaperones and Protection from Neurodegenerative Diseases, First Edition.
Edited by Stephan N. Witt.

to environmental stresses (elevated temperature, oxidative stress, anoxia, ethanol, heavy metals, and chemical denaturants), and even assist in protein degradation. Some molecular chaperones are expressed constitutively, whereas others are induced by various stresses. The molecular chaperones induced in response to heat stress were discovered first, and they were named heat shock proteins (HSP), and this phenomenon was termed the *heat shock response* (Ritossa, 1962). Heat shock proteins are now often referred to as *molecular chaperones*. The common feature of inducible and constitutive molecular chaperones is that they transiently associate with substrate proteins, often via an exposed hydrophobic segment, and by a cycle of binding and release the substrate protein eventually adopts its native, active conformation. Many but not all chaperones requires ATP binding and hydrolysis to fuel this cycle.

In bacteria, there are two classes of protein chaperones: (i) The DnaK/DnaJ/GrpE chaperone machine functions in an ATP-dependent manner to inhibit and even reverse protein misfolding and aggregation (Bukau and Horwich, 1998; Slepenkov and Witt, 2002). DnaK, in coordination with its co-chaperone DnaJ, binds to exposed hydrophobic regions of cellular proteins (such as those found during translation, translocation, or misfolding), and through repeated cycles of binding and release promotes the refolding of misfolded proteins as well as the disaggregation of aggregated proteins. The functional homolog of DnaK/DnaJ in eukaryotes is Hsp70/Hsp40. (ii) The GroEL/GroES chaperone machine functions in an ATP-dependent manner to capture and sequester partially unfolded or kinetically trapped intermediates within a central cavity defined by GroEL subunits where it promotes their folding (Hartl and Hayer-Hartl, 2002). The functional homolog of GroEL is TriC (TCP-1 ring complex). As TriC is insufficiently abundant to play a major role in the folding of most proteins, the Hsp90 family of chaperones evolved to fill this niche. Hsp90 is an abundant cytosolic protein that interacts with intermediately folded proteins, preventing their aggregation. Although it lacks the ability of Hsp70 to refold denatured proteins, Hsp90 is highly efficient in preventing protein misfolding and aggregation, and is the eukaryotic cell's central chaperone involved in cell growth, signaling, and proliferation.

12.1.2 Hsp70

Expressed both constitutively and at elevated levels in response to stress, members of the Hsp70 chaperone family promote protein folding, reverse misfolding and aggregation, and are even involved in the unfolding and translocation of polypeptides into cellular compartments such as mitochondria and chloroplasts. The typical pathway of nascent chain folding in *Escherichia coli* initially involves the participation of DnaK and DnaJ. These chaperones have access to the growing polypeptide after ~50 residues, but their contribution (and the subsequent one of the GroEL/GroES chaperones) is primarily posttranslational. By contrast, in eukaryotes, Hsp70/Hsp40-mediated nascent chain folding occurs more at the cotranslational level, permitting the proper folding of sequential domains

(Netzer and Hartl, 1997) that are typically seen in eukaryotic proteins. Cotranslational folding decreases the formation of aggregation-sensitive intermediates; thus, the Hsp70 system is sufficient to maintain the folding of most cytosolic proteins. Hsp70 binds exposed hydrophobic segments of its substrate proteins via its substrate-binding domain (SBD) (Zhu *et al.*, 1996); such binding is regulated by ATP hydrolysis-induced conformational changes in the N-terminal nucleotide-binding domain (NBD; also known as the *ATPase domain*), stimulated by Hsp40 (see Figure 12.1a). Substrate release requires binding of ATP to the NBD of Hsp70. Nucleotide-exchange factor proteins, like GrpE (*E. coli*) or Bag1 (human), promote the release of adenosine diphosphate (ADP) from the Hsp70 chaperone's NBD. ATP binding triggers a global conformational change that is transmitted from the NBD to the SBD and lid. The result is that the lid rotates away from the SBD, the SBD changes conformation, and the substrate rapidly dissociates. Following this, substrates enter a new cycle of binding and release or fold into their native conformation (reviewed in Nollen and Morimoto, 2002; Slepenkov and Witt, 2002).

12.1.3 Hsp90

Although not required for the folding of most polypeptides, Hsp90 plays an important role in the maturation of regulatory molecules and others whose folding is inherently problematic (Nathan *et al.*, 1997). Many of these client proteins fall into two classes—transcription factors and signaling kinases. Such proteins tend to be short-lived and prone to alternative conformations. Indeed, common to Hsp90 client proteins is the fact that they are regulated, either negatively or positively, by chaperone interaction. Underscoring the important role played by

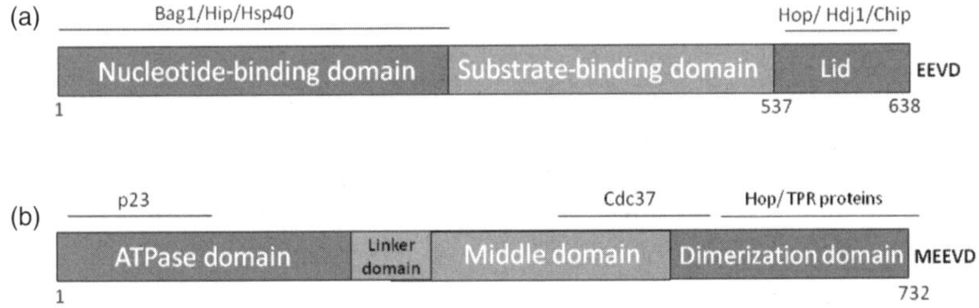

Figure 12.1 Schematic linear representation of the domain structure of the Hsp70 and Hsp90 chaperones (a and b, respectively). Binding sites of several important co-chaperones are indicated (composite of sites of interaction for both mammalian and yeast proteins), as are the C-terminal motifs characteristic of cytosolic Hsp70 and Hsp90 isoforms. For Hsp70, the nucleotide-binding domain binds ADP or ATP; the substrate-binding domain (SBD) binds segments of unfolded proteins or short peptides. In the ADP-bound state, the lid encapsulates and protects the bound substrate in the SBD; this prevents diffusion of the substrate out of the binding site. For Hsp90, evidence exists for the presence of substrate-binding sites within each of the three principal domains.

Hsp90 client proteins is the striking observation that when *Drosophila* Hsp90 is either mutated or pharmacologically impaired, phenotypic variation affecting most adult structures is produced (Rutherford and Lindquist, 1998). This observation raises the possibility that Hsp90 acts as a "capacitor" for morphological evolution by dampening the effects of mutations that destabilize proper protein folding. Such mutations can therefore remain silent and accumulate under neutral conditions (Rutherford and Lindquist, 1998).

Hsp90 has a modular structure in which the two well-conserved domains, an N-terminal ATP-binding domain and a C-terminal dimerization domain, are separated by a charged linker of variable length (illustrated in Figure 12.1b). Although binding of the client by Hsp90 occurs independently of ATP, its release requires ATP hydrolysis and is enhanced by the co-chaperone p23, which couples ATPase activity to polypeptide dissociation. The Hsp90–substrate binding and release cycle is also regulated by sequential interactions between Hsp90 and a second co-chaperone, Hop. Hop induces a conformational change in the ATPase domain of Hsp90 that inhibits ATPase activity. Hop dissociation results in a conformational change that permits ATP binding, transient dimerization of the N-termini, association with p23 and several immunophilins, and finally ATP hydrolysis. The latter opens the molecular clamp formed by the N-termini to release the substrate (reviewed in Nollen and Morimoto, 2002; Young and Hartl, 2000).

12.1.4 Heat Shock Factor — Master Regulator of the Eukaryotic Heat Shock Response

As alluded to above, the heat shock response is a universal response of organisms to thermal, chemical, and oxidative stress. In prokaryotes as well as eukaryotes, it is largely regulated at the transcriptional level. In *E. coli*, the stress responsive transcription factor, σ^{32}, is a promoter-specific subunit of RNA polymerase (RNAP). σ^{32} directs RNAP to the promoters of HSP-encoding genes, as well as others whose products permit increased cell survival in response to high temperature (reviewed in Guisbert *et al.*, 2008). In eukaryotes, proteotoxic conditions activate a single transcriptional activator, heat shock factor 1 (HSF1), a highly conserved, sequence-specific DNA-binding protein. In higher eukaryotes, its activation typically involves five steps: (i) interconversion of an inactive cytoplasmic monomer to a biologically active homotrimer; (ii) nuclear accumulation; (iii) DNA binding to heat shock response elements (HSEs) consisting of tandem inverted repeats of the pentameric sequence, AGAAN; (iv) hyperphosphorylation; and (v) transcriptional activation of target genes (reviewed in Wu, 1995).

In the yeast *Saccharomyces cerevisiae*, HSF1 is constitutively localized in the nucleus (Singh *et al.*, 2006) where a small population ($<10\%$) of molecules exist in a pre-trimerized, DNA-bound state (Giardina and Lis, 1995; Gross *et al.*, 1990; McDaniel *et al.*, 1989). Upon heat shock, the remaining HSF1 molecules bind cooperatively to HSEs (Erkine *et al.*, 1999). *Drosophila* HSF1 (dHSF1) is likewise found in the nucleus under control conditions (Westwood *et al.*, 1991) and similarly shows detectable DNA binding prior to heat shock (Boehm *et al.*, 2003). Nonetheless, following heat shock, the predominantly nucleoplasmic, monomeric dHSF1 is induced to form trimers that translocate to chromosomal loci, where the molecules remain stably bound to DNA, firing multiple rounds of RNA polymerase II (Pol II) transcription without undergoing detectable turnover (Yao *et al.*, 2006). In mammalian cells, HSF1 is found in both cytoplasmic and nuclear compartments prior to stress (Sarge *et al.*, 1993; Sheldon and Kingston, 1993).

HSF1 contains a highly conserved core, consisting of DNA-binding and trimerization domains (illustrated in Fig. 12.2a for yeast HSF1 [yHSF1] and in Fig. 12.2b for human HSF1, [hHSF1]). The DNA-binding domain adopts a "winged" helix– turn–helix structure (Harrison *et al.*, 1994; Vuister *et al.*, 1994). Interestingly, the "wing" structure does not bind DNA, but does contribute to the binding of the first trimer (Littlefield and Nelson, 1999) and may also increase the cooperativity between adjacent trimers that is seen both *in vitro* and *in vivo* (Amin *et al.*, 1994; Erkine *et al.*, 1999). The DNA-binding domain is in the $\alpha + \beta$ class of helix–turn–helix proteins, similar to that in the bacterial catabolite

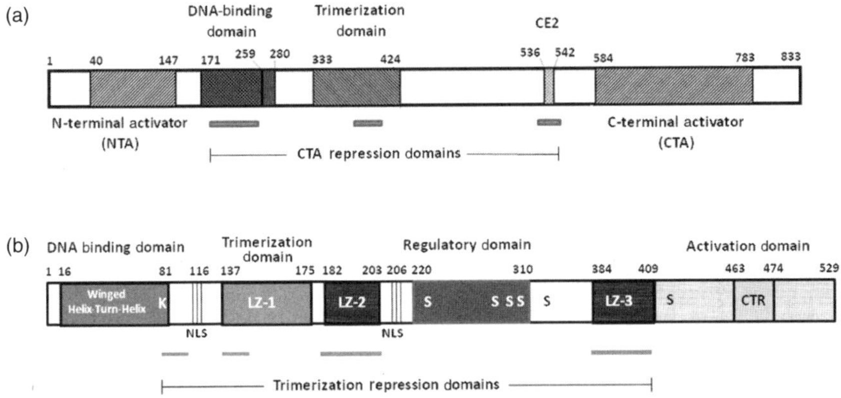

Figure 12.2 Domain maps of heat shock transcription factors from yeast and humans. (a) *S. cerevisiae* HSF1. Repression domains conferring temperature-sensitive regulation are indicated (the recognition helix within the DNA-binding domain (DBD), hydrophobic heptad repeats within the trimerization domain, and a heptapeptide sequence termed *CE2*.; all three act to repress the C-terminal activator). A second leucine zipper region, whose function is unknown, maps to the C-terminal activator (residues 628–667; not shown). A flexible linker of 21 amino acids, shown in dark blue and required for yHSF1 function, joins the DBD to the trimerization domain (Flick *et al.,* 1994). (b) *H. sapiens* HSF1. The indicated repression domains confer temperature-sensitive regulation and antagonize trimerization. S, serine residues implicated in hHSF1 regulation: Ser303, Ser307, and Ser308 are constitutively phosphorylated and repress hHSF1 activity, whereas Ser230, Ser326, and Ser419 are inducibly phosphorylated and linked to its activation. K, Lys80 residue subject to inducible acetylation/deacetylation that regulates the DNA-binding activity of hHSF1. NLS, nuclear localization signal; CTR, C-terminal repeat. (*A full color version of this figure appears in the color plate section.*)

activator protein (reviewed in Wu, 1995). The oligomerization domain is conserved in all HSF1 proteins and consists of hydrophobic heptad repeats ("leucine zippers") that form helical coiled-coil structures (reviewed in Wu, 1995). Most leucine zipper proteins form homo- or heterodimers, whereas the trimerization domain of HSF1 forms a highly unusual triple stranded, α-helical coiled-coil structure (Sorger and Nelson, 1989). yHSF1 additionally contains two transcriptional activation domains, one located at the N-terminus and the other at the C-terminus (Figure 12.2a). The N-terminal activator is critical for the transient heat shock response, whereas the C-terminal activator is important for sustained stimuli, and its deletion results in lethality at elevated temperatures (Nieto-Sotelo *et al.*, 1990; Sorger, 1990). Each activator exhibits an intrinsic constitutive activity that is repressed under non-heat-shock conditions by one or more masking domains. How these masking domains function is unclear. One possibility is that they physically interact with activation domains (Chen and Parker, 2002); another is that they serve as binding sites for corepressors such as molecular chaperones.

The core DNA-binding/trimerization domain structure of hHSF1 is similar to that of its yeast counterpart. However, hHSF1 contains additional domains, including a regulatory domain within which are located three sites of constitutive phosphorylation (Ser303, Ser307, Ser308); these appear to be important for negative regulation and may be targeted by molecular chaperones. On the other hand, sites of inducible phosphorylation (Ser230, Ser326, Ser419) promote hHSF1 activity (reviewed in Westerheide and Morimoto, 2005). Moreover, hHSF1 is negatively regulated by acetylation of a critical lysine residue located within the DNA-binding domain (Lys80). Deacetylation of Lys80 by the nicotinamide adenine dinucleotide (NAD^+)-dependent deacetylase SIRT1 is required to permit stable binding of hHSF1 to DNA and concomitant trans-activation of target *HSP* genes (Westerheide *et al.*, 2009).

Genomewide localization analysis in *S. cerevisiae* has identified >160 genes as HSF1 targets. These genes encode proteins involved in protein folding, degradation, and trafficking; energy generation; maintenance of cell integrity; cell signaling; and transcription (Hahn *et al.*, 2004). Moreover, yHSF1 is an essential protein, and its deletion results in lethality even in cells maintained at low temperatures (Sorger and Pelham, 1988). The essential nature of this protein arises from the fact that it is required to stimulate, at minimum, the constitutive transcription of Hsp70 and Hsp90-encoding genes (Gross *et al.*, 1990; McDaniel *et al.*, 1989; Park and Craig, 1989). One of the ways that it does this is by maintaining the promoter regions of such genes in an open chromatin conformation (Erkine *et al.*, 1996; Gross *et al.*, 1993; Venturi *et al.*, 2000).

Observations such as these suggest that a critical activity of HSF1 is to remodel chromatin structure. Indeed, heat-shock-activated yHSF1 rapidly recruits chromatin-modification complexes to the genes that it regulates, including the ATP-dependent remodeling complexes Swi/Snf, RSC, and Isw1 (Erkina *et al.*, 2008, 2010; Shivaswamy and Iyer, 2008; Zhao *et al.*, 2005). Likewise, hHSF1 recruits the human ATP-remodeling complex Brg1 (ortholog of yeast Swi/Snf) in response to thermal shock (de la Serna *et al.*, 2000). yHSF1 additionally triggers the rapid and simultaneous recruitment of histone acetyl transferases (HATs) and histone deacetylase complexes (HDACs) to *HSP* gene promoters and their linked coding regions (Kremer and Gross, 2009). Synchronous recruitment of activating and repressing activities may serve to fine-tune the heat shock transcriptional response. Histone modification enzymes are similarly recruited to the stress-activated human *HSP70* gene (Thomson *et al.*, 2004), suggesting that chromatin remodeling/nucleosome modification is a central, conserved activity of HSF1. In addition to removal of promoter-associated nucleosomes, activated HSF1 triggers the disassembly of nucleosomes across heat shock gene-coding regions in both yeast and *Drosophila* (Petesch and Lis, 2008; Zhao *et al.*, 2005). In the case of yeast, eviction of nucleosomes is kinetically linked to the presence of elongating Pol II (Zhao *et al.*, 2005). In the case of *Drosophila*, nucleosomes are lost across the entire *HSP70* locus in an initial wave that precedes Pol II transcription; this dramatic remodeling requires poly(ADP)-ribose polymerase or its activity (Petesch and Lis, 2008).

HSF1 additionally plays an important role in promoting the assembly of the transcription preinitiation complex, as both Pol II and TATA-binding protein (TBP) are recruited in an HSF1-dependent manner (Erkina *et al.*, 2008; Sekinger and Gross, 2001; Zhao *et al.*, 2005). The ability of yHSF1 to stimulate activated transcription has been shown to be dependent on two general transcription factors (GTFs)—TFIIE and Mediator (Lee *et al.*, 1999; Sakurai and Fukasawa, 1999; Singh *et al.*, 2006). In yeast, Mediator dampens basal *HSP* gene transcription and yet enhances heat shock-induced transcription, thus significantly increasing the dynamic range of heat shock gene expression (Singh *et al.*, 2006). Therefore, HSF1 partners with a combination of chromatin- and non-chromatin-associated factors to transcriptionally regulate its target *HSP* genes.

12.1.5 Chaperone-Mediated Regulation of HSF1 Function

Several lines of evidence, both genetic and biochemical, argue for the existence of a negative feedback loop in the regulation of HSF1, whereby products of genes activated by HSF1 downregulate its activity. Early studies suggested a potential negative regulatory role for Hsp70. For example, genetic analysis in yeast indicated that the Hsp70 protein, Ssa1, negatively regulated its own transcription via an upstream regulatory sequence (Stone and Craig, 1990). Consistent with this, mouse HSF1 was shown to associate with Hsp70 in whole cell lysates; addition of ATP and other hydrolyzable nucleotides resulted in the dissociation of the Hsp70–HSF1 complex (Abravaya *et al.*, 1992). Furthermore, Hsp70 blocked the acquisition of DNA-binding activity by HSF1; such an inhibitory effect was abolished by ATP (Abravaya *et al.*, 1992). Hsp70 was also found to associate with the activation domain of hHSF1 (Shi *et al.*, 1998). Together, these data suggest that Hsp70 can block the activation of HSF1 by preventing its binding to DNA; Hsp70 (and its co-chaperone Hsp40) may additionally suppress the ability of HSF1 to transactivate target genes during recovery from stress. Parallel observations have been reported for the regulation of σ^{32}: the major Hsp70 chaperone, DnaK, and its co-chaperones DnaJ and GrpE bind σ^{32} and inactivate it, thereby coupling σ^{32} function to the cellular protein folding state (reviewed in Guisbert *et al.*, 2008).

Several eukaryotes have very low levels of endogenous Hsp70 and yet are capable of suppressing HSF1 in the absence of stress (reviewed in Voellmy and Boellmann, 2007). Therefore, another chaperone may have a more universal role, and indeed more recent work has indicated that Hsp90 possesses a central, negative role in HSF1 regulation. In a landmark study, Voellmy and colleagues demonstrated that an Hsp90-containing HSF1 complex is present in the unstressed human cell and dissociates during stress (Zou *et al.*, 1998). Immunodepletion of Hsp90, but not Hsp70 (nor the closely related Hsc70 protein), nor the co-chaperones Hsp40, Hop, Hip, or the cyclophilin CyP40, dramatically activated the hHSF1 monomer to trimer transition in whole cell extracts and, concomitantly, the DNA-binding activity of hHSF1 (Zou *et al.*, 1998). Two additional studies found hHSF1 in complexes with Hsp90, the immunophilin FKBP52, and the co-chaperone p23 (Guo *et al.*, 2001; Nair *et al.*, 1996). Mutations in the hHSF1 regulatory domain that increased transcriptional competence (see Figure 12.2b; residues 220–310) impaired the assembly of hHSF1 into Hsp90-containing complexes (Guo *et al.*, 2001). Moreover, the naturally occurring compound, geldanamycin (GA), activates hHSF1 by virtue of its ability to impair Hsp90 activity. It does this by competing with ATP for binding to the ATP site, thus preventing completion of the interaction cycle by blocking p23 binding (Zou *et al.*, 1998). GA treatment of mouse cells significantly delays disassembly of HSF1 trimers after a heat shock and restores stress-induced *HSP* gene expression (Winklhofer *et al.*, 2001). These results are consistent with the idea that mammalian HSF1 is repressed in the absence of stress by an Hsp90 complex, and that in response to stress or pharmacological inhibition of Hsp90, this complex dissociates, resulting in HSF1 activation.

Comparable experiments carried out in *Xenopus* oocytes microinjected with either antibodies or Hsp90 (or treated with GA) led to similar findings and conclusions (Ali *et al.*, 1998). Immunodepletion of not only Hsp90 but also p23 enhanced HSF1 DNA-binding activity in frog oocytes (Bharadwaj *et al.*, 1999). Likewise, in both yeast and *Drosophila*, there is genetic evidence for a negative role of Hsp90 in HSF1 regulation under stress and nonstress conditions. In yeast, this is based on the increased expression of HSF1-regulated genes in genetic contexts in which the function of Hsp90 is reduced or that of Cpr7, a CyP40-type cyclophilin required for full Hsp90 function, eliminated. Hsp90 and Cpr7 function synergistically to repress gene expression from HSF1-dependent promoters (Duina *et al.*, 1998). Consistent with this, yHSF1 is retained by affinity columns of purified yeast Hsp90 (cited in Duina *et al.*, 1998). In *Drosophila*, HSF1 appears to be similarly inhibited by Hsp90 and its co-chaperones based on the fact that small inhibitory RNA (siRNA) depletion of Hsp90, Hsp70, and Hsp40 increases the DNA-binding activity of dHSF1; co-depletion of two of these proteins produced synergistic results (Marchler and Wu, 2001). Taken together, the results from mammalian, amphibian, insect, and fungal systems argue that Hsp90, by itself and/or associated with co-chaperones, is the most important repressor of HSF1. Interesting—but as yet unproven—is the possibility that Hsp90 may act on DNA-bound HSF1 trimers to induce their disassembly and that of the associated transcription initiation complex, such as previously shown for nuclear hormone receptors (Freeman and Yamamoto, 2002). If this turns out to be the case, it would enable the HSF1-regulated transcriptional machinery to directly detect changes in the intracellular protein folding environment.

12.1.6 Harnessing Chaperones in the Treatment of Human Disease

12.1.6.1 Neurodegenerative Disease. Given that HSF1 coordinately activates the expression of multiple protein chaperones and other cytoprotective genes, HSF1 activation could be of potential therapeutic value in the treatment of neurodegenerative diseases associated with protein misfolding such as Alzheimer's disease (AD), Parkinson's disease (PD), amyotrophic lateral sclerosis, and Huntington's disease (HD). Indeed, the critical event in these neurodegenerative diseases is a toxic gain-of-function mutation associated with the appearance of oligomers and other aggregates consisting of β-amyloid protein, α-synuclein, superoxide dismutase, and huntingtin, respectively. Of relevance, the conversion of HSF1 to the high-affinity DNA-binding homotrimeric state is not robust in neuronal cells. Although the mechanism is unclear, this could underlie the selective sensitivity of neuronal cells for disease even though misfolded proteins are expressed in all tissues (reviewed in Morimoto, 2008).

Manipulation of Hsp90 function using natural compounds therefore represents a potentially powerful therapeutic approach. As proof of principle, Bonini and colleagues tested this idea in a *Drosophila melanogaster* mutant that had been genetically engineered to overexpress α-synuclein, a protein implicated in the etiology of PD. Earlier work demonstrated that overexpressed Hsp70 could prevent dopaminergic neuronal loss associated with α-synuclein (Auluck and Bonini, 2002). When such α-synuclein-expressing flies were aged at 28°C (a stressful temperature) and additionally treated with GA, protection from α-synuclein toxicity occurred (Auluck *et al.*, 2005). However, similarly treated flies that also contained the hsf^4 allele (encoding an HSF1 mutant protein whose function is abolished at 28°C) evinced evidence of α-synuclein toxicity irrespective of GA treatment (Auluck *et al.*, 2005). These results suggest that GA-mediated HSF1 activation is capable of ameliorating α-synuclein toxicity in a model organism, and may do so via elevated expression of chaperones such as Hsp70 and Hsp40.

Other naturally occurring compounds that might offer potential benefit in the treatment of neurodegenerative disease via HSF1 activation include salicylates, resveratrol, and celastrol. Salicylates are widely prescribed anti-inflammatory drugs that affect the activity of both HSF1 and NF-κB; in the case of HSF1, sodium salicylate stimulates the ability of the protein to trimerize and bind DNA, although HSF1 does not activate the target gene transcription (Giardina and Lis, 1995; Jurivich *et al.*, 1992). Thus, activation of transcription requires a subsequent stimulus, which theoretically could be elevated tissue temperature associated with fever. This might contribute to the effectiveness of sodium salicylate (aspirin) in treating inflammation.

Resveratrol, a bioactive compound in red wine, is a potent inducer of sirtuins, an evolutionarily conserved family of NAD^+-dependent lysine deacetylases. Its founding member, the budding yeast protein Sir2 (Imai *et al.*, 2000), was originally shown to be an important mediator of the silent chromatin structure through its ability to deacetylate N-terminal lysine residues on nucleosomal histones (Braunstein *et al.*, 1993; Hoppe *et al.*, 2002). Such nucleosomal deacetylation is thought to contribute to the structural integrity of both the nucleolus and telomeres, thereby enhancing longevity (Dang *et al.*, 2009; Guarente, 2000). The human ortholog, SIRT1, regulates the function of not only histones but also other key molecules in the nucleus including HSF1 (Westerheide *et al.*, 2009), as discussed above. Although not formally demonstrated, it is thus conceivable that chronic exposure to low doses of resveratrol could slow neurodegeneration via SIRT1 activation of HSF1. On the other hand, an excess of nicotinamide, a water-soluble vitamin that is part of the vitamin B group, might contribute to neurodegeneration through its ability to inhibit sirtuins. Nicotinamide potently inhibits HSF1 activation of heat shock genes in human cells, presumably by preventing deacetylation of Lys80 within the protein's DNA-binding domain (Westerheide *et al.*, 2009) (see Section 12.1.4).

Celastrol, a quinone methide triterpene and active component from Chinese herbal medicine, specifically activates hHSF1 at multiple levels—including activation of DNA binding, hyperphosphorylation, and transcriptional activation (Westerheide *et al.*, 2004). How celastrol does this is unclear, but it is unlikely to work by inducing global protein denaturation; indeed, it elicits no interference with the chaperone function of Hsp70 in an *in vitro* assay (Westerheide *et al.*, 2004). Therefore, celastrol might be useful as a therapeutic in the treatment of neurodegenerative disease.

12.1.6.2 Cancer. The potential of GA and related compounds as a therapeutic in the treatment of human cancer has also been investigated. The rationale for this arises from the fact that increased Hsp70 and Hsp90 levels are a hallmark of cancer (reviewed in Buchner, 1999; Westerheide and Morimoto, 2005). Horikoshi and colleagues reported that GA, as well as a derivative, 17-AAG, induced cell cycle arrest of tumor cells because of proteasomal degradation of the mitogenic signaling proteins PI3 kinase, IKK, NF-κB, and others found within the Ras/Raf/MEK/ERK MAPK pathways (Fukuyo *et al.*, 2010). This striking result may be due, at least in part, to the fact that Hsp90 in tumor cells has 100 times the affinity for GA derivatives than Hsp90 in normal cells, presumably because of the large multi-chaperone complexes present with higher ATPase activity. GA derivatives, in combination with radiotherapy, have also been tried in animal models, and this appears to represent a promising approach. GA selectively abrogates the G2/M checkpoint by inducing ubiquitin-dependent degradation of

Chk1 and by downregulating CDC25C phosphatase activity. Abrogation of the G2/M checkpoint renders cells more sensitive to radiation-induced DNA damage (reviewed in Fukuyo *et al.*, 2010). Note that in these cases, GA may be acting primarily to promote proteasomal degradation of Hsp90 client proteins (see, e.g., Thomas *et al.*, 2006) as opposed to activating HSF1, which would lead to the elevated expression of all chaperones. Efficacy of GA treatment in cancer may therefore stem from the opposing roles of Hsp70 and Hsp90 in the mammalian cell, in which Hsp90 stabilizes its client proteins, whereas Hsp70 promotes their ubiquitylation (Thomas *et al.*, 2006).

Other compounds have also been tested as potential chemotherapeutics. Radicicol, a monocyclic lactone antibiotic that inhibits Hsp90 function by interacting with N-terminal ATPase pocket similar to GA, induces apoptosis in cultured human cells. Yet, in animals, radicicol lacked activity, possibly because of instability as a result of α-, β-, and γ-unsaturated carbonyl groups that are highly reactive to nucleophiles (reviewed in Fukuyo *et al.*, 2010).

A caveat of therapeutics that either directly or indirectly activates hHSF1 is that this protein is a powerful, multifaceted enhancer of oncogenesis (Dai *et al.*, 2007). Therefore, HSF1 activation via functional depletion of Hsp90 could have unintended, and adverse, consequences. HSF1 knockout protects mice from tumors induced by mutations in either oncogenes or tumor suppressor genes. In cultured mouse or human cells, HSF1 facilitates transformation to the malignant state by inducing the expression of genes that contribute to cell proliferation, survival, protein synthesis, and glucose metabolism (Dai *et al.*, 2007). Therefore, stimulation of HSF1 activity in tumor cells, as a consequence of inactivation of Hsp90 (or by any other mechanism), represents a potentially serious drawback to any therapeutic that targets this transcription factor.

12.2 GENE THERAPY FOR NEURODEGENERATIVE DISEASES

The concept of gene therapy was first developed within the field of genetics in the early twentieth century, and the feasibility of the concept was more vividly envisioned when man harnessed the ability to engineer recombinant DNA in bacteria in the 1950s and 1960s (Wolff and Lederberg, 1994). Although progress has been made over the past 30-year period of basic and clinical gene therapy research, it remains unclear if gene therapy will evolve to be a therapeutic option. Gene therapy clinical trials for diseases of the central nervous system (CNS) have been positive so far in terms of safety and tolerability of the treatments, although clear-cut efficacy is still elusive. Perhaps, the example with the greatest preclinical and clinical rationale is for specific diseases of the eye where gene therapy can improve and even restore vision (Simonelli *et al.*, 2010). Gene therapy is worth considering for conditions that (i) involve single gene defects in metabolism and other disease-causing familial mutations; (ii) are life threatening; or (iii) are orphans with respect to efficacious drugs or surgeries. Neurodegenerative diseases may involve inherited disease-causing mutations, are deadly, and are typically without effective medication. In 30 years more, it is hard not to foresee the emergence of gene-based therapeutics when conventional medications and surgeries are ineffective.

Methods of gene transfer are viral or nonviral. The latter may involve naked DNA or DNA complexed with chemicals such as cationic lipids. Although it remains difficult to approach the efficiency of viruses that have evolved to infect host cells and express their genes or RNA, nonviral methods have the potential to be safer and less restricted to innate viral tropism. Manipulating the viral genome by removing genes and adding mammalian genes has led to many types of specific recombinant viruses. In this process, transcomplementation is necessary to propagate the recombinant viruses, by providing the necessary viral gene products in trans for efficient packaging, which are then purified away during preparation of the recombinant vector. For gene transfer to the brain, recombinant herpes simplex virus was first used (Federoff *et al.*, 1992), followed by adenovirus (Le Gal La Salle *et al.*, 1993), and later by lentivirus (Naldini *et al.*, 1996) and adeno-associated virus (AAV) (Kaplitt *et al.*, 1994; McCown *et al.*, 1996). Each system has unique properties and tropisms that can have advantages for specific niche diseases—for example, the high packaging capacity of herpes, or the ability to integrate into chromosomes for lentivirus, which is necessary for prolonged expression in dividing cells.

Although the unique properties of different viral and nonviral gene transfer vectors may offer specific therapeutic utility, progress in the CNS in both preclinical and clinical work has been dominated by AAV, owing to its efficiency and safety margin. The first recombinant AAV vector was introduced in 1984 (Hermonat and Muzyczka, 1984), and the first example of AAV gene transfer in the brain of an animal was in 1994 (Kaplitt *et al.*, 1994). Since then, the widespread use of the AAV system has rapidly expanded in preclinical studies optimizing gene transfer efficiency through improvements in promoter, specific AAV subtype, and purification method (reviewed in Klein *et al.*, 2009). The first case of recombinant AAV gene transfer to the human brain was in 2003 in a PD clinical trial (Kaplitt *et al.*, 2007). There have been three separate trials for PD using three different gene strategies, owing to the fact that such a circumscribed portion of the brain causes motor symptomatology (Christine *et al.*, 2009; Herzog *et al.*, 2009; Kaplitt *et al.*, 2007). The other well-documented clinical trials using AAV in the CNS are for Alzheimer's, Canavan, and Batten diseases (Mandel and Burger, 2004; Worgall *et al.*, 2008). Several distinct gene modification strategies can be used: (i) gene replacement; (ii) overexpression of a specific disease-modifying gene product; (iii) overexpression of a nonspecific survival factor such as a neurotrophic factor; (iv) knockdown of a disease-related protein; (v) cancer cell suicide; and (vi) regulatable expression (outlined in Manfredsson and Mandel, 2010).

12.2.1 Gene Therapy Paradigms

Neurodegenerative diseases are often thought of as protein misfolding disorders, which could perhaps be corrected by chaperones. Molecular chaperones could act specifically on detoxifying neuropathological substrate proteins, or perhaps in a less specific manner by inducing a heat shock response. Specific protein lesions are associated with (and are diagnostically defining for) specific neurodegenerative diseases, such as (i) microtubule-associated protein tau neurofibrillary tangles in AD and other "tauopathy" diseases such as progressive supranuclear palsy; (ii) α-synuclein Lewy bodies in PD and dementia with Lewy bodies; (iii) polyglutamine repeat inclusions in HD; and (iv) cytoplasmic Tar DNA-binding protein-43 (TDP-43) inclusions in amyotrophic lateral sclerosis. Such lesions are typically ubiquitinated, which suggests that the host cell may be targeting the abnormal aggregates for the ubiquitin–proteasome triage system. Many other proteins in addition to ubiquitin can be colocalized to specific aggregates, such as the molecular chaperone heat shock proteins (Braak *et al.*, 2001; Uryu *et al.*, 2006). This supports the hypothesis that refolding of misfolded proteins could be a cellular response to neurodegenerative processes, or that the sequestration (and inactivation) of chaperones in lesions could cause further formation of the aberrant aggregates. In either case, an elevation in levels and replacement of functional chaperones within a cell that is prone to make aggregates could therefore be protective, which constitutes the premise for studying neuroprotection conferred by chaperones.

The preclinical rationale for neurodegenerative disease-modifying properties of heat shock proteins is vast, because of the colocalization of heat shock proteins in neuropathological lesions, as mentioned (Braak *et al.*, 2001; Uryu *et al.*, 2006), and the many studies showing the protective effects of molecular chaperone gene delivery, as reviewed in other chapters of this book. *Drosophila* models were used to assay Hsp70 effects on either α-synuclein or polyglutamine-induced neuronal loss by Bonini's group (Auluck and Bonini, 2002; Warrick *et al.*, 1999). Protection of the dopaminergic neurons of the substantia nigra has also been documented when Hsp70 is expressed with AAV in rodents using specific lesioning techniques such as the neurotoxin 1-methyl-4-phenyl-1,2,3,6-tetrahydropyridine (MPTP) (Dong *et al.*, 2005) or overexpression of a substrate protein for parkin ubiquitination called CDCrel-1 (Jung *et al.*, 2008).

In another model of dopaminergic neuron degeneration in rats, using mutant α-synuclein overexpression, coexpression of a chaperone known to resolve aggregated proteins in yeast, Hsp104, protected nigral dopamine neurons from α-synuclein-induced cell loss (Lo Bianco *et al.*, 2008) (see Chapter 7). In the same study looking at *in vitro* fibrillization of α-synuclein, Hsp104 and the chaperones Hsp70, Hsp40, and Hdj2 promoted disassembly of α-synuclein oligomers and fibrils (Lo Bianco *et al.*, 2008). Regarding Alzheimer's research, Hsp104 can also block the assembly of purified amyloid precursor protein cleavage products into β-amyloid (Vashist *et al.*, 2010), suggesting that it could prevent the formation of amyloid plaques in an animal. Regarding Huntington's research, crossing a transgenic mouse line for Hsp104 with a transgenic mouse line for polyglutamine pathology reduced the pathological aggregates and increased life span (Vacher *et al.*, 2005). Hsp104 holds potential as a protective gene to be delivered via gene therapy into patients with a variety of different neurodegenerative diseases.

Hsp90 and other heat shock proteins are also therapeutic targets. Both in AD brains and tau transgenic mice, Hsp90 levels inversely correlate with tau pathology, and direct association of Hsp90 with tau, which promotes nonpathological tau (increased solubility, less phosphorylation, and more microtubule interaction), has been demonstrated (Dou *et al.*, 2003). Knockout of Hsp70 or Hsp110 in mice leads to tau hyperphosphorylation and neurodegeneration, suggesting more chaperone-based strategies to counteract tau pathogenesis (Eroglu *et al.*, 2010). Other chaperones for tau such as Pin1, CHIP, and BAG2 have also been shown to detoxify or degrade pathological tau (Carrettiero *et al.*, 2009; Lu *et al.*, 1999; Petrucelli *et al.*, 2004), supporting them as rational targets for tau disease modification, potentially via gene delivery. Regarding amyotrophic lateral sclerosis research, the small heat shock protein HspB8 has been shown to promote the removal of disease-related aggregates of superoxide dismutase or TDP-43 (Crippa *et al.*, 2010). Another pathway that has been implicated in neurodegeneration is chaperone-mediated autophagy for lysosomal destruction of pathological aggregates (Koga and Cuervo, 2010). For example, in *Caenorhabditis elegans*, the autophagy-related enzymes, ATG7 and VPS41, reduced α-synuclein-induced neuronal loss when overexpressed and enhanced the neuron loss when their expression was blocked (Hamamichi *et al.*, 2008).

Further refinement of gene transfer technology could, one day, lead to pinpoint expression only in the affected cells, perhaps involving optogenetics where neurons can be made sensitive to specific wavelengths of light via gene transfer of specific opsin photopigments (Zhang *et al.*, 2010). The innate problems in targeting the gene transfer could be overcome by stimulating only specific cells with light. Another hurdle is the sufficiently widespread CNS expression for diseases with widespread pathology such as AD and amyotrophic lateral sclerosis, and the need for intracranial injections. Recent advances in AAV technology have made peripheral to central gene delivery to the nervous system possible (Foust *et al.*, 2009; McCarty *et al.*, 2009), which has now made new disease-modifying strategies possible for the neurobiologist.

The greatest rationale for a gene therapy strategy involving molecular chaperones would be a neurodegenerative disease in which a molecular chaperone is mutated, such as valosin-containing protein (VCP) or survival motor neuron (SMN). (Similar to yeast Hsp104 and human torsinA, VCP is a member of the AAA+ ATPase family of proteins; see Chapters 7 and 11.) Mutations in VCP are associated with inclusion body myopathy associated with Paget disease of the bone and frontotemporal dementia (Boeddrich *et al.*, 2006; Gitcho *et al.*, 2009). The pathological lesions in this disease are cytoplasmic TDP-43 inclusions as in amyotrophic lateral sclerosis, and mutant VCP inhibits autophagy functions (Ju *et al.*, 2009) and promotes the cytoplasmic accumulation of TDP-43 (Ritson *et al.*, 2010), which is normally a nuclear protein. Adding back the fully functional VCP chaperone could potentially restore normal autophagy and clear the cytoplasm of TDP-43. Another putative chaperone is SMN, which is involved in the assembly of small nuclear ribonucleoproteins. Mutations of SMN are linked to spinal muscular atrophy, a rapidly progressing motor neuron disease (Burghes and Beattie, 2009), for which there is no effective medication. Diseases involving mutations in the chaperones VCP or SMN would therefore meet the criteria to consider gene therapy to restore the full chaperone functionality that has been compromised in inherited forms of neurodegenerative disease.

12.3 OSMOLYTES AND PHARMACOLOGICAL CHAPERONES

Osmolytes and pharmacological chaperones are low molecular mass organic molecules that promote protein folding, but by different mechanisms. Osmolytes do not bind to proteins, are nonspecific, and thus are needed at high concentrations. Because osmolytes do not bind to proteins, they are not true chaperones; nevertheless, they are intriguing molecules, perhaps with therapeutic potential. Pharmacological chaperones are ligands of specific proteins, and since one ligand binds to one protein, the binding is specific. The focus in this section is on osmolytes and pharmacological chaperones and how they can be used as treatments for various neurodegenerative diseases.

12.3.1 Osmolytes

Osmolytes are low molecular mass neutral organic molecules, which are synthesized by bacteria, yeast, plants, and animals to protect proteins from harsh conditions that cause denaturation, such as high osmolarity, desiccation, and accumulation of certain metabolites (Yancey *et al.*, 1982; Yancey and Somero, 1979). Selective pressures have resulted in diverse evolutionarily distant organisms synthesizing similar osmolytes for protection from harsh environments (Somero, 1986). Osmolytes often occur at very high concentrations in cells (0.1–1 M), they do not bind to proteins, and they have a unique ability to stabilize native folded proteins (and destabilize potentially toxic unfolded proteins) in response to rapid changes in the external and internal milieu. Sugars, polyalcohols, amino acids, and modified amino acids are osmolytes (Figure 12.3). Almost every modern biochemistry textbook describes the use of sugars and glycerol as stabilizers of purified proteins; yet surprisingly little is known about their mechanisms of action. Because of the high osmolyte concentrations that must be used to achieve protein stability, many, but not all, of these compounds cannot be used therapeutically on humans. As described below, one osmolyte—trehalose—holds great potential to break up toxic protein aggregates in diseases such as Alzheimer's, Huntington's, and Parkinson's.

12.3.1.1 Osmolyte Mechanism. To understand how osmolytes work, it is important to review the thermodynamics of protein folding. For the simple two-state protein folding reaction $U \Leftrightarrow F$, where U and F stand for unfolded and folded, respectively, suppose the free energy of folding $\Delta G_{U \rightarrow F}$ is negative indicating that the unfolded to folded reaction is spontaneous. There are two components of $\Delta G_{U \rightarrow F}$; one is the enthalpy ($\Delta H_{U \rightarrow F}$) and the other is the entropy ($\Delta S_{U \rightarrow F}$), and each of these terms has two components, one from the protein and one from the solvent. What drives folding? Is it the change in enthalpy, entropy, or

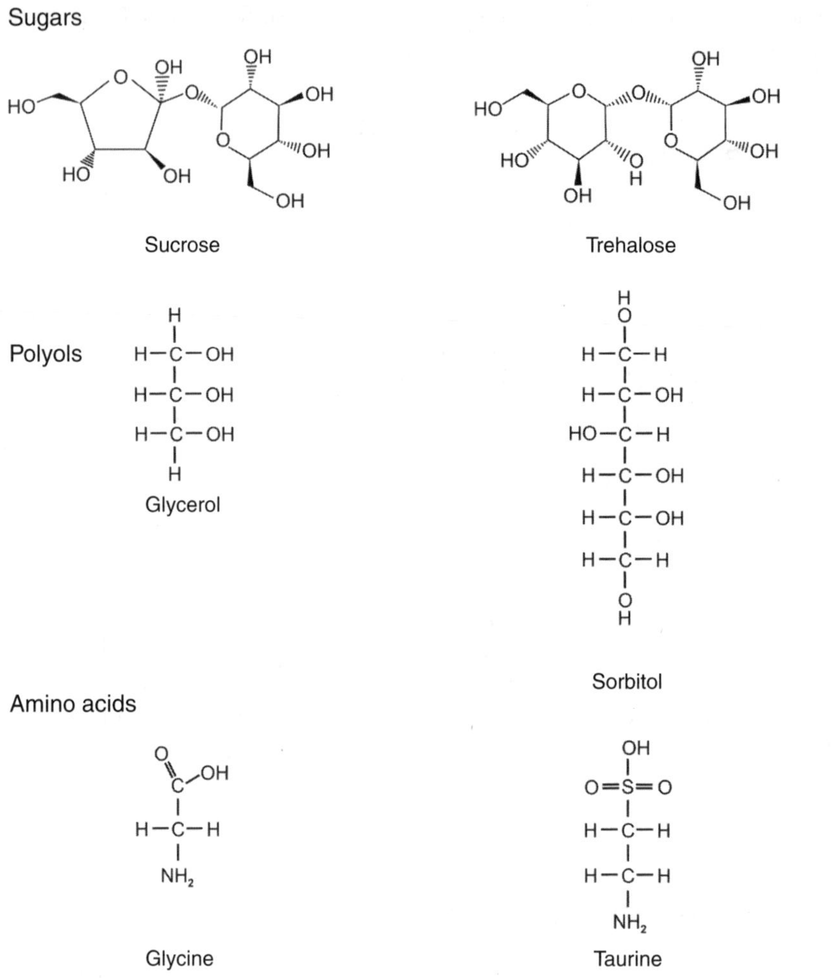

Figure 12.3 Osmolytes. Trehalose, a disaccharide of glucose, is a very effective osmolyte against huntingtin and α-synuclein aggregation, whereas sucrose, a dissacharide comprised of glucose and fructose, is not.

both? Considerable attention has been devoted to this problem, and the consensus is that a favorable change in the entropy of the solvent ($\Delta S_{U\rightarrow F[solvent]}>0$) drives folding. Most proteins contain amino acids with hydrophobic side chains such as leucine, isoleucine, valine, and phenylalanine. In aqueous solution, an unfolded protein's hydrophobic side chains, which have no polar groups to hydrogen bond with water molecules, stick out into and disrupt the hydrogen bonding network of water. To compensate for the loss of hydrogen bonding, water molecules are thought to form ordered shells (like a clathrate) around the hydrophobic sidechains (Donald Voet, 2004), which results in a decrease in solvent entropy. When the protein chain folds and the hydrophobic side chains are buried in the interior of the protein, water molecules no longer must form ordered shells around hydrophobic side chains, and this large, favorable increase in solvent entropy is the thermodynamic driving force for folding. This process—whereby a nonpolar molecule minimizes its contact with water—is often referred to as the *hydrophobic effect*. An increase in solvent entropy underlies the hydrophobic effect.

Given that osmolytes do not bind to proteins, how do osmolytes, which have been selected for through evolution, stabilize proteins from unfolding? Bolen and Baskakov (2001) have proposed that osmolytes stabilize proteins from denaturation because of an exclusion phenomenon, that is, an osmolyte has an unfavorable interaction with the surface of a native protein and an even less favorable interaction with the peptide backbone of the denatured/unfolded protein. The inability of an osmolyte to interact with the surface of a native protein in aqueous solution increases the free energy of the native state compared to the free energy of the native-state-absent osmolyte. Since in the unfolded state, there is far more surface area than in the folded state, and since osmolytes fail to interact with the peptide backbone, the free energy of the unfolded state of a protein in an aqueous solution of osmolyte is much larger than the free energy of the unfolded-state-absent osmolyte. In short, osmolytes stabilize the native state of proteins because of the dramatic destabilization of the unfolded state (Arakawa *et al.*, 2006). Bolen refers to this osmolyte effect on protein stability as the "osmophobic effect," and postulates that the osmophobic effect is an additional force to be considered that drives protein folding. Taking the hydrophobic effect and the osmophobic effect together, Bolen and Baskakov pointed out that for both of these effects "an unfavorable interaction between a structural component of the protein and a component of solution is responsible for the force itself."

Plants and microorganisms subject to water stress have evolved to synthesize unique small molecule osmolytes to stabilize their proteins. Neurons are not subject to water stress and consequently do not synthesize osmolytes. Neurons and other cells are subject to proteotoxic stress, however, which is defined here as the formation of toxic protein aggregates that kill neurons. Examples of proteotoxic amyloidogenic aggregates are Aβ peptides (AD), α-synuclein (PD and various synucleinopathies), polyQ (HD and various ataxias), tau (tauopathies), and prions. Because in most cases proteotoxic aggregates form in neurons late in life, typically after reproduction, there is no selective pressure for humans to have neurons that synthesize novel molecules that protect cells from toxic protein aggregates. It is even possible that those genes in humans that ensure robust reproductive fitness may promote aging and consequent formation of proteotoxic aggregates. In the following sections, we discuss intriguing examples of how osmolytes are being used to protect different types of cells and organisms from toxic protein aggregates. It will become evident that the mechanism(s) by which osmolytes protect cells is a subject of debate, and that the osmophobic effect may not be the primary protective effect. Although human neurons do not express osmolytes, osmolytes may be a unique therapy for some human neurodegenerative diseases. This is an example of how studying microorganisms and plants can bring insight into the human condition.

12.3.1.2 Trehalose Protects RNase A from Thermal Denaturation In Vitro. Trehalose, which is a disaccharide of glucose, is an osmolyte that is synthesized by many microorganisms in response to stress (Figure 12.3). Trehalose protects cells from desiccation, stabilizes proteins during lyophilization (Carpenter *et al.*, 1993), and protects against thermal denaturation in aqueous solution (Carninci *et al.*, 1998). It is useful to briefly describe a typical thermal denaturation experiment of a protein. The typical experiment commences with the protein of interest in a test tube at a relatively low temperature, say, 20°C, where 100% of the protein molecules are folded—one then monitors a readout such as an absorbance signal or a circular dichroism signal as the temperature is slowly increased. A plot of the signal versus temperature is the protein's melting curve, and the temperature at which 50% of the protein is folded and 50% denatured is the midpoint temperature (T_m). Because protein folding–unfolding is cooperative, a typical melting curve exhibits a sharp transition between the folded state and the unfolded state. An alternative experiment is where at a constant temperature, a chemical denaturant like guanidinium hydrochloride or urea is gradually added to the protein solution, and unfolding is monitored either by changes in the absorbance or circular dichroism signal. In this type of melting experiment, the denaturant concentration at which 50% of the protein is folded and 50% denatured is the midpoint concentration ($[\text{urea}]_m$). Melting curves are often used to evaluate the effect of additives such as osmolytes on protein stability, and such an experiment revealed how remarkably good a stabilizer trehalose is.

The analysis of the effect of trehalose (1–2 M) on the thermal denaturation of five different purified proteins (RNase A, lysozyme, cytochrome c, α-chymotrypsinogen, and trypsin inhibitor) in aqueous solution was conducted by Kaushik and Bhat (2003). Trehalose at 2 M increased the transition temperature of RNase A by 18.2°C (T_m : 38.2°C → 56.5°C), which is a very large increase in thermal stability. Trehalose increased the T_m of RNase A better than any other sugar or polyol ever reported. Similarly large increases in T_m induced by trehalose were also found for the other proteins. Although the effect is quite large, one must consider the high concentration of trehalose needed to achieve these effects. On the other hand, the concentration of trehalose needed to affect protein stability may depend on the sequence of the protein of interest, as shown below.

12.3.1.3 Trehalose and Huntington's Disease. In 2004, Tanaka *et al.* demonstrated the potential of trehalose for treating HD (Tanaka *et al.*, 2004). HD is a fatal polyQ (Q = the amino acid glutamine) disorder that causes neurodegeneration in humans. Few treatments exist for this disease. Like other polyQ diseases, age of onset and severity are inversely related to the length of the polyQ segment attached to the N-terminus of the huntingtin protein. The polyQ sequence promotes the aggregation of huntingtin, and the resultant proteotoxic aggregates kill cells, although the mechanism of cell death remains unknown. Scientists worldwide are trying to find therapeutics that can disrupt proteotoxic polyQ–huntingtin aggregates. Tanaka *et al.* developed a unique screen for compounds that disrupt polyQ aggregates.

The protein myoglobin, which does not aggregate at 37°C, was engineered to aggregate via the addition of a 35 glutamine repeat to its C-terminus (Mb-Q35) (Tanaka *et al.*, 2004). Mb-Q35 aggregates readily form at 37°C, and the aggregation is easily monitored by following absorbance at 550 nm. About 200 compounds were tested for their ability to inhibit the formation of Hb-Q35 aggregates, and the best inhibitors were trehalose and an *N*-acetylgalactosamine tetramer (GalNAcGT). Tanaka *et al.* also showed that disaccharides in general have a propensity to decrease polyglutamine-induced aggregation, and that trehalose was the most active. To deduce the effect of trehalose on the stability of wild-type Mb versus Mb-Q35, the midpoint of the unfolding transition was determined. Instead of using heat to disrupt protein structure, the denaturant guanidine hydrochloride was used. Although trehalose at 0.1 mM had no effect on the stability of wild-type myoglobin, this low concentration significantly increased the stability of Mb-Q35 to resist denaturation by guanidinium hydrochloride. It was concluded that trehalose inhibits the aggregation of Mb-Q35 by "interacting with the expanded polyglutamine region."

One might wonder how such a low concentration of trehalose is sufficient to affect the stability of Mb-Q35 whereas 2 M trehalose is needed to affect the stability of RNAase A (see above). One possibility is that the size of the protein is important, and that larger proteins simply need higher concentrations of trehalose to achieve an effect. In the case of Mb-Q35, the aggregation domain consists of only 35 glutamine residues, and thus it is likely that a much lower concentration of trehalose is needed to disrupt its aggregation.

Trehalose was subsequently tested in a transgenic mouse model of HD (Tanaka *et al.*, 2004). A variety of disease parameters, including brain atrophy, formation of intranuclear aggregates, motor function, and survival were evaluated for these transgenic HD mice treated with or without trehalose. Overall, trehalose reduced brain atrophy, decreased the number of intranuclear inclusions, improved motor function, and increased mean survival by 10%. This groundbreaking study demonstrated that compared to untreated animals, trehalose improved a variety of symptoms associated with HD and even increased survival.

12.3.1.4 Trehalose Induces Autophagy Via an mTOR-Independent Path. An unexpected twist in the story of trehalose came about shortly following the publication of the protective effect of trehalose in the mouse model of HD. It was reported that although trehalose can inhibit toxic aggregation via its function as an osmolyte, the *in vivo* function of trehalose may be somewhat more complicated. Indeed, trehalose appears to be a novel inducer of autophagy, and the protective function of trehalose in cell and animal models of proteinopathies may thus be due to its ability to induce autophagy rather than its ability to stabilize proteins. Note that several studies have reported that mutant α-synucleins, huntingtin fragments, and other polyQ-containing proteins are taken up by autophagy

(Berger *et al.*, 2006; Ravikumar *et al.*, 2002, 2004; Sarkar *et al.*, 2005; Webb *et al.*, 2003).

Autophagy is a highly regulated cellular process that serves to rid cells of proteins, protein aggregates, and even whole organelles by the transfer of these proteins and organelles to the vacuole (in yeast) and the lysosome (in mammalian cells) (He and Klionsky, 2009; Klionsky, 2007; Klionsky *et al.*, 2007). There are three forms of autophagy, and these are macroautophagy, microautophagy, and chaperone-mediated autophagy (see Chapter 3). The autophagy discussed here is macroautophagy, which herein is referred to as *autophagy*. The process of autophagy occurs when proteins, protein aggregates, or whole organelles are encapsulated in a lipid bilayer called an *autophagosome*. The autophagosome then transits to and merges with the vacuole or lysosome, where the autophagosome and its contents are proteolytically degraded. This is one way for the cells to get rid of damaged proteins as well as a way to recycle amino acids. Autophagy is negatively regulated by the master cellular nutrient sensor, the protein kinase mTOR (mammalian target of rapamycin). One way to induce autophagy in mammalian cells is to use the drug rapamycin. Rapamycin inhibits mTOR kinase activity, which in turn triggers autophagy. Because rapamycin triggers the clearance of potentially toxic proteotoxic aggregates, it is considered to be a powerful candidate drug for various neurodegenerative disease, but unfortunately rapamycin has many unwanted side effects, such as downregulating protein synthesis. A global search is underway for drugs that induce autophagy but are less toxic than rapamycin. One idea is that drugs that induce autophagy via an mTOR-independent pathway might not affect protein synthesis, and thus they would be better tolerated on a long-term basis by humans.

The seminal work showing that trehalose can induce mTOR-independent autophagy was conducted using COS-7 cells, which are non-neuronal, and SK-N-SH cells, which are a neuronal precursor (Sarkar *et al.*, 2007a). The polyQ protein, enhanced green fluorescent protein (EGFP)-tagged HDQ74, was expressed in each of these cells, and aggregates could be readily detected by western blotting. Strikingly, trehalose decreased the amounts of both aggregated and soluble EGFP-HDQ74 and increased cell viability. Other sugars such as sucrose, raffinose, and sorbitol did not affect aggregates or cell viability, and thus the effect was specific to trehalose. Trehalose also enhanced the clearance of A30P and A53T α-synuclein, which are mutant forms of the protein α-synuclein that are thought to trigger early–onset PD. Thus, trehalose may also have therapeutic potential for PD.

To deduce the mechanism by which the polyQ–huntingtin fragment is cleared from cells, the clearance of the huntingtin fragment was investigated using inhibitors of autophagy (3-methyl adenine, 3-MA) and the proteasome (lactacystin). Note that each of these drugs increased the aggregation and toxicity of EGFP-HDQ74 in COS-7 cells, indicating that this mutant protein is cleared from cells by both routes. By using each of these drugs in combination with trehalose, it was revealed that trehalose specifically stimulates autophagy. Additional experiments revealed that trehalose does not inhibit the kinase activity of mTOR; thus, trehalose triggers autophagy via an mTOR-independent pathway. The authors of this study suggested that the beneficial effects of trehalose in the mouse model of HD (Tanaka *et al.*, 2004) were probably due to clearance of the aggregates via autophagy. This finding was a major advance in the field and it stimulated work to uncover by various screening processes novel inducers of mTOR-independent autophagy (Floto *et al.*, 2007; Sarkar *et al.*, 2007b). This could be the emergence of a whole new field, and it happened because of work focused on trehalose, a "simple" disaccharide osmolyte synthesized by yeast and other microbes!

Trehalose has remarkable effects on proteins: it stabilizes proteins via the osmolyte effect and it triggers mTOR-independent autophagy, which results in the clearance of toxic protein aggregates. Trehalose and other novel inducers of mTOR-independent autophagy may be used in the near future to clear aggregates of polyQ-mutants in HD and α-synuclein and its mutants in PD. Trehalose is also being used to thwart the aggregation of amyloidogenic peptides involved in AD, as described below.

12.3.1.5 Trehalose and Alzheimer's Disease — An Emerging Area. The amyloid beta protein (Aβ) is associated with AD. Certain cleavage products of the Aβ protein, specifically Aβ(1-40) and Aβ(1-42), are particularly aggregation prone, and such aggregates can disrupt membranes and kill cells. The use of trehalose is an emerging area of AD research. One study analyzed the ability of trehalose to inhibit Aβ40 and Aβ42 aggregation in the test tube, and found that a concentration of trehalose as low as 50 mM completely inhibited the aggregation of Aβ40 and partially inhibited (50%) the aggregation of the more hydrophobic peptide Aβ42 (Liu *et al.*, 2005). Being more hydrophobic and more prone to aggregation, Aβ42 is considered to be more toxic than Aβ40. The same researchers also reported that 50 mM trehalose significantly dissolved the preformed aggregates of Aβ40 but had much less of an effect on the preformed aggregates of Aβ42. In a cytotoxic assay, the preformed aggregates of Aβ40 were toxic to human neuroblastoma SH-SY5Y cells, whereas Aβ40 coincubated with 50 mM trehalose was not toxic. On the other hand, Aβ42 was toxic to SH-SY5Y cells whether it had been coincubated with trehalose or not. This study was intriguing and opened the door to other studies (De Bona *et al.*, 2009; Liu *et al.*, 2009; Qi *et al.*, 2009; Reddy *et al.*, 2009; Sarkar and Rubinsztein, 2008). This is such a new area that it is impossible to know whether trehalose will have value as a therapeutic agent for AD. One thing is for sure, studying compounds such as trehalose stimulates the emergence of new ideas and this leads to the design, synthesis, and testing of new compounds. This is how research goes forward and how treatments for diseases are found.

12.3.2 Lipid Storage Diseases and Pharmacological Chaperones

The following sections describe the biology of lipid storage diseases and various treatments, focusing on the molecular basis by which pharmacological chaperones alleviate the progression of Gaucher's disease (GD) and Fabry's disease.

Pharmacological chaperones have a huge potential to be used therapeutically for various diseases, including some neurodegenerative diseases that are associated with proteins that transit through the endoplasmic reticulum (ER), and in fact some of these molecules are in clinical trials (for reviews see Cohen and Kelly, 2003; Morello *et al.*, 2000; Welch and Brown, 1996; Yu *et al.*, 2007).

The ER is a cellular compartment that is composed of a lipid bilayer, which separates the inside of the compartment (lumen) from the cytosol. Receptors, proteins destined for export, and proteins that will reside in the plasma or lysosomal membranes transit through the ER and into the Golgi apparatus, where specific proteins are sorted into specific vesicles, which are transported to specific sites or exocytosed. The ER possesses a sophisticated protein quality control system consisting of membrane-bound protein sensors that detect misfolded and unfolded proteins, and improperly folded proteins are caught, ejected through the ER membrane into the cytosol, and degraded by the proteasome. This response is called the unfolded protein response (UPR) (Malhotra and Kaufman, 2007; Ron and Walter, 2007), and the subsequent proteasomal degradation is called endoplasmic reticulum associated degradation (ERAD) (Vembar and Brodsky, 2008).

Lipid storage diseases are a group of inherited metabolic diseases, and a subset of these storage diseases involves mutations in various lysosomal acid hydralase enzymes that degrade glycosphingolipids, and hence these diseases are referred to as *sphingolipidoses* (Futerman and van Meer, 2004; Kolter and Sandhoff, 2006). The various lysosomal hydralases are proteins that use the ER to transit to their destination, the lysosome. A mutation in a lysosomal hydrolase results in the slow accumulation of glycosphingolipids in lysosomes, and this sometimes results in neurodegeneration. Mutations in lysosomal hydrolases typically occur in the active site or distal to the active site (Grace *et al.*, 1994; Premkumar *et al.*, 2005; Sawkar *et al.*, 2002). Mutations in the active site of a lysosomal hydralase cause a decrease in the activity of the enzyme, but typically the mutant enzyme traffics properly from the ER to the lysosome. Mutations distal to the active site that fail to affect enzymatic activity are problematic because often such mutations cause misfolding of the mutant at pH 7.0 (ER) but not at pH 4.5 (lysosome). Such misfolded mutant hydralases are retro-translocated out of the ER and degraded by ERAD. It is often the case that the mutation that prevents the enzyme from transiting through the ER (~pH 7) has no effect on the activity of the enzyme in its natural compartment, the lysosome (~pH 4.5). As will be shown, the beauty of pharmacological chaperones is that they rescue misfolded mutant enzymes in the ER, which in turn enables them to transit to their proper destination, the lysosome, where they function properly. Other therapies are also discussed.

12.3.2.1 Gaucher's Disease. GD, which is the most common lipid storage disease, is an inherited metabolic disease involving a deficiency in the lysosomal hydralase enzyme called acid-β-glucosidase (glucosylceramidase or glucocerebrosidase). Acid-β-glucosidase is coded for by the gene *GBA1,* and there are approximately 300 mutations reported in this gene (Hruska *et al.*, 2008). GD is an autosomal recessive disease, meaning that individuals who are homozygous for mutations in *GBA1* have the disease, whereas individuals who are heterozygous for mutations in *GBA1* are usually asymptomatic; but interestingly, such heterozygotes can have up to a 20-fold higher incidence of PD (Bras *et al.*, 2008; Bultron *et al.*; Sidransky *et al.*, 2009; Varkonyi *et al.*, 2003). The reason for this higher incidence of PD is not known. GD patients usually have oversized macrophages in the reticuloendothelium system, anemia, enlargement of different organs, bone damage, and sometimes CNS involvement. Without treatment, there are high rates of morbidity and mortality. Type 1 GD is the most common form, accounting for 95% of all cases, and its symptoms include anemia, hepatomegaly, splenomegaly, thrombocytopenia, and bone disease. Type 2 is lethal and presents in infants and affects the brain, liver, lungs, and spleen. Type 3 has neurological involvement and includes convulsions, ataxia, and dementia. Understanding a disease like this requires a combination of genetics, cell biology, biochemistry, and structural studies of the crystallized enzyme and its various mutants.

The lysosome is an acidic cellular compartment that serves to enzymatically degrade and recycle lipids, including glycosphingolipids and proteins. The glycosphingolipid glucosylceramide is synthesized in the ER and transits through the ER to its destination, the plasma membrane, where it is an essential structural component of the membrane along with cholesterol and phospholipids. The synthesis of various lipids and their deposition in various cellular membranes is balanced by their breakdown or catabolism in lysosomes. Plasma membrane lipids recycle to the lysosome for degradation via the endocytic pathway. Once the endocytic vesicles that contain glucosylceramide enter lysosomes, acid-β-glucosidase cleaves glucose from glucosylceramide to yield glucose and ceramide, and these two degradation products can escape from the lysosome to be recycled, whereas glucosylceramide cannot. A deficiency in acid-β-glucosidase activity or concentration results in a failure of this cleavage reaction and causes the progressive accumulation of glucosylceramide in lysosomes, and such an accumulation impairs the function of the lysosome and ultimately kills cells (Figure 12.4). Note that deficiencies in other acid hydralases cause inherited sphingolipidoses such as Fabry's disease, Farber's disease, GM2-gangliosidoses, Krabbe disease, Niemann–Pick disease, Sandhoff disease, Tay–Sachs disease, and others (Kolter and Sandhoff, 2006). There are several different possible therapies for GD, and three of them are discussed below.

12.3.2.2 Enzyme Replacement Therapy (ERT). Enzyme replacement therapy (ERT) is the current standard treatment for patients with GD types 1 and 3 (Elstein and Zimran, 2009; Martins *et al.*, 2009). ERT entails the intravenous infusion of purified human recombinant acid-β-glucosidase (Cerezyme® or imiglucerase, Genzyme Inc.) into patients. In general, the drug is well tolerated and effective at reducing and maintaining liver and spleen volumes. Approximately 4000 patients worldwide have been treated with this drug. Because Cerezyme fails to cross the blood–brain barrier, neurological symptoms in type 3 patients are unresolved. Given that GD patients must take Cerezyme lifelong on a biweekly basis, and given the cost of this treatment at $200,000

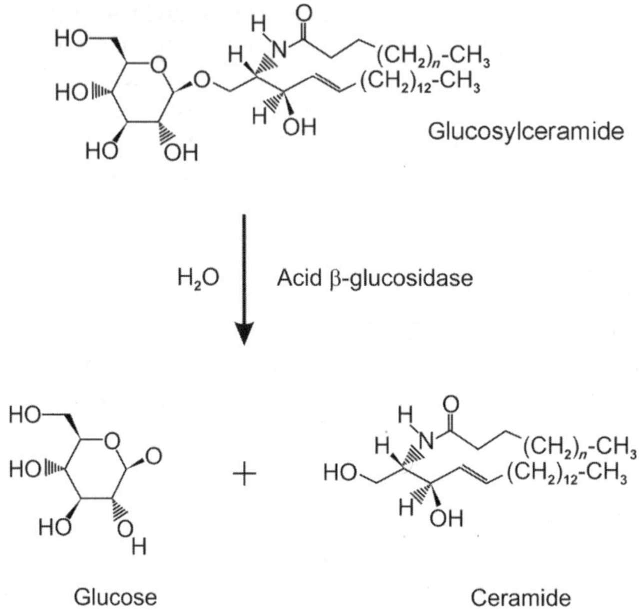

Figure 12.4 The biochemistry of Gaucher's disease. Glucosylceramide is cleaved into glucose and ceramide within the lysosome compartment by the enzyme, acid β-glucosidase. Mutations in this enzyme cause an accumulation of glucosylceramide in lysosomes, which in turn causes multiple cell and organ defects. The two cleavage products, glucose and ceramide, are recycled.

per year, this therapy is not a viable option for many GD patients. A worldwide disruption of the production of Cerezyme occurred in June 2009 because of viral contamination of the cells from which Cerezyme is synthesized, that is, Chinese hamster ovary (CHO) cells (Hollak *et al.*, 2010; Pollack, 2010). Genzyme stopped production and only about 20% of the expected production for 2009 became available, and production for 2010 was halted. This disruption in supply of a drug that is so vital to many people underscores the importance of identifying small molecules that might replace ERT. Scientists are working on finding less costly treatments for this disease, and pharmacological chaperones are an interesting potential treatment.

12.3.2.3 Substrate Reduction Therapy (SRT). The root cause of GD is an imbalance between the rates of glucosylceramide synthesis and degradation. Because the rate of glucosylceramide degradation is very low in GD patients, a logical therapeutic approach is to slow down the rate of glucosylceramide synthesis. This can be accomplished using a competitive inhibitor of glucosylceramide synthase (GlcT-1; UDP-glucose:*N*-acylsphingosine D-glucosyltransferase), which is the enzyme that synthesizes glucosylceramide. This approach is referred to as substrate reduction therapy (SRT) (Aerts *et al.*, 2006; Wennekes *et al.*, 2009). SRT is based on a landmark paper published in 1994, in which

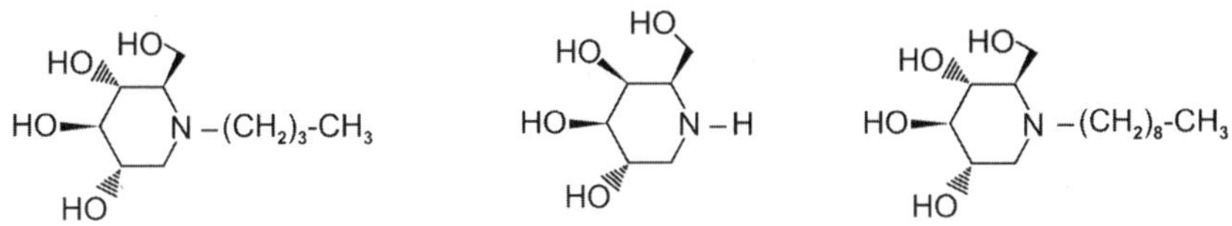

NB-DNJ **inhibits glucosylceramide synthase** **Gaucher**

DGJ **inhibits α-gal A** **Fabry**

NN-DNJ **inhibits acid-β-glucosidase** **Gaucher**

Figure 12.5 Drugs for Gaucher's and Fabry's diseases. NB-DNJ (*N*-butyldeoxynojirimycin, Zavesca) inhibits the enzyme glucosylceramide synthase. This drug decreases the level of glucosylceramide in cells and is used for GD. DGJ (1-deoxy-galactonojirimycin) is a pharmacological chaperone that binds to α-galactosidase A. The drug is used for Fabry's disease. NN-DNJ (*N*-(*n*-nonyl)deoxynojirimycin) is a pharmacological chaperone that binds to acid β-glucosidase. This drug has potential for GD.

Platt and colleagues showed that the imino sugar *N*-butyldeoxynojirimycin (NB-DNJ) (Figure 12.5) potently inhibits glucosylceramide synthase (Platt *et al.*, 1994). When this discovery was made, NB-DNJ had already been under evaluation in the clinic as a possible therapy for acquired immunodeficiency syndrome (AIDS). NB-DNJ (miglustat, Zavesca) is currently approved for use in type 1 Gaucher patients in the United States and other countries, but in general, it is only used for patients for whom ERT is not an option (Aerts *et al.*, 2006; Hollak *et al.*, 2010). Whether this drug crosses the blood–brain barrier is not known.

12.3.2.4 Pharmacological Chaperones. A pharmacological chaperone is a small molecule "whose function is to assist a protein to fold properly and enter its normal processing pathway smoothly" (Fan *et al.*, 1999). The finding that a small molecule can act like a protein chaperone and assist the folding of a protein was a landmark discovery, one that has far-reaching potential for novel drugs that can thwart neurodegenerative diseases. It is useful to briefly describe the study that set forth this concept, and to do this, we will briefly discuss Fabry's disease, which is an inherited metabolic disease involving a deficiency in lysosomal α-galactosidase A (α-Gal A). The use of pharmacological chaperones can of course be extended to similar lipidoses such as GD, as will be shown.

Fan and colleagues reported in 1999 (Fan *et al.*, 1999) that the competitive inhibitor of α-Gal A called 1-deoxy-galactonojirimycin (DGJ) (Figure 12.5), which is a ceramide analog, had properties that were not at all attributable to it being a competitive inhibitor. DGJ was tested on lymphoblasts obtained from Fabry patients that contained either the R301Q or Q279E mutations, which are mutations associated with a late-onset form of Fabry that affects the heart.

Incubating the lymphoblasts with 20 μM DGJ resulted in an 8- to 10-fold increase in α-Gal A activity (up to about 45% of normal activity) that persisted for four days after withdrawal of the drug; whereas, a DGJ concentration greater than 20 μM prevented the increase in activity. Additional *in vivo* experiments revealed that the R301Q mutant localized to the lysosome when cells were treated with DGJ, whereas the mutant enzyme localized to the Golgi without DGJ. *In vitro* experiments showed that DGJ inhibited the unfolding of purified R301Q at pH 7, the pH of the ER compartment.

It was proposed that DGJ, when used at subinhibitory concentrations, promotes the folding and trafficking of mutant α-Gal A molecules through the ER. Specifically, DGJ was proposed to bind to the active site of mutant α-Gal A in a pH-dependent manner: DGJ binds to the mutant much tighter at pH 7.0 (ER pH) than at pH 4.5 (lysosomal pH). Note that evidence for a pH dependence for the binding of a nojirimycin derivative to another glycosidase supports this idea (Dale *et al.*, 1985). Tight binding of DGJ to the active site of a mutant α-Gal A, say, for example, R301Q, is thought to induce the folding of the misfolded domain or domains distal to the active site, and the properly folded mutant enzyme–DGJ complex then successfully transits through the ER to the lysosome. On route to the lysosome, DGJ dissociates, and once in the lysosome, the mutant enzyme maintains its structure (the mutations causes misfolding at pH 7.0 not 4.5) and displays normal enzymatic activity. The gist is that a pharmacological chaperone is a competitive inhibitor that is used at low concentrations, and such low concentrations promote the folding and trafficking of some mutant enzymes through the ER.

Some GD patients, especially those with mutations in acid-β-glucosidase that affect folding at pH 7.0, may also benefit from pharmacological chaperones. Specifically, a derivative of nojirimycin was recently shown to act as a pharmacological chaperone for a GD-associated mutant (N370S) of acid-β-glucosidase (Sawkar *et al.*, 2002). Sawkar and colleagues demonstrated that incubating patient-derived fibroblasts for nine days with a subinhibitory concentration (10 μM) of *N*-(*n*-nonyl)deoxynojirimycin (NN-DNJ) (Figure 12.5) led to a twofold increase in N370S acid-β-glucosidase activity, which persisted for several days after the withdrawal of the drug. The N370S mutant is the most common mutation associated with GD. *In vitro* experiments with purified wild-type acid-β-glucosidase revealed that NN-DNJ protected the enzyme in a dose-dependent manner from heat denaturation. The authors of this study proposed that NN-DNJ functions as a pharmacological chaperone and rescues misfolding of the N370S mutant in the ER, which in turn leads to the proper trafficking of the mutant to the lysosome, where it displays normal enzymatic activity. Although the N370S mutant does not typically cause neurological complications, the use of chemical chaperones to promote the proper folding and trafficking of mutant enzymes through the ER in neurons should also work, as long as the pharmacological chaperone can cross the blood–brain barrier.

12.4 SUMMARY/CONCLUSIONS

The chapters in this book have examined the various roles that protein chaperones play in protecting neurons from toxic protein aggregates that trigger cell death. A potential problem with protein chaperones (or any other protective proteins) as a therapy is that they are costly to prepare and often fail to cross the blood–brain barrier. Hydrophobic, small molecules are much more likely to cross the blood-brain barrier than high molecular mass proteins. If small molecule activators of HSF1 can be engineered to cross the blood-brain barrier, they would activate the heat shock response in neurons and thereby increase the intracellular levels of endogenous protein chaperones. A recent study has identified small organic molecules as activators of hHSF1. One of these activators, termed *HSF1A*, elevates chaperone expression and suppresses protein misfolding and cell death in poly-Q-expressing neuronal precursor cells, and protects against cytotoxicity in a fly model of neurodegeneration (Neef *et al.*, 2010).

Alternatively, gene therapy can introduce a protective gene of interest into a patient, and genes can be delivered directly into the brain. As viral vectors improve, gene therapy may be used to treat more neurodegenerative diseases. But clearly, there are high hurdles. Pharmacological chaperones, similar to protein chaperones, bind to a target protein and promote its folding, but because they are hydrophobic in nature and are of low molecular mass, pharmacological chaperones have a far greater potential than protein chaperones to cross the blood–brain barrier. We predict that in the near future pharmacological chaperones, or perhaps even osmolytes, will be used as treatments for not only lipid storage diseases but also for diseases like Alzheimer's, Huntington's, and Parkinson's.

REFERENCES

Abravaya K, Myers MP, Murphy SP, Morimoto RI. The human heat shock protein hsp70 interacts with HSF, the transcription factor that regulates heat shock gene expression. *Genes Dev* 1992;**6**:1153–1164.

Aerts JM, Hollak CE, Boot RG, Groener JE, Maas M. Substrate reduction therapy of glycosphingolipid storage disorders. *J Inherit Metab Dis* 2006;**29**:449–456.

Ali A, Bharadwaj S, O'Carroll R, Ovsenek N. Hsp90 interacts with and regulates the activity of heat shock factor 1 in *Xenopus* oocytes. *Mol Cell Biol* 1998;**18**:4949–4960.

Amin J, Fernandez M, Ananthan J, Lis JT, Voellmy R. Cooperative binding of heat shock transcription factor to the Hsp70 promoter *in vivo* and *in vitro*. *J Biol Chem* 1994;**269**:4804–4811.

Arakawa T, Ejima D, Kita Y, Tsumoto K. Small molecule pharmacological chaperones: from thermodynamic stabilization to pharmaceutical drugs. *Biochim Biophys Acta* 2006;**1764**:1677–1687.

Auluck PK, Bonini NM. Pharmacological prevention of Parkinson disease in *Drosophila*. *Nat Med* 2002;**8**:1185–1186.

Auluck PK, Meulener MC, Bonini NM. Mechanisms of suppression of {alpha}-synuclein neurotoxicity by geldanamycin in *Drosophila*. *J Biol Chem* 2005;**280**:2873–2878.

Berger Z, Ravikumar B, Menzies FM, Oroz LG, Underwood BR, Pangalos MN, Schmitt I, Wullner U, Evert BO, O'Kane CJ, *et al.* Rapamycin alleviates toxicity of different aggregate-prone proteins. *Hum Mol Genet* 2006;**15**:433–442.

Bharadwaj S, Ali A, Ovsenek N. Multiple components of the Hsp90 chaperone complex function in regulation of heat shock factor 1 *in vivo*. *Mol Cell Biol* 1999;**19**:8033–8041.

Boeddrich A, Gaumer S, Haacke A, Tzvetkov N, Albrecht M, Evert BO, Muller EC, Lurz R, Breuer P, Schugardt N, *et al.* An arginine/lysine-rich motif is crucial for VCP/p97-mediated modulation of ataxin-3 fibrillogenesis. *EMBO J* 2006;**25**:1547–1558.

Boehm AK, Saunders A, Werner J, Lis JT. Transcription factor and polymerase recruitment, modification, and movement on dhsp70 *in vivo* in the minutes following heat shock. *Mol Cell Biol* 2003;**23**:7628–7637.

Bolen DW, Baskakov IV. The osmophobic effect: natural selection of a thermodynamic force in protein folding. *J Mol Biol* 2001;**310**:955–963.

Braak H, Del Tredici K, Sandmann-Kiel D, Rub U, Schultz C. Nerve cells expressing heat-shock proteins in Parkinson's disease. *Acta Neuropathol (Berl)* 2001;**102**:449–454.

Bras J, Singleton A, Cookson MR, Hardy J. Emerging pathways in genetic Parkinson's disease: potential role of ceramide metabolism in Lewy body disease. *FEBS J* 2008;**275**:5767–5773.

Braunstein M, Rose AB, Holmes SG, Allis CD, Broach JR. Transcriptional silencing in yeast is associated with reduced nucleosome acetylation. *Genes Dev* 1993;**7**:592–604.

Buchner J. Hsp90 & Co.—a holding for folding. *Trends Biochem Sci* 1999;**24**:136–141.

Bukau B, Horwich AL. The Hsp70 and Hsp60 chaperone machines. *Cell* 1998;**92**:351–366.

Bultron G, Kacena K, Pearson D, Boxer M, Yang R, Sathe S, Pastores G, Mistry PK. The risk of Parkinson's disease in type 1 Gaucher disease. *J Inherit Metab Dis.* 2010;**33**:167–173.

Burghes AH, Beattie CE. Spinal muscular atrophy: why do low levels of survival motor neuron protein make motor neurons sick? *Nat Rev Neurosci* 2009;**10**:597–609.

Carninci P, Nishiyama Y, Westover A, Itoh M, Nagaoka S, Sasaki N, Okazaki Y, Muramatsu M, Hayashizaki Y. Thermostabilization and thermoactivation of thermolabile enzymes by trehalose and its application for the synthesis of full length cDNA. *Proc Natl Acad Sci U S A* 1998;**95**:520–524.

Carpenter JF, Prestrelski SJ, Arakawa T. Separation of freezing- and drying-induced denaturation of lyophilized proteins using stress-specific stabilization. I. Enzyme activity and calorimetric studies. *Arch Biochem Biophys* 1993;**303**:456–464.

Carrettiero DC, Hernandez I, Neveu P, Papagiannakopoulos T, Kosik KS. The cochaperone BAG2 sweeps paired helical filament-insoluble tau from the microtubule. *J Neurosci* 2009;**29**:2151–2161.

Chen T, Parker CS. Dynamic association of transcriptional activation domains and regulatory regions in *Saccharomyces cerevisiae* heat shock factor. *Proc Natl Acad Sci U S A* 2002;**99**:1200–1205.

Christine CW, Starr PA, Larson PS, Eberling JL, Jagust WJ, Hawkins RA, VanBrocklin HF, Wright JF, Bankiewicz KS, Aminoff MJ. Safety and tolerability of putaminal AADC gene therapy for Parkinson disease. *Neurology* 2009;**73**:1662–1669.

Cohen FE, Kelly JW. Therapeutic approaches to protein-misfolding diseases. *Nature* 2003;**426**:905–909.

Crippa V, Sau D, Rusmini P, Boncoraglio A, Onesto E, Bolzoni E, Galbiati M, Fontana E, Marino M, Carra S, *et al.* The small heat shock protein B8 (HspB8) promotes autophagic removal of misfolded proteins involved in amyotrophic lateral sclerosis (ALS). *Hum Mol Genet* 2010;**19**:3440–3456.

Dai C, Whitesell L, Rogers AB, Lindquist S. Heat shock factor 1 is a powerful multifaceted modifier of carcinogenesis. *Cell* 2007;**130**:1005–1018.

Dale MP, Ensley HE, Kern K, Sastry KA, Byers LD. Reversible inhibitors of beta-glucosidase. *Biochemistry* 1985;**24**:3530–3539.

Dang W, Steffen KK, Perry R, Dorsey JA, Johnson FB, Shilatifard A, Kaeberlein M, Kennedy BK, Berger SL. Histone H4 lysine 16 acetylation regulates cellular lifespan. *Nature* 2009;**459**:802–807.

De Bona P, Giuffrida ML, Caraci F, Copani A, Pignataro B, Attanasio F, Cataldo S, Pappalardo G, Rizzarelli E. Design and synthesis of new trehalose-conjugated pentapeptides as inhibitors of Abeta(1-42) fibrillogenesis and toxicity. *J Pept Sci* 2009;**15**:220–228.

Donald Voet JGV. *Biochemistry*. Volume One, 3rd ed. New York: John Wiley & Sons, Inc.; 2004.

Dong Z, Wolfer DP, Lipp HP, Bueler H. Hsp70 gene transfer by adeno-associated virus inhibits MPTP-induced nigrostriatal degeneration in the mouse model of Parkinson disease. *Mol Ther* 2005;**11**:80–88.

Dou F, Netzer WJ, Tanemura K, Li F, Hartl FU, Takashima A, Gouras GK, Greengard P, Xu H. Chaperones increase association of tau protein with microtubules. *Proc Natl Acad Sci U S A* 2003;**100**:721–726.

Duina AA, Kalton HM, Gaber RF. Requirement for Hsp90 and a CyP-40-type cyclophilin in negative regulation of the heat shock response. *J Biol Chem* 1998;**273**:18974–18978.

Elstein D, Zimran A. Review of the safety and efficacy of imiglucerase treatment of Gaucher disease. *Biologics* 2009;**3**:407–417.

Erkina TY, Tschetter PA, Erkine AM. Different requirements of the SWI/SNF complex for robust nucleosome displacement at promoters of heat shock factor and Msn2- and Msn4-regulated heat shock genes. *Mol Cell Biol* 2008;**28**:1207–1217.

Erkina TY, Zou Y, Freeling S, Vorobyev VI, Erkine AM. Functional interplay between chromatin remodeling complexes RSC, SWI/SNF and ISWI in regulation of yeast heat shock genes. *Nucleic Acids Res* 2010;**38**:1441–1449.

Erkine AM, Adams CC, Diken T, Gross DS. Heat shock factor gains access to the yeast *HSC82* promoter independently of other sequence-specific factors and antagonizes nucleosomal repression of basal and induced transcription. *Mol Cell Biol* 1996;**16**:7004–7017.

Erkine AM, Magrogan SF, Sekinger EA, Gross DS. Cooperative binding of heat shock factor to the yeast *Hsp82* promoter *in vivo* and *in vitro*. *Mol Cell Biol* 1999;**19**:1627–1639.

Eroglu B, Moskophidis D, Mivechi NF. Loss of Hsp110 leads to age-dependent tau hyperphosphorylation and early accumulation of insoluble amyloid-beta. *Mol Cell Biol* 2010;**30**:4626–4643.

Fan JQ, Ishii S, Asano N, Suzuki Y. Accelerated transport and maturation of lysosomal alpha-galactosidase A in Fabry lymphoblasts by an enzyme inhibitor. *Nat Med* 1999;**5**:112–115.

Federoff HJ, Geschwind MD, Geller AI, Kessler JA. Expression of nerve growth factor *in vivo* from a defective herpes simplex virus 1 vector prevents effects of axotomy on sympathetic ganglia. *Proc Natl Acad Sci U S A* 1992;**89**:1636–1640.

Flick KE, Gonzalez L, Harrison CJ, Nelson HCM. Yeast heat shock transcription factor contains a flexible linker between the DNA-binding and trimerization domains. *J Biol Chem* 1994;**269**:12475–12481.

Floto RA, Sarkar S, Perlstein EO, Kampmann B, Schreiber SL, Rubinsztein DC. Small molecule enhancers of rapamycin-induced TOR inhibition promote autophagy, reduce toxicity in Huntington's disease models and enhance killing of mycobacteria by macrophages. *Autophagy* 2007;**3**:620–622.

Foust KD, Nurre E, Montgomery CL, Hernandez A, Chan CM, Kaspar BK. Intravascular AAV9 preferentially targets neonatal neurons and adult astrocytes. *Nat Biotechnol* 2009;**27**:59–65.

Freeman BC, Yamamoto KR. Disassembly of transcriptional regulatory complexes by molecular chaperones. *Science* 2002;**296**:2232–2235.

Fukuyo Y, Hunt CR, Horikoshi N. Geldanamycin and its anti-cancer activities. *Cancer Lett* 2010;**290**:24–35.

Futerman AH, van Meer G. The cell biology of lysosomal storage disorders. *Nat Rev Mol Cell Biol* 2004;**5**:554–565.

Giardina C, Lis JT. Dynamic protein-DNA architecture of a yeast heat shock promoter. *Mol Cell Biol* 1995;**15**:2737–2744.

Gitcho MA, Strider J, Carter D, Taylor-Reinwald L, Forman MS, Goate AM, Cairns NJ. VCP mutations causing frontotemporal lobar degeneration disrupt localization of TDP-43 and induce cell death. *J Biol Chem* 2009;**284**:12384–12398.

Grace ME, Newman KM, Scheinker V, Berg-Fussman A, Grabowski GA. Analysis of human acid beta-glucosidase by site-directed mutagenesis and heterologous expression. *J Biol Chem* 1994;**269**:2283–2291.

Gross DS, Adams CC, Lee S, Stentz B. A critical role for heat shock transcription factor in establishing a nucleosome-free region over the TATA-initiation site of the yeast *Hsp82* heat shock gene. *EMBO J* 1993;**12**:3931–3945.

Gross DS, English KE, Collins KW, Lee S. Genomic footprinting of the yeast *Hsp82* promoter reveals marked distortion of the DNA helix and constitutive occupancy of heat shock and TATA elements. *J Mol Biol* 1990;**216**:611–631.

Guarente L. Sir2 links chromatin silencing, metabolism, and aging. *Genes Dev* 2000;**14**:1021–1026.

Guisbert E, Yura T, Rhodius VA, Gross CA. Convergence of molecular, modeling, and systems approaches for an understanding of the *Escherichia coli* heat shock response. *Microbiol Mol Biol Rev* 2008;**72**:545–554.

Guo Y, Guettouche T, Fenna M, Boellmann F, Pratt WB, Toft DO, Smith DF, Voellmy R. Evidence for a mechanism of repression of heat shock factor 1 transcriptional activity by a multichaperone complex. *J Biol Chem* 2001;**276**:45791–45799.

Hahn JS, Hu Z, Thiele DJ, Iyer VR. Genome-wide analysis of the biology of stress responses through heat shock transcription factor. *Mol Cell Biol* 2004;**24**:5249–5256.

Hamamichi S, Rivas RN, Knight AL, Cao S, Caldwell KA, Caldwell GA. Hypothesis-based RNAi screening identifies neuroprotective genes in a Parkinson's disease model. *Proc Natl Acad Sci U S A* 2008;**105**:728–733.

Harrison CJ, Bohm AA, Nelson HCM. Crystal structure of the DNA binding domain of the heat shock transcription factor. *Science* 1994;**263**:224–227.

Hartl FU, Hayer-Hartl M. Molecular chaperones in the cytosol: from nascent chain to folded protein. *Science* 2002;**295**:1852–1858.

He C, Klionsky DJ. Regulation mechanisms and signaling pathways of autophagy. *Annu Rev Genet* 2009;**43**:67–93.

Hermonat PL, Muzyczka N. Use of adeno-associated virus as a mammalian DNA cloning vector: transduction of neomycin resistance into mammalian tissue culture cells. *Proc Natl Acad Sci U S A* 1984;**81**:6466–6470.

Herzog CD, Brown L, Gammon D, Kruegel B, Lin R, Wilson A, Bolton A, Printz M, Gasmi M, Bishop KM, *et al.* Expression, bioactivity, and safety 1 year after adeno-associated viral vector type 2-mediated delivery of neurturin to the monkey nigrostriatal system support cere-120 for Parkinson's disease. *Neurosurgery* 2009;**64**:602–612. (discussion 612–603).

Hollak CE, vom Dahl S, Aerts JM, Belmatoug N, Bembi B, Cohen Y, Collin-Histed T, Deegan P, van Dussen L, Giraldo P, *et al.* Force Majeure: therapeutic measures in response to restricted supply of imiglucerase (Cerezyme) for patients with Gaucher disease. *Blood Cells Mol Dis* 2010;**44**:41–47.

Hoppe GJ, Tanny JC, Rudner AD, Gerber SA, Danaie S, Gygi SP, Moazed D. Steps in assembly of silent chromatin in yeast: Sir3-independent binding of a Sir2/Sir4 complex to silencers and role for Sir2-dependent deacetylation. *Mol Cell Biol* 2002;**22**:4167–4180.

Hruska KS, LaMarca ME, Scott CR, Sidransky E. Gaucher disease: mutation and polymorphism spectrum in the glucocerebrosidase gene (GBA). *Hum Mutat* 2008;**29**:567–583.

Imai S-I, Armstrong CM, Kaeberlein M, Guarente L. Transcriptional silencing and longevity protein Sir2 is an NAD-dependent histone deacetylase. *Nature* 2000;**403**:795–800.

Ju JS, Fuentealba RA, Miller SE, Jackson E, Piwnica-Worms D, Baloh RH, Weihl CC. Valosin-containing protein (VCP) is required for autophagy and is disrupted in VCP disease. *J Cell Biol* 2009;**187**:875–888.

Jung AE, Fitzsimons HL, Bland RJ, During MJ, Young D. Hsp70 and constitutively active HSF1 mediate protection against CDCrel-1-mediated toxicity. *Mol Ther* 2008;**16**:1048–1055.

Jurivich D, Sistonen L, Kroes R, Morimoto R. Effect of sodium salicylate on the human heat shock response. *Science* 1992;**255**:1243–1245.

Kaplitt MG, Feigin A, Tang C, Fitzsimons HL, Mattis P, Lawlor PA, Bland RJ, Young D, Strybing K, Eidelberg D, *et al.* Safety and tolerability of gene therapy with an adeno-associated virus (AAV) borne GAD gene for Parkinson's disease: an open label, phase I trial. *Lancet* 2007;**369**:2097–2105.

Kaplitt MG, Leone P, Samulski RJ, Xiao X, Pfaff DW, O'Malley KL, During MJ. Long-term gene expression and phenotypic correction using adeno-associated virus vectors in the mammalian brain. *Nat Genet* 1994;**8**:148–154.

Kaushik JK, Bhat R. Why is trehalose an exceptional protein stabilizer? An analysis of the thermal stability of proteins in the presence of the compatible osmolyte trehalose. *J Biol Chem* 2003;**278**:26458–26465.

Klein RL, Wang DB, King MA. Versatile somatic gene transfer for modeling neurodegenerative disease. *Neurotox Res*. 2009;**16**:329–342.

Klionsky DJ. Autophagy: from phenomenology to molecular understanding in less than a decade. *Nat Rev Mol Cell Biol* 2007;**8**:931–937.

Klionsky DJ, Cuervo AM, Seglen PO. Methods for monitoring autophagy from yeast to human. *Autophagy* 2007;**3**:181–206.

Koga H, Cuervo AM. Chaperone-mediated autophagy dysfunction in the pathogenesis of neurodegeneration. *Neurobiol Dis* 2010;doi:10.1016/j.nbd.2010.07.006.

Kolter T, Sandhoff K. Sphingolipid metabolism diseases. *Biochim Biophys Acta* 2006;**1758**:2057–2079.

Kremer SB, Gross DS. SAGA and Rpd3 chromatin modification complexes dynamically regulate heat shock gene structure and expression. *J Biol Chem* 2009;**284**:32914–32931.

Le Gal La Salle G, Robert JJ, Berrard S, Ridoux V, Stratford-Perricaudet LD, Perricaudet M, Mallet J. An adenovirus vector for gene transfer into neurons and glia in the brain. *Science* 1993;**259**:988–990.

Lee D-k, Kim S, Lis JT. Different upstream transcriptional activators have distinct co-activator requirements. *Genes Dev* 1999;**13**:2934–2939.

Littlefield O, Nelson HC. A new use for the 'wing' of the 'winged' helix-turn-helix motif in the HSF-DNA cocrystal. *Nat Struct Biol* 1999;**6**:464–470.

Liu R, Barkhordarian H, Emadi S, Park CB, Sierks MR. Trehalose differentially inhibits aggregation and neurotoxicity of beta-amyloid 40 and 42. *Neurobiol Dis* 2005;**20**:74–81.

Liu FF, Ji L, Dong XY, Sun Y. Molecular insight into the inhibition effect of trehalose on the nucleation and elongation of amyloid beta-peptide oligomers. *J Phys Chem B* 2009;**113**:11320–11329.

Lo Bianco C, Shorter J, Regulier E, Lashuel H, Iwatsubo T, Lindquist S, Aebischer P. Hsp104 antagonizes alpha-synuclein aggregation and reduces dopaminergic degeneration in a rat model of Parkinson disease. *J Clin Invest* 2008;**118**:3087–3097.

Lu PJ, Wulf G, Zhou XZ, Davies P, Lu KP. The prolyl isomerase Pin1 restores the function of Alzheimer-associated phosphorylated tau protein. *Nature* 1999;**399**:784–788.

Malhotra JD, Kaufman RJ. The endoplasmic reticulum and the unfolded protein response. *Semin Cell Dev Biol* 2007;**18**:716–731.

Mandel RJ, Burger C. Clinical trials in neurological disorders using AAV vectors: promises and challenges. *Curr Opin Mol Ther* 2004;**6**:482–490.

Manfredsson FP, Mandel RJ. Development of gene therapy for neurological disorders. *Discov Med* 2010;**9**:204–211.

Marchler G, Wu C. Modulation of *Drosophila* heat shock transcription factor activity by the molecular chaperone DROJ1. *EMBO J* 2001;**20**:499–509.

Martins AM, Valadares ER, Porta G, Coelho J, Semionato Filho J, Pianovski MA, Kerstenetzky MS, Montoril Mde F, Aranda PC, Pires RF, *et al.* Recommendations on diagnosis, treatment, and monitoring for Gaucher disease. *J Pediatr* 2009;**155**:S10–S18.

McCarty DM, DiRosario J, Gulaid K, Muenzer J, Fu H. Mannitol-facilitated CNS entry of rAAV2 vector significantly delayed the neurological disease progression in MPS IIIB mice. *Gene Ther* 2009;**16**:1340–1352.

McCown TJ, Xiao X, Li J, Breese GR, Samulski RJ. Differential and persistent expression patterns of CNS gene transfer by an adeno-associated virus (AAV) vector. *Brain Res* 1996;**713**:99–107.

McDaniel D, Caplan AJ, Lee MS, Adams CC, Fishel BR, Gross DS, Garrard WT. Basal-level expression of the yeast *Hsp82* gene requires a heat shock regulatory element. *Mol Cell Biol* 1989;**9**:4789–4798.

Morello JP, Petaja-Repo UE, Bichet DG, Bouvier M. Pharmacological chaperones: a new twist on receptor folding. *Trends Pharmacol Sci* 2000;**21**:466–469.

Morimoto RI. Proteotoxic stress and inducible chaperone networks in neurodegenerative disease and aging. *Genes Dev* 2008;**22**:1427–1438.

Nair SC, Toran EJ, Rimerman RA, Hjermstad S, Smithgall TE, Smith DF. A pathway of multi-chaperone interactions common to diverse regulatory proteins: estrogen receptor, Fes tyrosine kinase, heat shock transcription factor Hsf1, and the aryl hydrocarbon receptor. *Cell Stress Chaperones* 1996;**1**:237–250.

Naldini L, Blomer U, Gage FH, Trono D, Verma IM. Efficient transfer, integration, and sustained long-term expression of the transgene in adult rat brains injected with a lentiviral vector. *Proc Natl Acad Sci U S A* 1996;**93**:11382–11388.

Nathan DF, Vos MH, Lindquist S. *In Vivo* functions of the *Saccharomyces cerevisiae* Hsp90 chaperone. *Proc Natl Acad Sci U S A* 1997;**94**:12949–12956.

Neef DW, Turski ML, Thiele DJ. Modulation of heat shock transcription factor 1 as a therapeutic target for small molecule intervention in neurodegenerative disease. *PLoS Biol* 2010;**8**:e1000291.

Netzer WJ, Hartl FU. Recombination of protein domains facilitated by co-translational folding in eukaryotes. *Nature* 1997;**388**:343–349.

Nieto-Sotelo J, Wiederrecht G, Okuda A, Parker CS. The yeast heat shock transcription factor contains a transcriptional activation domain whose activity is repressed under nonshock conditions. *Cell* 1990;**62**:807–817.

Nollen EA, Morimoto RI. Chaperoning signaling pathways: molecular chaperones as stress-sensing 'heat shock' proteins. *J Cell Sci* 2002;**115**:2809–2816.

Park H-O, Craig EA. Positive and negative regulation of basal expression of a yeast Hsp70 gene. *Mol Cell Biol* 1989;**9**:2025–2033.

Petesch SJ, Lis JT. Rapid, transcription-independent loss of nucleosomes over a large chromatin domain at Hsp70 loci. *Cell* 2008;**134**:74–84.

Petrucelli L, Dickson D, Kehoe K, Taylor J, Snyder H, Grover A, De Lucia M, McGowan E, Lewis J, Prihar G, *et al.* CHIP and Hsp70 regulate tau ubiquitination, degradation and aggregation. *Hum Mol Genet* 2004;**13**:703–714.

Platt FM, Neises GR, Dwek RA, Butters TD. N-butyldeoxynojirimycin is a novel inhibitor of glycolipid biosynthesis. *J Biol Chem* 1994;**269**:8362–8365.

Pollack A. Genzyme Drug Shortage Leaves Users Feeling Betrayed. *New York Times (New York)* 2010.

Premkumar L, Sawkar AR, Boldin-Adamsky S, Toker L, Silman I, Kelly JW, Futerman AH, Sussman JL. X-ray structure of human acid-beta-glucosidase covalently bound to conduritol-B-epoxide. Implications for Gaucher disease. *J Biol Chem* 2005;**280**:23815–23819.

Qi W, Zhang A, Good TA, Fernandez EJ. Two disaccharides and trimethylamine N-oxide affect Abeta aggregation differently, but all attenuate oligomer-induced membrane permeability. *Biochemistry* 2009;**48**:8908–8919.

Ravikumar B, Duden R, Rubinsztein DC. Aggregate-prone proteins with polyglutamine and polyalanine expansions are degraded by autophagy. *Hum Mol Genet* 2002;**11**:1107–1117.

Ravikumar B, Vacher C, Berger Z, Davies JE, Luo S, Oroz LG, Scaravilli F, Easton DF, Duden R, O'Kane CJ, *et al.* Inhibition of mTOR induces autophagy and reduces toxicity of polyglutamine expansions in fly and mouse models of Huntington disease. *Nat Genet* 2004;**36**:585–595.

Reddy AS, Izmitli A, de Pablo JJ. Effect of trehalose on amyloid beta (29–40)-membrane interaction. *J Chem Phys* 2009;**131**:085101.

Ritossa F. A new puffing pattern induced by temperature shock and DNP in *Drosophila*. *Experientia* 1962;**18**:571–573.

Ritson GP, Custer SK, Freibaum BD, Guinto JB, Geffel D, Moore J, Tang W, Winton MJ, Neumann M, Trojanowski JQ, *et al.* TDP-43 mediates degeneration in a novel *Drosophila* model of disease caused by mutations in VCP/p97. *J Neurosci* 2010;**30**:7729–7739.

Ron D, Walter P. Signal integration in the endoplasmic reticulum unfolded protein response. *Nat Rev Mol Cell Biol* 2007;**8**:519–529.

Rutherford S, Lindquist S. Hsp90 as a capacitor for morphological evolution. *Nature* 1998;**396**:336–342.

Sakurai H, Fukasawa T. Activator-specific requirement for the general transcription factor IIE in yeast. *Biochem Biophys Res Commun* 1999;**261**:734–739.

Sarge KD, Murphy SP, Morimoto RI. Activation of heat shock gene transcription by heat shock factor 1 involves oligomerization, acquisition of DNA-binding activity, and nuclear localization and can occur in the absence of stress. *Mol Cell Biol* 1993;**13**:1392–1407.

Sarkar S, Davies JE, Huang Z, Tunnacliffe A, Rubinsztein DC. Trehalose, a novel mTOR-independent autophagy enhancer, accelerates the clearance of mutant huntingtin and alpha-synuclein. *J Biol Chem* 2007a;**282**:5641–5652.

Sarkar S, Floto RA, Berger Z, Imarisio S, Cordenier A, Pasco M, Cook LJ, Rubinsztein DC. Lithium induces autophagy by inhibiting inositol monophosphatase. *J Cell Biol* 2005;**170**:1101–1111.

Sarkar S, Perlstein EO, Imarisio S, Pineau S, Cordenier A, Maglathlin RL, Webster JA, Lewis TA, O'Kane CJ, Schreiber SL, *et al.* Small molecules enhance autophagy and reduce toxicity in Huntington's disease models. *Nat Chem Biol* 2007b;**3**:331–338.

Sarkar S, Rubinsztein DC. Small molecule enhancers of autophagy for neurodegenerative diseases. *Mol Biosyst* 2008;**4**:895–901.

Sawkar AR, Cheng WC, Beutler E, Wong CH, Balch WE, Kelly JW. Chemical chaperones increase the cellular activity of N370S beta-glucosidase: a therapeutic strategy for Gaucher disease. *Proc Natl Acad Sci U S A* 2002;**99**:15428–15433.

Sekinger EA, Gross DS. Silenced chromatin is permissive to activator binding and PIC recruitment. *Cell* 2001;**105**:403–414.

de la Serna IL, Carlson KA, Hill DA, Guidi CJ, Stephenson RO, Sif S, Kingston RE, Imbalzano AN. Mammalian SWI-SNF complexes contribute to activation of the *hsp70* gene. *Mol Cell Biol* 2000;**20**:2839–2851.

Sheldon LA, Kingston RE. Hydrophobic coiled-coil domains regulate the subcellular localization of human heat shock factor 2. *Genes Dev* 1993;**7**:1549–1558.

Shi Y, Mosser DD, Morimoto RI. Molecular chaperones as HSF1-specific transcriptional repressors. *Genes Dev* 1998;**12**:654–666.

Shivaswamy S, Iyer VR. Stress-dependent dynamics of global chromatin remodeling in yeast: a dual role for SWI/SNF in the heat shock stress response. *Mol Cell Biol* 2008;**28**(7): 2221–2234.

Sidransky E, Nalls MA, Aasly JO, Aharon-Peretz J, Annesi G, Barbosa ER, Bar-Shira A, Berg D, Bras J, Brice A, *et al.* Multicenter analysis of glucocerebrosidase mutations in Parkinson's disease. *N Engl J Med* 2009;**361**:1651–1661.

Simonelli F, Maguire AM, Testa F, Pierce EA, Mingozzi F, Bennicelli JL, Rossi S, Marshall K, Banfi S, Surace EM, *et al.* Gene therapy for Leber's congenital amaurosis is safe and effective through 1.5 years after vector administration. *Mol Ther* 2010;**18**:643–650.

Singh H, Erkine AM, Kremer SB, Duttweiler HM, Davis DA, Iqbal J, Gross RR, Gross DS. A functional module of yeast Mediator that governs the dynamic range of heat shock gene expression. *Genetics* 2006;**172**:2169–2184.

Slepenkov SV, Witt SN. The unfolding story of the *Escherichia coli* Hsp70 DnaK: is DnaK a holdase or an unfoldase? *Mol Microbiol* 2002;**45**:1197–1206.

Somero GN. Protons, osmolytes, and fitness of internal milieu for protein function. *Am J Physiol* 1986;**251**:R197–R213.

Sorger PK. Yeast heat shock factor contains separable transient and sustained response transcriptional activators. *Cell* 1990;**62**:793–805.

Sorger PK, Nelson HCM. Trimerization of a yeast transcriptional activator via a coiled-coil motif. *Cell* 1989;**59**:807–813.

Sorger PK, Pelham HRB. Yeast heat shock factor is an essential DNA-binding protein that exhibits temperature-dependent phosphorylation. *Cell* 1988;**54**:855–864.

Stone DE, Craig EA. Self-regulation of 70-kilodalton heat shock proteins in *Saccharomyces cerevisiae*. *Mol Cell Biol* 1990;**10**:1622–1632.

Tanaka M, Machida Y, Niu S, Ikeda T, Jana NR, Doi H, Kurosawa M, Nekooki M, Nukina N. Trehalose alleviates polyglutamine-mediated pathology in a mouse model of Huntington disease. *Nat Med* 2004;**10**:148–154.

Thomas M, Harrell JM, Morishima Y, Peng HM, Pratt WB, Lieberman AP. Pharmacologic and genetic inhibition of hsp90-dependent trafficking reduces aggregation and promotes degradation of the expanded glutamine androgen receptor without stress protein induction. *Hum Mol Genet* 2006;**15**:1876–1883.

Thomson S, Hollis A, Hazzalin CA, Mahadevan LC. Distinct stimulus-specific histone modifications at *Hsp70* chromatin targeted by the transcription factor heat shock factor-1. *Mol Cell* 2004;**15**:585–594.

Uryu K, Richter-Landsberg C, Welch W, Sun E, Goldbaum O, Norris EH, Pham CT, Yazawa I, Hilburger K, Micsenyi M, *et al.* Convergence of heat shock protein 90 with ubiquitin in filamentous alpha-synuclein inclusions of alpha-synucleinopathies. *Am J Pathol* 2006;**168**:947–961.

Vacher C, Garcia-Oroz L, Rubinsztein DC. Overexpression of yeast hsp104 reduces polyglutamine aggregation and prolongs survival of a transgenic mouse model of Huntington's disease. *Hum Mol Genet* 2005;**14**:3425–3433.

Varkonyi J, Rosenbaum H, Baumann N, MacKenzie JJ, Simon Z, Aharon-Peretz J, Walker JM, Tayebi N, Sidransky E. Gaucher disease associated with parkinsonism: four further case reports. *Am J Med Genet A* 2003;**116**:348–351.

Vashist S, Cushman M, Shorter J. Applying Hsp104 to protein-misfolding disorders. *Biochem Cell Biol* 2010;**88**:1–13.

Vembar SS, Brodsky JL. One step at a time: endoplasmic reticulum-associated degradation. *Nat Rev Mol Cell Biol* 2008;**9**:944–957.

Venturi CB, Erkine AM, Gross DS. Cell cycle-dependent binding of yeast heat shock factor to nucleosomes. *Mol Cell Biol* 2000;**20**:6435–6448.

Voellmy R, Boellmann F. Chaperone regulation of the heat shock protein response. *Adv Exp Med Biol* 2007;**594**:89–99.

Vuister GW, Kim SJ, Orosz A, Marquardt J, Wu C, Bax A. Solution structure of the DNA-binding domain of *Drosophila* heat shock transcription factor. *Nat Struct Biol* 1994;**1**:605–614.

Warrick JM, Chan HY, Gray-Board GL, Chai Y, Paulson HL, Bonini NM. Suppression of polyglutamine-mediated neurodegeneration in *Drosophila* by the molecular chaperone HSP70. *Nat Genet* 1999;**23**:425–428.

Webb JL, Ravikumar B, Atkins J, Skepper JN, Rubinsztein DC. Alpha-synuclein is degraded by both autophagy and the proteasome. *J Biol Chem* 2003;**278**:25009–25013.

Welch WJ, Brown CR. Influence of molecular and chemical chaperones on protein folding. *Cell Stress Chaperones* 1996;**1**:109–115.

Wennekes T, van den Berg RJ, Boot RG, van der Marel GA, Overkleeft HS, Aerts JM. Glycosphingolipids–nature, function, and pharmacological modulation. *Angew Chem Int Ed Engl* 2009;**48**:8848–8869.

Westerheide SD, Anckar J, Stevens SM, Sistonen L, Morimoto RI Jr. Stress-inducible regulation of heat shock factor 1 by the deacetylase SIRT1. *Science* 2009;**323**:1063–1066.

Westerheide SD, Bosman JD, Mbadugha BN, Kawahara TL, Matsumoto G, Kim S, Gu W, Devlin JP, Silverman RB, Morimoto RI. Celastrols as inducers of the heat shock response and cytoprotection. *J Biol Chem* 2004;**279**:56053–56060.

Westerheide SD, Morimoto RI. Heat shock response modulators as therapeutic tools for diseases of protein conformation. *J Biol Chem* 2005;**280**:33097–33100.

Westwood JT, Clos J, Wu C. Stress-induced oligomerization and chromosomal relocalization of heat-shock factor. *Nature* 1991;**353**:822–827.

Winklhofer K, Reintjes A, Hoener M, Voellmy R, Tatzelt J. Geldanamycin restores a defective heat shock response *in vivo*. *J Biol Chem* 2001;**276**:45160–45167.

Wolff JA, Lederberg J. An early history of gene transfer and therapy. *Hum Gene Ther* 1994;**5**:469–480.

Worgall S, Sondhi D, Hackett NR, Kosofsky B, Kekatpure MV, Neyzi N, Dyke JP, Ballon D, Heier L, Greenwald BM, *et al.* Treatment of late infantile neuronal ceroid lipofuscinosis by CNS administration of a serotype 2 adeno-associated virus expressing CLN2 cDNA. *Hum Gene Ther* 2008;**19**:463–474.

Wu C. Heat shock transcription factors: structure and regulation. *Ann Rev Cell Dev Biol* 1995;**11**:441–469.

Yancey PH, Clark ME, Hand SC, Bowlus RD, Somero GN. Living with water stress: evolution of osmolyte systems. *Science* 1982;**217**:1214–1222.

Yancey PH, Somero GN. Counteraction of urea destabilization of protein structure by methylamine osmoregulatory compounds of elasmobranch fishes. *Biochem J* 1979;**183**:317–323.

Yao J, Munson KM, Webb WW, Lis JT. Dynamics of heat shock factor association with native gene loci in living cells. *Nature* 2006;**442**:1050–1053.

Young JC, Hartl FU. Polypeptide release by Hsp90 involves ATP hydrolysis and is enhanced by the co-chaperone p23. *EMBO J* 2000;**19**:5930–5940.

Yu Z, Sawkar AR, Kelly JW. Pharmacologic chaperoning as a strategy to treat Gaucher disease. *FEBS J* 2007;**274**:4944–4950.

Zhang F, Gradinaru V, Adamantidis AR, Durand R, Airan RD, de Lecea L, Deisseroth K. Optogenetic interrogation of neural circuits: technology for probing mammalian brain structures. *Nat Protoc* 2010;**5**:439–456.

Zhao J, Herrera-Diaz J, Gross DS. Domain-wide displacement of histones by activated heat shock factor occurs independently of Swi/Snf and is not correlated with RNA polymerase II density. *Mol Cell Biol* 2005;**25**:8985–8999.

Zhu X, Zhao X, Burkholder WF, Gragerov A, Ogata CM, Gottesman ME, Hendrickson WA. Structural analysis of substrate binding by the molecular chaperone DnaK. *Science* 1996;**272**:1606–1614.

Zou J, Guo Y, Guettouche T, Smith DF, Voellmy R. Repression of heat shock transcription factor HSF1 activation by Hsp90 (Hsp90 complex) that forms a stress-sensitive complex with HSF1. *Cell* 1998;**94**:471–480.

INDEX

Protein Chaperones and Protection from Neurodegenerative Diseases, First Edition.
Edited by Stephan N. Witt.